Student Solutions Manual

Nancy S. Boudreau

Bowling Green State University

Statistics ELEVENTH EDITION

McClave | Sincich

PEARSON

Prentice Hall

Upper Saddle River, NJ 07458

Editorial Director: Christine Hoag
Editor-in-Chief, Mathematics & Statistics: Deirdre Lynch
Print Supplement Editor: Joanne Wendelken
Senior Managing Editor: Linda Mihatov Behrens
Associate Managing Editor: Bayani Mendoza de Leon
Project Manager, Production: Raegan Keida Heerema
Supplement Cover Manager: Paul Gourhan
Supplement Cover Designer: Victoria Colotta
Operations Specialist: Ilene Kahn
Senior Operations Supervisor: Diane Peirano

© 2009 Pearson Education, Inc.
Pearson Prentice Hall
Pearson Education, Inc.
Upper Saddle River, NJ 07458

Printed in the United States of America

10 9 8 7 6 5 4 3 2

ISBN-13: 978-0-13-513281-4

ISBN-10:   0-13-513281-9

Pearson Education Ltd., *London*
Pearson Education Singapore, Pte. Ltd.
Pearson Education, Canada, Ltd.
Pearson Education—Japan
Pearson Education Australia Pty, Ltd.
Pearson Education North Asia, Ltd.
Pearson Educación de Mexico, S.A. de C.V.
Pearson Education Malaysia, Pte. Ltd.
Pearson Education Upper Saddle River, New Jersey

Contents

Preface

This solutions manual is designed to accompany the text, *Statistics*, Tenth Edition, by James T. McClave and Terry Sincich. It provides answers to most odd-numbered exercises for each chapter in the text. Other methods of solution may also be appropriate; however, the author has presented one that she believes to be most instructive to the beginning Statistics student. The student should first attempt to solve the assigned exercises without help from this manual. Then, if unsuccessful, the solution in the manual will clarify points necessary to the solution. The student who successfully solves an exercise should still refer to the manual's solution. Many points are clarified and expanded upon to provide maximum insight into and benefit from each exercise.

Instructors will also benefit from the use of this manual. It will save time in preparing presentations of the solutions and possibly provide another point of view regarding their meaning.

Some of the exercises are subjective in nature and thus omitted from the Answer Key at the end of *Statistics*, Tenth Edition. The subjective decisions regarding these exercises have been made and are explained by the author. Solutions based on these decisions are presented; the solution to this type of exercise is often most instructive. When an alternative interpretation of an exercise may occur, the author has often addressed it and given justification for the approach taken.

I would like to thank Kelly Barber for creating the art work and for typing this work.

Nancy S. Boudreau
Bowling Green State University
Bowling Green, Ohio

Statistics, Data, and Statistical Thinking

1.1   Statistics is a science that deals with the collection, classification, analysis, and interpretation of information or data. It is a meaningful, useful science with a broad, almost limitless scope of applications to business, government, and the physical and social sciences.

1.3   The first element of inferential statistics is the population of interest. The population is a set of existing units. The second element is one or more variables that are to be investigated. A variable is a characteristic or property of an individual population unit. The third element is the sample. A sample is a subset of the units of a population. The fourth element is the inference about the population based on information contained in the sample. A statistical inference is an estimate, prediction, or generalization about a population based on information contained in a sample. The fifth and final element of inferential statistics is the measure of reliability for the inference. The reliability of an inference is how confident one is that the inference is correct.

1.5   Quantitative data are measurements that are recorded on a meaningful numerical scale. Qualitative data are measurements that are not numerical in nature; they can only be classified into one of a group of categories.

1.7   A population is a set of existing units such as people, objects, transactions, or events. A sample is a subset of the units of a population.

1.9   An inference without a measure of reliability is nothing more than a guess. A measure of reliability separates statistical inference from fortune telling or guessing. Reliability gives a measure of how confident one is that the inference is correct.

1.11   The data consisting of the classifications A, B, C, and D are qualitative. These data are nominal and thus are qualitative. After the data are input as 1, 2, 3, and 4, they are still nominal and thus qualitative. The only differences between the two data sets are the names of the categories. The numbers associated with the four groups are meaningless.

1.13   a.   The experimental units for this study are the 15 earthquake sites.

       b.   From 1940 to 1995, there were many more than 15 earthquakes. Thus, these 15 earthquakes represent a sample.

       c.   There are 3 variables in this problem. The variable 'type of ground motion' has 3 levels (short, long, or forward directive). Thus, this variable is qualitative. The variable 'earthquake magnitude' is measured on a Richter scale which results in a meaningful number. Thus this variable is quantitative. The variable peak ground acceleration is measured in feet per second and is quantitative.

1.15   a.   The population of interest is all citizens of the United States.

       b.   The variable of interest is the view of each citizen as to whether the United States was justified in invading Iraq or not. It is qualitative.

c. The sample is the 2000 individuals selected for the poll.

d. The inference of interest is to estimate the proportion of all citizens who believe the United States was justified in invading Iraq.

e. The method of data collection is a survey.

f. It is not very likely that the sample will be representative of the population of all citizens of the United States. By selecting phone numbers at random, the sample will be limited to only those people who have telephones. Also, many people share the same phone number, so each person would not have an equal chance of being contacted. Another possible problem is the time of day the calls are made. If the calls are made in the evening, those people who work in the evening would not be represented.

1.17 a. Whether the data collected on the chief executive officers at the 500 largest U. S. companies is a population or a sample depends on what one is interested in. If one is only interested in the information from the CEO's of the 500 largest U.S. companies, then this data form a population. If one is interested in the information on CEO's from all U.S. firms, then this data would form a sample.

b. 1. The industry type of the CEO's company is a qualitative variable. The industry type is a name.

   2. The CEO's total compensation is a meaningful number. Thus, it is a quantitative variable.

   3. The CEO's total compensation over the previous five years is quantitative.

   4. The number of company stock shares (millions) held is a meaningful number. Thus, it is a quantitative variable.

   5. The CEO's age is a meaningful number. Thus, it is a quantitative variable.

   6. The CEO's efficiency rating is a meaningful number. Thus, it is a quantitative variable.

1.19 a. The town where the sample was collected is a name and thus, is a qualitative variable.

b. The type of water supply is a category. This is a qualitative variable.

c. The acidic level is a meaningful number. Thus, this is a quantitative variable.

d. The turbidity level is measured on a numerical scale and is a quantitative variable.

e. The temperature is measured on a numerical scale and is a quantitative variable.

f. The number of fecal coliforms per 100 milliliters is measured on a numerical scale and is thus, a quantitative variable.

g.	The free chlorine-residual (milligrams per liter) is measured on a numerical scale and is a quantitative variable.

h.	The presence of hydrogen sulphide is measured on a categorical scale (yes or no). This is a qualitative variable.

1.21	a.	Length of maximum span can take on values such as 15 feet, 50 feet, 75 feet, etc. Therefore, it is quantitative.

b.	The number of vehicle lanes can take on values such as 2, 4, etc. Therefore, it is quantitative.

c.	The answer to this item is "yes" or "no," which are not numeric. Therefore, it is qualitative.

d.	Average daily traffic could take on values such as 150 vehicles, 3,579 vehicles, 53,295 vehicles, etc. Therefore, it is quantitative.

e.	Condition can take on values "good," "fair," or "poor," which are not numeric. Therefore, it is qualitative.

f.	The length of the bypass or detour could take on values such as 1 mile, 4 miles, etc. Therefore, it is quantitative.

g.	Route type can take on values "interstate," U.S.," "state," "county," or "city," which are not numeric. Therefore, it is qualitative.

1.23	a.	The data collection method used by the cancer researchers is a designed experiment. Those participating in the study were randomly assigned to one of the two screening methods.

b.	The experimental units in this study are the 50,000 smokers.

c.	The variable measured in this study is the age at which the scanning method first detects a tumor. Since age is represented by a meaningful number, it is quantitative.

d.	The population of interest is all smokers in the U.S. The sample of interest is the set of 50,000 smokers participating in the study.

e.	The inference of interest is the difference in mean age at which each of the scanning methods first detects a tumor.

1.25   a.   The data collection method used by the researchers was a designed experiment. Half of the boxers received the massage and half did not.

        b.   The experimental units are the amateur boxers.

        c.   There are two variables measured on the boxers – heart rate and blood lactate level. Both of these variables are measured on a numeric scale and thus, are quantitative.

        d.   The inferences drawn from the analysis are: there is no difference in the mean heart rates between the two groups of boxers (those receiving massage and those not receiving massage) and there is no difference in the mean blood lactate levels between the two groups of boxers. Thus, massage did not affect the recovery rate of the boxers.

        e.   No. Only amateur boxers were used in the experiment. Thus, all inferences relate only to boxers.

1.27   a.   The data for this study were collected from a designed experiment. The subjects in the study were randomly divided into two groups - one group received the drug and the other group received a placebo.

        b.   This study involves interential statistics. The goal of the research was to see if the drug was effective in reducing blood loss of burn patients who undergo skin replacement surgery. The researchers are not particularly interested in the outcomes for the patients in the study, but rather the outcomes for all possible burn patients.

        c.   The population of interest to the researchers is all possible burn patients. The sample is the 14 burn patients who participated in the study.

1.29   a.   The population of interest to the psychologist is the set of all male soccer players between the ages of 14 and 29 who played up to five times per week.

        b.   The variables of interest are the frequency of "headers" and the IQ of each soccer player.

        c.   The average number of headers per game per player can take on values such as 0, 1, 2, etc. Therefore, it is quantitative. IQ can take on values such as 85, 100, 103, etc. Therefore, it is quantitative.

        d.   The sample is the set of 60 male soccer players between the ages of 14 and 29 who played up to five times per week.

        e.   The inference made by the psychologist is that "heading" the ball in soccer lowers players' IQs.

        f.   There could be several reasons why the inference could be misleading. First, "heading" the ball is a skill. It is possible that those players who are more athletically inclined are not as intellectually gifted as those players who are not as athletically inclined. Also, it appears that the data were collected only once per player. The only way one could conclude that "heading" the ball lowers players' IQs is by collecting data on each player over a period of time and seeing if each player's IQ decreases.

1.31 a. The population of interest to the pollsters is the set of all Americans.

   b. The variable of interest is whether or not the person believes the Nazi extermination of the Jews happened. Since the views are not numeric, the variable is qualitative.

   c. The sample is the set of 1000 adults and high school students who were asked the question.

   d. The method of data collection is by survey.

   e. The pollsters inferred that one in five Americans believe that it is possible that the Holocaust never happened.

   f. The reliability of this inference is very suspect. The question presented to the sample is not a "yes," "no" question. It asks one to decide if it seems possible or if it seems impossible. Therefore, an answer of "yes" to this question is totally meaningless

Methods for Describing
Set of Data

2.1   The class frequency is the number of observations in the data set falling in a particular class. The class relative frequency is the class frequency divided by the total number of observations in the data set. The class percentage is the class relative frequency multiplied by 100.

2.3   In a bar graph, a bar or rectangle is drawn above each class of the qualitative variable corresponding to the class frequency or class relative frequency. In a Pareto diagram, the bars of the bar graph are arranged in order from the largest to the smallest from left to right.

2.5   a.   To find the frequency for each class, count the number of times each letter occurs. The frequencies for the three classes are:

Class	Frequency
X	8
Y	9
Z	3
Total	20

   b.   The relative frequency for each class is found by dividing the frequency by the total sample size. The relative frequency for the class X is 8/20 = .40. The relative frequency for the class Y is 9/20 = .45. The relative frequency for the class Z is 3/20 = .15.

Class	Frequency	Relative Frequency
X	8	.40
Y	9	.45
Z	3	.15
Total	20	1.00

   c.   The frequency bar chart is:

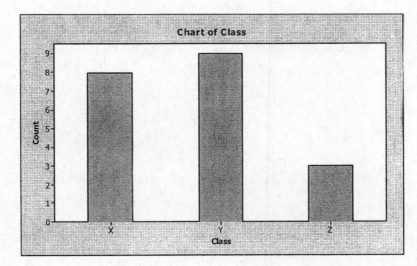

d.   The pie chart for the relative frequency distribution is:

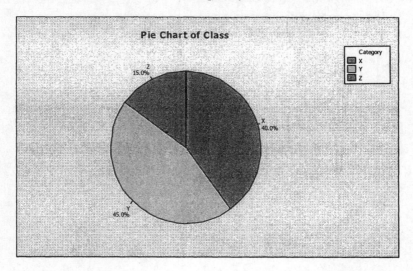

2.7   a.   The proportion of books at reading level 1 is found by dividing the number of books at level 1 by the total number of books, or 39/266 = .147.

b.   For reading level 2, the proportion of books is 76/266 = .286. For reading level 3, the proportion of books is 50/266 = .188. For reading level 4, the proportion of books is 87/266 = .327. For reading level 5, the proportion of books is 11/266 = .041. For reading level 6, the proportion of books is 3/266 = .011.

c.   The sum of the proportions is .147 + .286 + .188 + .327 + .041 + .011 = 1.000.

d.   Using MINITAB, a bar graph of the data is:

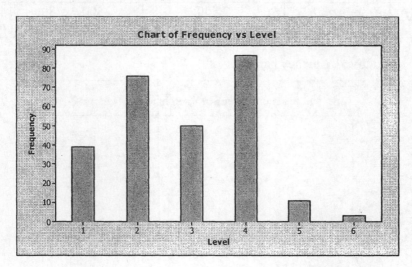

e. Using MINITAB, the Pareto diagram is:

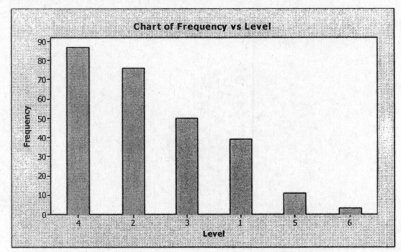

From this diagram, the reading level that occurs the most frequently is level 4.

2.9 a. To construct a relative frequency table for data, we must find the relative frequency for each species of rhinos. To find the relative frequency, divide the frequency by the total population size, 17,800. The relative frequency for African Black rhinos is 3,610/17,800 = .203. The rest of the relative frequencies are found in a similar manner and are reported in the table.

Rhino species	Population Estimate	Relative Frequency
African Black	3.610	3610 / 17800 = .203
African White	11,330	11330 / 17800 = .637
(Asian) Sumatran	300	300 / 17800 = .017
(Asian) Javan	60	60 / 17800 = .003
(Asian) Indian	2,500	2500 / 17800 = .140
Total	**17,800**	**1.000**

b. The relative frequency bar chart is:

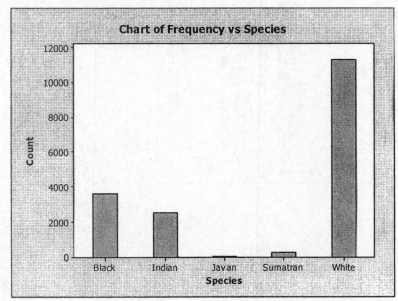

c. The proportion of rhinos that are African is (.203 + .637) = .840.

   The proportion of rhinos that are Asian is (.017 + .003 + .140) = .160.

2.11 a. From the summary table, the proportion of melt ponds that had landfast ice is 38.89 / 100 = .3889.

   b. Yes. From the summary table 17.46% of the melt ponds had first-year ice.

   c. Using MINITAB, the Pareto diagram is:

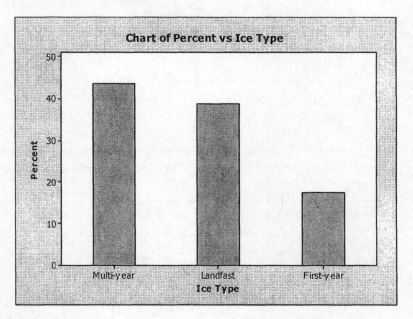

   The most commonly found type of ice is multi-year ice, followed by landfast ice. The least frequently found type of ice is first-year ice.

2.13 a. The data collection method was a survey.

   b. Since the data were numbers (percentage of US labor and materials), the variable is quantitative. Once the data were collected, they were grouped into 4 categories.

c.   Using MINITAB, a pie chart of the data is:

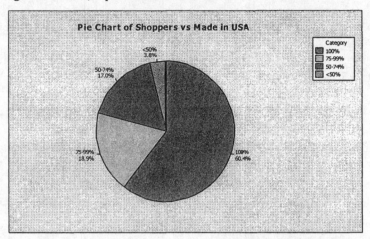

About 60% of those surveyed believe that "Made in USA" means 100% US labor and materials.

2.15   First, construct a frequency table. There are 5 categories of highest degree obtained. The relative frequencies are found by dividing the frequencies by the total number of CEO's in the sample, which is 40. The table is:

Highest Degree Obtained	Frequency	Relative Frequency
None	2	2 / 40 = .05
Bachelors	8	8 / 40 = .20
MBA	20	20 / 40 = .50
Masters	4	4 / 40 = .10
PhD	2	2 / 40 = .05
Law	4	4 / 40 = .10
Total	**40**	**1.00**

Using MINITAB, a pie chart of the data is:

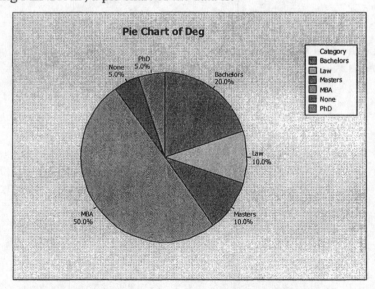

It appears that most CEO's have some type of advanced degree. Of the 40 CEO's studied, 30 or 75% had some type of advanced degree.

2.17   a.   There were 1,470 responses that were missing. In addition, 14 responses were 8 = Don't know and 7 responses were 9 = Missing. The missing values were not included, but those responding with an 8 were kept. Therefore, there were only 1333 useable responses. The frequency table is:

Response	Frequency	Relative Frequency
1	450	450/1333 = .338
2	627	627/1333 = .470
3	219	219/1333 = .164
4	23	23/1333 = .017
8	14	14/1333 = .011
Totals	**1333**	**1.000**

b.   Using MINITAB, the pie chart for the data is:

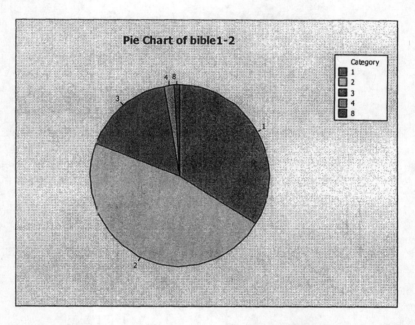

c.   The response with the highest frequency is 2, 'the Bible is the inspired word of God but not everything is to be taken literally'. Almost 47% of the respondents selected this answer. About one-third of the respondents answered 1, 'the Bible is the actual word of God and is to be taken literally'. Very few (1.7%) of the respondents chose response 4, 'the Bible has some other origin' and response 8 (1.1%), 'Don't know'.

2.19   a.   Using MINITAB, pie charts for Well Class, Aquifer, and Detect MTBE are:

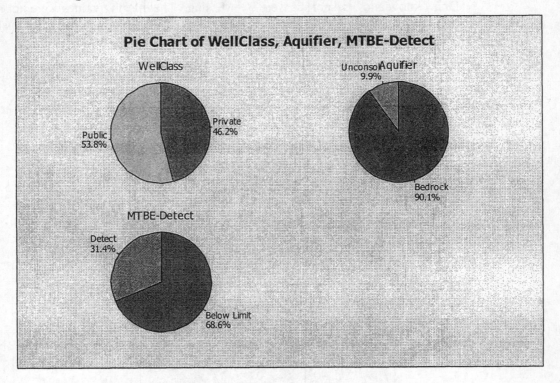

b.   Using MINITAB, the side-by-side bar charts to compare the proportions of contaminated wells for private and public wells classes are:

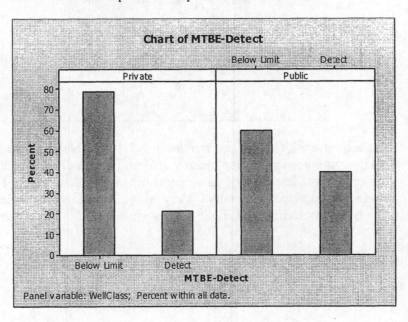

c.   Using MINITAB, the side-by-side bar charts to compare the proportions of contaminated wells for bedrock and unconsolidated aquifers are:

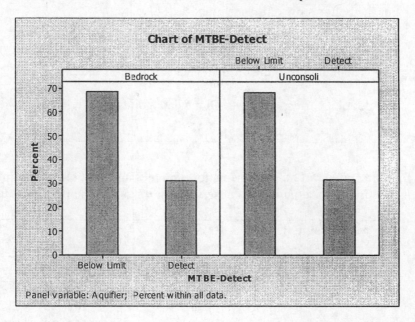

d.   For the pie charts in part **a**, a little more than half of the wells are public (53.8%) and a little less than half are private (46.2%). Most of the aquifers are bedrock (90.1%) and very few are unconsolidated (9.9%). About two-thirds of the wells are not contaminated (below limit – 68.6%) and about one-third are contaminated (detect – 31.4%).

For part **b**, a larger proportion of public wells are contaminated than private wells.

For part **c**, about the same proportion of bedrock and unconsolidated aquifers are contaminated.

2.21   In a dot plot, the values are placed on the horizontal axis. The numerical value of each measurement is located on the horizontal axis by a dot above the respective value. When values repeat, the dots are placed one above the other. In a stem-and-leaf display, the stems are the left-most digits of the measurements in the data set. They are arranged vertically from the smallest to the largest. The leaves are the right-most digit of the measurements in the data set. Each leaf is placed beside the appropriate stem. For each stem, the leaves are arranged in order from the smallest to the largest..

2.23   In a histogram, a class interval is a range of numbers above which the frequency of the measurements or relative frequency of the measurements is plotted.

2.25 a. The original data set has $1 + 3 + 5 + 7 + 4 + 3 = 23$ observations.

b. For the bottom row of the stem-and-leaf display:

  The stem is 0.
  The leaves are 0, 1, 2.
  The numbers in the original data set are 0, 1, and 2.

c. The dot plot corresponding to all the data points is:

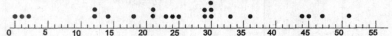

2.27 To find the number of measurements for each measurement class, multiply the relative frequency by the total number of observations, $n = 500$. The frequency table is:

Measurement Class	Relative Frequency	Frequency
.5 – 2.5	.10	500(.10) = 50
2.5 – 4.5	.15	500(.15) = 75
4.5 – 6.5	.25	500(.25) = 125
6.5 – 8.5	.20	500(.20) = 100
8.5 – 10.5	.05	500(.05) = 25
10.5 – 12.5	.10	500(.10) = 50
12.5 – 14.5	.10	500(.10) = 50
14.5 – 16.5	.05	500(.05) = 25
		500

The frequency histogram is:

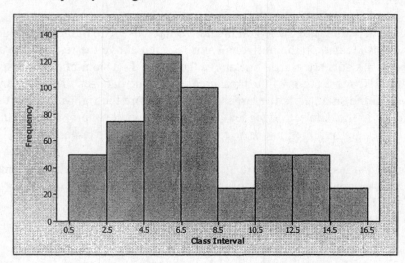

2.29 a. From the graph, the approximate percentage of aftershocks measuring between 1.5 and 2.5 is $6\% + 14\% + 19\% + 15\% + 11\% + 8\% = 73\%$.

b. From the graph, the approximate percentage of aftershocks measuring greater than 3 is $3\% + 2\% + 1.5\% + 1.25\% + .75\% + .5\% + .30\% = 9.3\%$.

2.31   a.   Using MINITAB, the stem-and-leaf display is as follows.

Character Stem-and-Leaf Display

```
Stem-and-leaf of No. Book  N  = 14
Leaf Unit = 1.0

    1      1  6
    5      2  0124
    6      2  8
   (3)     3  044
    5      3  9
    4      4  002
    1      4
    1      5  3
```

b.   The leaves that correspond to students who earned an "A" grade are underlined in the graph above. Those students who earned A's tended to read the most books.

2.33   Using MINITAB, a bar chart of the West Nile Virus cases in California counties is:

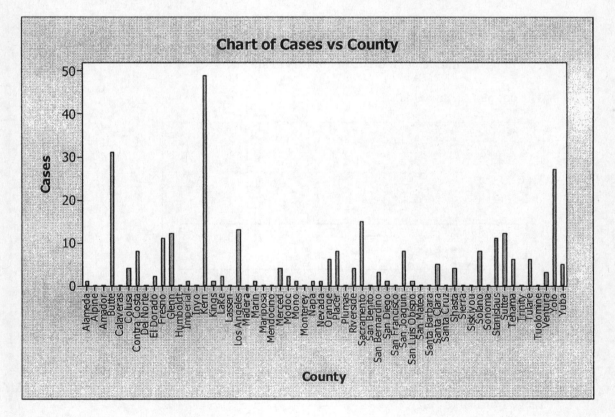

2.35   a.   Using MINITAB, the stem-and-leaf display for the duration is:

Character Stem-and-Leaf Display

```
Stem-and-leaf of Duration   N  = 18
Leaf Unit = 0.10

     7      1  0000000
     8      2  0
    (2)     3  00
     8      4  000
     5      5
     5      6
     5      7
     5      8
     5      9  0
     4     10  0
     3     11  0
     2     12  00
```

b.   The leaves that correspond to sit-ins where at least one arrest was made are underlined in the graph above. The pattern revealed does not support the theory that sit-ins of longer duration are more likely to lead to arrests.

2.37   The dot plot for these data is:

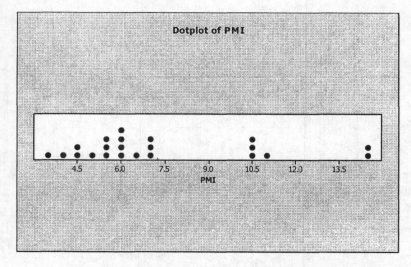

From the dot plot, most of the PMI's range from 3 to 7.5 (16 of the 22). The rest of the PMI's range from 10 to 15.

2.39   a.   Using Minitab, the stem-and-leaf display for the data is:

```
Stem-and-Leaf of LOSS        N = 19
Leaf Unit = 1.0

   8        0  11234559
  (3)       1  123
   8        2  00
   6        3  9
   5        4  ⑥
   4        5  ⑥
   3        6  ①  ⑤
   1        7
   1        8
   1        9
   1       10  ⓪
```

   b.   The numbers circled on the display in part **a** are associated with the eclipses of Saturnian satellites.

   c.   Since the five largest numbers are associated with eclipses of Saturnian satellites, it is much more likely that the greater light loss is associated with eclipses rather than occults.

2.41   a.   Using MINITAB, dot plots for the responses for the 3 groups are:

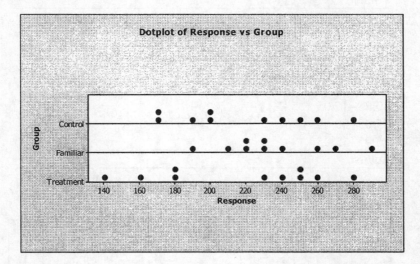

From the three dot plots, it appears that there is not much difference in the response rate between the Control group and the Treatment group. Both range from approximately 165 to 275. (The Treatment group did have a response of 139.) The response rate of the Familiarity group appears to be higher than for the other two groups. The response rate ranges from approximately 190 to 290.

b. First, we will construct frequency tables for each of the groups. We will use the following classes: 20.5–25.05, 25.05–30.05, 30.05–35.05, 35.05–40.05, 40.0545.05, 45.05–50.05, and 50.05–55.05. The frequencies and relative frequencies (found by dividing the frequencies by the sample size) are listed in the following tables:

Control Group:

Measurement Class	Frequency	Relative Frequency
20.05–25.05	1	.1
25.05–30.05	6	.6
30.05–35.05	2	.2
35.05–40.05	1	.1
Totals	10	1.0

Treatment Group:

Measurement Class	Frequency	Relative Frequency
25.05–30.05	1	.1
30.05–35.05	4	.4
35.05–40.05	3	.3
40.05–45.05	2	.2
Totals	10	1.0

Familiarity Group:

Measurement Class	Frequency	Relative Frequency
35.05–40.05	1	.1
40.05–45.05	4	.4
45.05–50.05	4	.4
50.05–55.05	1	.1
Totals	10	1.0

The relative frequency histograms for the three groups are as follows:

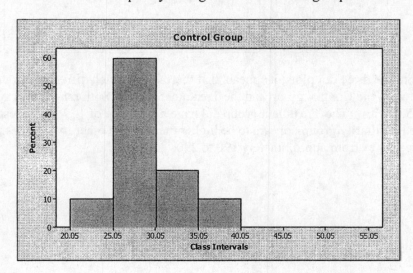

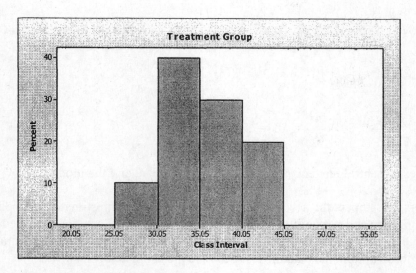

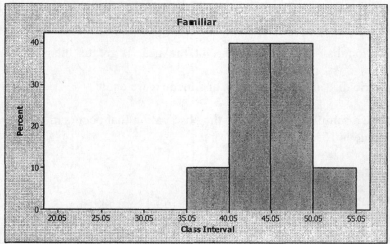

The three relative frequency histograms of the accuracy scores for the three groups indicate that the Familiarity group had the highest accuracy rate, the Treatment group had the second highest accuracy rate, and the Control group had the lowest accuracy rate.

2.43   a. $\sum x = 3 + 8 + 4 + 5 + 3 + 4 + 6 = 33$

    b. $\sum x^2 = 3^2 + 8^2 + 4^2 + 5^2 + 3^2 + 4^2 + 6^2 = 175$

    c. $\sum (x - 5)^2 = (3 - 5)^2 + (8 - 5)^2 + (4 - 5)^2 + (5 - 5)^2 + (3 - 5)^2 + (4 - 5)^2$
            $+ (6 - 5)^2 = 20$

    d. $\sum (x - 2)^2 = (3 - 2)^2 + (8 - 2)^2 + (4 - 2)^2 + (5 - 2)^2 + (3 - 2)^2 + (4 - 2)^2$
            $+ (6 - 2)^2 = 71$

    e. $\left(\sum x \right)^2 = (3 + 8 + 4 + 5 + 3 + 4 + 6)^2 = 33^2 = 1089$

2.45 a. $\sum x = 6 + 0 + (-2) + (-1) + 3 = 6$

b. $\sum x^2 = 6^2 + 0^2 + (-2)^2 + (-1)^2 + 3^2 = 50$

c. $\sum x^2 - \dfrac{\left(\sum x\right)^2}{5} = 50 - \dfrac{6^2}{5} = 50 - 7.2 = 42.8$

2.47 Three measures of central tendency are the mean, the median, and the mode.

2.49 The two factors that impact the accuracy of the sample mean as an estimate of the population mean are sample size and variability or spread of the data. For the sample size, the larger the sample size, the more accurate the estimate. For variability, all other factors remaining constant, the more variable the data the less accurate the estimate of the mean.

2.51 a. For a distribution that is skewed to the left, the mean is less than the median.

b. For a distribution that is skewed to the right, the mean is greater than the median.

c. For a symmetric distribution, the mean and median are equal.

2.53 Assume the data are a sample. The mode is the observation that occurs most frequently. For this sample, the mode is 15, which occurs 3 times.

The sample mean is:

$$\bar{x} - \frac{\sum x}{n} = \frac{18 + 10 + 15 + 13 + 17 + 15 + 12 + 15 + 18 + 16 + 11}{11} = \frac{160}{11} = 14.545$$

The median is the middle number when the data are arranged in order. The data arranged in order are: 10, 11, 12, 13, 15, 15, 15, 16, 17, 18, 18. The middle number is the 6th number, which is 15.

2.55 a. $\bar{x} = \dfrac{\sum x}{n} = \dfrac{85}{10} = 8.5$

b. $\bar{x} = \dfrac{400}{16} = 25$

c. $\bar{x} = \dfrac{35}{45} = .78$

d. $\bar{x} = \dfrac{242}{18} = 13.44$

2.57   a.   The mean is $\bar{x} = \dfrac{\sum x}{n} = \dfrac{443}{14} = 31.643$.

To find the median, the data must be arranged in order. In the problem, the data are already arranged in order. The are a total of 14 observations, which is an even number. The median is the average of the middle 2 numbers which are 30 and 34. The median is $\dfrac{30+34}{2} = 32$. The mode is the observation appearing the most. In this data set, the mode is 34 and 40 because each appears 2 times in the data set.

   b.   Since the mean and the median are almost the same, the distribution of the data set is approximately symmetric. This can be verified by the stem-and-leaf display of Exercise 2.31.

2.59   a.   The mean of the data is $\bar{x} = \dfrac{\sum x}{n} = \dfrac{11.77}{8} = 1.4713$.

   b.   The median is the average of the middle two numbers once the data are arranged in order. The data arranged in order are:

   1.37, 1.41, 1.42, 1.48, 1.50, 1.51, 1.53, 1.55

The middle two numbers are 1.48 and 1.50. The median is $\dfrac{1.48+1.50}{2} = 1.49$

   c.   Since the mean is less than the median, the data are somewhat skewed to the left.

2.61   a.   The sample mean radioactivity level is:

$$\bar{x} = \frac{\sum x}{n} = \frac{-43.75}{9} = -4.861$$

The median is the middle observation once they have been ordered. The 5th observation is −4.85. Thus the median is −4.85.

The mode is −5.00.

   b.   The average radioactivity level is −4.861. Half of the radioactivity levels are less than −4.85 and half are greater. The modal observation occurred 2 times.

2.63   a.   The mean response was 5.87. The average rating (across the 15 respondents) of the statement "The Training Game is a great way for students to understand the animal's perspective during training" was 5.87. Since the highest rating the statement could receive was 7, this indicated that on the average, the students rather strongly agreed with this statement.

The mode response was 6. More students rated the statement with a 6 than any other rating. A rating of 6 indicated a rather strong agreement to the statement.

b. Since the mode is only slightly larger than the mean, there probably is no skewness present. If any skewness exists, the data would probably be skewed to the left since the mode is larger than the mean and because the mode is very close to the largest value possible.

2.65 a. The first variable is gender. It has only two values which are not numerical, so it is qualitative. The next variable is group. There are three groups which are not numerical, so group is qualitative. The next variable is DIQ. This variable is measured on a numerical scale, so it is quantitative. The last variable is percent of pronoun errors. This variable is measured on the numerical scale, so it is quantitative.

b. In order to compute numerical descriptive measures, the data must have numbers associated with them. Qualitative variables do not have meaningful numbers associated with them, so one cannot compute numerical measures.

c. The mean of the DIQ scores for the SLI children is:

$$\bar{x} = \frac{\sum x}{n} = \frac{86 + 86 + 94 + \cdots + 95}{10} = \frac{936}{10} = 93.6$$

The median is the average of the middle two numbers after they have been arranged in order: 84, 86, 86, 87, 89, 94, 95, 98, 107, 110.

The median is $\frac{89 + 94}{2} = \frac{183}{2} = 91.5$

The mode is the value with the highest frequency. Since 86 occurred twice and no other value occurred more than once, the mode is 86.

d. The mean of the DIQ scores for the YND children is:

$$\bar{x} = \frac{\sum x}{n} = \frac{110 + 92 + 92 + \cdots + 92}{10} = \frac{953}{10} = 95.3$$

The median is the average of the middle two numbers after they have been arranged in order: 86, 90, 90, 92, 92, 92, 96, 100, 105, 110.

The median is $\frac{92 + 92}{2} = \frac{184}{2} = 92$

The mode is the value with the highest frequency. Since 92 occurred three times and no other value occurred more than twice, the mode is 92.

e. The mean of the DIQ scores for the OND children is:

$$\bar{x} = \frac{\sum x}{n} = \frac{110 + 113 + 113 + \cdots + 98}{10} = \frac{1019}{10} = 101.9$$

The median is the average of the middle two numbers after they have been arranged in order: 87, 92, 94, 95, 98, 108, 109, 110, 113, 113.

The median is $\frac{98 + 108}{2} = \frac{206}{2} = 103$

The mode is the value with the highest frequency. Since 113 occurred twice and no other value occurred more than once, the mode is 113.

f. Of the three groups, the SLI group had the lowest mean DIQ score (93.6), the YND group had a slightly higher mean DIQ score (95.3), while the OND group had the highest mean DIQ score (101.9). Thus, the SLI and the YND groups appear to be fairly similar with regard to DIQ, while the OND group appears to be much higher.

Of the three groups, the SLI group had the lowest median DIQ score (91.5), the YND group had a slightly higher median DIQ score (92), while the OND group had the highest median DIQ score (103). Thus, again, the SLI and the YND groups appear to be fairly similar with regard to DIQ, while the OND group appears to be much higher. Of the three groups, the SLI group had the lowest mode DIQ score (86), the YND group had a slightly higher mode DIQ score (92), while the OND group had the highest mode IDQ score (113). Thus, again, the SLI and the YND groups appear to be fairly similar with regard to DIQ, while the OND group appears to be much higher.

For the SLI group, the mean is larger than the median and the median is larger than the mode. For the YND group, the mean is greater than the median, but the mode is equal to the median. For the OND group, the mean is smaller than the median and the median is smaller than the mode. If only the means were used, the three groups would appear to have similar distributions – the means range from 93.6 to 101.9. If the medians were used, the range gets larger – the observations range from 91.5 to 103. If the modes were used, the range is quite different – the modes range from 86 to 113. Thus, it appears that all three "centers" should be computed fro each group.

g. **YND children**: The mean percentage of pronoun errors is:

$$\bar{x} = \frac{\sum x}{n} = \frac{94.40 + 19.05 + \ldots + 0}{10} = \frac{468.8}{10} = 46.88$$

The median is the average of the middle 2 numbers once the data have been arranged in order: 0, 0, 18.75, 19.05, 32.43, 55.00, 62.50, 86.67, 94.40, 100.00.

The median is $\frac{32.43 + 55.00}{2} = \frac{87.43}{2} = 43.715$

The mode is the value with the highest frequency. Since 0 occurs 2 times, 0 is the mode.

SLI children: The mean percentage of pronoun errors is:

$$\bar{x} = \frac{\sum x}{n} = \frac{60.00 + 40.00 + \ldots + 0}{10} = \frac{301.71}{10} = 30.171$$

The median is the average of the middle 2 numbers once the data have been arranged in order: 0, 0, 0, 27.27, 31.58, 33.33, 40.00, 42.86, 60.00, 66.67.

The median is $\frac{31.58 + 33.33}{2} = \frac{64.91}{2} = 32.455$

The mode is the value with the highest frequency. Since 0 occurs 3 times, 0 is the mode.

OND children: The mean percentage of pronoun errors is:

$$\bar{x} = \frac{\sum x}{n} = \frac{0 + 0 + \ldots + 0}{10} = \frac{0}{10} = 0$$

The median is the average of the middle 2 numbers once the data have been arranged in order: 0, 0, 0, 0, 0, 0, 0, 0, 0, 0,.

The median is $\frac{0 + 0}{2} = \frac{0}{2} = 0$

The mode is the value with the highest frequency. Since 0 occurs 10 times, 0 is the mode.

2.67   a.   The mean cylinder power measurements for student #11 is:

$$\bar{x} = \frac{\sum x}{n} = \frac{-.08 + (-.06) + \ldots + (-.16)}{25} = \frac{-3.86}{25} = -.1544$$

The median is the middle number once the data have been arranged in order:
−1.07, −.21, −.20, −.17, −.17, −.17, −.16, −.16, −.16, −.15, −.12, −.12, −.11, −.10,
−.09, −.09, −.09, −.08, −.08, −.07, −.07, −.06, −.06, −.06, −.04

The median is −.11.

The mode is the value with the highest frequency. Since −.17, −.16, −.09, −.06 each occur 3 times, all are modes.

From the printout, the mean is −0.1544 and the median is −0.11. (All the numbers in the table are negative numbers.) Since these two values are very close together, both are good representatives of the middle of the data set.

b.   The outlier is −1.07.

c.   After deleting −1.07, the mean is:

$$\bar{x} = \frac{\sum x}{n} = \frac{(-.08) + (-.06) + (-.15) + \cdots + (-.16)}{24} = \frac{-2.79}{24} = -.11625$$

The median is the average of the middle two numbers once the data are arranged in order.

The middle two numbers are 10 and 11. The median is $\frac{(-.10) + (-.11)}{2} = -.105$

The mean changes from −.1544 to −.11625 while the median changes from −.11 to −.105. The mean changes much more than the median when the outlier is removed.

2.69   The range of a data set is the difference between the largest and smallest measurements.

2.71   The sample variance is the sum of the squared deviations from the sample mean divided by the sample size minus 1. The population variance is the sum of the squared deviations from the population mean divided by the population size.

2.73   If the standard deviation increases, this implies that the data are more variable.

2.75   a.   Range = 42 − 37 = 5

$$s^2 = \frac{\sum x^2 - \frac{\left(\sum x\right)^2}{n}}{n-1} = \frac{7935 - \frac{199^2}{5}}{5-1} = 3.7 \qquad s = \sqrt{3.7} = 1.92$$

b.   Range = 100 − 1 = 99

$$s^2 = \frac{\sum x^2 - \frac{\left(\sum x\right)^2}{n}}{n-1} = \frac{25,795 - \frac{303^2}{9}}{9-1} = 1,949.25 \qquad s = \sqrt{1,949.25} = 44.15$$

c.   Range = 100 − 2 = 98

$$s^2 = \frac{\sum x^2 - \frac{\left(\sum x\right)^2}{n}}{n-1} = \frac{20,033 - \frac{295^2}{8}}{8-1} = 1,307.84 \qquad s = \sqrt{1,307.84} = 36.16$$

2.77   This is one possibility for the two data sets.

Data Set 1: 1, 1, 2, 2, 3, 3, 4, 4, 5, 5
Data Set 2: 1, 1, 1, 1, 1, 5, 5, 5, 5, 5

$$\bar{x}_1 = \frac{\sum x}{n} = \frac{1+1+2+2+3+3+4+4+5+5+}{10} = \frac{30}{10} = 3$$

$$\bar{x}_2 = \frac{\sum x}{n} = \frac{1+1+1+1+1+5+5+5+5+5}{10} = \frac{30}{10} = 3$$

Therefore, the two data sets have the same mean. The variances for the two data sets are:

$$s_1^2 = \frac{\sum x^2 - \frac{\left(\sum x\right)^2}{n}}{n-1} = \frac{110 - \frac{30^2}{10}}{9} = \frac{20}{9} = 2.2222$$

$$s_2^2 = \frac{\sum x^2 - \frac{\left(\sum x\right)^2}{n}}{n-1} = \frac{130 - \frac{30^2}{10}}{9} = \frac{40}{9} = 4.4444$$

The dot diagram for the two data sets are shown below.

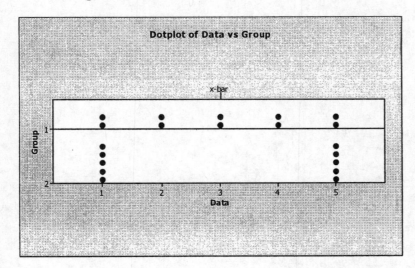

2.79   a.   Range = 3 − 0 = 3

$$s^2 = \frac{\sum x^2 - \frac{\left(\sum x\right)^2}{n}}{n-1} = \frac{15 - \frac{7^2}{5}}{5-1} = 1.3 \qquad s = \sqrt{1.3} = 1.1402$$

b. After adding 3 to each of the data points,

Range $= 6 - 3 = 3$

$$s^2 = \dfrac{\sum x^2 - \dfrac{\left(\sum x\right)^2}{n}}{n-1} = \dfrac{102 - \dfrac{22^2}{5}}{5-1} = 1.3 \qquad s = \sqrt{1.3} = 1.1402$$

c. After subtracting 4 from each of the data points,

Range $= -1 - (-4) = 3$

$$s^2 = \dfrac{\sum x^2 - \dfrac{\left(\sum x\right)^2}{n}}{n-1} = \dfrac{39 - \dfrac{(-13)^2}{5}}{5-1} = 1.3 \qquad s = \sqrt{1.3} = 1.1402$$

d. The range, variance, and standard deviation remain the same when any number is added to or subtracted from each measurement in the data set.

2.81 a. For those students who earned A, the range is $53 - 24 = 29$.

$$\text{The variance is } s^2 = \dfrac{\sum x^2 - \dfrac{\left(\sum x\right)^2}{n}}{n-1} = \dfrac{11{,}482 - \dfrac{296^2}{8}}{7} = \dfrac{530}{7} = 75.7143$$

The standard deviation is $s = \sqrt{s^2} = \sqrt{75.7143} = 8.701$.

b. For those students who earned a B or C, the range is $40 - 16 = 24$.

$$\text{The variance is } s^2 = \dfrac{\sum x^2 - \dfrac{\left(\sum x\right)^2}{n}}{n-1} = \dfrac{3{,}965 - \dfrac{147^2}{6}}{5} = \dfrac{363.5}{5} = 72.7$$

The standard deviation is $s = \sqrt{s^2} = \sqrt{72.7} = 8.526$.

c. The students who received A's have a more variable distribution of the number of books read. The range, variance, and standard deviation for this group are greater than the corresponding values for the B-C group.

2.83 a. Range $= 1.55 - 1.37 = .18$

b. $$s^2 = \dfrac{\sum x^2 - \dfrac{\left(\sum x\right)^2}{n}}{n-1} = \dfrac{17.3453 - \dfrac{11.77^2}{8}}{8-1} = .0041$$

c. $s = \sqrt{.0041} = .064$

d. If the standard deviation of the daily ammonia levels during the morning drive-time is 1.45 ppm (compared to .064 ppm in the afternoon drive-time), then the morning drive-time has more variable ammonia levels.

2.85 a. Range $= 13 - 1 = 12$

$$s^2 = \frac{\sum x^2 - \frac{\left(\sum x\right)^2}{n}}{n-1} = \frac{498 - \frac{80^2}{20}}{20-1} = \frac{178}{19} = 9.3684$$

$$s = \sqrt{s^2} = \sqrt{9.3584} = 3.061$$

b. Dropping the largest measurement:

Range $= 9 - 1 = 8$

$$s^2 = \frac{\sum x^2 - \frac{\left(\sum x\right)^2}{n}}{n-1} = \frac{329 - \frac{67^2}{19}}{19-1} = \frac{92.7368421}{18} = 5.1520$$

$$s = \sqrt{s^2} = \sqrt{5.1520} = 2.270$$

By dropping the largest observation from the data set, the range decreased from 12 to 8, the variance decreased from 9.3684 to 5.1520 and the standard deviation decreased from 3.061 to 2.270.

c. Dropping the largest and smallest measurements:

Range $= 9 - 1 = 8$

$$s^2 = \frac{\sum x^2 - \frac{\left(\sum x\right)^2}{n}}{n-1} = \frac{328 - \frac{66^2}{18}}{18-1} = \frac{86}{17} = 5.0588$$

$$s = \sqrt{s^2} = \sqrt{5.0588} = 2.249$$

By dropping the largest and smallest observations from the data set, the range decreased from 12 to 8, the variance decreased from 9.3684 to 5.0588 and the standard deviation decreased from 3.061 to 2.249.

2.87   a.   The unit of measurement of the variable of interest is dollars (the same as the mean and standard deviation). Based on this, the data are quantitative.

    b.   Since no information is given about the shape of the data set, we can only use Chebyshev's rule.

        $900 is 2 standard deviations below the mean, and $2100 is 2 standard deviations above the mean. Using Chebyshev's rule, at least 3/4 of the measurements (or $3/4 \times 200 = 150$ measurements) will fall between $900 and $2100.

        $600 is 3 standard deviations below the mean and $2400 is 3 standard deviations above the mean. Using Chebyshev's rule, at least 8/9 of the measurements (or $8/9 \times 200 \times 178$ measurements) will fall between $600 and $2400.

        $1200 is 1 standard deviation below the mean and $1800 is 1 standard deviation above the mean. Using Chebyshev's rule, nothing can be said about the number of measurements that will fall between $1200 and $1800.

        $1500 is equal to the mean and $2100 is 2 standard deviations above the mean. Using Chebyshev's rule, at least 3/4 of the measurements (or $3/4 \times 200 = 150$ measurements) will fall between $900 and $2100. It is possible that all of the 150 measurements will be between $900 and $1500. Thus, nothing can be said about the number of measurements between $1500 and $2100.

2.89   According to the Empirical Rule:

    a.   Approximately 68% of the measurements will be contained in the interval $\bar{x} - s$ to $\bar{x} + s$.

    b.   Approximately 95% of the measurements will be contained in the interval $\bar{x} - 2s$ to $\bar{x} + 2s$.

    c.   Essentially all the measurements will be contained in the interval $\bar{x} - 3s$ to $\bar{x} + 3s$.

2.91   Using Chebyshev's rule, at least 8/9 of the measurements will fall within 3 standard deviations of the mean. Thus, the range of the data would be around 6 standard deviations. Using the Empirical Rule, approximately 95% of the observations are within 2 standard deviations of the mean. Thus, the range of the data would be around 4 standard deviations. We would expect the standard deviation to be somewhere between Range/6 and Range/4.

For our data, the range = 760 − 135 = 625.

The Range/6 = 625/6 = 104.17 and Range/4 = 625/4 = 156.25.

Therefore, I would estimate that the standard deviation of the data set is between 104.17 and 156.25.

It would not be feasible to have a standard deviation of 25. If the standard deviation were 25, the data would span 625/25 = 25 standard deviations. This would be extremely unlikely.

2.93   a. $\bar{x} = \dfrac{\sum x}{n} = \dfrac{16,040}{169} = 94.91$

$$s^2 = \frac{\sum x^2 - \dfrac{\left(\sum x\right)^2}{n}}{n-1} = \frac{1,526,288 - \dfrac{16,040^2}{169}}{169-1} = 23.2837$$

$s = \sqrt{s^2} = \sqrt{23.2837} = 4.83$

b. $\bar{x} \pm s \Rightarrow 94.91 \pm 4.83 \Rightarrow (90.08,\ 99.74)$

$\bar{x} \pm 2s \Rightarrow 94.91 \pm 2(4.83) \Rightarrow 94.91 \pm 9.66 \Rightarrow (85.25,\ 104.57)$

$\bar{x} \pm 3s \Rightarrow 94.91 \pm 3(4.83) \Rightarrow 94.91 \pm 14.49 \Rightarrow (80.42,\ 109.40)$

c.   There are 137 out of 169 observations in the first interval. This is (137/169)*100% = 81.1%. There are 165 out of 169 observations in the second interval. This is (165/169)*100% = 97.6%. There are 166 out of 169 observations in the second interval. This is (166/169)*100% = 98.2%.

The percentages for the first 2 intervals are somewhat larger than we would expect using the Empirical Rule. The Empirical Rule indicates that approximately 68% of the observations will fall within 1 standard deviation of the mean. It also indicates that approximately 95% of the observations will fall within 2 standard deviations of the mean. Chebyshev's Theorem says that at least ¾ or 75% of the observations will fall within 2 standard deviations of the mean and at least 8/9 or 88.8% of the observations will fall within 3 standard deviations of the mean. It appears that our observed percentages agree with Chebyshev's Theorem better that the Empirical Rule.

2.95   a.   The mean is 79 and the standard deviation is 23. Thus, a measurement of 102 is 1 standard deviation above the mean. If nothing is known about the shape of the distribution, we must use Chebyshev's Rule to describe the distribution. Using this, we know nothing about the number of observations within 1 standard deviation of the mean, and thus, know nothing about the percentage of observations less than 102.

b.   If we assume that the distribution is mound-shaped, we can use the Empirical Rule to describe the distribution. If the distribution is mound-shaped, then it is symmetric. Thus, we know that half of the observations will be below 79. We know that about 68% of the observations are within 1 standard deviation of the mean. Therefore, about 34% of the observations will be between the mean and 1 standard deviation above the mean, or between 79 and 102. Thus, the percentage of observations less than 102 is about 50% + 34% = 84%.

2.97 For parts a and b, we must use Chebyshev's Rule to describe the data sets. For both the handrubbers and the handwashers, the standard deviation is greater than the mean. Since no observations can be less than 0, we know that the smallest observation in the data sets is less than 1 standard deviation from the mean. Thus, the distributions cannot be symmetric, but rather skewed to the right. We cannot use the Empirical Rule.

a. Using Chebyshev's Rule, we know that at least $1 - \frac{1}{k^2}$ of the observations fall within k standard deviations of the mean. To find k, we use:

$$.95 = 1 - \frac{1}{k^2} \Rightarrow \frac{1}{k^2} = 1 - .95 \Rightarrow \frac{1}{k^2} = .05 \Rightarrow k^2 = 20 \Rightarrow k = 4.5$$

Thus, the interval $\bar{x} \pm 4.5s$ will contain at least $1 - \frac{1}{4.5^2} = .95$ of the observations.

For handrubbers: $\bar{x} \pm 4.5s \Rightarrow 35 \pm 4.5(59) \Rightarrow 35 \pm 265.5 \Rightarrow (0, \ 300.5)$

b. For handwashers: $\bar{x} \pm 4.5s \Rightarrow 69 \pm 4.5(106) \Rightarrow 69 \pm 477 \Rightarrow (0, \ 546)$

2.99 a. Since no information is given about the distribution of the velocities of the Winchester bullets, we can only use Chebyshev's rule to describe the data. We know that at least 3/4 of the velocities will fall within the interval:

$$\bar{x} \pm 2s \Rightarrow 936 \pm 2(10) \Rightarrow 936 \pm 20 \Rightarrow (916, 956)$$

Also, at least 8/9 of the velocities will fall within the interval:

$$\bar{x} \pm 3s \Rightarrow 936 \pm 3(10) \Rightarrow 936 \pm 30 \Rightarrow (906, 966)$$

b. Since a velocity of 1,000 is much larger than the largest value in the second interval in part **a**, it is very unlikely that the bullet was manufactured by Winchester.

2.101 a. Regardless of the shape of the distribution, most of the observations will fall within 3 standard deviations of the mean. Thus, for the SAT-Math scores, an interval likely to contain a student's change in score is:

$$\bar{x} \pm 3s \Rightarrow 19 \pm 3(65) \Rightarrow 19 \pm 195 \Rightarrow (-176, 214)$$

b. Regardless of the shape of the distribution, most of the observations will fall within 3 standard deviations of the mean. Thus, for the SAT-Verbal scores, an interval likely to contain a student's change in score is:

$$\bar{x} \pm 3s \Rightarrow 7 \pm 3(49) \Rightarrow 7 \pm 147 \Rightarrow (-140, 154)$$

c. For the SAT-Verbal, the maximum increase in scores is about 154 points. For the SAT-Math, the maximum increase in scores is approximately 214. Thus, a student is more likely to get a 140-point increase on the SAT-Math test.

2.103   Using the definition of a percentile:

	Percentile	Percentage Above	Percentage Below
a.	75th	25%	75%
b.	50th	50%	50%
c.	20th	80%	20%
d.	84th	16%	84%

2.105   For mound-shaped distributions, we can use the Empirical Rule. About 95% of the observations will fall within 2 standard deviations of the mean. Thus, about 95% of the measurements will have z-scores between -2 and 2.

2.107   We first compute z-scores for each x value.

a. $z = \dfrac{x - \mu}{\sigma} = \dfrac{100 - 50}{25} = 2$

b. $z = \dfrac{x - \mu}{\sigma} = \dfrac{1 - 4}{1} = -3$

c. $z = \dfrac{x - \mu}{\sigma} = \dfrac{0 - 200}{100} = -2$

d. $z = \dfrac{x - \mu}{\sigma} = \dfrac{10 - 5}{3} = 1.67$

The above z-scores indicate that the x value in part **a** lies the greatest distance above the mean and the x value of part **b** lies the greatest distance below the mean.

2.109   The percentile ranking for the age of 25 years in the distribution of all ages of licensed drivers stopped by police is $100 - 74 = 26$ percentile.

2.111   From Exercise 2.93, $\bar{x} = 94.91$ and $s = 4.83$.

a.   The z-score for the Nautilus Explorer is: $z = \dfrac{x - \bar{x}}{s} = \dfrac{78 - 94.91}{4.83} = -3.50$

The score for the Nautilus Explorer is 3.50 standard deviations below the mean for all the cruise ships.

b.   The z-score for the Rotterdam is: $z = \dfrac{x - \bar{x}}{s} = \dfrac{98 - 94.91}{4.83} = 0.64$

The score for the Rotterdam is 0.64 standard deviations above the mean for all the cruise ships.

2.113   a.   Using SAS, the output is:

```
                            EGG LENGTH                           1

                        The UNIVARIATE Procedure
                            Variable:  length

                                Moments

N                          130     Sum Weights              130
Mean                 60.6538462    Sum Observations        7885
Std Deviation        43.9861168    Variance          1934.77847
Skewness              2.1091362    Kurtosis          4.70026314
Uncorrected SS          727842     Corrected SS      249586.423
Coeff Variation      72.5199136    Std Error Mean    3.85783765

                     Basic Statistical Measures

           Location                      Variability

        Mean      60.65385    Std Deviation         43.98612
        Median    49.50000    Variance                  1935
        Mode      35.00000    Range                220.00000
                              Interquartile Range   32.00000


                   Tests for Location: Mu0=0

        Test            -Statistic-        -----p Value------

        Student's t     t  15.72224    Pr > |t|      <.0001
        Sign            M         65    Pr >= |M|     <.0001
        Signed Rank     S     4257.5    Pr >= |S|     <.0001


                       Quantiles (Definition 5)

                       Quantile       Estimate

                       100% Max         236.0
                       99%              218.0
                       95%              160.0
                       90%              122.0
                       75% Q3            67.0
                       50% Median        49.5
                       25% Q1            35.0
                       10%               23.0
                       5%                19.0
                       1%                16.0
                       0% Min            16.0
```

```
                    The UNIVARIATE Procedure
                        Variable:  length

                       Extreme Observations

        ----Lowest----              ----Highest---

        Value      Obs              Value      Obs

         16.0      102               195       128
         16.0       90               205       123
         17.0      101               216       122
         18.0      103               218       129
         18.5      100               236       130


                        Missing Values

                                  -----Percent Of-----
        Missing                                Missing
          Value       Count      All Obs           Obs

             .            2         1.52        100.00
```

The 10th percentile egg length is 23.

b. From the printout, $\bar{x} = 60.65$ and $s = 43.99$. The z-score corresponding to the Moas, P. australis bird species' egg length is:

$$z = \frac{x - \bar{x}}{s} = \frac{205 - 60.65}{43.99} = 3.28$$

The z-score for Moas, P. australis is 3.28 standard deviations above the mean. This is a fairly large value for a *z*-score. This indicates that the egg length for Moas, P. australis could be very unusual.

2.115 We will use SAS to compute the percentiles. The output is:

```
                    The UNIVARIATE Procedure
                        Variable:  LEFTEYE

                             Moments

    N                        25     Sum Weights               25
    Mean                 -0.1544    Sum Observations       -3.86
    Std Deviation     0.19676721    Variance          0.03871733
    Skewness          -4.5220839    Kurtosis          21.7019614
    Uncorrected SS        1.5252    Corrected SS        0.929216
    Coeff Variation    -127.4399    Std Error Mean    0.03935344


                    Basic Statistical Measures

          Location                       Variability

      Mean     -0.15440      Std Deviation            0.19677
      Median   -0.11000      Variance                 0.03872
      Mode     -0.17000      Range                    1.03000
                             Interquartile Range      0.08000

NOTE: The mode displayed is the smallest of 4 modes with a count of 3.
```

```
                Tests for Location: Mu0=0

       Test              -Statistic-       -----p Value------

       Student's t     t  -3.92342      Pr > |t|    0.0006
       Sign            M     -12.5      Pr >= |M|   <.0001
       Signed Rank     S    -162.5      Pr >= |S|   <.0001

                   Quantiles (Definition 5)

                   Quantile      Estimate

                   100% Max        -0.04
                   99%             -0.04
                   95%             -0.06
                   90%             -0.06
                   75% Q3          -0.08
                   50% Median      -0.11
                   25% Q1          -0.16
                   10%             -0.20
                   5%              -0.21
                   1%              -1.07
                   0% Min          -1.07

                   Extreme Observations

          ----Lowest----          ----Highest---

          Value      Obs          Value      Obs

          -1.07       7           -0.07       23
          -0.21      18           -0.06        2
          -0.20      11           -0.06       12
          -0.17      22           -0.06       21
          -0.17      17           -0.04       16
```

a. From the printout above, the 10th percentile is −0.20. Thus, 10% of the cylinder power measurements are below −0.20 and 90% are above −0.20.

b. From the printout above, the 95th percentile is −0.06. Thus, 95% of the cylinder power measurements are below −0.06 and 5% are above −0.06.

c. $z = \dfrac{x - \overline{x}}{s} = \dfrac{-1.07 - (-.1544)}{.19677} = -4.65$. A power measurement of −1.07 is 4.65 standard deviations below the mean. Since this z-score is so small, it is an extremely unlikely value to observe. The cylinder value of −1.07 is an extreme value.

2.117   a.   From the problem, $\mu = 2.7$ and $\sigma = .5$

$$z = \frac{x - \mu}{\sigma} \Rightarrow z\sigma = x - \mu \Rightarrow x = \mu + z\sigma$$

For $z = 2.0$, $x = 2.7 + 2.0(.5) = 3.7$

For $z = -1.0$, $x = 2.7 - 1.0(.5) = 2.2$

For $z = .5$, $x = 2.7 + .5(.5) = 2.95$

For $z = -2.5$, $x = 2.7 - 2.5(.5) = 1.45$

   b.   For $z = -1.6$, $x = 2.7 - 1.6(.5) = 1.9$

   c.   If we assume the distribution of GPAs is approximately mound-shaped, we can use the Empirical Rule.

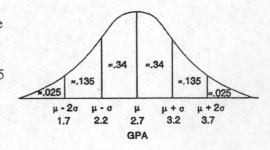

From the Empirical Rule, we know that $-.025$ or -2.5% of the students will have GPAs above 3.7 (with $z = 2$). Thus, the GPA corresponding to summa cum laude (top 2.5%) will be greater than 3.7 ($z > 2$).

We know that $-.16$ or 16% of the students will have GPAs above 3.2 ($z = 1$). Thus, the limit on GPAs for cum laude (top 16%) will be greater than 3.2 ($z > 1$).

We must assume the distribution is mound-shaped.

2.119   The 25th percentile is the observation with 25% of the observations below it and 75% of the observations above it. Another name for the 25th percentile is the lower quartile.

The 50th percentile is the observation with 50% of the observations below it and 50% of the observations above it. Another name for the 50th percentile is the median.

The 75th percentile is the observation with 75% of the observations below it and 25% of the observations above it. Another name for the 75th percentile is the upper quartile.

2.121   The hinges of a box plot are the upper quartile and the lower quartile (top and bottom of the rectangle).

2.123   To determine if the measurements are outliers, compute the z-score.

a. $z = \dfrac{x - \bar{x}}{s} = \dfrac{65 - 57}{11} = .727$ Since this z-score is less than 3 in magnitude, 65 is not an outlier.

b. $z = \dfrac{x - \bar{x}}{s} = \dfrac{21 - 57}{11} = -3.273$ Since this z-score is more than 3 in magnitude, 21 is an outlier.

c. $z = \dfrac{x - \bar{x}}{s} = \dfrac{72 - 57}{11} = 1.354$ Since this z-score is less than 3 in magnitude, 72 is not an outlier.

d. $z = \dfrac{x - \bar{x}}{s} = \dfrac{98 - 57}{11} = 3.727$ Since this z-score is more than 3 in magnitude, 98 is an outlier.

2.125   a.   The median is approximately 4.

c. Q_L (Lower quartile) is approximately 3 and Q_U (Upper quartile) is approximately 6.

d.   The interquartile range is $IQR = Q_U - Q_L \approx 6 - 3 = 3$.

e.   The data set is skewed to the right since the right whisker is longer than the left and there are outlying observations to the right.

f.   50% of the observations lie to the right of the median. 75% of the observations lie to the left of the upper quartile.

g.   The outliers in the data set are 12, 13, and 16.

2.127   a.   The approximate 25th percentile PASI score before treatment is 10. The approximate median before treatment is 15. The approximate 75th percentile PASI score before treatment is 28.

b.   The approximate 25th percentile PASI score after treatment is 2. The approximate median after treatment is 4. The approximate 75th percentile PASI score after treatment is 7.

c.   Since the 75th percentile after treatment is lower than the 25th percentile before treatment, it appears that the ichthyotherapy is effective in treating psoriasis.

2.129   a.   The z-score is $z = \dfrac{x - \bar{x}}{s} = \dfrac{3.3 - 7.3}{3.18} = -1.26$.

b.   A PMI score of 3.3 would not be considered an outlier. The z-score is -1.26. A z-score this small is not considered unusual.

2.131   a.   Using MINITAB, the side-by-side box plots are:

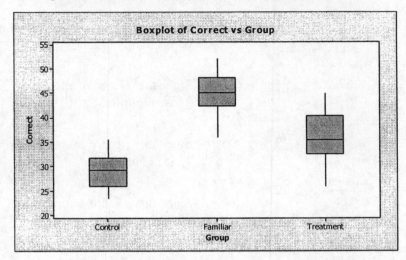

b.   The median for the Control group is approximately 29, the median for the Familiar group is approximately 45, and the median for the Treatment group is approximately 35. It appears that the learned breathing pattern strategy helped make the speech more understandable.

c.   The interquartile range for the Control group is from about 26 to 31 or 5 units, the interquartile range for the Familiar group is from about 42.5 to 47.5 or 5 units, and the interquartile range for the Treatment group is from about 32.5 to 40 or 7.5 units. The spread for the Treatment group is larger than the spreads for the other two groups.

d.   The standard deviations for the three groups agree fairly well with the corresponding interquartile ranges. The standard deviation for the Treatment group is the largest as is the interquartile range. The standard deviation for the Control group is the smallest, while the interquartile range for the Control group is tied with the Familiar group.

e.   Since there are no asterisks or observations outside the inner fences, there is no evidence that there are any outliers.

2.133   Using MINITAB, the boxplot is:

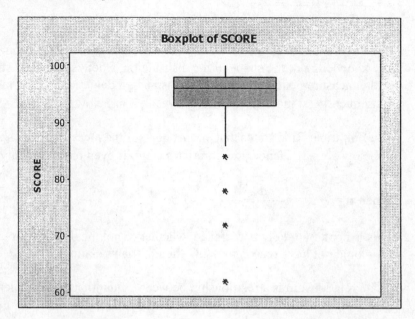

From the boxplot, there appears to be 4 outliers: 62, 72, 78, and 84.

b.   From Exercise 2.93, $\bar{x} = 94.91$ and $s = 4.83$. Since the data are skewed to the left, we will consider observations more than 3 standard deviations from the mean to be outliers. An observation with a z-score of 3 would have the value:

$$z = \frac{x - \bar{x}}{s} \Rightarrow 3 = \frac{x - 94.91}{4.83} \Rightarrow 3(4.83) = x - 94.91$$
$$\Rightarrow 14.49 = x - 94.91 \Rightarrow x = 109.4$$

An observation with a z-score of -3 would have the value:

$$z = \frac{x - \bar{x}}{s} \Rightarrow -3 = \frac{x - 94.91}{4.83} \Rightarrow -3(4.83) = x - 94.91$$
$$\Rightarrow -14.49 = x - 94.91 \Rightarrow x = 80.42$$

Observations greater than 109.40 or less than 80.42 would be considered outliers. Using this criterion, the following observations would be outliers: 62, 72, and 78.

c.   No, these methods do not agree. Using the boxplot, 4 observations were identified as outliers. Using the z-score method, only 3 observations were identified as outliers. Since the data are very highly skewed to the left, the z-score method may not be appropriate.

2.135 a. $z = \dfrac{x - \mu}{\sigma} = \dfrac{4 - 7}{1} = -3$

   b.   The z-score is low enough to suspect that the librarian's claim is incorrect. Even without any knowledge of the shape of the distribution, Chebyshev's rule states that at least 8/9 of the measurements will fall within 3 standard deviations of the mean (and, consequently, at most 1/9 will be above $z = 3$ or below $z = -3$).

   c.   The Empirical Rule states that almost none of the measurements should be above $z = 3$ or below $z = -3$. Hence, the librarian's claim is even more unlikely.

   d.   When $\sigma = 2$, $z = \dfrac{x - \mu}{\sigma} = \dfrac{4 - 7}{2} = -1.5$

   This is not an unlikely occurrence, whether or not the data are mound-shaped. Hence, we would not have reason to doubt the librarian's claim.

2.137 A bivariate relationship is a relationship between 2 quantitative variables.

2.139 Using MINITAB, the scatterplot is:

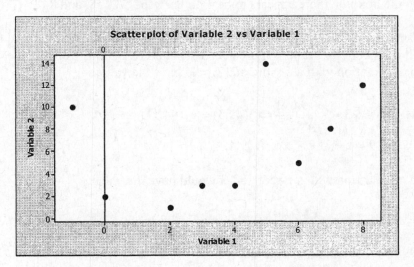

From the scatterplot, there does not appear to be much of a trend between variable 1 and variable 2. There is a slight positive linear trend - as variable 1 increases, variable 2 tends to increase. However, this relationship appears to be very weak.

2.141   Using MINITAB, a scatterplot of the data is:

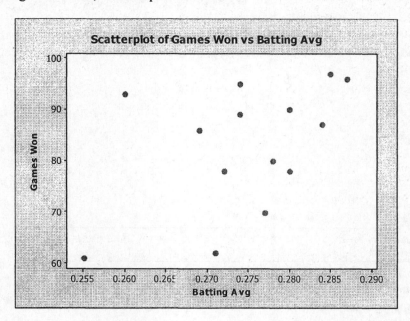

There appears to be a positive relationship between the number of games won and the team batting average. As the team batting average increases, the teams tend to win more games.

2.143   a.   Using MINITAB, the scatterplot is:

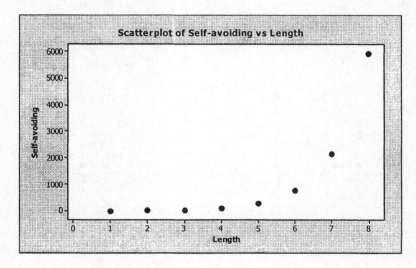

It appears that as length increases, the self-avoiding walks increase, but at an increasing rate.

b.   Using MINITAB, the scatterplot of the data is:

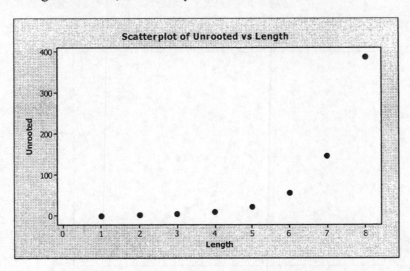

It appears that as the length increases, the unrooted walks also increase, but at an increasing rate.

2.145   a.   Using MINITAB, a scatterplot of the data is:

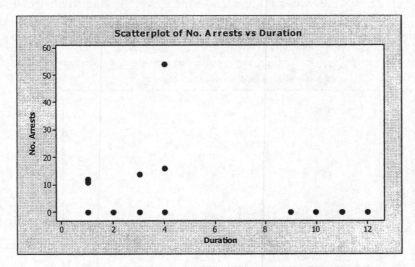

There does not appear to be much of a trend between the number of arrests and the duration of the sit-in.

b. Using MINITAB, a scatterplot of the data using only those observations where there was at least one arrest is:

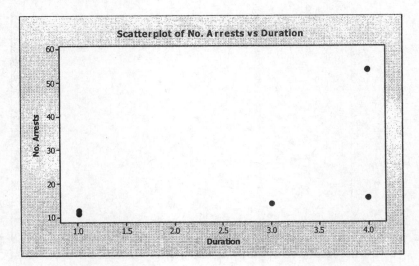

There appears to be a positive relationship between the number of arrests and the duration of the sit-in. As the duration increases, the number of arrests tend to increase.

c. Since there are only 5 observations used in the scatterplot in part **b**, the reliability is suspect.

2.147 a. Using MINITAB, a scatterplot of the data is:

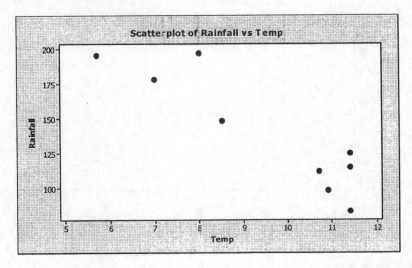

There appears to be a negative relationship between the annual rainfall and the maximum daily temperature. As the maximum daily temperature increases, the amount of annual rainfall tends to decrease.

b. Using MINITAB, a scatterplot of the relationship between annual rainfall and total plant cover is:

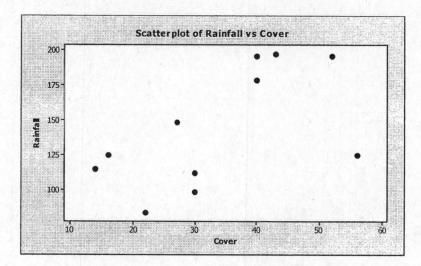

There appears to be a positive relationship between annual rainfall and total plant cover. As the total plant cover increases, the annual rainfall tends to increase.

c. Using MINITAB, a scatterplot of the relationship between annual rainfall and number of ant species is:

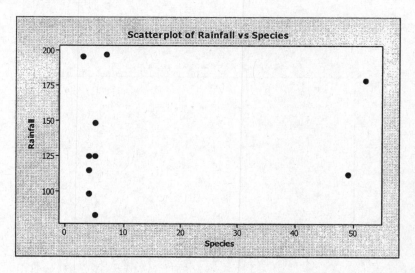

There does not appear to be a relationship between annual rainfall and number of ant species.

d. Using MINITAB, a scatterplot of the relationship between annual rainfall and species diversity index is:

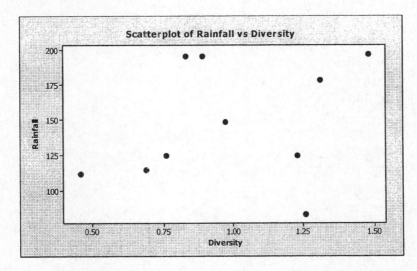

There appears to be a positive relationship between annual rainfall and species diversity index. As the species diversity index increases, the annual rainfall tends to increase.

2.149 Using MINITAB, a graph of the accuracy scores versus the driving distance is as follows:

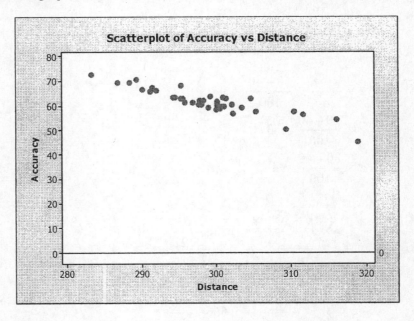

Yes, the golfer's concern is valid. From the graph, as the driving distance increases, the accuracy decreases.

2.151 The median is preferred to the mean as a measure of central tendency when the data set is skewed either right or left. If the data set has some extreme observations, the median better measure of central tendency than the mean.

2.153 A stem-and-leaf display is generally preferred over a histogram when the data set is relatively small.

2.155   One technique for distorting information on a graph is by stretching the vertical axis by starting the vertical axis somewhere above 0.

2.157   a. $z = \dfrac{x - \mu}{\sigma} = \dfrac{50 - 60}{10} = -1$

$z = \dfrac{70 - 60}{10} = 1$

$z = \dfrac{80 - 60}{10} = 2$

b. $z = \dfrac{x - \mu}{\sigma} = \dfrac{50 - 60}{5} = -2$

$z = \dfrac{70 - 60}{5} = 2$

$z = \dfrac{80 - 60}{5} = 4$

c. $z = \dfrac{x - \mu}{\sigma} = \dfrac{50 - 40}{10} = 1$

$z = \dfrac{70 - 40}{10} = 3$

$z = \dfrac{80 - 40}{10} = 4$

d. $z = \dfrac{x - \mu}{\sigma} = \dfrac{50 - 40}{100} = .1$

$z = \dfrac{70 - 40}{100} = .3$

$z = \dfrac{80 - 40}{100} = 4$

2.159   a. $s^2 = \dfrac{\sum x^2 - \dfrac{\left(\sum x\right)^2}{n}}{n - 1} = \dfrac{246 - \dfrac{63^2}{22}}{22 - 1} = 3.1234$

b. $s^2 = \dfrac{\sum x^2 - \dfrac{\left(\sum x\right)^2}{n}}{n - 1} = \dfrac{666 - \dfrac{106^2}{25}}{25 - 1} = 9.0233$

c. $s^2 = \dfrac{\sum x^2 - \dfrac{\left(\sum x\right)^2}{n}}{n - 1} = \dfrac{76 - \dfrac{11^2}{7}}{7 - 1} = 9.7857$

2.161 a. $\sum x = 4 + 6 + 6 + 5 + 6 + 7 = 34$

$\sum x^2 = 4^2 + 6^2 + 6^2 + 5^2 + 6^2 + 7^2 = 198$

$\bar{x} = \dfrac{\sum x}{n} = \dfrac{34}{6} = 5.67$

$s^2 = \dfrac{\sum x^2 - \dfrac{\left(\sum x\right)^2}{n}}{n-1} = \dfrac{198 - \dfrac{34^2}{6}}{6-1} = \dfrac{5.3333}{5} = 1.0667$

$s = \sqrt{1.0667} = 1.03$

b. $\sum x = -1 + 4 + (-3) + 0 + (-3) + (-6) = -9$

$\sum x^2 = (-1)^2 + 4^2 + (-3)^2 + 0^2 + (-3)^2 + (-6)^2 = 71$

$\bar{x} = \dfrac{\sum x}{n} = \dfrac{-9}{6} = -\$1.5$

$s^2 = \dfrac{\sum x^2 - \dfrac{\left(\sum x\right)^2}{n}}{n-1} = \dfrac{71 - \dfrac{(-9)^2}{6}}{6-1} = \dfrac{57.5}{5} = 11.5$ dollars squared

$s = \sqrt{11.5} = \$3.39$

c. $\sum x = \dfrac{3}{5} + \dfrac{4}{5} + \dfrac{2}{5} + \dfrac{1}{5} + \dfrac{1}{16} = 2.0625$

$\sum x^2 = \left(\dfrac{3}{5}\right)^2 + \left(\dfrac{4}{5}\right)^2 + \left(\dfrac{2}{5}\right)^2 + \left(\dfrac{1}{5}\right)^2 + \left(\dfrac{1}{16}\right)^2 = 1.2039$

$\bar{x} = \dfrac{\sum x}{n} = \dfrac{2.0625}{5} = .4125\%$

$s^2 = \dfrac{\sum x^2 - \dfrac{\left(\sum x\right)^2}{n}}{n-1} = \dfrac{1.2039 - \dfrac{2.0625^2}{5}}{5-1} = \dfrac{.3531}{4} = .0883\%$ squared

$s = \sqrt{.0883} = .30\%$

d.   (a)   Range $= 7 - 4 = 3$

(b)   Range $= \$4 - (\$-6) = \$10$

(c)   Range $= \dfrac{4}{5}\% - \dfrac{1}{16}\% = \dfrac{64}{80}\% - \dfrac{5}{80}\% = \dfrac{59}{80\%} = .7375\%$

2.163   Using MINITAB, the scatterplot is:

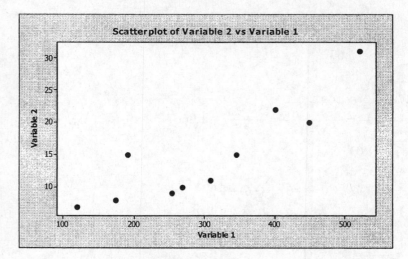

From the scatterplot, it appears that there is a trend. As variable 1 increases, variable 2 also tends to increase.

2.165   a.   The type of graph is a pie chart.

        b.   Breast cancer is treated most often. Breast cancer consists of 17% of all cancers treated at Moffitt.

        c.   Of all Moffitt's patients, 13% are treated for melanoma, 3% are treated for lymphoma, and 3% are treated for leukemia. Altogether, 13% + 3% + 3% = 19% of all patients at Moffitt are treated for one of the three types of cancer.

2.167   a.   The sample mean is 603.7. This is the average of the 98 observations. The standard deviation is StDev = 185.4. We would expect most of the observations to fall within 2 standard deviations of the mean. The minimum value is 216. This is the smallest observation in the data set. The Q1 value is 475. This is the 25th percentile. Twenty-five percent of all observations in the data set are less than or equal to 475. The median is 605.0. Half of the observations in the data set are above 605 and half are below. The Q3 value is 724.3. This is the 75th percentile. Seventy-five percent of all observations in the data set are less than or equal to 724.3. The maximum value is 1240.0. This is the value of the largest observation in the data set.

        b.   The z-score is: $z = \dfrac{x - \bar{x}}{s} = \dfrac{408 - 603.7}{185.4} = -1.06$

              A head-injury rating of 408 is less than the mean head-injury rating. It is a little more than one standard deviation below the mean.

2.169   a.   Using MINITAB, a bar chart is:

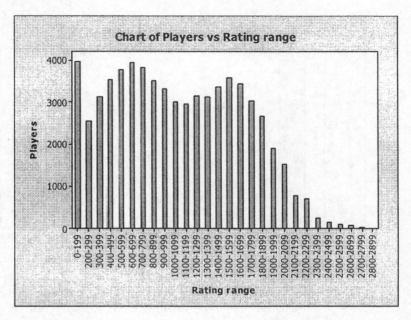

b.   The total number of players with ratings of 2,000 or higher is 1 + 13 + 66 + 87 + 133 + 231 + 691 + 783 + 1516 = 3521. The percentage of players is found by dividing the number of players with ratings of 2,000 or higher by the total number of players which is 65,455 and then multiplying by 100%. The percentage is (3,521 / 65,455)*100% = 5.38%.

c.   The average rating is 1068. This is a measure of central tendency or a rating of an average player.

2.171   a.   To compute the relative frequencies for each category, divide the frequency by the total sample size. For category "Violence to others," the relative frequency is 373/562 = .66. The rest of the relative frequencies are found in a similar manner and are displayed in the table.

Media Coverage	Number of Items	Relative Frequency
Violence to others	373	.66
Sympathetic	102	.18
Harm to self	71	.13
Comic images	12	.02
Criticism of definitions	4	.01
Total	562	1.00

b.   The relative frequency bar graph is:

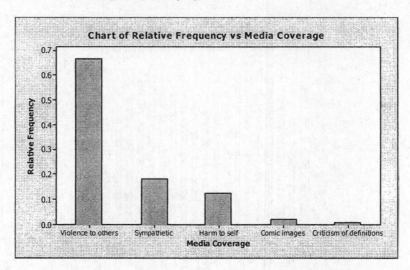

c.   Most of the items related to mental health were classified as "violence to others." The next two most reported classifications were "sympathetic " and "harm to self."

2.173   a.   Using MINITAB, the Pareto diagram is:

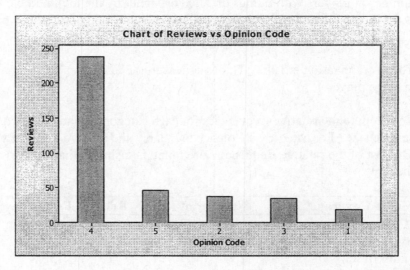

The opinion that occurred most often was "favorable/recommended" with 238 responses. The total number of responses was $19 + 37 + 35 + 238 + 46 = 375$. The proportion of books receiving a "favorable/recommended" opinion is $238/375 = .635$.

b.   Books receiving either a 4 (favorable/recommended or a 5 (outstanding/significant) were reviewed as favorable and recommended for purchase. The total number of books receiving a rating of 4 or 5 is $238 + 46 = 284$. The proportion of books receiving these ratings is $284/375 = .757$. This proportion is more than .75 of 75%. Thus, the statement made is correct.

2.175 a. To display the status, we use a pie chart. From the pie
chart, we see that 58% of the Beanie babies are retired
and 42% are current.

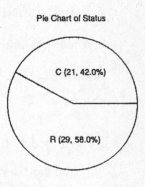

b. Using Minitab, a histogram of the values is:

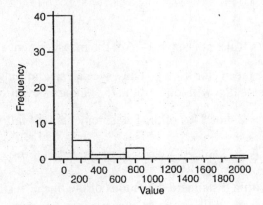

Most (40 of 50) Beanie babies have values less than $100. Of the remaining 10, 5 have
values between $100 and $300, 1 has a value between $300 and $500, 1 has a value
between $500 and $700, 2 have values between $700 and $900, and 1 has a value
between $1900 and $2100.

c. A plot of the value versus the age of the Beanie Baby is as follows:

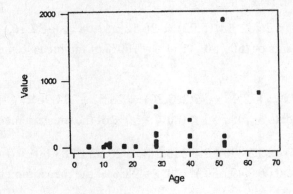

From the plot, it appears that as the age increases, the value tends to increase.

d.   Using MINITAB, the descriptive statistics are:

Descriptive Statistics: Age, Value

```
Variable    N   N*    Mean   SE Mean   StDev   Minimum      Q1   Median      Q3   Maximum
Age        50   0    26.52      2.25   15.92      5.00   12.00    28.00   40.C0     64.00
Value      50   0    128.9      44.3   313.2      10.0    15.0     26.5    65.0    1900.0
```

According to Chebyshev's Rule, at least $1-\dfrac{1}{k^2}$ of the measurements will fall within k standard deviations of the mean (for k > 1). Thus, we can say nothing about the percentage of measurements that will fall within .75 standard deviations of the mean. At least $1-\dfrac{1}{2.5^2}=1-.16=.84$ or 84% of the measurements will fall within 2.5 standard deviations of the mean. At least $1-\dfrac{1}{4^2}=1-.0625=.9375$ or 93.75% of the measurements will fall within 4 standard deviations of the mean.

e.   For the interval $\bar{x}\pm.75s \Rightarrow 26.52\pm.75(15.92)\Rightarrow 26.52\pm11.94 \Rightarrow (14.58,\ \ 38.46)$, there are 22 measurements or (22 / 50)*100% = 44% of the measurements.

For the interval $\bar{x}\pm 2.5s \Rightarrow 26.52\pm 2.5(15.92)\Rightarrow 26.52\pm39.8 \Rightarrow (-13.28,\ \ 66.32)$, there are 50 measurements or (50 / 50)*100% = 100% of the measurements. This is at least 84%.

For the interval $\bar{x}\pm 4s \Rightarrow 26.52\pm 4(15.92)\Rightarrow 26.52\pm63.68 \Rightarrow (-37.16,\ \ 90.20)$, there are 50 measurements or (50 / 50)*100% = 100% of the measurements. This is at least 93.75%.

f.   For the interval $\bar{x}\pm.75s \Rightarrow 128.90\pm.75(313.20)\Rightarrow 26.52\pm 234.90 \Rightarrow (-106.00,\ \ 363.80)$, there are 45 measurements or (45 / 50)*100% = 90% of the measurements.

For the interval $\bar{x}\pm 2.5s \Rightarrow 128.90\pm 2.5(313.20)\Rightarrow 26.52\pm 783.00 \Rightarrow (-654.10,\ \ 911.90)$, there are 49 measurements or (49 / 50)*100% = 98% of the measurements. This is at least 84%.

For the interval $\bar{x}\pm 4s \Rightarrow 128.90\pm 4(313.20)\Rightarrow 26.52\pm 1,252.80 \Rightarrow (-1,123.90,\ \ 1,381.70)$, there are 49 measurements or (49 / 50)*100% = 98% of the measurements. This is at least 93.75%.

2.177 A relative frequency bar graph is used to depict the data:

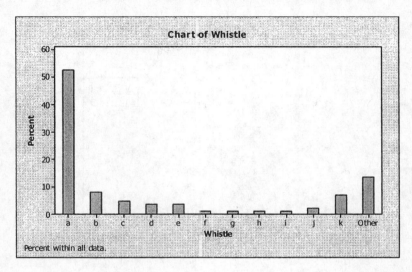

From the bar graph, over half of the whistle types were "Type a." The next most frequent category was the "Other types" with a relative frequency of about .14. Whistle types b and k were the next most frequent. None of the other whistle types had relative frequencies higher than .05.

2.179 a. Using MINITAB, the boxplot is:

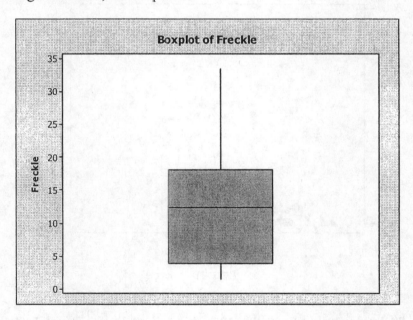

From the boxplot, there are no outliers detected.

b. Since there are no outliers, we cannot find the z-score for the points identified as outliers.

2.181 a. For each transect, three variables were measured. The number of seabirds found is quantitative. The length of the transect is also quantitative. Whether or not the transect was in an oiled area is qualitative.

b.   The experimental unit is the transect.

c.   A pie chart of the oiled and unoiled areas is:

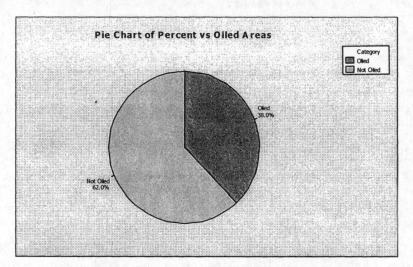

d.   Using MINITAB, a scattergram of the data is:

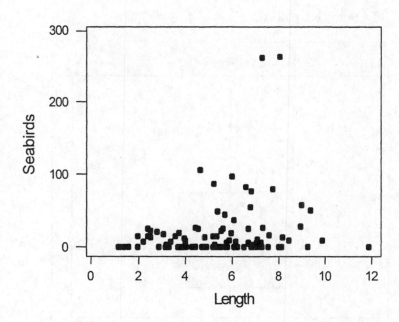

e.   The mean density for the unoiled area is 3.27, while the mean for the oiled area is 3.495. The median for the unoiled area is .89 and is .70 for the oiled area. These are both fairly similar.

f.   Using Chebyshev's Theorem, at least 75% of the observations will fall within 2 standard deviations of the mean. This interval for unoiled areas would be:

$$\bar{x} \pm 2s \Rightarrow 3.27 \pm 2(6.7) \Rightarrow 3.27 \pm 13.4 \Rightarrow (-10.13, 16.67)$$

g. Using Chebyshev's Theorem, at least 75% of the observations will fall within 2 standard deviations of the mean. This interval for oiled areas would be:

$$\bar{x} \pm 2s \Rightarrow 3.495 \pm 2(5.968) \Rightarrow 3.495 \pm 11.936 \Rightarrow (-8.441, 15.431)$$

h. From the above two intervals, we know that at least 75% of the observations for the unoiled area will fall between -10.31 and 16.67 and at most 25% of the observations will fall above 15.431 for the oiled areas. Thus, the unoiled areas would be more likely to have a seabird density of 16.

2.183 a. Using Minitab, a stem-and-leaf display of the data is:

```
Stem-and-leaf of VELOCITY      N = 51
Leaf Unit = 100

   1    18  4
   3    18  79
  12    19  001112444
  18    19  566788
  20    20  12
  21    20  7
  21    21
  23    21  99
  (5)   22  11344
  23    22  5666777777889
  10    23  001222344
   1    23
   1    24
   1    24  9
```

b. From this stem-and-leaf display, it is fairly obvious that there are two different distributions since there are two groups of data.

c. Since there appears to be two distributions, we will compute two sets of numerical descriptive measures. We will call the group with the smaller velocities A1775A and the group with the larger velocities A1775B.

For A1775A:

$$\bar{x} = \frac{\sum x}{n} = \frac{408,707}{21} = 19,462.2$$

$$s^2 = \frac{\sum x^2 - \dfrac{\left(\sum x\right)^2}{n}}{n-1} = \frac{7,960,019,531 - \dfrac{408,707^2}{21}}{21-1} = 283,329.3$$

$$s = \sqrt{283,329.3} = 532.29$$

For A1775B:

$$\bar{x} = \frac{\sum x}{n} = \frac{685,154}{30} = 22,838.5$$

$$s^2 = \frac{\sum x^2 - \dfrac{\left(\sum x\right)^2}{n}}{n-1} = \frac{15,656,992,942 - \dfrac{685,154^2}{30}}{30-1} = 314,694.83$$

$$s = \sqrt{314,694.83} = 560.98$$

d.	To determine which of the two clusters this observation probably belongs to, we will compute z-scores for this observation for each of the two clusters.

For A1775A:
$$z = \frac{20,000 - 19,462.2}{532.29} = 1.01$$

Since this z-score is so small, it would not be unlikely that this observation came from this cluster.

For A1775B:

$$z = \frac{20,000 - 22,838.5}{560.98} = -5.06$$

Since this z-score is so large (in magnitude), it would be very unlikely that this observation came from this cluster.

Thus, this observation probably came from the cluster A1775A.

2.185	First, compute the z-score for 1.80: $z = \dfrac{x - \mu}{\sigma} = \dfrac{1.80 - 2.00}{.08} = -2.5$

If the data are actually mound-shaped, it would be very unusual (less than 2.5%) to observe a batch with 1.80% zinc phosphide if the true mean is 2.0%. Thus, if we did observe 1.8%, we would conclude that the mean percent of zinc phosphide in today's production is probably less than 2.0%.

2.187	a.	The variable "Days in Jail Before Suicide" is measured on a numerical scale, so it is quantitative. The variables "Marital Status", "Race", "Murder/Manslaughter Charge", and "Time of Suicide" are not measured on a numerical scale, so they are all qualitative. The variable "Year" is measured on a numerical scale, so it is quantitative.

b.   Using MINITAB, the pie chart for the data is:

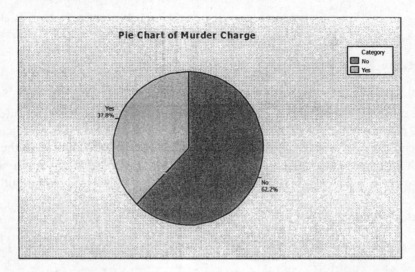

Suicides are more likely to be committed by inmates charged with lesser crimes than by inmates charged with murder/manslaughter. Of the suicides reported, 62.2% are committed by those convicted of a lesser charge.

c.   Using MINITAB, the pie chart for the data is:

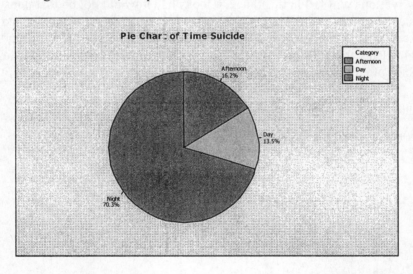

Suicides are much more likely to be committed at night than any other time. Of the suicides reported, 70.3% were committed at night.

d. Using MINITAB, the descriptive statistics are:

Variable	N	Mean	Median	TrMean	StDev	SE Mean
Jail	37	41.4	15.0	30.3	66.7	11.0

Variable	Minimum	Maximum	Q1	Q3
Jail	1.0	309.0	4.0	41.5

The mean length of time an inmate spent in jail before committing suicide is 41.4 days.
The median length of time an inmate spent in jail before committing suicide is 15 days.
Since the mean is much larger than the median, the data are skewed to the right. Most
of those committing suicide, commit it within 15 days of arriving in jail. However,
there are a few inmates who spend many more days in jail before committing suicide.

e. First, compute the z-score associated with 200 days:

$$z = \frac{x - \bar{x}}{s} = \frac{200 - 41.4}{66.7} = 2.38$$

Using Chebyshev's rule, we know that at most $1/2^2 = 1/4$ of the observations will fall
more than 2 standard deviations from the mean. Also, we know at most $1/3^2 = 1/9$ of
the observations will fall more than 3 standard deviations from the mean. The score 200
is 2.38 standard deviations from the mean. Since we know that at most 1/4 of the
observations will fall in this range, it looks like it would not be that unusual to see
someone commit suicide after 200 days. However, if we look at the data, of the 37
observations, there are only 2 observations of 200 or larger. This proportion is $2/37 =$
.054. Using this information, it would be rather unusual for an inmate to commit
suicide after 200 days.

f. Using MINITAB, the stem-and-leaf plot of the data is:

```
Stem-and-leaf of Year      N = 37
Leaf Unit = 1.0

  1    196  7
  5    196  8889
 11    197  000001
 13    197  23
 16    197  455
 18    197  66
 (2)   197  99
 17    198  000111
 11    198  233
  8    198  55555
  3    198  77
  1    198  9
```

From the stem-and-leaf plot, it does not appear that the number of suicides have
decreased over time.

2.189 a. Due to the "elite" superstars, the salary distribution is skewed to the right. Since this
implies that the median is less than the mean, the players' association would want to use
the median.

b. The owners, by the logic of part **a**, would want to use the mean.

Probability

3.1   An experiment is an act or process of observation that leads to a single outcome that cannot be predicted with certainty.

3.3   A sample space of an experiment is a collection of all its sample points.

3.5   Two probability rules for sample points are:

    1.   All sample point probabilities **must** lie between 0 and 1, i.e. $0 \leq p_i \leq 1$.

    2.   The probabilities of all sample points within a sample space **must** sum to 1, i.e., $\sum p_i = 1$.

3.7   The probability of an event A is calculated by summing the probabilities of the sample points in the sample space for A.

3.9   a.   Since the probabilities must sum to 1,

$$P(E_3) = 1 - P(E_1) - P(E_2) - P(E_4) - P(E_5) = 1 - .1 - .2 - .1 - .1 = .5$$

    b. $P(E_3) = 1 - P(E_3) - P(E_2) - P(E_4) - P(E_5)$
$$\Rightarrow 2P(E_3) = 1 - .1 - .2 - .1 \Rightarrow 2P(E_3) = .6 \Rightarrow P(E_3) = .3$$

    c. $P(E_3) = 1 - P(E_1) - P(E_2) - P(E_4) - P(E_5) = 1 - .1 - .1 - .1 - .1 = .6$

3.11 $P(A) = P(1) + P(2) + P(3) = .05 + .20 + .30 = .55$

    $P(B) = P(1) + P(3) + P(5) = .05 + .30 + .15 = .50$

    $P(C) = P(1) + P(2) + P(3) + P(5) = .05 + .20 + .30 + .15 = .70$

3.13   a. $\dbinom{5}{2} = \dfrac{5!}{2!(5-2)!} = \dfrac{5!}{2!3!} = \dfrac{5 \cdot 4 \cdot 3 \cdot 2 \cdot 1}{2 \cdot 1 \cdot 3 \cdot 2 \cdot 1} = 10$

    b. $\dbinom{6}{3} = \dfrac{6!}{3!(6-3)!} = \dfrac{6!}{3!3!} = \dfrac{6 \cdot 5 \cdot 4 \cdot 3 \cdot 2 \cdot 1}{3 \cdot 2 \cdot 1 \cdot 3 \cdot 2 \cdot 1} = 20$

    c. $\dbinom{20}{5} = \dfrac{20!}{5!(20-5)!} = \dfrac{20!}{5!15!} = \dfrac{20 \cdot 19 \cdot 18 \cdots 2 \cdot 1}{5 \cdot 4 \cdot 3 \cdot 2 \cdot 1 \cdot 15 \cdot 14 \cdot 13 \cdots 2 \cdot 1} = 15{,}504$

3.15 a. If we denote the marbles as B_1, B_2, R_1, R_2, R_3, then the ten equally likely sample points in the sample space would be:

$$S: \begin{bmatrix} (B_1,B_2),(B_1,R_1),(B_1,R_2),(B_1,R_3),(B_2,R_1) \\ (B_2,R_2),(B_2,R_3),(R_1,R_2),(R_1,R_3),(R_2,R_3) \end{bmatrix}$$

Notice that order is ignored, as the only concern is whether or not a marble is selected.

b. Each of these ten would be equally likely, implying that each occurs with a probability 1/10.

c. $P(A) = \dfrac{1}{10}$ $P(B) = 6\left(\dfrac{1}{10}\right) = \dfrac{6}{10} = \dfrac{3}{5}$ $P(C) = 3\left(\dfrac{1}{10}\right) = \dfrac{3}{10}$

3.17 Use the relative frequency as an estimate for the probability. Thus,
P(physically assaulted) = 600/12,000 = .05.

3.19 a. Let A = {chicken passes inspection with fecal contamination}. $P(A) = 1 / 100 = .01$.

b. Yes. The relative frequency of passing inspection with fecal contamination is 306/32,075 = .0095 ≈ .01.

3.21 a. Define the following event:

A: {Beech tree is damaged by fungi}

$$P(A) = \frac{49}{188} = .261$$

b. There would be 3 sample points for this experiment: trunk, leaves, and branch. We can convert the percentages to proportions and use these to estimate the probabilities.

Sample Points	Probabilities
Trunk	.85
Leaves	.10
Branch	.05
Total	1.00

3.23 a. There are 4 possible outcomes for this experiment: Interview, Observation + Participation, Observation Only, and Grounded Theory.

b. The 4 sample points are not equally likely to occur because the different outcomes do not occur at the same rate.

c. We will use the relative frequencies to assign probabilities to the sample points.
P(Interview) = 5,079 / 7,506 = .677
P(Observation + Participation) = 1,042 / 7,506 = .139
P(Observation Only) = 848 / 7,506 = .113
P(Grounded Theory) = 537 / 7,506 = .072

d. (Interview or Grounded Theory) = (5,079 + 537) / 7,506 = 5,616 / 7,506 = .748

3.25　a.　There are a total of $\binom{5}{3} = \dfrac{5!}{3!(5-3)!} = \dfrac{5!}{3!2!} = \dfrac{5\cdot4\cdot3\cdot2\cdot1}{3\cdot2\cdot1\cdot2\cdot1} = 10$ ways to get 3-grill displays.

These 10 display combinations are:

1, 2, 3	1, 3, 4	2, 3, 4	3, 4, 5
1, 2, 4	1, 3, 5	2, 3, 5	
1, 2, 5	1, 4, 5	2, 4, 5	

However, since grill #2 must be selected, there are only 6 possibilities:

1, 2, 3	2, 3, 4
1, 2, 4	2, 3, 5
1, 2, 5	2, 4, 5

b.　We can estimate the probabilities by using the relative frequency for each sample point. The relative frequency is found by dividing the frequency by the total sample size of 124. These estimates are contained in the following table:

Grill Display Combination	Number of Students	Probability
1-2-3	35	.282
1-2-4	8	.065
1-2-5	42	.339
2-3-4	4	.032
2-3-5	1	.008
2-4-5	34	.274
TOTAL	124	1.0000

c.　Of the 6 sample points, only 3 of them contain Grill #1.

P(Grill #1 chosen) = P(1-2-3 or 1-2-4 or 1-2-5) = .282 + .065 + .339 = .686.

3.27　a.　There are a total of $\binom{8}{2} = \dfrac{8!}{2!(8-2)!} = \dfrac{8!}{2!6!} = \dfrac{8\cdot7\cdot6\cdot5\cdot4\cdot3\cdot2\cdot1}{2\cdot1\cdot6\cdot5\cdot4\cdot3\cdot2\cdot1} = 28$ possible Quinella bets.

b.　If all the players are of equal ability, then the probability of getting any combination is 1/28.

3.29　The total number of pairs of bullets is a combination of 1,837 bullets taken 2 at a time or:

$$\binom{1,837}{2} = \frac{1,837!}{2!\,1,835!} = \frac{1,837(1,836)}{2} = 1,686,366$$

The probability of finding a match or a false positive is 693 / 1,686,366 = .000411.

This probability is very small. The confidence in the FBI's forensic evidence should be very high.

3.31   a. $\binom{6}{2} = \dfrac{6!}{2!(6-2)!} = \dfrac{6!}{2!4!} = \dfrac{6 \cdot 5 \cdot 4 \cdot 3 \cdot 2 \cdot 1}{2 \cdot 1 \cdot 4 \cdot 3 \cdot 2 \cdot 1} = 15$

        b. $\binom{6}{3} = \dfrac{6!}{3!(6-3)!} = \dfrac{6!}{3!3!} = \dfrac{6 \cdot 5 \cdot 4 \cdot 3 \cdot 2 \cdot 1}{3 \cdot 2 \cdot 1 \cdot 3 \cdot 2 \cdot 1} = 20$

        c. $\binom{6}{4} = \dfrac{6!}{4!(6-4)!} = \dfrac{6!}{4!2!} = \dfrac{6 \cdot 5 \cdot 4 \cdot 3 \cdot 2 \cdot 1}{4 \cdot 3 \cdot 2 \cdot 1 \cdot 2 \cdot 1} = 15$

        d. $\binom{6}{5} = \dfrac{6!}{5!(6-5)!} = \dfrac{6!}{5!1!} = \dfrac{6 \cdot 5 \cdot 4 \cdot 3 \cdot 2 \cdot 1}{5 \cdot 4 \cdot 3 \cdot 2 \cdot 1 \cdot 1} = 6$

        e.   In addition to the combinations computed in parts a – d, we need to find the number of 0-drug combinations, 1-drug combination, and 6-drug combinations.

        $\binom{6}{0} = \dfrac{6!}{0!(6-0)!} = \dfrac{6!}{0!6!} = \dfrac{6 \cdot 5 \cdot 4 \cdot 3 \cdot 2 \cdot 1}{1 \cdot 6 \cdot 5 \cdot 4 \cdot 3 \cdot 2 \cdot 1} = 1$

        $\binom{6}{1} = \dfrac{6!}{1!(6-1)!} = \dfrac{6!}{1!5!!} = \dfrac{6 \cdot 5 \cdot 4 \cdot 3 \cdot 2 \cdot 1}{1 \cdot 5 \cdot 4 \cdot 3 \cdot 2 \cdot 1} = 6$

        $\binom{6}{6} = \dfrac{6!}{6!(6-6)!} = \dfrac{6!}{6!0!} = \dfrac{6 \cdot 5 \cdot 4 \cdot 3 \cdot 2 \cdot 1}{6 \cdot 5 \cdot 4 \cdot 3 \cdot 2 \cdot 1 \cdot 1} = 1$

        The total number of ways the 6 drugs can be combined is $15 + 20 + 15 + 6 + 1 + 6 + 1 = 64$.

3.33   The union of 2 events A and B is the event that occurs if either A or B or both occur on a single performance of the experiment.

3.35   The complement of an event A is the event that A does not occur – that is, the event consisting of all sample points that are not in event A.

3.37   The Additive Rule of Probability is: The probability of the union of events A and B is the sum of the probability of events A and B minus the probability of the intersection of events A and B, that is $P(A \cup B) = P(A) + P(B) - P(A \cap B)$.

3.39   The Additive Rule of Probability for mutually exclusive events is: If 2 events A and B are mutually exclusive, the probability of the union of A and B equals the sum of the probabilities of A and B; that is, $P(A \cup B) = P(A) + P(B)$.

3.41 a. A: {*HHH, HHT, HTH, THH, TTH, THT, HTT*}
    B: {*HHH, TTH, THT, HTT*}
    $A \cup B$: {*HHH, HHT, HTH, THH, TTH, THT, HTT*}
    A^c: {*TTT*}
    $A \cap B$: {*HHH, TTH, THT, HTT*}

b. If the coin is fair, then each of the 8 possible outcomes are equally likely, with probability 1/8.

$$P(A) = \frac{7}{8} \qquad P(B) = \frac{4}{8} = \frac{1}{2} \qquad P(A \cup B) = \frac{7}{8}$$

$$P(A^c) = \frac{1}{8} \qquad P(A \cap B) = \frac{4}{8} = \frac{1}{2}$$

c. $P(A \cup B) = P(A) + P(B) - P(A \cap B) = \frac{7}{8} + \frac{1}{2} - \frac{1}{2} = \frac{7}{8}$

d. No. $P(A \cap B) = \frac{1}{2}$ which is not 0.

3.43 a. $P(A) = P(E_1) + P(E_2) + P(E_3) + P(E_5) + P(E_6) = \frac{1}{5} + \frac{1}{5} + \frac{1}{5} + \frac{1}{20} + \frac{1}{10} = \frac{15}{20} = \frac{3}{4}$

b. $P(B) = P(E_2) + P(E_3) + P(E_4) + P(E_7) = \frac{1}{5} + \frac{1}{5} + \frac{1}{20} + \frac{1}{5} = \frac{13}{20}$

c. $P(A \cup B) = P(E_1) + P(E_2) + P(E_3) + P(E_4) + P(E_5) + P(E_6) + P(E_7)$
    $= \frac{1}{5} + \frac{1}{5} + \frac{1}{5} + \frac{1}{20} + \frac{1}{20} + \frac{1}{10} + \frac{1}{5} = 1$

d. $P(A \cap B) = P(E_2) + P(E_3) = \frac{1}{5} + \frac{1}{5} = \frac{2}{5}$

e. $P(A^c) = 1 - P(A) = 1 - \frac{3}{4} = \frac{1}{4}$

f. $P(B^c) = 1 - P(B) = 1 - \frac{13}{20} = \frac{7}{20}$

g. $P(A \cup A^c) = P(E_1) + P(E_2) + P(E_3) + P(E_4) + P(E_5) + P(E_6) + P(E_7)$
    $= \frac{1}{5} + \frac{1}{5} + \frac{1}{5} + \frac{1}{20} + \frac{1}{20} + \frac{1}{10} + \frac{1}{5} = 1$

h. $P(A^c \cap B) = P(E_4) + P(E_7) = \frac{1}{20} + \frac{1}{5} = \frac{5}{20} = \frac{1}{4}$

3.45   a. $P(A) = .50 + .10 + .05 = .65$

b. $P(B) = .10 + .07 + .50 + .05 = .72$

c. $P(C) = .25$

d. $P(D) = .05 + .03 = .08$.

e. $P(A^c) = .25 + .07 + .03 = .35$ (Note: $P(A^c) = 1 - P(A) = 1 - .65 = .35$)

f. $P(A \cup B) = P(B) = .10 + .07 + .50 + .05 = .72$

g. $P(A \cap B) = P(A) = .50 + .10 + .05 = .65$

h.   Two events are mutually exclusive if they have no sample points in common or if the probability of their intersection is 0.

$P(A \cap B) = .50 + .10 + .05 = .65$. Since this is not 0, A and B are not mutually exclusive.
$P(A \cap C) = 0$. Since this is 0, A and C are mutually exclusive.

$P(A \cap D) = .05$. Since this is not 0, A and D are not mutually exclusive.

$P(B \cap C) = 0$. Since this is 0, B and C are mutually exclusive.

$P(B \cap D) = .05$. Since this is not 0, B and D are not mutually exclusive.

$P(C \cap D) = 0$. Since this is 0, C and D are mutually exclusive.

3.47   a.   The sample points of this experiment are the locations where toxic chemical incidents occurred. They are:

School laboratory, In Transit, Chemical plant, Non-chemical plant, and Other.

b.   Reasonable probabilities would be the percents of the incidents changed to proportions.

P(School laboratory) = .06, P(In Transit) = .26, P(Chemical plant) = .21,
P(Non-chemical plant) = .35, P(Other) = .12

c. P(School laboratory) = .06

d. P(Chemical plant or Non-chemical plant) = .21 + .35 = .56

e. P(Not occur In Transit) = 1 − P(In Transit) = 1 − .26 = .74

3.49   a.   A Venn Diagram that illustrates the results of the gene profiling analysis is:

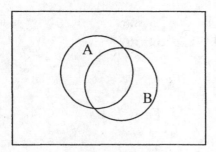

    b.   From the problem, we know $P(A) = .41$, $P(B) = .42$, and $P(A \cap B) = .40$
          $P(A \cup B) = P(A) + P(B) - P(A \cap B) = .41 + .42 - .40 = .43$.

    c. $P(\text{Neither}) = P(A \cup B)^C = 1 - P(A \cup B) = 1 - .43 = .57$

3.51   a.   The event $A \cap B$ is the event the outcome is black and odd. The event is $A \cap B$: {11, 13, 15, 17, 29, 31, 33, 35}

    b.   The event $A \cup B$ is the event the outcome is black or odd or both. The event $A \cup B$ is {2, 4, 6, 8, 10, 11, 13, 15, 17, 20, 22, 24, 26, 28, 29, 31, 33, 35, 1, 3, 5, 7, 9, 19, 21, 23, 25, 27}

    c.   Assuming all events are equally likely, each has a probability of 1/38.

$$P(A) = 18\left(\frac{1}{38}\right) = \frac{18}{38} = \frac{9}{19}$$

$$P(B) = 18\left(\frac{1}{38}\right) = \frac{18}{38} = \frac{9}{19}$$

$$P(A \cap B) = 8\left(\frac{1}{38}\right) = \frac{8}{38} = \frac{4}{19}$$

$$P(A \cup B) = 28\left(\frac{1}{38}\right) = \frac{28}{38} = \frac{14}{19}$$

$$P(C) = 18\left(\frac{1}{38}\right) = \frac{18}{38} = \frac{9}{19}$$

    d.   The event $A \cap B \cap C$ is the event the outcome is odd and black and low. The event $A \cap B \cap C$ is {11, 13, 15, 17}.

    e. $P(A \cup B) = P(A) + P(B) - P(A \cap B) = \dfrac{9}{19} + \dfrac{9}{19} - \dfrac{4}{19} = \dfrac{14}{19}$

    f. $P(A \cap B \cap C) = 4\left(\dfrac{1}{38}\right) = \dfrac{4}{38} = \dfrac{2}{19}$

g.   The event $A \cup B \cup C$ is the event the outcome is odd or black or low. The event $A \cup B \cup C$ is:

$$\{1, 2, 3, \dots, 29, 31, 33, 35\}$$

or

$$\{\text{All simple events except } 00, 0, 30, 32, 34, 36\}$$

h. $P(A \cup B \cup C) = 32\left(\dfrac{1}{38}\right) = \dfrac{32}{38} = \dfrac{16}{19}$

3.53   a.   Define the following events:

F: {Fight}
N: {No fight}
W: {Initiator Wins}
T: {No Clear Winner}
L: {Initiator Loses}

$P(F \cap W) = 26 / 167 = .156$

b. $P(N) = 103 / 167 = .617$

c. $P(T) = 35 / 167 = .210$

d. $P(F \cup L) = (26 + 23 + 15 + 11) / 167 = 75 / 167 = .449$

e.   Yes. $P(T \cap L) = 0$.

3.55   a.   The sample points for this experiment are:

(PTW-R, Jury), (PTW-R, Judge), (PTW-A/D, Jury), (PTW-A/D, Judge),
(DTW-R, Jury), (DTW-R, Judge), (DTW-A/D, Jury), (DTW-A/D, Judge)

b. $P(A) = \dfrac{1,465}{2,143} = .684$

c. $P(B) = \dfrac{265}{2,143} = .124$

d.   No. $P(A \cap B) = \dfrac{194}{2,143} = .091$. Since this is not 0, events A and B are not mutually exclusive.

e. $P(A^c) = 1 - P(A) = 1 - .684 = .316$.

f. $P(A \cup B) = P(A) + P(B) - P(A \cap B) = .684 + .124 - .091 = .717$

g. $P(A \cap B) = \dfrac{194}{2,143} = .091$

3.57   There are a total of 6 x 6 x 6 = 216 possible outcomes to the throwing of 3 fair dice. A partition is a set of 3 numbers that sum to a particular number.

There are 6 partitions for a sum of 9 on 3 dice:

$$(1, 2, 6), (1, 3, 5), (1, 4, 4), (2, 2, 5), (2, 3, 4), (3, 3, 3).$$

However, some of these partitions can be obtained in more than one way if we let the order of the numbers (corresponding to particular die) matter. The total set of possible outcomes that sum to 9, where order matters is:

(1, 2, 6),	(1, 3, 5),	(1, 4, 4),	(2, 2, 5),	(2, 3, 4),	(3, 3, 3),
(1, 6, 2),	(1, 5, 3),	(4, 1, 4),	(2, 5, 2),	(2, 4, 3),	
(2, 1, 6),	(3, 1, 5),	(4, 4, 1),	(5, 2, 2),	(3, 2, 4),	
(2, 6, 1),	(3, 5, 1),			(3, 4, 2),	
(6, 1, 2),	(5, 1, 3),			(4, 2, 3),	
(6, 2, 1),	(5, 3, 1),			(4, 3, 2)	

There are a total of 25 sample points. Thus, the probability of getting a sum of 9 is 25 / 216 = .116.
There are also 6 partitions for a sum of 10 on 3 dice:

$$(1, 3, 6), (1, 4, 5), (2, 2, 6), (2, 3, 5), (2, 4, 4), (3, 3, 4).$$

Again, some of these partitions can be obtained in more than one way if we let the order of the numbers (corresponding to particular die) matter. The total set of possible outcomes that sum to 10, where order matters is:

(1, 3, 6),	(1, 4, 5),	(2, 2, 6),	(2, 3, 5),	(2, 4, 4),	(3, 3, 4),
(1, 6, 3),	(1, 5, 4),	(2, 6, 2),	(2, 5, 3),	(4, 2, 4),	(3, 4, 3),
(3, 1, 6),	(4, 1, 5),	(6, 2, 2),	(3, 2, 5),	(4, 4, 2),	(4, 3, 3),
(3, 6, 1),	(4, 5, 1),		(3, 5, 2),		
(6, 1, 3),	(5, 1, 4),		(5, 2, 3),		
(6, 3, 1),	(5, 4, 1),		(5, 3, 2)		

There are a total of 27 sample points. Thus, the probability of getting a sum of 10 is 27 / 216 = .125.

3.59   The Multiplicative Rule of Probability for two independent events is:

$$P(A \cap B) = P(A)P(B)$$

3.61   The Multiplicative Rule of Probability is:

$$P(A \cap B) = P(A)P(B \mid A), \text{ or equivalently } P(A \cap B) = P(B)P(A \mid B)$$

3.63   a. $P(A \mid B) = \dfrac{P(A \cap B)}{P(B)} = \dfrac{.1}{.2} = .5$

        b. $P(B \mid A) = \dfrac{P(A \cap B)}{P(A)} = \dfrac{.1}{.4} = .25$

        c.   No, events A and B are not independent. If A and B were independent, then $P(A \mid B) = P(A)$. For this problem, $P(A \mid B) = .5$ and $P(A) = .4$.

3.65   a. $P(A \cap B) = P(A)\,P(B) = .4(.2) = .08$

        b. $P(A \mid B) = \dfrac{P(A \cap B)}{P(B)} = \dfrac{.08}{.2} = .4$

        c. $P(A \cup B) = P(A) + P(B) - P(A \cap B) = .4 + .2 - .08 = .52$

3.67   a. $P(A) = P(E_1) + P(E_2) + P(E_3) = .1 + .1 + .2 = .4$
              $P(B) = P(E_2) + P(E_3) + P(E_5) = .1 + .2 + .1 = .4$
              $P(A \cap B) = P(E_2) + P(E_3) = .1 + .2 = .3$

        b. $P(E_1 \mid A) = \dfrac{P(E_1 \cap A)}{P(A)} = \dfrac{.1}{.4} = .25$

              $P(E_2 \mid A) = \dfrac{P(E_2 \cap A)}{P(A)} = \dfrac{.1}{.4} = .25$

              $P(E_3 \mid A) = \dfrac{P(E_3 \cap A)}{P(A)} = \dfrac{.2}{.4} = .5$

              We are given that $P(E_1) = .1$, $P(E_2) = .1$, and $P(E_3) = .2$
              Thus, $P(E_1) = P(E_2)$ and $P(E_3) = 2P(E_1)$
              From above, $P(E_1 \mid A) = P(E_2 \mid A)$ and $P(E_3 \mid A) = 2P(E_1 \mid A)$

              $P(E_1 \mid A) + P(E_2 \mid A) + P(E_3 \mid A) = .25 + .25 + .50 = 1.00$

        c.   Using the sum of the conditional probabilities,
              $$P(B \mid A) = P(E_2 \mid A) + P(E_3 \mid A) = .25 + .5 = .75$$
              Using the formula,

              $$P(B \mid A) = \dfrac{P(A \cap B)}{P(A)} = \dfrac{.3}{.4} = .75$$

3.69   The 36 possible outcomes obtained when tossing two dice are listed below:

$$(1, 1)\ (1, 2)\ (1, 3)\ (1, 4)\ (1, 5)\ (1, 6)$$
$$(2, 1)\ (2, 2)\ (2, 3)\ (2, 4)\ (2, 5)\ (2, 6)$$
$$(3, 1)\ (3, 2)\ (3, 3)\ (3, 4)\ (3, 5)\ (3, 6)$$
$$(4, 1)\ (4, 2)\ (4, 3)\ (4, 4)\ (4, 5)\ (4, 6)$$
$$(5, 1)\ (5, 2)\ (5, 3)\ (5, 4)\ (5, 5)\ (5, 6)$$
$$(6, 1)\ (6, 2)\ (6, 3)\ (6, 4)\ (6, 5)\ (6, 6)$$

A: $\{(1, 2), (1, 4), (1, 6), (2, 1), (2, 3), (2, 5), (3, 2), (3, 4), (3, 6), (4, 1), (4, 3),$
$(4, 5), (5, 2), (5, 4), (5, 6), (6, 1), (6, 3), (6, 5)\}$

B: $\{(3, 6), (4, 5), (5, 4), (5, 6), (6, 3), (6, 5), (6, 6)\}$

$A \cap B$: $\{(3, 6), (4, 5), (5, 4), (5, 6), (6, 3), (6, 5)\}$

If A and B are independent, then $P(A)P(B) = P(A \cap B)$.

$$P(A) = \frac{18}{36} = \frac{1}{2} \qquad P(B) = \frac{7}{36} \qquad P(A \cap B) = \frac{6}{36} = \frac{1}{6}$$

$P(A)P(B) = \frac{1}{2} \cdot \frac{7}{36} = \frac{7}{72} \neq \frac{1}{6} = F(A \cap B)$. Thus, A and B are not independent.

3.71   Let W_1 and W_2 represent the two white chips, R_1 and R_2 represent the two red chips, and B_1 and B_2 represent the two blue chips. The sample space is:

W_1W_2	W_2R_1	R_1B_1
W_1R_1	W_2R_2	R_1B_2
W_1R_2	W_2B_1	R_2B_1
W_1B_1	W_2B_2	R_2B_2
W_1B_2	R_1R_2	B_1B_2

Assuming each event is equally likely, each event will have a probability of 1/15.

Then, $P(A) = P(W_1W_2) + P(R_1R_2) + P(B_1B_2) = 3\left(\frac{1}{15}\right) = \frac{3}{15} = \frac{1}{5}$

$$P(B) = P(R_1R_2) = \frac{1}{15}$$

$P(C) = P(W_1W_2) + P(W_1R_1) + P(W_1R_2) + P(W_1B_1) + P(W_1B_2) + P(W_2R_1)$
$$+ P(W_2R_2) + P(W_2B_1) + P(W_2B_2) + P(R_1R_2) + P(R_1B_1) + P(R_1B_2)$$
$$+ P(R_2B_1) + P(R_2B_2)$$

$$= 14\left(\frac{1}{15}\right) = \frac{14}{15}$$

$$P(A \cap B) = P(R_1R_2) = \frac{1}{15}$$

$$P(A^c) = 1 - P(A) = 1 - \frac{1}{5} = \frac{4}{5}$$

$$P(A^c \cap B) = 0$$

$$P(B \cap C) = P(R_1R_2) = \frac{1}{15}$$

$$P(A \cap C) = P(W_1W_2) + P(R_1R_2) = 2\left(\frac{1}{15}\right) = \frac{2}{15}$$

$$P(A^c \cap C) = P(W_1R_1) + P(W_1R_2) + P(W_1B_1) + P(W_1B_2) + P(W_2R_1)$$
$$+ P(W_2R_2) + P(W_2B_1) + P(W_2B_2) + P(R_1B_1) + P(R_1B_2)$$
$$+ P(R_2B_1) + P(R_2B_2)$$

$$= 12\left(\frac{1}{15}\right) = \frac{12}{15} = \frac{4}{5}$$

$$P(B \mid A) = \frac{P(A \cap B)}{P(A)} = \frac{\frac{1}{15}}{\frac{1}{5}} = \frac{1}{3}$$

$$P(B \mid A^c) = \frac{P(A^c \cap B)}{P(A^c)} = \frac{0}{\frac{4}{5}} = 0$$

$$P(B \mid C) = \frac{P(B \cap C)}{P(C)} = \frac{\frac{1}{15}}{\frac{14}{15}} = \frac{1}{14}$$

$$P(A \mid C) = \frac{P(A \cap C)}{P(C)} = \frac{\frac{2}{15}}{\frac{14}{15}} = \frac{1}{7}$$

$$P(C \mid A^c) = \frac{P(A^c \cap C)}{P(A^c)} = \frac{\frac{4}{5}}{\frac{4}{5}} = 1$$

3.73   Define the following events:

C: {Speeding is cause of fatal crash}
M: {Missing curve is cause of fatal crash}

From the problem, $P(C) = .3$ and $P(C \cap M) = .12$

$$P(M \mid C) = \frac{P(C \cap M)}{P(C)} = \frac{.12}{.30} = .40$$

3.75   a.   Define the following event:

           A: {Psychic picks box with crystal}
           If psychic is just guessing, $P(A) = 1/10 = .1$

      b. P(Psychic guesses correct at least once in 7 trials)
           $= 1 - P$(Psychic does not guess correct in 7 trials)

           $= 1 - P(A^c \cap A^c \cap A^c \cap A^c \cap A^c \cap A^c \cap A^c) = 1 - .9(.9)(.9)(.9)(.9)(.9)$
           $= 1 - .478 = .522$ (This assumes that the trials are independent)

      c.   If the psychic is just guessing, then P(Psychic does not guess correct in 7 trials) $= .478$.

           Thus, if the person was not a psychic (and was merely guessing) we would expect the person to guess wrong in all seven trials about half the time. This would not be a rare event. Thus, this outcome would support the notion that the person is guessing.

3.77   Define the following events:

      A: {Winner is from National League}

      B: {Winner is from American League}

      C: {Winner is from Eastern Division}

      D: {Winner is from Central Division}

      E: {Winner is from Western Division}

      a. $P(C \mid B) = \dfrac{P(C \cap B)}{P(B)} = \dfrac{7/16}{10/16} = \dfrac{7}{10} = .7$

      b. $P(A \mid D) = \dfrac{P(A \cap D)}{P(D)} = \dfrac{2/16}{4/16} = \dfrac{2}{4} = .5$

      c. $P(E^C \mid A) = \dfrac{P(E^C \cap A)}{P(A)} = \dfrac{3/16}{6/16} = \dfrac{3}{6} = .5$

3.79   Define the following event:

      R: {Fish is red snapper}

      From the problem, $P(R^C) = .77$

      a. $P(R) = 1 - P(R^C) = 1 - .77 = .23$

      b. P(At least one customer is served red snapper) $= 1 - P$(No customers are served red snapper) $= 1 - P(R^C \cap R^C \cap R^C \cap R^C \cap R^C) = 1 - .77(.77)(.77)(.77)(.77) = 1 - .271 = .729$

3.81 a. P(Species extinct) = 38 /132 = .2879

b. If 9 species are chosen and all are extinct, then for the 10th pick, there are only 132 − 9 = 123 species to pick from, of which 38 − 9 = 29 are extinct. P(Species extinct on 10th selection | first 9 species are extinct) = 29 / 123 = .236.

3.83 Define the following events:

I: {Intruder}
N: {No intruder}
A: {System A sounds alarm}
B: {System B sounds alarm}

a. From the problem:

$P(A|I) = .9.$

$P(B|I) = .95$

$P(A|N) = .2$

$P(B|N) = .1$

b. We know that events A and B are independent. Thus,

$P(A \cap B|I) = P(A \mid I)\, P(B|I) = .9(.95) = .855$

c. $P(A \cap B|N) = P(A \mid N)\, P(B|N) = .2(.1) = .02$

d. $P(A \cup B|I) = P(A \mid I) + P(B|I) - P(A \cap B|I) = .9 + .95 - .855 = .995$

3.85 Define the following events:

A: {Player is white}

B: {Player is black}

C: {Player is a guard}

D: {Player is a forward}

E: {Player is a center}

a. $P(E|A) = \dfrac{P(E \cap A)}{P(A)} = \dfrac{28/368}{84/368} = \dfrac{28}{84} = .333$

b. $P(E|B) = \dfrac{P(E \cap B)}{P(B)} = \dfrac{34/368}{284/368} = \dfrac{34}{284} = .120$

c.   Events A and E are independent if $P(E|A) = P(E)$. From part a, $P(E|A) = .333$.

$P(E) = 62 / 368 = .168$

Since $P(E|A) \neq P(E)$, the events are not independent.

d.   Since the events {White player} and {Center} are not independent, this supports the theory of "stacking". Also, from part a, we found that $P(\text{Center}|\text{White}) = .333$ and from part b, $P(\text{Center}|\text{Black}) = .120$. Thus, the probability of playing Center depends on race. Again, this supports the theory of "stacking".

3.87 $P(A|B) = 0$

3.89   a.   The 8 outcomes are:

A and C win first round matches and A wins final (A defeats B, C defeats D, A defeats C)
A and C win first round matches and C wins final (A defeats B, C defeats D, C defeats A)
A and D win first round matches and A wins final (A defeats B, D defeats C, A defeats D)
A and D win first round matches and D wins final (A defeats B, D defeats C, D defeats A)
B and C win first round matches and B wins final (B defeats A, C defeats D, B defeats C)
B and C win first round matches and C wins final (B defeats A, C defeats D, C defeats B)
B and D win first round matches and B wins final (B defeats A, D defeats C, B defeats D)
B and D win first round matches and D wins final (B defeats A, D defeats C, D defeats B)

b.   If all the players are of equal ability, then each of the above outcomes are equally likely. Each outcome has a probability of 1/8 of occurring. $P(A \text{ wins}) = 2/8 = .25$.

c.   There are 2 outcomes where A wins the tournament:

A and C win first round matches and A wins final (A defeats B, C defeats D, A defeats C)
A and D win first round matches and A wins final (A defeats B, D defeats C, A defeats D)

$P(A \text{ wins}) = P(A \text{ defeats } B \cap C \text{ defeats } D \cap A \text{ defeats } C)$
$\qquad\qquad\quad + P(A \text{ defeats } B \cap D \text{ defeats } C \cap A \text{ defeats } D)$

$\qquad\quad = P(A \text{ defeats } B)P(C \text{ defeats } D)P(A \text{ defeats } C)$
$\qquad\qquad\quad + P(A \text{ defeats } B)P(D \text{ defeats } C)P(A \text{ defeats } D)$

$\qquad\quad = .9(.4)(.7) + .9(1 - .4)(.6) = .252 + .324 = .576$

3.91   a.   Define the following events:

W: {Player wins the game Go}
F: {Player plays first (black stones)}

$P(W \cap F) = 319/577 = .553$

b. $P(W \cap F \mid CA) = 34/34 = 1$
$P(W \cap F \mid CB) = 69/79 = .873$
$P(W \cap F \mid CC) = 66/118 = .559$
$P(W \cap F \mid BA) = 40/54 = .741$
$P(W \cap F \mid BB) = 52/95 = .547$
$P(W \cap F \mid BC) = 27/79 = .342$
$P(W \cap F \mid AA) = 15/28 = .536$
$P(W \cap F \mid AB) = 11/51 = .216$
$P(W \cap F \mid AC) = 3/39 = .077$

c. There are three combinations where the player with the black stones (first) is ranked higher than the player with the white stones: CA, CB, and BA.

$P(W \cap F \mid CA \cup CB \cup BA) = (34 + 69 + 40)/(34 + 79 + 54) = 143/167 = .856$

d. There are three combinations where the players are of the same level: CC, BB, and AA.

$P(W \cap F \mid CC \cup BB \cup AA) = (66 + 52 + 15)/(118 + 95 + 28) = 133/241 = .552$

3.93 a. The total number of outcomes is a combination of 14 numbers taken 4 at a time.

$$\binom{14}{4} = \frac{14!}{4!(14-4)!} = \frac{14!}{4!10!} = \frac{14 \cdot 13 \cdot 12 \cdots 2 \cdot 1}{4 \cdot 3 \cdot 2 \cdot 1 \cdot 10 \cdot 9 \cdots 2 \cdot 1} = 1{,}001$$

b. The probability that any team obtains the first pick in the draft is that team's number of combinations divided by 1000. The probabilities are:

NBA Lottery Team	Number of Combinations	Probability
Worst Record	250	.250
2nd worst record	200	.200
3rd worst record	157	.157
4th worst record	120	.120
5th worst record	89	.089
6th worst record	64	.064
7th worst record	44	.044
8th worst record	29	.029
9th worst record	18	.018
10th worst record	11	.011
11th worst record	7	.007
12th worst record	6	.006
13th worst record	5	.005
Total	1000	1.000

c. Define the following events:

W_{i1}: {i^{th} worst team gets 1st pick}
W_{12}: {Worst team gets 2nd pick}

If the 2nd worst team obtains the number 1 pick, then that team is removed from the group that could obtain the number 2 pick. Thus, the 200 combinations assigned to the team with the 2nd worst record are removed from the available combinations. Now, there are only 800 possible combinations. The probability that the team with the worst record will get the 2nd pick in the draft given that the team with the 2nd worst record gets the first pick would be $P(W_{12} | W_{21}) = 250 / 800 = .313$.

d. If the 3rd worst team obtains the number 1 pick, then that team is removed from the group that could obtain the number 2 pick. Thus, the 157 combinations assigned to the team with the 3rd worst record are removed from the available combinations. Now, there are only $1000 - 157 = 843$ possible combinations. The probability that the team with the worst record will get the 2nd pick in the draft given that the team with the 3rd worst record gets the first pick would be $P(W_{12} | W_{31}) = 250 / 843 = .297$.

e. We need to find $P(W_{12} | W_{11}^c) = \dfrac{P(W_{12} \cap W_{11}^c)}{P(W_{11}^c)} = \dfrac{P(W_{12})}{1 - P(W_{11})}$

Now, $P(W_{11}^c) = 1 - P(W_{11}) = 1 - .250 = .750$

$$P(W_{12}) = P(W_{12} | W_{21})P(W_{21}) + P(W_{12} | W_{31})P(W_{31})$$
$$+ P(W_{12} | W_{41})P(W_{41}) + \cdots + + P(W_{12} | W_{131})P(W_{131})$$

To aid in the computation, we will use the following table:

NBA Lottery Team	Number of Combinations	Probability $P(W_{i1})$	$P(W_{12} | W_{i1})$	$P(W_{12} | W_{i1})P(W_{i1})$
Worst Record	250	.250		
2nd worst record	200	.200	.313	.0626
3rd worst record	157	.157	.297	.0466
4th worst record	120	.120	.284	.0341
5th worst record	89	.089	.274	.0244
6th worst record	64	.064	.267	.0171
7th worst record	44	.044	.262	.0115
8th worst record	29	.029	.257	.0075
9th worst record	18	.018	.255	.0046
10th worst record	11	.011	.253	.0028
11th worst record	7	.007	.252	.0018
12th worst record	6	.006	.252	.0015
13th worst record	5	.005	.251	.0013
Total	1000	1.000		.2158

Thus,

$$P(W_{12}) = P(W_{12} \mid W_{21})P(W_{21}) + P(W_{12} \mid W_{31})P(W_{31})$$
$$+ P(W_{12} \mid W_{41})P(W_{41}) + \cdots + + P(W_{12} \mid W_{13\ 1})P(W_{13\ 1}) = .2158$$

Finally, $P(W_{12} \mid W_{11}^c) = \dfrac{P(W_{12} \cap W_{11}^c)}{P(W_{11}^c)} = \dfrac{P(W_{12})}{1 - P(W_{11})} = \dfrac{.2158}{.750} = .288$

3.95   A representative sample exhibits characteristics typical of those possessed by the population of interest. The most common way to obtain a representative sample is to use a random sample.

3.97   A blind study is one where the participants in the study do not know what treatment they are receiving.

3.99   a. $\dbinom{600}{3} = \dfrac{600!}{(3!)(597!)} = \dfrac{(600)(599)(598)}{(1)(2)(3)} = 35,820,200$

      b.   All sets are equally likely in random sampling, so that the probability of any particular sample is:

$$\frac{1}{35,820,200}$$

      c.   Repeated samples would be highly unlikely.

3.101 a.   Decide on a starting point on the random number table. Then take the first n numbers reading down, and this would be the sample. Group the digits on the random number table into groups of 7 (for part **b**) or groups of 4 (for part **c**). Eliminate any duplicates and numbers that begin with zero since they are not valid telephone numbers.

      b.   Starting in Row 6, column 5, take the first 10 seven-digit numbers reading down. The telephone numbers are:

                     277-5653
                     988-7231
                     188-7620
                     174-5318
                     530-6059
                     709-9779
                     496-2669
                     889-7433
                     482-3752
                     772-3313

      c.   Starting in Row 10, column 7, take the first 5 four-digit numbers reading down. The 5 telephone numbers are:

                     373-3886
                     373-5686
                     373-1866
                     373-3632
                     373-6768

3.105   First, number the intersections from 1 to 5000. Using the random number table, select a
        starting point that contains 4 digits. Following either the row or column, select successive 4
        digit numbers until 50 different 4 digit numbers between 0001 and 5000 are selected. The
        intersections corresponding to these 50 four digit numbers are selected for sampling.

        The second method requires the rows to be numbered from 00 to 99 and the columns to be
        numbered from 1 to 50. Using the random number table, select a starting point that contains
        2 digits. Following either the row or column, select successive 2 digit numbers in sets of 2.
        The first 2 digit number will correspond to the row number. The second 2 digit number must
        be between 1 and 50 and corresponds to the column number. This procedure is followed until
        50 pairs of 2 digit numbers are selected that correspond to 50 different intersections.

3.107   Suppose a basketball team has a total of 10 players. There are 5 distinct positions on the team
        – center, forward 1, forward 2, shooting guard, and point guard. How many different ways
        could a coach arrange the starting lineup, assuming that all players could play all positions?

        We can apply the Permutations Rule to find the number of combinations.

        $$P_n^N = N(N-1)(N-2)\cdots(N-n+1) = \frac{N!}{(N-n)!} = \frac{10!}{(10-5)!}$$

        $$= \frac{10!}{5!} = \frac{10\cdot 9\cdots 1}{5\cdot 4\cdot 3\cdot 2\cdot 1} = 10\cdot 9\ 8\cdot 7\cdot 6 = 30,240.$$

3.109   The Combinations Rule does not take order into account, while the Permutations Rule does
        take order into account.

3.111   Since the number of possible outcomes from tossing a coin is always two on each flip, the
        multiplicative rule would yield a product of a series of 2's.

        a. $(2)(2) = 2^2 = 4$

        b. $(2)(2)(2) = 2^3 = 8$

        c. $(2)(2)(2)(2)(2) = 2^5 = 32$

        d. 2^n

3.113   a. $\binom{7}{3} = \frac{7!}{(3!)(4!)} = \frac{7\cdot 6\cdot 5}{3\cdot 2\cdot 1} = 35$

        b. $\binom{6}{2} = \frac{6!}{(2!)(4!)} = \frac{6\cdot 5}{2\cdot 1} = 15$

        c. $\binom{30}{2} = \frac{30!}{(2!)(28!)} = \frac{30\cdot 29}{2\cdot 1} = 435$

        d. $\binom{10}{8} = \frac{10!}{(8!)(2!)} = \frac{10\cdot 9}{2\cdot 1} = 45$

        e. $\binom{q}{r} = \frac{q!}{r!(q-r!)}$

3.115   a.   Since order is not important, the number of choices is:

$$\binom{8}{5} = \frac{8!}{5!3!} = 56$$

b.   Order is important because it is necessary to distinguish between the positions (guard, forward, and center). However, the two players chosen as guards and the two forwards are indistinguishable once they are chosen. That is, choosing player #1 as the first guard and player #3 as the second guard results in the same team as choosing player #3 for the first guard and player #1 for the second guard. Denote the five positions to be filled as G_1, G_2, F_1, F_2, and C. The number of distinct teams, assuming each position was distinct is:

$$P_5^8 = \frac{8!}{3!} = 8(7)(6)(5)(4) = 6720$$

However, since G_1 and G_2 are indistinguishable, this number must be divided by 2, or $6720/2 = 3360$ possible teams. Similarly, F_1 and F_2 are indistinguishable, so that there are in fact only $3360/2 = 1680$ possible teams.

c.   In this situation, all five positions are distinct. Thus, there are $P_5^8 = 6720$ possible teams.

3.117   a.   To visit all 4 cities and then return home, the order of the visits is important. The number of different orders is a permutation of 4 things taken 4 at a time or

$$P_4^4 = \frac{4!}{(4-4)!} = \frac{4 \cdot 3 \cdot 2 \cdot 1}{1} = 24$$

b.   If the salesperson visits B first, then he only has 2 choices for the next city (D or E), since B and C are not connected. Once in D or E, he has 2 choices for the next city (if in D, can choose C or E; if in E, can choose D or C). Once in the third city, there is only one choice for the fourth city. The possible trips with B first are $2 \times 2 \times 1 = 4$: BDEC, BDCE, BEDC, BECD.

Similarly, if city C is the first city, there are 2 choices for the second city (D or E), 2 choices for the third, and one for the fourth. Thus, there are $2 \times 2 \times 1 = 4$ choices: CDEB, CDBE, CEDB, CEBD.

If city D is the first chosen, then to visit all four cities exactly once, city B or C must be chosen next. The third city must be E, and the fourth city the only one left. Thus, there are $2 \times 1 \times 1 = 2$ choices: DBEC, DCEB.

Similarly, if city E is the first chosen, then B or C must be the second city, D must be the third, and the fourth city must be the one remaining. Thus, there are $2 \times 1 \times 1 = 2$ choices: EBDC, ECDB.

The total number of routes are $4 + 4 + 2 + 2 = 12$.

3.119 a. The total number of distributors that make up Elaine's group is $6(5) = 30$.

b. The total number of distributors that make up Elaine's group now is $6(5)(7)(5) = 1,050$.

3.121 a. From the problem, we have three students and three hospitals. Suppose we consider the three hospitals as three positions. Using the permutations rule, there are

$$P = P_3^3 = \frac{3!}{(3-3)!} = \frac{3 \cdot 2 \cdot 1}{1} = 6 \text{ different assignments possible.}$$

These assignments are:

$$[S_1H_A \quad S_2H_B \quad S_3H_C] \quad [S_1H_A \quad S_3H_B \quad S_2H_C]$$
$$[S_2H_A \quad S_1H_B \quad S_3H_C] \quad [S_2H_A \quad S_3H_B \quad S_1H_C]$$
$$[S_3H_A \quad S_1H_B \quad S_2H_C] \quad [S_3H_A \quad S_2H_B \quad S_1H_C]$$

b. Assuming that all of the above assignments are equally likely, the probability of any one of them is $1/6$. The probability that student #1 is assigned to hospital B is $2/6 = 1/3$.

3.123 a. There are a total of $2 \times 3 \times 3 = 18$ maintenance organization alternatives.

b. If all alternatives are equally likely, then each has a probability of $1/18$ of being selected. If 4 of these alternatives are feasible, then P(feasible alternative) $= 4/18 = .222$.

3.125 The professor will choose 5 of 10 questions. Of these 10 questions, the student has prepared answers for 7, and has not prepared the other three. The 5 questions on the exam can be chosen in

$$\binom{10}{5} = \frac{10!}{5!5!} = 252 \text{ ways}$$

a. If the student has prepared all 5 exam questions, the professor has chosen 5 from the 7 prepared questions. This can be done in

$$\binom{7}{5} = 21 \text{ ways}$$

Hence, the desired probability is $21/252$.

b. Define the following events.

A: {The student is prepared for no questions}
B: {The student is prepared for 1 question}
C: {The student is prepared for 2 questions}

A will occur if the professor picks 0 questions from the 7 prepared and 5 from the 3 unprepared. This is impossible, as is B. C will occur if the professor picks 2 questions from the 7 prepared and 3 questions from the 3 unprepared. This can happen in

$$\binom{7}{2}\binom{3}{3} = 21(1) = 21 \text{ ways}$$

Hence, $P(\text{prepared for less than 3}) = P(A) + P(B) + P(C) = 0 + 0 + 21/252 = 21/252$

c. The professor must pick 4 from the 7 prepared questions and 1 from the remaining 3 questions. This can happen in

$$\binom{7}{4}\binom{3}{1} = 35(3) = 105 \text{ ways}$$

Hence, $P(\text{prepared for exactly } 4) = 105/252$

3.127 a. The number of hands possible is:

$$\binom{52}{5} = \frac{52!}{5!47!} = \frac{52 \cdot 51 \cdot 50 \cdot 49 \cdot 48}{5 \cdot 4 \cdot 3 \cdot 2 \cdot 1} = 2,598,960$$

b. Every set of five cards would be equally likely. Therefore, the probability of any one type of hand can be calculated by finding the number of ways of obtaining that particular type of hand and dividing by the total number of hands possible. For example,

$$P(A) = \frac{\text{Number of ways of obtaining a flush}}{\text{Number of hands possible}}$$

To determine the number of flushes possible, it is necessary to recognize that there would be the same number of flushes in spades as hearts, diamond, or clubs. If we can determine the number of ways of obtaining a flush in any particular suit, then four times that amount would be the number of flushes possible. However, since there are 13 cards of each suit, it follows that the number of flushes for any particular suit is:

$$\binom{13}{5} = \frac{13!}{5!8!} = 1,287$$

Then the total number of flushes possible would be:

$$4 \cdot 1,287 = 5,148$$

Then $P(A) = \dfrac{5,148}{2,598,960} \approx .002$

c. To compute $P(B)$, it is necessary to recognize that there are 10 different sets of cards' values which result in straights:

$$\{A, 2, 3, 4, 5\}$$
$$\{2, 3, 4, 5, 6\}$$
$$\{3, 4, 5, 6, 7\}$$
$$\{4, 5, 6, 7, 8\}$$
$$\{5, 6, 7, 8, 9\}$$
$$\{6, 7, 8, 9, 10\}$$
$$\{7, 8, 9, 10, J\}$$
$$\{8, 9, 10, J, Q\}$$
$$\{9, 10, J, Q, K\}$$
$$\{10, J, Q, K, A\}$$

We can determine the total number of possible straights by determining the number of ways that any one particular type of straight could occur and then multiply this amount by 10. But the number of ways of obtaining any one straight, say $\{10, J, Q, K, A\}$, is done easily by the multiplicative rule. Since there would be one 10 from four to be selected, and one jack from four to be selected, etc., the number of ways of obtaining a straight of the form $\{10, J, Q, K, A\}$ is:

$$4 \cdot 4 \cdot 4 \cdot 4 \cdot 4 = 1024$$

There would then be $10 \cdot 1024 = 10,240$ possible straights, so that $P(B) = \dfrac{10,240}{2,598,960}$

$\approx .00394$

d.   There are only four straight flushes of each type. For example, a straight flush of the type $\{10, J, Q, K, A\}$ can occur in spades, hearts, diamonds, and clubs. Since there are 10 types of straights, there are $4 \cdot 10 = 40$ possible straight flushes.

Thus, $P(A \cap B) = \dfrac{40}{2,598,960} = .0000154$

3.129   If events A and B are independent, then the probability that event B occurs does not depend on whether event A has occurred or not. If events A and B are independent, then $P(B \mid A) = P(B)$.

3.131   a. $P(B_1 \cap A) = P(A \mid B_1)P(B_1) = .3(.75) = .225$

b. $P(B_2 \cap A) = P(A \mid B_2)P(B_2) = .5(.25) = .125$

c. $P(A) = P(B_1 \cap A) + P(B_2 \cap A) = .225 + .125 = .35$

d. $P(B_1 \mid A) = \dfrac{P(B_1 \cap A)}{P(A)} = \dfrac{.225}{.35} = .643$

e. $P(B_2 \mid A) = \dfrac{P(B_2 \cap A)}{P(A)} = \dfrac{.125}{.35} = .357$

3.133   If A is independent of B_1, B_2, and B_3, then $P(A \mid B_1) = P(A) = .4$.

Then $P(B_1 \mid A) = \dfrac{P(A \mid B_1)P(B_1)}{P(A)} = \dfrac{.4(.2)}{.4} = .2$

3.135   Define the following events:

T: {Athlete illegally uses testosterone}
P: {Test for testosterone is positive}
N: {Athlete does not illegally use testosterone}

a. $P(P \mid T) = \dfrac{50}{100} = .50$

b. $P(P^c \mid N) = 1 - P(P \mid N) = 1 - \dfrac{9}{900} = 1 - .01 = .99$

c. $P(T \mid P) = \dfrac{P(P \mid T)P(T)}{P(P)}$.

Now, $P(P) = P(P \mid N)P(N) + P(P \mid T)P(T) = .01(.9) + .5(.1) = .009 + .05 = .059$

$P(T \mid P) = \dfrac{P(P \mid T)P(T)}{P(P)} = \dfrac{.5(.1)}{.059} = \dfrac{.05}{.059} = .847$

3.137 a. $P(E_1 \mid error) = \dfrac{P(E_1 \cap error)}{P(error)}$

$= \dfrac{P(error \mid E_1)P(E_1)}{P(error \mid E_1)P(E_1) + P(error \mid E_2)P(E_2) + P(error \mid E_3)P(E_3)}$

$= \dfrac{.01(.30)}{.01(.30) + .03(.20) + .02(.50)} = \dfrac{.003}{.003 + .006 + .01} = \dfrac{.003}{.019} = .158$

b. $P(E_2 \mid error) = \dfrac{P(E_2 \cap error)}{P(error)}$

$= \dfrac{P(error \mid E_2)P(E_2)}{P(error \mid E_1)P(E_1) + P(error \mid E_2)P(E_2) + P(error \mid E_3)P(E_3)}$

$= \dfrac{.03(.20)}{.01(.30) + .03(.20) + .02(.50)} = \dfrac{.006}{.003 + .006 + .01} = \dfrac{.006}{.019} = .316$

c. $P(E_3 \mid error) = \dfrac{P(E_3 \cap error)}{P(error)}$

$= \dfrac{P(error \mid E_3)P(E_3)}{P(error \mid E_1)P(E_1) + P(error \mid E_2)P(E_2) + P(error \mid E_3)P(E_3)}$

$= \dfrac{.02(.50)}{.01(.30) + .03(.20) + .02(.50)} = \dfrac{.01}{.003 + .006 + .01} = \dfrac{.01}{.019} = .526$

d. If there was a serious error, the probability that the error was made by engineer 3 is .526. This probability is higher than for any of the other engineers. Thus engineer #3 is most likely responsible for the error.

3.139   From Exercise 3.83, we defined the following events:

   I: {Intruder}
   N: {No intruder}
   A: {System A sounds alarm}
   B: {System B sounds alarm}
   Also, from Exercise 3.81, $P(A \cap B \mid I) = .855$ and $P(A \cap B \mid N) = .02$.

   Thus, $P(A \cap B) = P(A \cap B \mid I)P(I) + P(A \cap B \mid N)P(N) = .855(.4) + .02(.6)$
   $= .342 + .012 = .354$

   $P(I \cap A \cap B) = P(A \cap B \mid I)P(I) = .855(.4) = .342$

   $$P(I \mid A \cap B) = \frac{P(I \cap A \cap B)}{P(A \cap B)} = \frac{.342}{.354} = .966$$

3.141   a. $A \cup B$

   b. B^c

   c. $A \cap B$

   d. $A^c \mid B$

3.143   a.   Since events A and B are mutually exclusive, $P(A \cap B) = 0$.

   $$P(A \mid B) = \frac{P(A \cap B)}{P(B)} = \frac{0}{.3} = 0$$

   b.   If Events A and B are independent, then $P(A \mid B) = P(A)$. From part **a**, we know $P(A \mid B) = 0$. However, we also know $P(A) = .2$. Thus, events A and B are not independent.

3.145   We know $P(A \mid B) = \dfrac{P(A \cap B)}{P(B)}$. Thus, $P(B) = \dfrac{P(A \cap B)}{P(A \mid B)} = \dfrac{.4}{.8} = .5$.

3.147   a.

(1, *H*)	(2, 1)	(3, *H*)	(4, 1)	(5, *H*)	(6, 1)
(1, *T*)	(2, 2)	(3, *T*)	(4, 2)	(5, *T*)	(6, 2)
	(2, 3)		(4, 3)		(6, 3)
	(2, 4)		(4, 4)		(6, 4)
	(2, 5)		(4, 5)		(6, 5)
	(2, 6)		(4, 6)		(6, 6)

b. Each simple event is an intersection of two independent events. Each simple event whose first element is 1, 3, or 5 has probability

$$\left(\frac{1}{6}\right)\left(\frac{1}{2}\right) = \frac{1}{12}$$

while each simple event whose first element is 2, 4, or 6 has probability

$$\left(\frac{1}{6}\right)\left(\frac{1}{6}\right) = \frac{1}{36}$$

c. $P(A) = P\{(1, H), (3, H), (5, H)\} = \dfrac{1}{12} + \dfrac{1}{12} + \dfrac{1}{12} = \dfrac{3}{12} = \dfrac{1}{4}$

$P(B) = P\{(1, H), (1, T), (3, H), (3, T), (5, H), (5, T)\} = \dfrac{6}{12} = \dfrac{1}{2}$

d. A^c: {all except $(1, H)$, $(3, H)$, and $(5, H)$}
   B^c: {$(2, 1), (2, 2), (2, 3), (2, 4), (2, 5), (2, 6), (4, 1), (4, 2), (4, 3), (4, 4), (4, 5), (4, 6),$
         $(6, 1), (6, 2), (6, 3), (6, 4), (6, 5), (6, 6)$}
   $A \cap B$: {$(1, H), (3, H), (5, H)$}
   $A \cup B$: {$(1, H), (1, T), (3, H), (3, T), (5, H), (5, T)$}

e. $P(A^c) = 1 - P(A) = 1 - \dfrac{1}{4} = \dfrac{3}{4}$

$P(B^c) = 1 - P(B) = 1 - \dfrac{1}{2} = \dfrac{1}{2}$

$P(A \cap B) = P(A) = \dfrac{1}{4}$

$P(A \cup B) = P(A) + P(B) - P(A \cap B) = \dfrac{1}{4} + \dfrac{1}{2} - \dfrac{1}{4} = \dfrac{1}{2}$

$P(A \mid B) = \dfrac{P(A \cap B)}{P(B)} = \dfrac{1/4}{1/2} = \dfrac{1}{2}$

$P(B \mid A) = \dfrac{P(A \cap B)}{P(A)} = \dfrac{1/4}{1/4} = 1$

f. $P(A \cap B) \neq 0$, so that A and B are not mutually exclusive.
   $P(A \mid B) \neq P(A)$, so that A and B are not independent.

3.149 a. Because events A and B are independent, we have:

$$P(A \cap B) = P(A)P(B) = (.3)(.1) = .03$$

Thus, $P(A \cap B) \neq 0$, and the two events cannot be mutually exclusive.

b. $P(A \mid B) = \dfrac{P(A \cap B)}{P(B)} = \dfrac{.03}{.1} = .3$ $P(B \mid A) = \dfrac{P(A \cap B)}{P(A)} = \dfrac{.03}{.1} = .1$

c. $P(A \cup B) = P(A) + P(B) - P(A \cap B) = .3 + .1 - .03 = .37$

3.151   a. $6! = 6 \cdot 5 \cdot 4 \cdot 3 \cdot 2 \cdot 1 = 720$

b. $\binom{10}{9} = \dfrac{10!}{9!(10-9)!} = \dfrac{10 \cdot 9 \ 8 \cdot \ \cdots \ \cdot 1}{9 \cdot 8 \cdot 7 \cdot \ \cdots \ \cdot 1 \cdot 1} = 10$

c. $\binom{10}{1} = \dfrac{10!}{1!(10-1)!} = \dfrac{10 \cdot 9 \cdot 8 \cdot \ \cdots \ \cdot 1}{1 \cdot 9 \cdot 8 \cdot \ \cdots \ \cdot 1} = 10$

d. $P_2^6 = \dfrac{6!}{(6-2)!} = \dfrac{6 \cdot 5 \cdot 4 \cdot 3 \cdot 2 \cdot 1}{4 \cdot 3 \cdot 2 \cdot 1} = 30$

e. $\binom{6}{3} = \dfrac{6!}{3!(6-3)!} = \dfrac{6 \cdot 5 \cdot 4 \cdot 3 \cdot 2 \cdot 1}{3 \cdot 2 \cdot 1 \cdot 3 \cdot 2 \cdot 1} = 20$

f. $0! = 1$

g. $P_4^{10} = \dfrac{10!}{(10-4)!} = \dfrac{10 \cdot 9 \cdot 8 \cdot \ \cdots \ \cdot 1}{6 \cdot 5 \cdot 4 \cdot 3 \cdot 2 \cdot 1} = 5040$

h. $P_2^{50} = \dfrac{50!}{(50-2)!} = \dfrac{50 \cdot 49 \cdot 48 \cdot \ \cdots \ \cdot 1}{48 \cdot 47 \cdot \ \cdots \ \cdot 1} = 2450$

3.153   From the problem, to find probabilities, we must convert the percents to proportions by dividing by 100. Thus, $P(1 \text{ star}) = 0$, $P(2 \text{ stars}) = .0408$, $P(3 \text{ stars}) = .1735$, $P(4 \text{ stars}) = .6020$, and $P(5 \text{ stars}) = .1837$.

a.   False. The probability of an event cannot be 4. The probability of an event must be between 0 and 1.

b.   True. $P(4 \text{ or } 5 \text{ stars}) = P(4 \text{ stars}) + P(5 \text{ stars}) = .6020 + .1837 = .7857$.

c.   True. No cars have a rating of 1 star, thus $P(1 \text{ star}) = 0$.

d.   False. $P(2 \text{ stars}) = .0408$ and $P(5 \text{ stars}) = .1837$. Since .0408 us smaller than .1837, the car has a better chance of having a 5-star rating than a 2-star rating.

3.155   Suppose we define the following event:

A = {eighth-grader scores above 655 on mathematics assessment test}

Then the probability that a randomly selected eighth-grader has a score of 655 or below on the mathematics assessment test is:

$P(A^c) = 1 - P(A) = 1 - .05 = .95$.

3.157   First, define the following events:

A: {Driver was African American}
B: {Driver exceeded the speed limit by at least 5 mph}
C: {Driver was stopped for speeding}

a. $P(A) = .14$
$P(B) = .98$
$P(A \mid B) = .15$
$P(A \mid C) = .35$

b. $P(A \cap B) = P(A \mid B) \, P(B) = .15(.98) = .147$

c. $P(A^c \mid B) = 1 - P(A \mid B) = 1 - .15 = .85$

d. $P(A^c \mid C) = 1 - P(A \mid C) = 1 - .35 = .65$

e.   If events A and B are independent, then $P(A \mid B) = P(A)$. $P(A \mid B) = .15$ and $P(A) = .14$.
For this example, $P(A \mid B) \neq P(A)$, but they are very close to each other. To a close
approximation, A and B are independent.

f.   If African Americans are not stopped more often for speeding than expected, then
events A and C should be independent. We already know from part e that events A and
B are essentially independent - i.e. the proportion of the speeders who are African
American is equal to the proportion of African Americans.

Are events A and C independent? If they are, then $P(A \mid C) = P(A)$. From the above,
$P(A \mid C) = .35$ and $P(A) = .14$. Thus, events A and C are not independent. This means
that the proportion of those stopped for speeding who are African American is not equal
to the proportion of African Americans in the population. Since the $P(A \mid C) > P(A)$,
this implies that on this stretch of highway, African Americans are stopped for speeding
at a higher rate than would be expected.

3.159   a.   The sample space is the listing of all possible nearshore bar conditions. The sample
space is {Single, shore parallel; Other; Planar}.

b.   Define the following events:

SP: {Single, shore parallel}
P: {Planar}
O: {Other}

Assuming that all the sample points are equally likely, $P(SP) = 2 / 6 = .333$.
$P(O) = 2 / 6 = .333$. $P(P) = 2 / 6 = .333$

c. $P(P \cup SP) = 4 / 6 = .667$

d.   The sample space is the listing of all possible beach conditions. The sample
space is {No dunes/flat; Bluff/scarp; Single dune; Not observed}.

e.   Define the following events:

N: {No dunes/flat}
B: {Bluff/scarp}
D: {Single dune}
NO: {Not observed}

Assuming that all the sample points are equally likely, $P(N) = 2 / 6 = .333$.
$P(B) = 1 / 6 = .167$. $P(D) = 2 / 6 = .333$. $P(NO) = 1/6 = .167$.

f. $P(N^C) = 1 - P(N) = 1 - .333 = .667$

3.161   a.   The sample points would be the possible answers to the question. Thus, the sample points would be: Total population, Agricultural change, Presence of industry, Growth, and Population concentration.

b.   Reasonable probabilities would correspond to the proportions given in the pie chart. These would be:

P(Total population) $= .18$, P(Agricultural change) $= .05$, P(Presence of industry) $= .27$, P(growth) $= .05$, and P(Population concentration) $= .45$.

c.   Sample points related to population are Total population and Population concentration. $P(\text{population-related}) = P(\text{Total population}) + P(\text{Population concentration}) = .18 + .45 = .63$.

3.163   a. $P(\text{Most salient role is spouse}) = 424 / 1{,}102 = .385$

b. $P(\text{Most salient role is parent or grandparent})$
$= 269 / 1{,}102 + 148 / 1{,}102 = .244 + .134 = .378$

c. $P(\text{Most salient role does not involve a spouse or relative})$
$= 73 / 1{,}102 + 59 / 1{,}102 + 34 / 1{,}102 + 36 / 1{,}102 = .066 + .054 + .031 + .033 = .184$

3.165   a.   Define the following event:

F_i: {Player makes a foul shot on ith attempt} $P(F_i) = .8$
$P(F_i') = 1 - P(F_i) = 1 - .8 = .2$

The event "the player scores on both shots" is $F_1 \cap F_2$. If the throws are independent, then:

$$P(F_1 \cap F_2) = P(F_1)P(F_2) = .8(.8) = .64$$

The event "the player scores on exactly one shot" is:

$$(F_1 \cap F_2^c) \cup (F_1^c \cap F_2) = P(F_1 \cap F_2^c) + P(F_1^c \cap F_2)$$
$$= P(F_1)P(F_2^c) + P(F_1^c)P(F_2)$$
$$= .8(.2) + .2(.8) = .16 + .16 = .32$$

The event "the player scores on neither shot" is $F_1^c \cap F_2^c$.

$$P(F_1^c \cap F_2^c) = P(F_1^c)P(F_2^c) = .2(.2) = .04$$

b.   We know $P(F_1) = .8$, $P(F_2 | F_1) = .9$, and $P(F_2 | F_1^c) = .7$

The probability the player scores on both shots is:

$$P(F_1 \cap F_2) = P(F_2 | F_1)P(F_1) = .9(.8) = .72$$

The probability the player scores on exactly one shot is:

$$
\begin{aligned}
P(F_1 \cap F_2^c) + P(F_1^c \cap F_2) &= P(F_2^c | F_1)P(F_1) + P(F_2 | F_1^c)P(F_1^c) \\
&= [1 - P(F_2 | F_1)]P(F_1) + P(F_2 | F_1^c)P(F_1^c) \\
&= (1 - .9)(.8) + .7(.2) = .08 + .14 = .22
\end{aligned}
$$

The probability the player scores on neither shot is:

$$P(F_1^c \cap F_2^c) = P(F_2^c | F_1^c)P(F_1^c) = [1 - P(F_2 | F_1^c)]P(F_1^c) = (1 - .7)(.2) = .06$$

c.   Two consecutive foul shots are probably dependent. The outcome of the second shot probably depends on the outcome of the first.

3.167   Define the following events:

A: {Seed carries single spikelets}
B: {Seed carries paired spikelets}
C: {Seed produces ears with single spikelets}
D: {Seed produces ears with paired spikelets}

From the problem, $P(A) = .4$, $P(B) = .6$, $P(C | A) = .29$, $P(D | A) = .71$, $P(C | B) = .26$, and $P(D | B) = .74$.

a. $P(A \cap C) = P(C | A)P(A) = .29(.4) = .116$

b. $P(D) = P(A \cap D) + P(B \cap D) = P(D | A)P(A) + P(D | B)P(B) = .71(.4) + .74(.6)$
$= .284 + .444 = .728$

3.169   Define the following events:

A_1: {Component 1 in System A works properly}
A_2: {Component 2 in System A works properly}
A_3: {Component 3 in System A works properly}
B_1: {Component 1 in System B works properly}
B_2: {Component 2 in System B works properly}
B_3: {Component 3 in System B works properly}
B_4: {Component 4 in System B works properly}
A: {System A works properly}
B: {System B works properly}
C: {Subsystem C works properly}
D: {Subsystem D works properly}

$P(A_1) = 1 - P(A_1^c) = 1 - .12 = .88$ $P(B_1) = 1 - P(B_1^c) = 1 - .1 = .9$

$P(A_2) = 1 - P(A_2^c) = 1 - .09 = .91$ $P(B_2) = 1 - P(B_2^c) = 1 - .1 = .9$

$P(A_3) = 1 - P(A_3^c) = 1 - .11 = .89$ $P(B_3) = 1 - P(B_3^c) = 1 - .1 = .9$

$P(B_4) = 1 - P(B_4^c) = 1 - .1 = .9$

a. $P(A) = P(A_1 \cap A_2 \cap A_3) = P(A_1)P(A_2)P(A_3) = .88(.91)(.89) = .7127$
    (since the three components operate independently)

b. $P(A^c) = 1 - P(A) = 1 - .7127 - .2873$

c. Subsystem C works properly if both components B_1 and B_2 work properly.

    $$P(C) = P(B_1 \cap B_2) = P(B_1)P(B_2) = .9(.9) = .81$$
    (since the 2 components operate independently)

    Similarly, subsystem D works properly if both components B_3 and B_4 work properly.

    $$P(D) = P(B_3 \cap B_4) = P(B_3)P(B_4) = .9(.9) = .81$$

    System B operates properly if either subsystem C or subsystem D operates properly.

    $$P(B) = P(C \cup D) = P(C) + P(D) - P(C \cap D) = P(C) + P(D) - P(C)P(D)$$
    $$= .81 + .81 - .81(.81) = .9639$$

d. The probability that at least one subsystem in System B fails is:

    $$P(C \cap D^c) + P(C^c \cap D) = P(C)P(D^c) + P(C^c)P(D)$$
    $$= .81(1 - .81) + (1 - .81)(.81) = .1539 + .1539 = .3078$$

e. System B fails to operate only if both subsystems C and D fail.

    $$P(B^c) = P(C^c \cap D^c) = P(C^c)P(D^c) = (1 - .81)(1 - .81) = .0361$$

f. If System B operates correctly 99% of the time, then it fails 1% of the time. The probability that one of the subsystems fails is $1 - .81 = .19$. The probability that n subsystems fail is $.19^n$. Thus, we must find n such that

    $$.19^n \leq .01 \implies n \geq 3$$

3.171 Define the following event:

D: {Chip is defective}

From the Exercise, $P(S_1) = .15$, $P(S_2) = .05$, $P(S_3) = .10$, $P(S_4) = .20$, $P(S_5) = .12$, $P(S_6) = .20$, and $P(S_7) = .18$. Also, $P(D|S_1) = .001$, $P(D|S_2) = .0003$, $P(D|S_3) = .0007$, $P(D|S_4) = .006$, $P(D|S_5) = .0002$, $P(D|S_6) = .0002$, and $P(D|S_7) = .001$.

a.   We must find the probability of each supplier given a defective chip.

$$P(S_1 | D) = \frac{P(S_1 \cap D)}{P(D)} =$$

$$\frac{P(D|S_1)P(S_1)}{P(D|S_1)P(S_1) + P(D|S_2)P(S_2) + P(D|S_3)P(S_3) + P(D|S_4)P(S_4) + P(D|S_5)P(S_5) + P(D|S_6)P(S_6) + P(D|S_7)P(S_7)}$$

$$= \frac{.001(.15)}{.00\overline{1}(.15) + .0003(.05) + .0007(.10) + .006(.20) + .0002(.12) + .0002(.02) + .001(.18)}$$

$$= \frac{.00015}{.00015 + .000015 + .00007 + .0012 + .000024 + .00004 + .00018} = \frac{.00015}{.001679} = .0893$$

$$P(S_2 | D) = \frac{P(S_2 \cap D)}{P(D)} = \frac{P(D|S_2)P(S_2)}{P(D)} = \frac{.0003(.05)}{.001679} = \frac{.000015}{.001679} = .0089$$

$$P(S_3 | D) = \frac{P(S_3 \cap D)}{P(D)} = \frac{P(D|S_3)P(S_3)}{P(D)} = \frac{.0007(.10)}{.001679} = \frac{.00007}{.001679} = .0417$$

$$P(S_4 | D) = \frac{P(S_4 \cap D)}{P(D)} = \frac{P(D|S_4)P(S_4)}{P(D)} = \frac{.006(.20)}{.001679} = \frac{.0012}{.001679} = .7147$$

$$P(S_5 | D) = \frac{P(S_5 \cap D)}{P(D)} = \frac{P(D|S_5)P(S_5)}{P(D)} = \frac{.0002(.12)}{.001679} = \frac{.000024}{.001679} = .0143$$

$$P(S_6 | D) = \frac{P(S_6 \cap D)}{P(D)} = \frac{P(D|S_6)P(S_6)}{P(D)} = \frac{.0002(.20)}{.001679} = \frac{.00004}{.001679} = .0238$$

$$P(S_7 | D) = \frac{P(S_7 \cap D)}{P(D)} = \frac{P(D|S_7)P(S_7)}{P(D)} = \frac{.001(.18)}{.001679} = \frac{.00018}{.001679} = .1072$$

Of these probabilities, .7147 is the largest. This implies that if a failure is observed, supplier number 4 was most likely responsible.

b. If the seven suppliers all produce defective chips at the same rate of .0005, then $P(D|S_i)$ =.0005 for all $i = 1, 2, 3, \ldots 7$ and $P(D) = .0005$.

For any supplier i, $P(S_i \cap D) = P(D|S_i)P(S_i) = .0005 P(S_i)$ and

$$P(S_i \mid D) = \frac{P(S_i \cap D)}{P(D)} = \frac{P(D|S_i)P(S_i)}{.0005} = \frac{.0005P(S_i)}{.0005} = P(S_i)$$

Thus, if a defective is observed, then it most likely came from the supplier with the largest proportion of sales (probability). In this case, the most likely supplier would be either supplier 4 or supplier 6. Both of these have probabilities of .20.

3.173 a. The odds in favor of Smarty Jones are $\dfrac{1}{3}$ to $\left(1-\dfrac{1}{3}\right)$ or $\dfrac{1}{3}$ to $\dfrac{2}{3}$ or 1 to 2

b. If the odds are 1 to 1, $P(\text{Smarty Jones will win}) = \dfrac{1}{1+1} = \dfrac{1}{2} = .5$

c. If the odds against Smarty Jones winning are 3 to 2, the odds for Smarty Jones winning are 2 to 3. The probability Smarty Jones will win is $\dfrac{2}{2+3} = \dfrac{2}{5} = .4$

3.175 Define the following event:

A: {Antigens match}

From the problem, $P(A) = .25$ and $P(A') = 1 - P(A) = 1 - .25 = .75$

a. The probability that one sibling has a match is $P(A) = .25$

b. The probability that all three will match is:

$$P(AAA) = P(A \cap A \cap A) = P(A)P(A)P(A) = .25^3 = .0156$$

c. The probability that none of the three match is:

$$P(A^c A^c A^c) = P(A^c \cap A^c \cap A^c) = P(A^c)P(A^c)P(A^c) = .75^3 = .4219$$

d. For this part, $P(A) = .001$ and $P(A^c) = 1 - P(A) = 1 - .001 = .999$

$P(A) = .001$

$P(AAA) = P(A \cap A \cap A) = P(A)P(A)P(A) = .001^3 = .000000001$

$P(A^c A^c A^c) = P(A^c \cap A^c \cap A^c) = P(A^c)P(A^c)P(A^c) = .999^3 = .9970$

3.177   We will work with the following events:

   A: {Woman is pregnant}
   B: {Pregnancy test is positive}

The given probabilities can be written as:

$$P(A) = .75, \ P(B|A) = .99, \ P(B|A^c) = .02$$

Then, $P(A|B) = \dfrac{P(A \cap B)}{P(B)} = \dfrac{P(B|A) \cdot P(A)}{P(A \cap B) + P(A^c \cap B)}$

$$= \dfrac{P(B|A) \cdot P(A)}{P(B|A) \cdot P(A) + P(B|A^c) \cdot P(A^c)}$$

$$= \dfrac{(.99)(.75)}{(.99)(.75) + (.02)(.25)} = \dfrac{.7425}{.7475} \approx .993$$

3.179   We first determine the number of simple events for this experiment. According to the partitions rule, the number of ways of dealing 52 cards so that each of the four players receives 13 cards is:

$$\frac{52!}{13!13!13!13!} = 5.3644738 \times 10^{28}$$

If the cards are well shuffled, then the probability of each simple event is:

$$\frac{1}{5.3644738} \text{ times } 10^{-28}$$

We now need to count the number of simple events that result in one player receiving all the diamonds, another all the hearts, another all the spades, and another all the clubs. To do this, first consider a particular *ordering* that yields the desired result:

   Player #1 receives 13 diamonds
   Player #2 receives 13 hearts
   Player #3 receives 13 spades
   Player #4 receives 13 clubs

This particular *ordered* event can happen in

$$\binom{13}{13}\binom{13}{13}\binom{13}{13}\binom{13}{13} = 1$$

way. However, the *ordering* of the players is not relevant to us.

Since there are $P_4^4 = \dfrac{4!}{(4-4)!} = 24$ possible orderings of the players, we conclude that the

probability of the rare event observed in Dubuque is:

$$\frac{24}{5.3644738 \times 10^{28}} \approx 4.4739 \times 10^{-28}$$

3.181   If each of 3 players uses a fair coin, the sample space is:

HHH   HTT
HHT   THT
HTH   TTH
THH   TTT

Since each event is equally likely, each event will have probability 1/8.

The probability of odd man out on the first roll is:

$P(HHT) + P(HTH) + P(THH) + P(HTT) + P(THT) + P(TTH)$
$= 1/8 + 1/8 + 1/8 + 1/8 + 1/8 + 1/8 = 6/8 = 3/4$

If one player uses a two-headed coin, the sample space will be (assume 3rd player has the two-headed coin):

HHH   THH
HTH   TTH

Since each event is equally likely, each will have probability 1/4.

The probability the 3rd player is odd man out is $P(TTH) = 1/4$.

From the first part, the probability the 3rd player is odd man out is $P(HHT) + P(TTH) = 1/8 + 1/8 = 1/4$. Thus, the two probabilities are the same.

3.183   If a coin is balanced, then $P(H) = P(T) = .5$. To find the probability of any sequence, you find the probability of the intersection of the simple events. Thus,

$P(HHHHHHHHHH) = P(H \cap H \cap H \cap H \cap H \cap H \cap H \cap H \cap H \cap H)$

$= P(H)P(H)P(H)P(H)P(H)P(H)P(H)P(H)P(H)P(H) = .5(.5)(.5)(.5)(.5)(.5)(.5)(.5)(.5)(.5)$

$= .00098$

Similarly, $P(HHTTHTTHHH) = P(H \cap H \cap T \cap T \cap H \cap T \cap T \cap H \cap H \cap H)$

$= P(H)P(H)P(T)P(T)P(H)P(T)P(T)P(H)P(H)P(H) = .5(.5)(.5)(.5)(.5)(.5)(.5)(.5)(.5)$

$= .00098$

Similarly, $P(TTTTTTTTTT) = P(T \cap T \cap T \cap T \cap T \cap T \cap T \cap T \cap T \cap T)$

$= P(T)P(T)P(T)P(T)P(T)P(T)P(T)P(T)P(T)P(T) = .5(.5)(.5)(.5)(.5)(.5)(.5)(.5)(.5)$

$= .00098$

Thus, any of these 3 sequences are equally likely. In fact, any specific sequence has the same probability.

Now, suppose we compute the probability of getting all heads or all tails:

P(All heads or all tails) = P(All heads $\cup$ All tails)

= P(All heads) + P(All tails) − P(All heads $\cap$ All tails) = .00098 + .00098 − 0 = .00196

Consequently, the probability of getting a mix of heads and tails is:

P(10 coin tosses result in a mix of heads and tails) = 1 − P(All heads or all tails)

= 1 − .00196 = .99804

So, even though any particular sequence has a probability of .00098, we just found out the probability of getting all heads or all tails is only .00196 and the probability of getting a mix of heads and tails is .99804. If we know that one of these sequences actually occurred, then we would conclude that it was probably the one with the mix of heads and tails because the probability of a mix is close to 1.

Discrete Random Variables

4.1   A random variable is a rule that assigns one and only one value to each simple event of an experiment.

4.3   a.   Since we can count the number of words spelled correctly, the random variable is discrete.

       b.   Since we can assume values in an interval, the amount of water flowing through the Hoover Dam in a day is continuous.

       c.   Since time is measured on an interval, the random variable is continuous.

       d.   Since we can count the number of bacteria per cubic centimeter in drinking water, this random variable is discrete.

       e.   Since we cannot count the amount of carbon monoxide produced per gallon of unleaded gas, this random variable is continuous.

       f.   Since weight is measured on an interval, weight is a continuous random variable.

4.5   a.   The reaction time difference is continuous because it lies within an interval.

       b.   Since we can count the number of violent crimes, this random variable is discrete.

       c.   Since we can count the number of near misses in a month, this variable is discrete.

       d.   Since we can count the number of winners each week, this variable is discrete.

       e.   Since we can count the number of free throws made per game by a basketball team, this variable is discrete.

       f.   Since distance traveled by a school bus lies in some interval, this is a continuous random variable.

4.7   The values x can assume are 0, 1, 2, 3, 4, ... Thus, x is a discrete random variable.

4.13   The probability distribution of a discrete random variable can be represented by a table, graph, or formula that specifies the probability associated with each possible value the random variable can assume.

4.15   a.   We know $\sum p(x) = 1$. Thus, $p(2) + p(3) + p(5) + p(8) + p(10) = 1$

$$\Rightarrow p(5) = 1 - p(2) - p(3) - p(8) - p(10) = 1 - .15 - .10 - .25 - .25 = .25$$

b. $P(x = 2 \text{ or } x = 10) = P(x = 2) + P(x = 10) = .15 + .25 = .40$

c. $P(x \le 8) = P(x = 2) + P(x = 3) + P(x = 5) + P(x = 8) = .15 + .10 + .25 + .25 = .75$

4.17   a. $P(x \le 12) = P(x = 10) + P(x = 11) + P(x = 12) = .2 + .3 + .2 = .7$

b. $P(x > 12) = 1 - P(x \le 12) = 1 - .7 = .3$

c. $P(x \le 14) = P(x = 10) + P(x = 11) + P(x = 12) + P(x = 13) + P(x = 14)$
$= .2 + .3 + .2 + .1 + .2 = 1$

d. $P(x = 14) = .2$

e. $P(x \le 11 \text{ or } x > 12) = P(x \le 11) + P(x > 12)$
$= P(x = 10) + P(x = 11) + P(x = 13) + P(x = 14)$
$= .2 + .3 + .1 + .2 = .8$

4.19   a.   The simple events are (where H = head, T = tail):

	HHH	HHT	HTH	THH	HTT	THT	TTH	TTT
x = # heads	3	2	2	2	1	1	1	0

b.   If each event is equally likely, then $P(\text{simple event}) = \dfrac{1}{n} = \dfrac{1}{8}$

$$p(3) = \frac{1}{8}, \; p(2) = \frac{1}{8} + \frac{1}{8} + \frac{1}{8} = \frac{3}{8}, \; p(1) = \frac{1}{8} + \frac{1}{8} + \frac{1}{8} = \frac{3}{8}, \text{ and } p(0) = \frac{1}{8}$$

c.

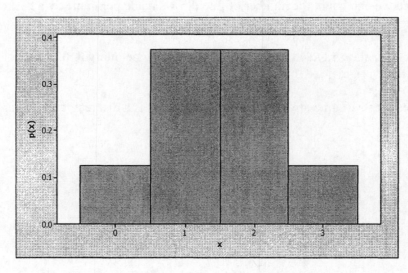

d. $P(x = 2 \text{ or } x = 3) = p(2) + p(3) = \dfrac{3}{8} + \dfrac{1}{8} = \dfrac{4}{8} = \dfrac{1}{2}$

4.21   a. $p(0) + p(1) + p(2) + p(3) + p(4) = .09 + .30 + .37 + .20 + .04 = 1.00$

       b. $P(x = 3 \text{ or } 4) = p(3) + p(4) = .20 + .04 = .24$

       c. $P(x < 2) = p(0) + p(1) = .09 + .30 = .39$

4.23   a. X is the sum of the grill numbers. To find the probability distribution for x, we divide the number of students choosing the particular grill display combination divided by the total number of students or 124. The table below shows the probability distribution for x.

Grill Display Combination	Number of Students	x	$p(x)$
1-2-3	35	6	35/124 = .282
1-2-4	8	7	8/124 = .065
1-2-5	42	8	42/124 = .339
2-3-4	4	9	4/124 = .032
2-3-5	1	10	1/124 = .008
2-4-5	34	11	34/124 = .274
Total	124		1.00

       b. $P(x > 10) = P(x = 11) = .274$.

4.25   Suppose we define the following events:

P: {Student scores perfect 1600}
N: {Student does not score perfect 1600}

From the text, $P(P) = 5/10,000 = .0005$. Thus, $P(N) = 1 - P(P) = 1 - .0005 = .9995$

       a.   If three students are randomly selected, the possible outcomes for the 3 students are as follows:
           PPP PPN PNP NPP PNN NPN NNP NNN

          Each of the sample points listed is really the intersection of independent events.

          Thus, $P(PPP) = P(P \cap P \cap P) = P(P)P(P)P(P) = .0005(.0005)(.0005) = .0000$

          $P(PPN) = P(P \cap P \cap N) = P(P)P(P)P(N) = .0005(.0005)(.9995) = .0000$
              $= P(PNP) = P(NPP)$

          $P(PNN) = P(P \cap N \cap N) = P(P)P(N)P(N) = .0005(.9995)(.9995) = .0005$
              $= P(NPN) = P(NNP)$

          $P(NNN) = P(N \cap N \cap N) = P(N)P(N)P(N) = .9995(.9995)(.9995) = .9985$

          Thus, $P(x = 0) = P(NNN) = .9985$
              $P(x = 1) = P(PNN) + P(NPN) + P(NNP) = .0005(3) = .0015$
              $P(x = 2) = P(PPN) + P(PNP) + P(NPP) = .0000(3) = .0000$
              $P(x = 3) = P(PPP) = .0000$

b. The graph of $p(x)$ is:

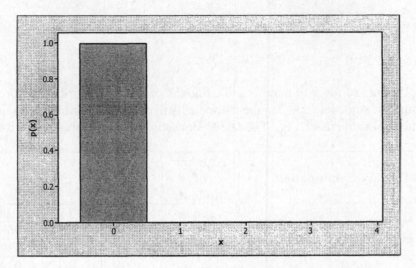

c. $P(x \le 1) = P(x = 0) + P(x = 1) = .9985 + .0015 = 1.0000$

4.27 Assigning points according to the directions is:

OUTCOME OF APPEAL	Number of cases	Points awarded, x
Plaintiff trial win – reversed	71	−1
Plaintiff trial win – affirmed/dismissed	240	5
Defendant trial win – reversed	68	−3
Defendant trial win – affirmed/dismissed	299	5
TOTAL	678	

To find the probabilities for x, we divide the frequencies by the total sample size. The probability distribution for x is:

x	p(x)
−3	68 / 678 = .100
−1	71 / 678 = .105
5	(240+299)/678 = .795
TOTAL	1.000

Using MINITAB, the graph of the probability distribution is:

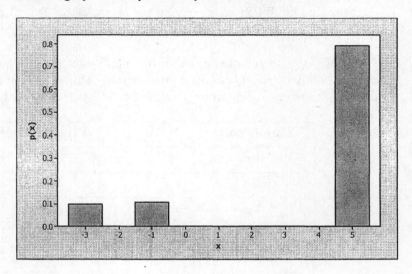

4.29   The sample space of the experiment would be:

   S: {BBB, BBG, BGB, GBB, GGB, GBG, BGG, GGG}

where B and G represent a boy and girl respectively. Since the female parent always donates an X chromosome, the gender of any child is determined by the chromosome donated by the father; an X chromosome donated by the father will produce a girl, a Y, a boy. Each child therefore has a .5 probability of being either gender (as might be expected). From this, it follows that each of the above simple events has a probability of 1/8. If we let z represent the number of male offspring, the probability distribution of z is:

z	0	1	2	3
$p(z)$	1/8	3/8	3/8	1/8

Then,

   $P(\text{At least one boy}) = P(z \geq 1) = 1 - P(z = 0) = 1 - 1/8 = 7/8$

4.31   a.   Each point in the system can have one of 2 status levels, "free" or "obstacle". Define the following events:

   A_F: {Point A is free} A_O: {Point A is obstacle}
   B_F: {Point B is free} B_O: {Point B is obstacle}
   C_F: {Point C is free} C_O: {Point C is obstacle}

   Thus, the sample points for the space are:

   $A_F B_F C_F$, $A_F B_F C_O$, $A_F B_O C_F$, $A_F B_O C_O$, $A_O B_F C_F$, $A_O B_F C_O$, $A_O B_O C_F$, $A_O B_O C_O$

b. Since it is stated that the probability of any point in the system having a "free" status is .5, the probability of any point having an "obstacle" status is also .5, Thus, the probability of each of the sample points above is $P(A_iB_iC_i) = .5(.5)(.5) = .125$.

The values of x, the number of free links in the system, for each sample point are listed below. A link is free if both the points are free. Thus, a link from A to B is free if A is free and B is free. A link from B to C is free if B is free and C is free.

Sample point	x	Probability
$A_FB_FC_F$	2	.125
$A_FB_FC_O$	1	.125
$A_FB_OC_F$	0	.125
$A_FB_OC_O$	0	.125
$A_OB_FC_F$	1	.125
$A_OB_FC_O$	0	.125
$A_OB_OC_F$	0	.125
$A_OB_OC_O$	0	.125

The probability distribution for x is:

x	Probability
0	.625
1	.250
2	.125

4.33 The expected value of a random variable, E(x), does not always equal a specific value of the random variable, x. The E(x) is the mean of the probability distribution and does not have to equal a specific value of x.

4.35 a. $\mu = E(x) = \sum xp(x) = 1(.2) + 2(.4) + 4(.2) + 10(.2) = .2 + .8 + .8 + 2 = 3.8$

b. $\sigma^2 = E[(x - \mu)^2] = \sum (x - \mu)^2 p(x)$
$$= (1 - 3.8)^2(.2) + (2 - 3.8)^2(.4) + (4 - 3.8)^2(.2) + (10 - 3.8)^2(.2)$$
$$= 1.568 + 1.296 + .008 + 7.688 = 10.56$$

c. $\sigma = \sqrt{10.59} = 3.2496$

d. The average value of x over many trials is 3.8.

e. No. The random variable can only take on values 1, 2, 4, or 10.

f. Yes. It is possible that μ can be equal to an actual value of x.

4.37   a.   It would seem that the mean of both would be 1 since they both are symmetric distributions centered at 1.

   b. $P(x)$ seems more variable since there appears to be greater probability for the two extreme values of 0 and 2 than there is in the distribution of y.

   c.   For x: $\mu = E(x) = \sum xp(x) = 0(.3) + 1(.4) + 2(.3) = 0 + .4 + .6 = 1$

$\sigma^2 = E[(x - \mu)^2] = \sum (x - \mu)^2 p(x)$

$= (0 - 1)^2(.3) + (1 - 1)^2(.4) + (2 - 1)^2(.3) = .3 + 0 + .3 = .6$

   For y: $\mu = E(y) = \sum yp(y) = 0(.1) + 1(.8) + 2(.1) = 0 + .8 + .2 = 1$

$\sigma^2 = E[(y - \mu)^2] = \sum (y - \mu)^2 p(y)$

$= (0 - 1)^2(.1) + (1 - 1)^2(.8) + (2 - 1)^2(.1) = .1 + 0 + .1 = .2$

   The variance for x is larger than that for y.

4.39   a.   To find the probabilities associated with each value of x, we divide the frequency associated with each value by the total sample size, 743. The probabilities appear in the table:

First Digit	Frequency of Occurrence	Probability
1	109	109 / 743 = .1467
2	75	75 / 743 = .1009
3	77	77 / 743 = .1036
4	99	99 / 743 = .1332
5	72	72 / 743 = .0969
6	117	117 / 743 = .1575
7	89	89 / 743 = .1198
8	62	62 / 743 = .0834
9	43	43 / 743 = 0579
Total	**743**	1.0000

   b. $\mu = E(x) = \sum xp(x) = 1(.1467) + 2(.1009) + 3(.1036) + 4(.1332) + 5(.0969)$
$+ 6(.1575) + 7(.1198) + 8(.0834) + 9(.0579) = 4.6485$

   c.   Over a very large number of trials, the average first significant digit is 4.6485.

4.41   a. $E(x) = \sum xp(x) = 0(.09) + 1(.30) + 2(.37) + 3(.20) + 4(.04) = 0 + .30 + .74 + .60 + .16 = 1.8$

   In a random samples of 4 homes, the average number of homes with high dust mite levels is 1.8.

b. $\sigma^2 = E[(x-\mu)^2] = \sum (x-\mu)^2 p(x)$

$= (0-1.8)^2\,(.09) + (1-1.8)^2(.30) + (2-1.8)^2(.37) + (3-1.8)^2(.20) + (4-1.8)^2(.04)$

$= .2916 + .1920 + .0148 + .2880 + .1936 = .98$

$\sigma = \sqrt{.98} = .9899$

c. $\mu \pm 2\sigma \Rightarrow 1.8 \pm 2(.9899) \Rightarrow 1.8 \pm 1.9798 \Rightarrow (-.1798, 3.7798)$

$P(-.1798 < x < 3.7798) = p(0) + p(1) + p(2) + p(3) = .09 + .30 + .37 + .20 = .96$

Chebyshev's Theorem says that the interval $\mu \pm 2\sigma$ will contain at least .75 of the data. The Empirical Rule says that approximately .95 of the data will be contained in the interval. Both Chebyshev's Theorem and the Empirical Rule fit the distribution.

4.43   a.   Define the following events:

M: {Miami Beach, FL}
C: {Coney Island, NY}
S: {Surfside, CA}
B: {Monmouth Beach, NJ}
O: {Ocean City, NJ}
L: {Spring Lake, NJ}

The possible pairs of beach hotspots that can be selected are:

MC, MS, MB, MO, ML, CS, CB, CO, CL, SB, SO, SL, BO, BL, OL

b. If we are just selecting 2 hotspots from 6, each of the 15 combinations are equally likely. Thus, each would have a probability of 1/15.

c. The value of x for each of the pairs is:

Pair	x	Pair	x
MC	0	CL	1
MS	0	SB	1
MB	1	SO	0
MO	0	SL	1
ML	1	BO	1
CS	0	BL	2
CB	1	OL	1
CO	0		

d. The probability distribution for x is:

x	$p(x)$
0	6/15
1	8/15
2	1/15

e. $\mu = E(x) = \sum xp(x) = 0(6/15) + 1(8/15) + 2(1/15) = 10/15 = .667$

    The average total number of hotspots with a planar nearshore bar condition in each pair is .667.

4.45   For a $5 bet, you will either win $5 or lose $5 (–$5). The probability distribution for the net winnings is:

x	$p(x)$
–5	20/38
5	18/38

$$\mu = E(x) = \sum xp(x) = -5\left(\frac{20}{38}\right) + 5\left(\frac{18}{38}\right) = -.263$$

Over a large number of trials, the average winning for a $5 bet on red is -$0.263.

4.47   Let x = bookie's earnings per dollar wagered. Then x can take on values $1 (you lose) and $–5 (you win). The only way you win is if you pick 3 winners in 3 games. If the probability of picking 1 winner in 1 game is .5, then $P(www) = p(w)p(w)p(w) = .5(.5)(.5) = .125$ (assuming games are independent).

Thus, the probability distribution for x is:

x	$p(x)$
$1	.875
$–5	.125

$$E(x) = \sum xp(x) = 1(.875) - 5(.125) = .875 - .625 = \$.25$$

On the average, the bookie will earn $.25 for each bet made.

4.49   The formula for p(x) for a binomial random variable with n = 7 and p = .2 is:

$$p(x) = \binom{n}{x} p^x q^{n-x} = \binom{7}{x} .2^x .8^{7-x} \quad (x = 0,1,2,...,7)$$

4.51   a. $\dfrac{6!}{2!(6-2)!} = \dfrac{6!}{2!4!} = \dfrac{6 \cdot 5 \cdot 4 \cdot 3 \cdot 2 \cdot 1}{(2 \cdot 1)(4 \cdot 3 \cdot 2 \cdot 1)} = 15$

     b. $\binom{5}{2} = \dfrac{5!}{2!(5-2)!} = \dfrac{5!}{2!3!} = \dfrac{5 \cdot 4 \cdot 3 \cdot 2 \cdot 1}{(2 \cdot 1)(3 \cdot 2 \cdot 1)} = 10$

     c. $\binom{7}{0} = \dfrac{7!}{0!(7-0)!} = \dfrac{7!}{0!7!} = \dfrac{7 \cdot 6 \cdot 5 \cdot 4 \cdot 3 \cdot 2 \cdot 1}{(1)(7 \cdot 6 \cdot 5 \cdot 4 \cdot 3 \cdot 2 \cdot 1)} = 1$

(Note: 0! = 1)

d. $\begin{pmatrix} 6 \\ 6 \end{pmatrix} = \dfrac{6!}{6!(6-6)!} = \dfrac{6!}{6!0!} = \dfrac{6 \cdot 5 \cdot 4 \cdot 3 \cdot 2 \cdot 1}{(6 \cdot 5 \cdot 4 \cdot 3 \cdot 2 \cdot 1)(1)} = 1$

e. $\begin{pmatrix} 4 \\ 3 \end{pmatrix} = \dfrac{4!}{3!(4-3)!} = \dfrac{4!}{3!1!} = \dfrac{4 \cdot 3 \cdot 2 \cdot 1}{(3 \cdot 2 \cdot 1)(1)} = 4$

4.53 a. $P(x=1) = \dfrac{5!}{1!4!}(.2)^1(.8)^4 = \dfrac{5 \cdot 4 \cdot 3 \cdot 2 \cdot 1}{(1)(4 \cdot 3 \cdot 2 \cdot 1)}(.2)^1(.8)^4 = 5(.2)^1(.8)^4 = .4096$

b. $P(x=2) = \dfrac{4!}{2!2!}(.6)^2(.4)^2 = \dfrac{4 \cdot 3 \cdot 2 \cdot 1}{(2 \cdot 1)(2 \cdot 1)}(.6)^2(.4)^2 = 6(.6)^2(.4)^2 = .3456$

c. $P(x=0) = \dfrac{3!}{0!3!}(.7)^0(.3)^3 = \dfrac{3 \cdot 2 \cdot 1}{(1)(3 \cdot 2 \cdot 1)}(.7)^0(.3)^3 = 1(.7)^0(.3)^3 = .027$

d. $P(x=3) = \dfrac{5!}{3!2!}(.1)^3(.9)^2 = \dfrac{5 \cdot 4 \cdot 3 \cdot 2 \cdot 1}{(3 \cdot 2 \cdot 1)(2 \cdot 1)}(.1)^3(.9)^2 = 10(.1)^3(.9)^2 = .0081$

e. $P(x=2) = \dfrac{4!}{2!2!}(.4)^2(.6)^2 = \dfrac{4 \cdot 3 \cdot 2 \cdot 1}{(2 \cdot 1)(2 \cdot 1)}(.4)^2(.6)^2 = 6(.4)^2(.6)^2 = .3456$

f. $P(x=1) = \dfrac{3!}{1!2!}(.9)^1(.1)^2 = \dfrac{3 \cdot 2 \cdot 1}{(1)(2 \cdot 1)}(.9)^1(.1)^2 = 3(.9)^1(.1)^2 = .027$

4.55 a. $\mu = np = 25(.5) = 12.5$

$\sigma^2 = np(1-p) = 25(.5)(.5) = 6.25$
$\sigma = \sqrt{\sigma^2} = \sqrt{6.25} = 2.5$

b. $\mu = np = 80(.2) = 16$

$\sigma^2 = np(1-p) = 80(.2)(.8) = 12.8$
$\sigma = \sqrt{\sigma^2} = \sqrt{12.8} = 3.578$

c. $\mu = np = 100(.6) = 60$

$\sigma^2 = np(1-p) = 100(.6)(.4) = 24$
$\sigma = \sqrt{\sigma^2} = \sqrt{24} = 4.899$

d. $\mu = np = 70(.9) = 63$

$\sigma^2 = np(1-p) = 70(.9)(.1) = 6.3$
$\sigma = \sqrt{\sigma^2} = \sqrt{6.3} = 2.510$

e. $\mu = np = 60(.8) = 48$

$$\sigma^2 = np(1-p) = 60(.8)(.2) = 9.6$$
$$\sigma = \sqrt{\sigma^2} = \sqrt{9.6} = 3.098$$

f. $\mu = np = 1,000(.04) = 40$

$$\sigma^2 = np(1-p) = 1,000(.04)(.96) = 38.4$$
$$\sigma = \sqrt{\sigma^2} = \sqrt{38.4} = 6.197$$

4.57   a. $P(x < 10) = P(x \le 9) = 0$

       b. $P(x \ge 10) = 1 - P(x \le 9) = 1 - .002 = .998$

       c. $P(x = 2) = P(x \le 2) - P(x \le 1) = .206 \ .069 = .137$

4.59 x is a binomial random variable with $n = 4$.

       a.   If the probability distribution of x is symmetric, $p(0) = p(4)$ and $p(1) = p(3)$.

Since $p(x) = \binom{n}{x} p^x q^{n-x}$ $x = 0, 1, \ldots, n,$

When $n = 4$,

$$\binom{4}{0} p^0 q^4 = \binom{4}{4} p^4 q^0$$

$$\Rightarrow \frac{4!}{0!4!} p^0 q^4 = \frac{4!}{4!0!} p^4 q^0$$
$$\Rightarrow q^4 = p^4$$
$$\Rightarrow p = q$$

Since $p + q = 1, p = .5$

Therefore, the probability distribution of x is symmetric when $p = .5$.

       b.   If the probability distribution of x is skewed to the right, then the mean is greater than the median. Therefore, there are more small values in the distribution (0, 1) than large values (3, 4). Therefore, p must be smaller than .5. Let $p = .2$ and the probability distribution of x will be skewed to the right.

       c.   If the probability distribution of x is skewed to the left, then the mean is smaller than the median. Therefore, there are more large values in the distribution (3, 4) than small values (0, 1). Therefore, p must be larger than .5. Let $p = .8$ and the probability distribution of x will be skewed to the left.

d. In part **a**, x is a binomial random variable with $n = 4$ and $p = .5$.

$$p(x) = \binom{4}{x}.5^x.5^{4-x} \qquad x = 0, 1, 2, 3, 4$$

$$p(0) = \binom{4}{0}.5^0.5^4 = \frac{4!}{0!4!}.5^4 = 1(.5)^4 = .0625$$

$$p(1) = \binom{4}{1}.5^1.5^3 = \frac{4!}{1!3!}.5^4 = 4(.5)^4 = .25$$

$$p(2) = \binom{4}{2}.5^2.5^2 = \frac{4!}{2!2!}.5^4 = 6(.5)^4 = .375$$

$p(3) = p(1) = .25$ (since the distribution is symmetric)
$p(4) = p(0) = .0625$

The probability distribution of x in tabular form is:

x	0	1	2	3	4
$p(x)$	.0625	.25	.375	.25	.0625

$$\mu = np = 4(.5) = 2$$

The graph of the probability distribution of x when $n = 4$ and $p = .5$ is as follows.

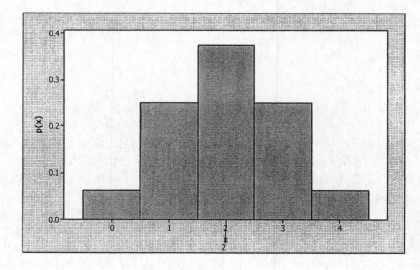

In part **b**, x is a binomial random variable with $n = 4$ and $p = .2$.

$$p(x) = \binom{4}{x} .2^x .8^{4-x} \qquad x = 0, 1, 2, 3, 4$$

$$p(0) = \binom{4}{0} .2^0 .8^4 = 1(1).8^4 = .4096$$

$$p(1) = \binom{4}{1} .2^1 .8^3 = 4(.2)(.8)^3 = .4096$$

$$p(2) = \binom{4}{2} .2^2 .8^2 = 6(.2)^2 (.8)^2 = .1536$$

$$p(3) = \binom{4}{3} .2^3 .8^1 = 4(.2)^3 (.8) = .0256$$

$$p(4) = \binom{4}{0} .2^4 .8^0 = 1(.2)^4 (1) = .0016$$

The probability distribution of x in tabular form is:

x	0	1	2	3	4
$p(x)$	.4096	.4096	.1536	.0256	.0016

$$\mu = np = 4(.2) = .8$$

The graph of the probability distribution of x when $n = 4$ and $p = .2$ is as follows:

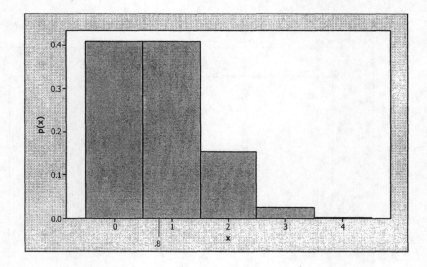

In part **c**, x is a binomial random variable with $n = 4$ and $p = .8$.

$$p(x) = \binom{4}{x}.8^x.2^{4-x} \quad x = 0, 1, 2, 3, 4$$

$$p(0) = \binom{4}{0}.8^0.2^4 = 1(1).2^4 = .0016$$

$$p(1) = \binom{4}{1}.8^1.2^3 = 4(.8)(.2)^3 = .0256$$

$$p(2) = \binom{4}{2}.8^2.2^2 = 6(.8)^2(.2)^2 = .1536$$

$$p(3) = \binom{4}{3}.8^3.2^1 = 4(.8)^3(.2) = .4096$$

$$p(4) = \binom{4}{4}.8^4.2^0 = 1(.8)^4(1) = .4096$$

The probability distribution of x in tabular form is:

x	0	1	2	3	4
$p(x)$	.0016	.0256	.1536	.4096	.4096

Note: The distribution of x when $n = 4$ and $p = .2$ is the reverse of the distribution of x when $n = 4$ and $p = .8$.

$$\mu = np = 4(.8) = 3.2$$

The graph of the probability distribution of x when $n = 4$ and $p = .8$ is as follows:

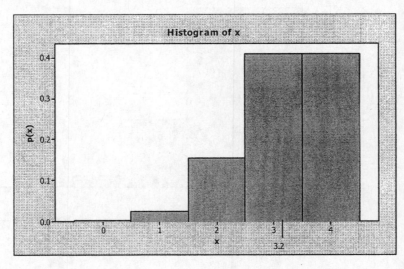

e. In general, when $p = .5$, a binomial distribution will be symmetric regardless of the value of n. When p is less than .5, the binomial distribution will be skewed to the right; and when p is greater than .5, it will be skewed to the left. (Refer to parts **a**, **b**, and **c**.)

4.61   a.   We will check the characteristics of a binomial random variable:

   1.   This experiment consists of $n = 20$ identical trials.

   2.   There are only 2 possible outcomes for each trial. SBIRS detects intruding object (S) or not (F).

   3.   The probability of S remains the same from trial to trial. In this case, $p = P(S) = .8$ for each trial.

   4.   The trials are independent. Whether SBIRS detects one intruding object should not effect whether it detects another.

   5. x = number of intruding objects detected by SBIRS in 20 trials.

   b. $n = 20$ and $p = P(S) = .8$

   c.   Using the Table II, Appendix A with $n = 20$ and $p = .8$,

   $P(x = 15) = P(x \le 15) - P(x \le 14) = .370 - .196 = .174$.

   d. $P(x \ge 15) = 1 - P(x \le 14) = 1 - .196 = .804$

   e. $\mu = E(x) = np = 20(.8) = 16$. The average number of intruding objects detected in 20 trials is 16.

4.63   a.   From the problem, x is a binomial random variable with $n = 10$ and $p = .8$.

   $P(x = 3) = P(x \le 3) - P(x \le 2) = .001 - .000 = .001$ (from Table II, Appendix A)

   b. $P(x \le 7) = .322$

   c. $P(x > 4) = 1 - P(x \le 4) = 1 - .006 = .994$

4.65   a.   Let x = number of beach trees damaged by fungi in 20 trials. Then x is a binomial random variable with $n = 20$ and $p = .25$.

$$P(x < 10) = P(x = 0) + P(x = 1) + \cdots + P(x = 9)$$

$$= \binom{20}{0}.25^0.75^{20} + \binom{20}{1}.25^1.75^{19} + \binom{20}{2}.25^2.75^{18} + \cdots + \binom{20}{9}.25^9.75^{11}$$

$$= .0032 + .0211 + .0669 + .1339 + .1897 + .2023 + .1686 + .1124 + .0609 + .0271$$

$$= .9861$$

b. $P(x > 15) = P(x = 16) + P(x = 17) + \cdots + P(x = 20)$

$$= \binom{20}{16}.25^{15}.75^4 + \binom{20}{17}.25^{17}.75^3 + \binom{20}{18}.25^{18}.75^2 + \cdots + \binom{20}{20}.25^{20}.75^0$$

$$= .000000356 + .000000027 + .000000001 + 0 + 0 = .000000384$$

4.67   a.   Let x = number of women out of 15 that have been abused. Then x is a binomial random variable with $n = 15$, S = woman has been abused, F = woman has not been abused, $P(S) = p = 1/3$ and $q = 1$ $p = 1 - 1/3 = 2/3$.

$$P(x \geq 4) = 1 - P(x < 4) = 1 - P(x = 0) - P(x = 1) - P(x = 2) - P(x = 3)$$
$$= 1 - \frac{15!}{0!15!}\left(\frac{1}{3}\right)^0\left(\frac{2}{3}\right)^{15} - \frac{15!}{1!14!}\left(\frac{1}{3}\right)^1\left(\frac{2}{3}\right)^{14} - \frac{15!}{2!13!}\left(\frac{1}{3}\right)^2\left(\frac{2}{3}\right)^{13}$$

$$- \frac{15!}{3!12!}\left(\frac{1}{3}\right)^3\left(\frac{2}{3}\right)^{12}$$

$$= 1 - .0023 - .0171 - .0599 - .1299 = .7908$$

b.   For $p = .1$,

$$P(x \geq 4) = 1 - P(x < 4) = 1 - P(x = 0) - P(x = 1) - P(x = 2) - P(x = 3)$$
$$= 1 - \frac{15!}{0!15!}.1^0.9^{15} - \frac{15!}{1!14!}.1^1.9^{14} - \frac{15!}{2!13!}.1^2.9^{13} - \frac{15!}{3!12!}.1^3.9^{12}$$
$$= 1 - .2059 - .3432 - .2669 - .1285 = .0555$$

c.   We sampled 15 women and actually found that 4 had been abused. If $p = 1/3$, the probability of observing 4 or more abused women is .7908. If $p = .1$, the probability of observing 4 or more abused women is only .0555. If $p = .1$, we would have seen a very unusual event because the probability is so small (.0555). If $p = 1/3$, we would have seen an event that was very common because the probability was very large (.7908). Since we normally do not see rare events, the true probability of abuse is probably close to 1/3.

4.69   a.   If the psychic is just guessing, she has 1 chance out of 10 of guessing the correct decision. Thus, $p = 1/10 = .1$.

       b.   Let x = number of correct decisions in 7 trials. Then x is a binomial random variable with $n = 7$ and $p = .1$. The expected number of correct decisions in 7 trials is

$$E(x) = np = 7(.1) = .7.$$

Thus, if the psychic is just guessing, we would expect her to make less than 1 correct decision in 7 trials.

       c. $P(x = 0) = \binom{7}{0}.1^0.9^7 = .4783$

       d.   Now, let $p = .5$. $P(x = 0) = \binom{7}{0}.5^0.5^7 = .0078$

       e.   Yes. If the psychic is just guessing, the probability that she makes no correct decisions is .4783. Thus, if the psychic, in fact, made no correct decisions, this would not be an unusual event if she is guessing. However, on the other hand, if she does have ESP ($p = .5$) and made no correct decisions, this would be a very unusual event ($P(x = 0) = .0078$). The evidence indicates the psychic probably is just guessing and does not have ESP.

4.71   a.   We must assume that the probability that a specific type of ball meets the requirements is always the same from trial to trial and the trials are independent. To use the binomial probability distribution, we need to know the probability that a specific type of golf ball meets the requirements.

       b.   For a binomial distribution,

$$\mu = np$$
$$\sigma = \sqrt{npq}$$

In this example, n = two dozen = $2 \cdot 12 = 24$.

            $p = .10$ (Success here means the golf ball *does not* meet standards.)
            $q = .90$
            $\mu = np = 24(.10) = 2.4$
            $\sigma = \sqrt{npq} = \sqrt{24(.10)(.90)} = 1.47$

       c.   In this situation,

            p = Probability of success
              = Probability golf ball *does* meet standards
              = .90
            $q = 1 - .90 = .10$
            $n = 24$
            $E(x) = \mu = np = 24(.90) = 21.60$
            $\sigma = \sqrt{npq} = \sqrt{24(.10)(.90)} = 1.47$ (Note that this is the same as in part **b**.)

4.73   The four characteristics of a Poisson random variable are:

   a.   The experiment consists of counting the number of times a certain event occurs during a given unit of time or in a given area or volume (or weight, distance, or any other unit of measurement).
   b.   The probability that an event occurs in a given unit of time, area, or volume is the same for all the units.
   c.   The number of events that occur in one unit of time, area, or volume is independent of the number that occur in other units.
   d.   The mean (or expected) number of events in each unit is denoted by the Greek letter lambda, λ.

4.75   For the Poisson probability distribution

$$p(x) = \frac{3^x e^{-3}}{x!} \qquad (x = 0, 1, 2, \ldots)$$

the value of λ is 3.

4.77   a.   In order to graph the probability distribution, we need to know the probabilities for the possible values of x. Using Table III of Appendix A with $\lambda = 3$:

$p(0) = .050$
$p(1) = P(x \le 1) - P(x = 0) = .199 - .050 = .149$
$p(2) = P(x \le 2) - P(x \le 1) = .423 - .199 = .224$
$p(3) = P(x \le 3) - P(x \le 2) = .647 - .423 = .224$
$p(4) = P(x \le 4) - P(x \le 3) = .815 - .647 = .168$
$p(5) = P(x \le 5) - P(x \le 4) = .916 - .815 = .101$
$p(6) = P(x \le 6) - P(x \le 5) = .966 - .916 = .050$
$p(7) = P(x \le 7) - P(x \le 6) = .988 - .966 = .022$
$p(8) = P(x \le 8) - P(x \le 7) = .996 - .988 = .008$
$p(9) = P(x \le 9) - P(x \le 8) = .999 - .996 = .003$
$p(10) \approx .001$

The probability distribution of x in graphical form is:

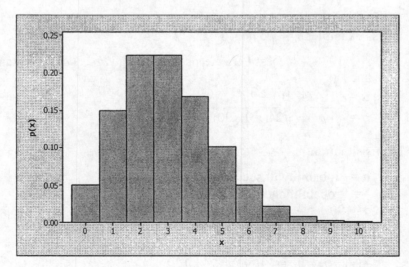

b. $\mu = \lambda = 3$

$\sigma^2 = \lambda = 3$

$\sigma = \sqrt{3} = 1.7321$

4.79 $\mu = \lambda = 1.5$

Using Table III of Appendix A:

a. $P(x \le 3) = .934$

b. $P(x \ge 3) = 1 - P(x \le 2) = 1 - .809 = .191$

c. $P(x = 3) = P(x \le 3) - P(x \le 2) = .934 - .809 = .125$

d. $P(x = 0) = .223$

e. $P(x > 0) = 1 - P(x = 0) = 1 - .223 = .777$

f. $P(x > 6) = 1 - P(x \le 6) = 1 - .999 = .001$

4.81 a. To graph the Poisson probability distribution with $\lambda = 3$, we need to calculate $p(x)$ for $x = 0$ to 10. Using Table III, Appendix A,

$p(0) = .050$
$p(1) = P(x \le 1) - P(x = 0) = .199 - .050 = .149$
$p(2) = P(x \le 2) - P(x \le 1) = .423 - .199 = .224$
$p(3) = P(x \le 3) - P(x \le 2) = .647 - .423 = .224$
$p(4) = P(x \le 4) - P(x \le 3) = .815 - .647 = .168$
$p(5) = P(x \le 5) - P(x \le 4) = .916 - .815 = .101$
$p(6) = P(x \le 6) - P(x \le 5) = .966 - .916 = .050$
$p(7) = P(x \le 7) - P(x \le 6) = .988 - .966 = .022$
$p(8) = P(x \le 8) - P(x \le 7) = .996 - .988 = .008$
$p(9) = P(x \le 9) - P(x \le 8) = .999 - .996 = .003$
$p(10) = P(x \le 10) - P(x \le 9) \approx 1 - .999 = .001$

The graph is shown here.

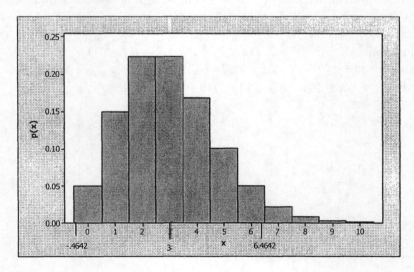

b. $\mu = \lambda = 3$

$\sigma = \sqrt{\lambda} = \sqrt{3} = 1.7321$

$\mu \pm 2\sigma \Rightarrow 3 \pm 2(1.7321) \Rightarrow (-.4642, 6.4642)$

c. $P(\mu - 2\sigma < x < \mu + 2\sigma) = P(-.4642 < x < 6.4642) = P(0 \le x \le 6)$
$$= P(x \le 6) = .966$$

4.83   a.   Using Table III, Appendix A, with $\lambda = 1$, $P(x = 0) = .368$.

b.   Using Table III, Appendix A, with $\lambda = 1$, $P(x > 1) = 1 - P(x \le 1) = 1 - .736 = .264$.

c.   Using Table III, Appendix A, with $\lambda = 1$, $P(x \le 2) = .920$.

4.85   a. $P(x = 0) = \dfrac{\lambda^0 e^{-\lambda}}{0!} = \dfrac{1.2^0 e^{-1.2}}{0!} = .3012$

b. $P(x = 1) = \dfrac{\lambda^1 e^{-\lambda}}{1!} = \dfrac{1.2^1 e^{-1.2}}{1!} = .3614$

c. $E(x) = \mu = \lambda = 1.2$

$\sigma^2 = \lambda = 1.2$

$\sigma = \sqrt{1.2} = 1.0954$

4.87   a. $P(x < 3) = P(x \le 2) = .125$ from Table III, Appendix A, with $\lambda = 5$.

b. $\mu = E(x) = \lambda = 5$. The average number of blocked calls during the peak hour is 3.

c. $P(x = 0) = .007$ from Table III, Appendix A, with $\lambda = 5$. The probability of no blocked calls in the peak hour is very small. Thus, it is not likely that there will be no blocked calls during the peak hour.

d.   Since the probability that $x = 0$ is so small, we would infer that we have either seen a very rare event or that the value of λ is not 5 but a value smaller than 5.

4.89   a.   Using Table III and $\lambda = 6.2$, $P(x = 2) = P(x \le 2) - P(x \le 1) = .054 - .015 = .039$
$P(x = 6) = P(x \le 6) - P(x \le 5) = .574 - .414 = .160$
$P(x = 10) = P(x \le 10) - P(x \le 9) = .949 - .902 = .047$

b. The plot of the distribution is:

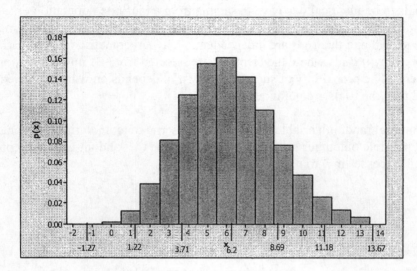

c. $\mu = \lambda = 6.2$, $\sigma = \sqrt{\lambda} = \sqrt{6.2} = 2.490$

$\mu \pm \sigma \Rightarrow 6.2 \pm 2.49 \Rightarrow (3.71, 8.69)$

$\mu \pm 2\sigma \Rightarrow 6.2 \pm 2(2.49) \Rightarrow 6.2 \pm 4.98 \Rightarrow (1.22, 11.18)$

$\mu \pm 3\sigma \Rightarrow 6.2 \pm 3(2.49) \Rightarrow 6.2 \pm 7.47 \Rightarrow (-1.27, 13.67)$

See the plot in part **b**.

d. First, we need to find the mean number of customers per hour. If the mean number of customers per 10 minutes is 6.2, then the mean number of customers per hour is $6.2(6) = 37.2 = \lambda$.

$\mu = \lambda = 37.2$ and $\sigma = \sqrt{\lambda} = \sqrt{37.2} = 6.099$

$\mu \pm 3\sigma \Rightarrow 37.2 \pm 3(6.099) \Rightarrow 37.2 \pm 18.297 \Rightarrow (18.903, 55.498)$

Using Chebyshev's Rule, we know at least 8/9 or 88.9% of the observations will fall within 3 standard deviations of the mean. The number 75 is way beyond the 3 standard deviation limit. Thus, it would be very unlikely that more than 75 customers entered the store per hour on Saturdays.

4.91 Let x = number of cars that will arrive within the next 30 minutes. If the average number of cars that arrive in 60 minutes is 10, the average number of cars that arrive in 30 minutes is $\lambda = 5$.

In order for no cars to be in line at closing time, no more than 1 car can arrive in the next 30 minutes.

$P(x \leq 1) = .040$ from Table III, Appendix A, with $\lambda = 5$.

Since this probability is so small, it is very likely that at least one car will be in line at closing time. The probability that at least one car will be in line at closing is $1 - .040 = .960$.

4.93   A binomial random variable is characterized by n trials where the trials are identical. In order for the trials to be identical, we have to sample from an infinite population or sample from a finite population with replacement. The probability of success on any trial remains constant from trial to trial and the trials are independent. A hypergeometric random variable is characterized by n trials where the elements are selected from a finite population without replacement. The probability of success on any trial depends on what happened on previous trials, and thus the trials are not independent.

Both a binomial random variable and a hypergeometric random variable are characterized by n trials, 2 possible outcomes on each trial (S or F), and the random variable represents the number of successes in n trials.

4.95   a. $P(x = 1) = \dfrac{\binom{r}{x}\binom{N-r}{n-x}}{\binom{N}{n}} = \dfrac{\binom{3}{1}\binom{5-3}{3-1}}{\binom{5}{3}} = \dfrac{\frac{3!}{1!2!}\frac{2!}{2!0!}}{\frac{5!}{3!2!}} = \dfrac{3(1)}{10} = .3$

b. $P(x = 3) = \dfrac{\binom{r}{x}\binom{N-r}{n-x}}{\binom{N}{n}} = \dfrac{\binom{3}{3}\binom{9-3}{5-3}}{\binom{9}{5}} = \dfrac{\frac{3!}{3!0!}\frac{6!}{2!4!}}{\frac{9!}{5!4!}} = \dfrac{1(15)}{126} = .119$

c. $P(x = 2) = \dfrac{\binom{r}{x}\binom{N-r}{n-x}}{\binom{N}{n}} = \dfrac{\binom{2}{2}\binom{4-2}{2-2}}{\binom{4}{2}} = \dfrac{\frac{2!}{2!0!}\frac{2!}{0!2!}}{\frac{4!}{2!2!}} = \dfrac{1(1)}{6} = .167$

d. $P(x = 0) = \dfrac{\binom{r}{x}\binom{N-r}{n-x}}{\binom{N}{n}} = \dfrac{\binom{2}{0}\binom{4-2}{2-0}}{\binom{4}{2}} = \dfrac{\frac{2!}{0!2!}\frac{2!}{2!0!}}{\frac{4!}{2!2!}} = \dfrac{1(1)}{6} = .167$

4.97   With $N = 12$, $n = 8$ and $r = 6$, x can take on values 2, 3, 4, 5, or 6.

a. $P(x = 2) = \dfrac{\binom{r}{x}\binom{N-r}{n-x}}{\binom{N}{n}} = \dfrac{\binom{6}{2}\binom{12-6}{8-2}}{\binom{12}{8}} = \dfrac{\frac{6!}{2!4!}\frac{6!}{6!0!}}{\frac{12!}{8!4!}} = \dfrac{15(1)}{495} = .030$

$P(x = 3) = \dfrac{\binom{r}{x}\binom{N-r}{n-x}}{\binom{N}{n}} = \dfrac{\binom{6}{3}\binom{12-6}{8-3}}{\binom{12}{8}} = \dfrac{\frac{6!}{3!3!}\frac{6!}{5!1!}}{\frac{12!}{8!4!}} = \dfrac{20(6)}{495} = .242$

$$P(x=4) = \frac{\binom{r}{x}\binom{N-r}{n-x}}{\binom{N}{n}} = \frac{\binom{6}{4}\binom{12-6}{8-4}}{\binom{12}{8}} = \frac{\frac{6!}{4!2!}\frac{6!}{4!2!}}{\frac{12!}{8!4!}} = \frac{15(15)}{495} = .455$$

$$P(x=5) = \frac{\binom{r}{x}\binom{N-r}{n-x}}{\binom{N}{n}} = \frac{\binom{6}{5}\binom{12-6}{8-5}}{\binom{12}{8}} = \frac{\frac{6!}{5!1!}\frac{6!}{3!3!}}{\frac{12!}{8!4!}} = \frac{6(20)}{495} = .242$$

$$P(x=6) = \frac{\binom{r}{x}\binom{N-r}{n-x}}{\binom{N}{n}} = \frac{\binom{6}{6}\binom{12-6}{8-6}}{\binom{12}{8}} = \frac{\frac{6!}{6!0!}\frac{6!}{2!4!}}{\frac{12!}{8!4!}} = \frac{1(15)}{495} = .030$$

The probability distribution of x in tabular form is:

x	$p(x)$
2	.030
3	.242
4	.455
5	.242
6	.030

b. $\mu = \dfrac{nr}{N} = \dfrac{8(6)}{12} = 4$

$\sigma^2 = \dfrac{r(N-r)n(N-n)}{N^2(N-1)} = \dfrac{6(12-6)8(12-8)}{12^2(12-1)} = \dfrac{1{,}152}{1{,}584} = .7273$

$\sigma = \sqrt{.7273} = .853$

c. $\mu \pm 2\sigma \Rightarrow 4 \pm 2(.853) \Rightarrow 4 \pm 1.706 \Rightarrow (2.294,\ 5.706)$

The graph of the distribution is:

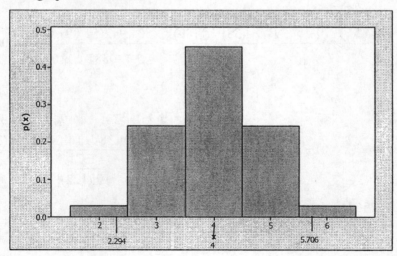

d. $P(2.294 < x < 5.706) = P(3 \leq x \leq 5) = .242 + .455 + .242 = .939$

4.99   For this problem, $N = 100$, $n = 10$, and $x = 4$.

a.   If the sample is drawn without replacement, the hypergeometric distribution should be used. The hypergeometric distribution requires that sampling be done without replacement.

b.   If the sample is drawn with replacement, the binomial distribution should be used. The binomial distribution requires that sampling be done with replacement.

4.101   Let x = number of men chosen in a sample of 4. For this problem, $N = 650$, $r = 3$, and $n = 4$.

$$P(x \geq 1) = 1 - P(x = 0) = 1 - \frac{\binom{3}{0}\binom{650-3}{4-0}}{\binom{650}{4}}$$

$$= 1 - \frac{\frac{3!}{0!3!}\frac{647!}{4!643!}}{\frac{650!}{4!646!}} = 1 - \frac{1(7,233,877,315)}{7,369,297,650} = 1 - .9816 = .0184$$

4.103   Let x = number of "clean" cartridges selected in 5 trials. For this problem, $N = 158$, $r = 122$, and $n = 5$.

$$P(x = 5) = \frac{\binom{r}{x}\binom{N-r}{n-x}}{\binom{N}{n}} = \frac{\binom{122}{5}\binom{36}{0}}{\binom{158}{5}} = \frac{\frac{122!}{5!\,117!}\frac{36!}{0!36!}}{\frac{158!}{5!\,153!}} = .2693$$

4.105   Let x = number of extinct bird species in $n = 10$ trials. Then x is a hypergeometric random variable with $N = 132$, $r = 38$, and $n = 10$.

a. $P(x = 5) = \dfrac{\binom{38}{5}\binom{132-38}{10-5}}{\binom{132}{10}} = \dfrac{\frac{38!}{5!(33)!}\cdot\frac{94!}{5!(89)!}}{\frac{132!}{10!(122)!}} = \dfrac{501,942(54,891,018)}{3.12058705x10^{14}} = .0883$

b. $P(x \leq 1) = P(x = 0) + P(x = 1) = \dfrac{\binom{38}{0}\binom{132-38}{10-0}}{\binom{132}{10}} = \dfrac{\frac{38!}{0!(38)!}\cdot\frac{94!}{10!(84)!}}{\frac{132!}{10!(122)!}} + \dfrac{\frac{38!}{1!(37)!}\cdot\frac{94!}{9!(85)!}}{\frac{132!}{10!(122)!}}$

$$= \frac{9.041256842x10^{12}}{3.12058705x10^{14}} + \frac{4.041973645x10^{13}}{3.12058705x10^{14}} = .0290 + .1295 = .1585$$

4.107 Let x = number of spoiled bottles in the sample of 3. Since the sampling will be done without replacement, x is a hypergeometric random variable with $N = 12$, $n = 3$, and $r = 1$.

$$P(x = 1) = \frac{\binom{r}{x}\binom{N-r}{n-x}}{\binom{N}{n}} = \frac{\binom{1}{1}\binom{12-1}{3-1}}{\binom{12}{3}} = \frac{\frac{1!}{1!0!}\frac{11!}{2!9!}}{\frac{12!}{3!9!}} = \frac{55}{220} = .25$$

4.109 Let x = number of grants awarded to the north side in 140 trials. The random variable x has a hypergeometric distribution with $N = 743$, $n = 140$, and $r = 601$.

a. $\mu = E(x) = \dfrac{nr}{N} = \dfrac{140(601)}{743} = 113.24$

$$\sigma^2 = \frac{r(N-r)n(N-n)}{N^2(N-1)} = \frac{601(743-601)140(743-140)}{743^2(743-1)} = 17.5884$$

$\sigma = \sqrt{17.5884} = 4.194$

b. If the grants were awarded at random, we would expect approximately 113 to be awarded to the north side. We observed 140. The z-score associated with 140 is:

$$z = \frac{x - \mu}{\sigma} = \frac{140 - 113.24}{4.194} = 6.38$$

Because this z-score is so large, it would be extremely unlikely to observe all 140 grants to the north side if they are randomly selected. Thus, we would conclude that the grants were not randomly selected.

4.111 a. Poisson

b. Binomial

c. Binomial

4.113 From Table II, Appendix A:

a. $P(x = 14) = P(x \le 14) - F(x \le 13) = .584 - .392 = .192$

b. $P(x \le 12) = .228$

c. $P(x > 12) = 1 - P(x \le 12) = 1 - .228 = .772$

d. $P(9 \le x \le 18) = P(x \le 18) - P(x \le 8) = .992 - .005 = .987$

e. $P(8 < x < 18) = P(x \le 17) - P(x \le 8) = .965 - .005 = .960$

f. $\mu = np = 20(.7) = 14$

$\sigma^2 = npq = 20(.7)(.3) = 4.2$, $\sigma = \sqrt{4.2} = 2.049$

g.　$\mu \pm 2\sigma \Rightarrow 14 \pm 2(2.049) \Rightarrow 14 \pm 4.098 \Rightarrow (9.902, 18.098)$

$$P(9.902 < x < 18.098) = P(10 \le x \le 18) = P(x \le 18) - P(x \le 9)$$
$$= .992 - .017 = .975$$

4.115　a.　$P(x = 1) = \binom{3}{1}(.1)^1(.9)^{3-1} = \dfrac{3!}{1!2!}(.1)^1(.9)^2 = 3(.1)(.81) = .243$

b.　$P(x = 4) = P(x \le 4) - P(x \le 3) = .238 - .107 = .131$ from Table II, Appendix A.

c.　$P(x = 0) = \binom{2}{0}(.4)^0(.6)^{2-0} = \dfrac{2!}{0!2!}(1)(.6)^2 = .36$

d.　$P(x = 4) = P(x \le 4) - P(x \le 3) = .969 - .812 = .157$ from Table II, Appendix A.

e.　$P(x = 12) = P(x \le 12) - P(x \le 11) = .184 - .056 = .128$ from Table II, Appendix A.

f.　$P(x = 8) = P(x \le 8) - P(x \le 7) = .954 - .833 = .121$ from Table II, Appendix A.

4.117　Using Table III, Appendix A,

a.　When $\lambda = 2, p(3) = P(x \le 3) - P(x \le 2) = .857 - .677 = .180$

b.　When $\lambda = 1, p(4) = P(x \le 4) - P(x \le 3) = .996 - .981 = .015$

c.　When $\lambda = .5, p(2) = P(x \le 2) - P(x \le 1) = .986 - .910 = .076$

4.119　a.　For this experiment, there are $n = 200$ smokers (n identical trials). For each smoker, there are 2 possible outcomes: $S =$ smoker enters treatment program and $F =$ smoker does not enter treatment program. The probability of S (smoker enters treatment program) is the same from trial to trial. This probability is $P(S) = p = .05$. $P(F) = 1 - P(S) = 1 - .05 = .95$. The trials are independent and $x =$ number of smokers entering treatment program in 200 trials. Thus, x is a binomial random variable.

b.　$p = P(S) = .05$. Of all the smokers, only .05 or 5% enter treatment programs.

c.　$E(x) = np = 200(.05) = 10$. For all samples of 200 smokers, the average number of smokers who enter a treatment program will be 10.

4.121   a.   In order to be a binomial random variable, the five characteristics must hold.

      1.   For this problem, there are 5 items scanned. We will assume that these 5 trials are identical.

      2.   For each item scanned, there are 2 possible outcomes: priced incorrectly (S) or priced correctly (F).

      3.   The probability of being priced incorrectly remains constant from trial to trial. For this problem, we will assume that the probability of being priced incorrectly is $P(S) = 1/30$ for each trial.

      4.   We will assume that whether one item is priced incorrectly is independent of any other.

      5.   The random variable x is the number of items priced incorrectly in 5 trials.

      Thus, x is a binomial random variable.

    b.   The estimate of p, the probability of an item being priced incorrectly is 1/30.

    c. $P(x = 1) = \binom{5}{1}(1/30)^1(29/30)^4 = .1455$

    d. $P(x \geq 1) = 1 - P(x = 0) = 1 - \binom{5}{0}(1/30)^0(29/30)^5 = 1 - .8441 = .1559$

4.123   The random variable x would have a binomial distribution:

    1. n identical trials. A sample of married women was taken and would be considered small compared to the entire population of married women, so the trials would be very close to identical.

    2.   There are only two possible outcomes for each trial. In this experiment, there are only two possible outcomes: the woman would marry the same man (S) or she would not (F).

    3.   The probability of Success stays the same from trial to trial. Here, $P(S) = .50$ and $P(F) = .50$. In reality, these probabilities would not be exactly the same from trial to trial, but rounded off to 4 decimal places, they would be the same.

    4.   The trials are independent. In this experiment, the trials would not be exactly independent because we would be sampling without replacement from a finite population. However, if the sample size is fairly small compared to the population size, the trials will be essentially independent.

    5.   The binomial random variable x would be the number of successes in n trials. For this experiment, x = the number of women out of those sampled who would marry the same man again.

    Thus, x would possess (approximately) a binomial distribution.

4.125   a.   In order for this to be a valid probability distribution, $\sum p(x) = 1$ and $0 \le p(x) \le 1$ for all values of x.

In this distribution, $0 \le p(x) \le 1$ for all values of x and $\sum p(x) = .051 + .099 + .093 + .635 + .122 = 1.000$. Thus, this is a valid probability distribution.

   b. $P(x = 1) = .051$

   c. $P(x \ge 4) = P(x = 4) + P(x = 5) = .635 + .122 = .757$

   d. $P(x = 2 \text{ or } x = 3) = P(x = 2) + P(x = 3) = .099 + .093 = .192$

   e. $E(x) = \mu = \sum xp(x) = 1(.051) + 2(.099) + 3(.093) + 4(.635) + 4(.122) = 3.678$

   f.   The average rating of all books rated is 3.678.

4.127   a. $\mu = E(x) = np = 800(.65) = 520$
        $\sigma = \sqrt{npq} = \sqrt{800(.65)(.35)} = 13.491$

   b.   Half of the 800 food items is 400. A value of $x = 400$ would have a z-score of:

$$z = \frac{x - \mu}{\sigma} = \frac{400 - 520}{13.491} = -8.895$$

Since the z-score associated with 400 items is so small (−8.895), it would be virtually impossible to observe less than half without any traces of pesticides if the 65% value was correct.

4.129   Let x = number of parents who yell at their child before, during, or after a meet in 20 trials. Then x has a binomial distribution with $n = 20$ and $p = .05$.

$P(x \ge 1) = 1 - P(x = 0) = 1 - .358 = .642$ (Using Table II, Appendix A)

4.131   a. $\sigma = \sqrt{\lambda} = \sqrt{4} = 2$

   b. $P(x > 10) = 1 - P(x \le 10) = 1 - .997 = .003$ from Table III, Appendix A, with $\lambda = 4$. Since the probability is so small (.003), it would be very unlikely that the plant would yield a value that would exceed the EPA limit.

4.133   Using Table II with $n = 25$ and $p = .8$:

   a. $P(x < 15) = P(x \le 14) = .006$

   b.   Since the probability of such an event is so small when $p = .8$, if less than 15 insects die we would conclude that the insecticide is not as effective as claimed.

4.135   a.   For this problem, we must assume that each couple has only one child. Let x = number of couples out of 15 who have a child with A-T. Then x is a binomial random variable with $n = 15$, S = couple has child with A-T, F = couple does not have a child with A-T, $P(S) = p = 1/5 = .20$ and $q = 1 - p = 1 - .20 = .80$.

$$P(x > 8) = 1 - P(x \le 8) = 1 - .999 = .001$$
(Using Table II, Appendix A, with $n = 15$ and $p = .2$)

b.   For $n = 10,000$,

$$P(x < 3,000) = P(x = 0) + P(x = 1) + \cdots + P(x = 2,999)$$

$$= \frac{10000!}{0!10000!} .20^0 .80^{10000} + \frac{10000!}{1!9999!} .20^1 .80^{9999} + \cdots + \frac{10000!}{2999!7001!} .20^{2999} .80^{7001}$$

$\mu = E(x) = np = 10,000(.20) = 2,000;\ \sigma^2 = npq = 10,000(.20)(.80) = 1,600$
$\sigma = \sqrt{1,600} = 40$

The value 3,000 has a z-score of:

$$z = \frac{x - \mu}{\sigma} = \frac{3,000 - 2,000}{40} = 25$$

This z-score is extremely large. Essentially all of the observations will have z-scores less than 25. Thus, $P(x < 3000) \approx 1.000$.

4.137   There will not be enough beds if x = the number of newly admitted patients exceeds 4. For $\lambda = 2.5$,

$$P(x > 4) = 1 - P(x \le 4)$$
$$= 1 - \{P(x = 0) + P(x = 1) + P(x = 2) + P(x = 3) + P(x = 4)\}$$
$$= 1 - \left(\frac{2.5^0 \cdot e^{-2.5}}{0!} + \frac{2.5^1 \cdot e^{-2.5}}{1!} + \frac{2.5^2 \cdot e^{-2.5}}{2!} + \frac{2.5^3 \cdot e^{-2.5}}{3!} + \frac{2.5^4 \cdot e^{-2.5}}{4!} \right)$$

$$= 1 - (.0821 + .2052 + .2565 + .2138 + .1336)$$
$$= 1 - .8912 = .1088$$

4.139   Let x = number of disasters in 25 trials. If NASA's assessment is correct, then x is a binomial random variable with $n = 25$ and $p = 1 / 60,000 = .00001667$. If the Air Force's assessment is correct, then x is a binomial random variable with $n = 25$ and $p = 1 / 35 = .02857$.

If NASA's assessment is correct, then the probability of no disasters in 25 missions would be:

$$P(x = 0) = \binom{25}{0} (1/60,000)^0 (59,999/60,000)^{25} = .9996$$

Thus, the probability of at least one disaster would be

$$P(x \ge 1) = 1 - P(x = 0) = 1 - .9996 = .0004$$

If the Air Force's assessment is correct, then the probability of no disasters in 25 missions would be:

$$P(x=0) = \binom{25}{0}(1/35)^0(34/35)^{25} = .4845$$

Thus, the probability of at least one disaster would be

$$P(x \geq 1) = 1 - P(x=0) = 1 - .4845 = .5155$$

One disaster actually did occur. If NASA's assessment were correct, it would be almost impossible for at least one disaster to occur in 25 trials. If the Air Force's assessment were correct, one disaster in 25 trials would not be an unusual event. Thus, the Air Force's assessment appears to be appropriate.

Continuous Random Variables

Chapter

5

5.1   A uniform random variable is a continuous random variable that appears to have equally likely outcomes over its range of possible values.

5.3   a.   For a uniform random variable,

$$f(x) = \begin{cases} \dfrac{1}{d-c} = \dfrac{1}{30-10} = \dfrac{1}{20} & 10 \leq x \leq 30 \\ 0 & \text{otherwise} \end{cases}$$

     b. $\mu = \dfrac{c+d}{2} = \dfrac{10+30}{2} = 20$

         $\sigma = \dfrac{d-c}{\sqrt{12}} = \dfrac{30-10}{\sqrt{12}} = 5.774$

     c.

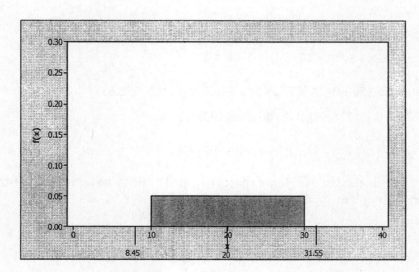

       $\mu \pm 2\sigma \Rightarrow 20 \pm 2(5.774) \Rightarrow 20 \pm 11.584 \Rightarrow (8.452, 31.548)$

5.5   a.   For a uniform random variable,

$$f(x) = \begin{cases} \dfrac{1}{d-c} = \dfrac{1}{4-2} = \dfrac{1}{2} & 2 \leq x \leq 4 \\ 0 & \text{otherwise} \end{cases}$$

     b. $\mu = \dfrac{c+d}{2} = \dfrac{2+4}{2} = 3$

         $\sigma = \dfrac{d-c}{\sqrt{12}} = \dfrac{4-2}{\sqrt{12}} = .577$

c. $\mu \pm \sigma \Rightarrow 3 \pm .577 \Rightarrow (2.423, 3.577)$

$$P(2.423 \leq x \leq 3.577) = (3.577 - 2.423)\frac{1}{2} = .577$$

d. $P(x > 2.78) = (4 - 2.78)\frac{1}{2} = .61$

e. $P(2.4 \leq x \leq 3.7) = (3.7 - 2.4)\frac{1}{2} = .65$

f. $P(x < 2) = 0$

5.7 For the uniform random variable,

$$f(x) = \begin{cases} \dfrac{1}{d-c} = \dfrac{1}{200-100} = \dfrac{1}{100} & 100 \leq x \leq 200 \\ 0 & \text{otherwise} \end{cases}$$

a. $\mu = \dfrac{c+d}{2} = \dfrac{100+200}{2} = 150$ $\qquad \sigma = \dfrac{d-c}{\sqrt{12}} = \dfrac{200-100}{\sqrt{12}} = 28.87$

$\mu \pm 2\sigma \Rightarrow 150 \pm 2(28.87) \Rightarrow 150 \pm 57.74 \Rightarrow (92.26, 207.74)$

$P(x < 92.26) + P(x > 207.74) = 0 + 0 = 0$

b. $\mu \pm 3\sigma \Rightarrow 150 \pm 3(28.87) \Rightarrow 150 \pm 86.61 \Rightarrow (63.39, 236.61)$

$P(63.39 < x < 236.61) = P(100 \leq x \leq 200) = 1$

c. $P(92.26 < x < 207.74) = P(100 \leq x \leq 200) = 1$

5.9 a. Let x be the number of anthrax spores detected. Then x has a uniform distribution on the interval from 0 to 10. Thus, $c = 0$ and $d = 10$.

$$f(x) = \begin{cases} \dfrac{1}{d-c} = \dfrac{1}{10-0} = \dfrac{1}{10} & \text{for } 0 \leq x \leq 10 \\ 0 & \text{otherwise} \end{cases}$$

$$P(x \leq 8) = (8-0)\left(\frac{1}{10}\right) = \frac{8}{10} = .8$$

b. $P(2 \leq x \leq 5) = (5-2)\left(\dfrac{1}{10}\right) = \dfrac{3}{10} = .3$

5.11 a. Suppose no bolt-on trace elements are used. Let x be a uniform random variable on the interval 260 to 290. From this, c = 260 and d = 290.

$$f(x) = \begin{cases} \dfrac{1}{d-c} = \dfrac{1}{290-260} = \dfrac{1}{30} & 260 \leq x \leq 290 \\ 0 & \text{otherwise} \end{cases}$$

$$P(280 \leq x \leq 284) = (284 - 280)\left(\dfrac{1}{30}\right) = 4\left(\dfrac{1}{30}\right) = .1333$$

Suppose bolt-on trace elements are used. Let y be a uniform random variable on the interval 278 to 285. From this, c = 278 and d = 285.

$$f(y) = \begin{cases} \dfrac{1}{d-c} = \dfrac{1}{285-278} = \dfrac{1}{7} & 278 \leq y \leq 285 \\ 0 & \text{otherwise} \end{cases}$$

$$P(280 \leq y \leq 284) = (284 - 280)\left(\dfrac{1}{7}\right) = 4\left(\dfrac{1}{7}\right) = .5714$$

b. Suppose no bolt-on trace elements are used.

$$P(x \leq 268) = (268 - 260)\left(\dfrac{1}{30}\right) = 8\left(\dfrac{1}{30}\right) = .2667$$

Suppose bolt-on trace elements are used.

$$P(x \leq 268) = (0)\left(\dfrac{1}{7}\right) = 0$$

5.13 Let x be cycle availability. Then x has a uniform distribution on the interval from 0 to 1. Thus, $c = 0$ and $d = 1$.

$$f(x) = \begin{cases} \dfrac{1}{d-c} = \dfrac{1}{1-0} = 1 & \text{for } 0 \leq x \leq 1 \\ 0 & \text{otherwise} \end{cases}$$

mean: $E(x) = \mu = \dfrac{c+d}{2} = \dfrac{0+1}{2} = .5$ The average cycle availability is .5.

standard deviation: $\sigma = \dfrac{d-c}{\sqrt{12}} = \dfrac{1-0}{\sqrt{12}} = .2887$

10th percentile: $P(x \leq k) = .10 \Rightarrow (k-0)(1) = .10 \Rightarrow k = .1$. 10% of all values of cycle availability are less than or equal to .10.

Lower quartile: $P(x \leq k) = .25 \Rightarrow (k-0)(1) = .25 \Rightarrow k = .25$. 25% of all values of cycle availability are less than or equal to .25.

Upper quartile: $P(x \leq k) = .75 \Rightarrow (k-0)(1) = .75 \Rightarrow k = .75$. 75% of all values of cycle availability are less than or equal to .75.

5.15 a. The amount dispensed by the beverage machine is a continuous random variable since it can take on any value between 6.5 and 7.5 ounces.

b. Since the amount dispensed is random between 6.5 and 7.5 ounces, x is a uniform random variable.

$$f(x) = \frac{1}{d-c} \quad (c \leq x \leq d)$$

$$\frac{1}{d-c} = \frac{1}{7.5-6.5} = \frac{1}{1} = 1$$

Therefore, $f(x) = \begin{cases} 1 & (6.5 \leq x \leq 7.5) \\ 0 & \text{otherwise} \end{cases}$

The graph is as follows:

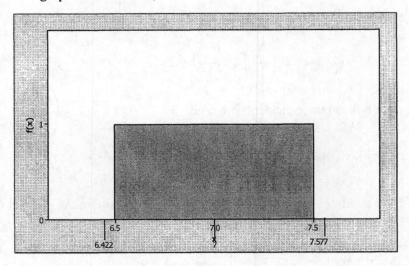

c. $\mu = \dfrac{c+d}{2} = \dfrac{6.5+7.5}{2} = \dfrac{14}{2} = 7$

$\sigma = \dfrac{d-c}{\sqrt{12}} = \dfrac{7.5-6.5}{\sqrt{12}} = .2887$

$\mu \pm 2\sigma \Rightarrow 7 \pm 2(.2887) \Rightarrow 7 \pm .5774 \Rightarrow (6.422, 7.577)$

d. $P(x \geq 7) = (7.5 - 7)(1) = .5$

e. $P(x < 6) = 0$

f. $P(6.5 \leq x \leq 7.25) = (7.25 - 6.5)(1) = .75$

g.   The probability that the next bottle filled will contain more than 7.25 ounces is:

$$P(x > 7.25) = (7.5 - 7.25)(1) = .25$$

The probability that the next 6 bottles filled will contain more than 7.25 ounces is:

$$P[(x > 7.25) \cap (x > 7.25) \cap (x > 7.25) \cap (x > 7.25) \cap (x > 7.25) \cap (x > 7.25)]$$
$$= [P(x > 7.25)]^6 = .25^6 = .0002$$

5.17   Let x = number of inches a gouge is from one end of the spindle. Then x has a uniform distribution with $f(x)$ as follows:

$$f(x) = \begin{cases} \dfrac{1}{d-c} = \dfrac{1}{18-0} = \dfrac{1}{18} & 0 \le x \le 18 \\ 0 & \text{otherwise} \end{cases}$$

In order to get at least 14 consecutive inches without a gouge, the gouge must be within 4 inches of either end. Thus, we must find:

$$P(x < 4) + P(x > 14) = (4 - 0)(1/18) + (18 - 14)(1/18)$$
$$= 4/18 + 4/18 = 8/18 = .4444$$

5.19   A normal distribution is a bell-shaped curve with the center of the distribution at μ.

5.21   If x has a normal distribution with mean μ and standard deviation σ, then the distribution of $z = \dfrac{x - \mu}{\sigma}$ is a normal with a mean of $\mu = 0$ and a standard deviation of $\sigma = 1$.

5.23   a. $P(-2.00 < z < 0) = .4772$
(from Table IV, Appendix A)

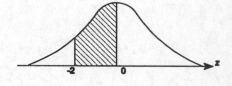

b. $P(-1.00 < z < 0) = .3413$
(from Table IV, Appendix A)

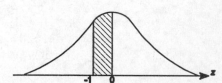

c. $P(-1.69 < z < 0) = .4545$
(from Table IV, Appendix A)

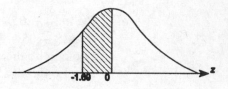

d. $P(-.58 < z < 0) = .2190$
(from Table IV, Appendix A)

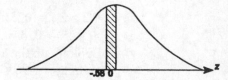

5.25   a. $P(z = 1) = 0$, since a single point does not have an area.

b. $P(z \le 1) = P(z \le 0) + P(0 < z \le 1)$
$$= A_1 + A_2$$
$$= .5 + .3413$$
$$= .8413$$
(Table IV, Appendix A)

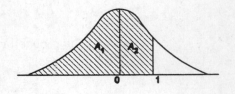

c. $P(z < 1) = P(z \le 1) = .8413$ (Refer to part **b**.)

d. $P(z > 1) = 1 - P(z \le 1) = 1 - .8413 = .1587$ (Refer to part **b**.)

e. $P(-1 \le z \le 1) = A_1 + A_2$
$$= .3413 + .3413$$
$$= .6826$$

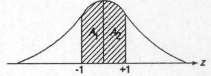

f. $P(-2 \le z \le 2) = A_1 + A_2$
$$= .4772 + .4772$$
$$= .9544$$

g. $P(-2.16 < z \le 0.55) = A_1 + A_2$
$$= .4846 + .2088$$
$$= .6934$$

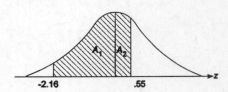

h. $P(-.42 < z < 1.96)$
$$= P(-.42 \le z \le 0) + P(0 \le z \le 1.96)$$
$$= A_1 + A_2$$
$$= .1628 + .4750$$
$$= .6378$$

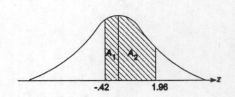

5.27   Using Table IV, Appendix A:

a. $P(z \ge z_0) = .05$
$A_1 = .5 - .05 = .4500$
Looking up the area .4500 in Table IV gives
$z_0 = 1.645$.

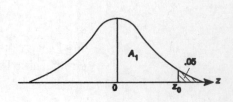

b. $P(z \ge z_0) = .025$
$A_1 = .5 - .025 = .4750$
Looking up the area .4750 in Table IV
gives $z_0 = 1.96$.

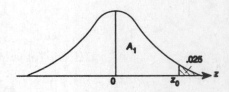

c. $P(z \le z_0) = .025$
$A_1 = .5 - .025 = .4750$
Looking up the area .4750 in Table IV gives
$z = 1.96$. Since z_0 is to the left of 0, $z_0 = 1.96$.

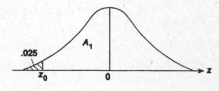

d. $P(z \ge z_0) = .10$
$A_1 = .5 - .1 = .4$
Looking up the area .4000 in Table IV
gives $z_0 = 1.28$.

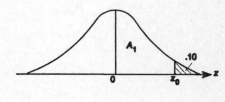

e. $P(z > z_0) = .10$
$A_1 = .5 - .1 = .4$
$z_0 = 1.28$ (same as in **d**)

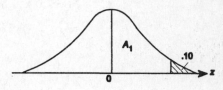

5.29 Using the formula $z = \dfrac{x - \mu}{\sigma}$ with $\mu = 25$ and $\sigma = 5$:

a. $z = \dfrac{25 - 25}{5} = 0$

b. $z = \dfrac{30 - 25}{5} = 1$

c. $z = \dfrac{37.5 - 25}{5} = 2.5$

d. $z = \dfrac{10 - 25}{5} = -3$

e. $z = \dfrac{50 - 25}{5} = 5$

f. $z = \dfrac{32 - 25}{5} = 1.4$

5.31 a. $P(x < \mu - 2\sigma) + P(x > \mu + 2\sigma) = P(z < -2) + P(z > 2)$
$\qquad\qquad\qquad = (.5 - .4772) + (.5 - .4772)$
$\qquad\qquad\qquad = 2(.5 - .4772) = .0456$
$\qquad\qquad\qquad\qquad$ (from Table IV, Appendix A)

$\qquad P(x < \mu - 3\sigma) + P(x > \mu + 3\sigma) = P(z < -3) + P(z > 3)$
$\qquad\qquad\qquad = (.5 - .4987) + (.5 - .4987)$
$\qquad\qquad\qquad = 2(.5 - .4987) = .0026$
$\qquad\qquad\qquad\qquad$ (from Table IV, Appendix A)

b. $P(\mu - \sigma < x < \mu + \sigma) = P(-1 < z < 1)$
$$= P(-1 < z < 0) + P(0 < z < 1)$$
$$= .3413 + .3413$$
$$= 2(.3413) = .6826$$

(from Table IV, Appendix A)

$P(\mu - 2\sigma < z < \mu + 2\sigma) = P(-2 < z < 2)$
$$= P(-2 < z < 0) + P(0 < z < 2)$$
$$= .4772 + .4772$$
$$= 2(.4772) = .9544$$

(from Table IV, Appendix A)

c. $P(x \le x_0) = .80$. Find x_0.

$$P(x \le x_0) = P\left(z \le \frac{x_0 - 300}{30}\right) = P(z \le z_0) = .80$$

$A_1 = .80 - .5 = .3000$

Looking up area .3000 in Table IV, $z_0 = .84$.

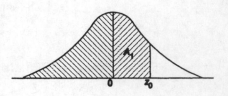

$$z_0 = \frac{x_0 - 300}{30} \Rightarrow .84 = \frac{x_0 - 300}{30} \Rightarrow x_0 = 325.2$$

$P(x \le x_0) = .10$. Find x_0.

$$P(x \le x_0) = P\left(z \le \frac{x_0 - 300}{30}\right) = P(z \le z_0) = .10$$

$A_1 = .50 - .10 = .4000$

Looking up area .4000 in Table IV, $z_0 = -1.28$.

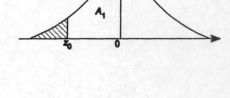

$$z_0 = \frac{x_0 - 300}{30} \Rightarrow -1.28 = \frac{x_0 - 300}{30} \Rightarrow x_0 = 261.6$$

5.33 a. $P(x < x_0) = .10 \Rightarrow P\left(z < \frac{x_0 - 30}{8}\right)$
$$= P(z < z_0) = .10$$
$A_1 = .5 - .10 = .4000$

Looking up the area .4000 in Table IV gives $z_0 = 1.28$. Since z_0 is to the left of 0, $z_0 = -1.28$.

$$z_0 = -1.28 = \frac{x_0 - 30}{8} \Rightarrow x_0 = 8(-1.28) + 30 = 19.76$$

b. $P(x < x_0) = .80 \Rightarrow P\left(z < \dfrac{x_0 - 30}{8} \right)$

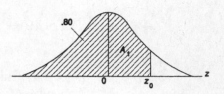

$$= P(z < z_0) = .80$$
$A_1 = .80 - .5 = .3000$
Looking up the area .3000 in Table IV gives $z_0 = 0.84$.

$$z_0 = 0.84 = \dfrac{x_0 - 30}{8} \Rightarrow x_0 = 8(0.84) + 30 = 36.72$$

c. $P(x > x_0) = .01 \Rightarrow P\left(z > \dfrac{x_0 - 30}{8} \right)$

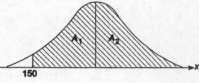

$$= P(z > z_0) = .01$$
$A_1 = .5 - .01 = .4900$
Looking up the area .4900 in Table IV gives $z_0 = 2.33$.

$$z_0 = 2.33 = \dfrac{x_0 - 30}{8} \Rightarrow x_0 = 8(2.33) + 30 = 48.64$$

5.35 The random variable x has a normal distribution with $\sigma = 25$.

We know $P(x > 150) = .90$. So, $A_1 + A_2 = .90$. Since $A_2 = .50$, $A_1 = .90 - .50 = .40$. Look up the area .40 in the body of Table IV; (take the closest value) $z_0 = -1.28$.

To find μ, substitute all the values into the z-score formula:

$$z = \dfrac{x - \mu}{\sigma}$$
$$\Rightarrow -1.28 = \dfrac{150 - \mu}{25}$$
$$\Rightarrow \quad \mu = 150 + 25(1.28) = 182$$

5.37 a. Let x = age of powerful women. Using Table IV, Appendix A,

$$P(55 < x < 60) = P\left(\dfrac{55 - 50}{5.3} < z < \dfrac{60 - 50}{5.3} \right) = P(.94 < z < 1.89)$$
$$= .4706 - .3264 = .1442$$

b. $P(48 < x < 52) = P\left(\dfrac{48 - 50}{5.3} < z < \dfrac{52 - 50}{5.3} \right) = P(-.38 < z < .38)$
$$= .1480 + .1480 = .2960$$

c. $P(x < 35) = P\left(z < \dfrac{35 - 50}{5.3} \right) = P(z < -2.83) = .5000 - .4977 = .0023$

d. $P(x > 40) = P\left(z > \dfrac{40 - 50}{5.3}\right) = P(z > -1.89) = .5000 + .4706 = .9706$

5.39   a.   Let x = transmission delay of an RSVP linked wireless device. Using Table IV, Appendix A,

$$P(x < 57) = P\left(z < \dfrac{57 - 48.5}{8.5}\right) = P(z < 1.00) = .5000 + .3413 = .8413$$

b. $P(40 < x < 60) = P\left(\dfrac{40 - 48.5}{8.5} < z < \dfrac{60 - 48.5}{8.5}\right) = P(-1.00 < z < 1.35)$
$$= .3413 + .4115 = .7528$$

5.41   a.   Let x = driver's head injury rating. Using Table IV, Appendix A,

$$P(500 < x < 700) = P\left(\dfrac{500 - 605}{185} < z < \dfrac{700 - 605}{185}\right) = P(-.57 < z < .51)$$
$$= .2157 + .1950 = .4107$$

b. $P(400 < x < 500) = P\left(\dfrac{400 - 605}{185} < z < \dfrac{500 - 605}{185}\right) = P(-1.11 < z < -.57)$
$$= .3665 - .2157 = .1508$$

c. $P(x < 850) = P\left(z < \dfrac{850 - 605}{185}\right) = P(z < 1.32) = .5 + .4066 = .9066$

d. $P(x > 1,000) = P\left(z > \dfrac{1,000 - 605}{185}\right) = P(z > 2.14) = .5 - .4838 = .0162$

e. $P(x > x_o) = .10 \Rightarrow P\left(z > \dfrac{x_o - 605}{185}\right) = .10 \Rightarrow z = 1.28 = \dfrac{x_o - 605}{185}$
$$\Rightarrow 236.8 = x_o - 605 \Rightarrow x_o = 841.8$$

5.43   Let x = height of women over 20 years old. We are given that $\mu = 64$ and $\sigma = 2.6$ and that the distribution of x is normal.

a.   Find x_L and x_U such that $P(x_L < x < x_U) = .50$.

First, we need to find z_0 such that $P(-z_0 < z < z_0) = .50$. Since we know that the center of the z distribution is 0, half of the area or $.50/2 = .25$ will be between $-z_0$ and 0 and half will be between 0 and z_0.

We look up .25 in the body of Table IV, Appendix A to find $z_0 = .67$.

We know

$$P(x_L < x < x_U) = P\left(\frac{x_L - 64}{2.6} < z < \frac{x_U - 64}{2.6} \right) = P(-.67 < z < .67) = .50$$

Thus,

$$\frac{x_L - 64}{2.6} = -.67 \quad \text{and} \quad \frac{x_U - 64}{2.6} = .67$$

$$\Rightarrow x_L = 64 - .67(2.6) = 64 - 1.74 = 62.26$$

$$\Rightarrow x_U = 64 + .67(2.6) = 64 + 1.74 = 65.74$$

b.   Find x_L and x_U such that $P(x_L < x < x_U) = .75$.

First, we need to find z_0 such that $P(-z_0 < z < z_0) = .75$. Since we know that the center of the z distribution is 0, half of the area or $.75/2 = .375$ will be between $-z_0$ and 0 and half will be between 0 and z_0.

We look up .375 in the body of Table IV, Appendix A to find $z_0 = 1.15$.

We know

$$P(x_L < x < x_U) = P\left(\frac{x_L - 64}{2.6} < z < \frac{x_U - 64}{2.6} \right) = P(-1.15 < z < 1.15) = .75$$

Thus,

$$\frac{x_L - 64}{2.6} = -1.15 \quad \text{and} \quad \frac{x_U - 64}{2.6} = 1.15$$

$$\Rightarrow x_L = 64 - 1.15(2.6) = 64 - 2.99 = 61.01$$

$$\Rightarrow x_U = 64 + 1.15(2.6) = 64 + 2.99 = 66.99$$

c.   Find x_L and x_U such that $P(x_L < x < x_U) = .90$.

First, we need to find z_0 such that $P(-z_0 < z < z_0) = .90$. Since we know that the center of the z distribution is 0, half of the area or $.90/2 = .45$ will be between $-z_0$ and 0 and half will be between 0 and z_0.

We look up .45 in the body of Table IV, Appendix A to find $z_0 = 1.645$.

We know

$$P(x_L < x < x_U) = P\left(\frac{x_L - 64}{2.6} < z < \frac{x_U - 64}{2.6} \right) = P(-1.645 < z < 1.645) = .90$$

Thus,

$$\frac{x_L - 64}{2.6} = -1.645 \quad \text{and} \quad \frac{x_U - 64}{2.6} = 1.645$$

$$\Rightarrow x_L = 64 - 1.645(2.6) = 64 - 4.28 = 59.72$$

$$\Rightarrow x_U = 64 + 1.645(2.6) = 64 + 4.28 = 68.28$$

d. Find x_L and x_U such that $P(x_L < x < x_U) = .95$.

First, we need to find z_0 such that $P(-z_0 < z < z_0) = .95$. Since we know that the center of the z distribution is 0, half of the area or $.95/2 = .475$ will be between $-z_0$ and 0 and half will be between 0 and z_0.

We look up .475 in the body of Table IV, Appendix A to find $z_0 = 1.96$.

We know

$$P(x_L < x < x_U) = P\left(\frac{x_L - 64}{2.6} < z < \frac{x_U - 64}{2.6}\right) = P(-1.96 < z < 1.96) = .95$$

Thus,

$$\frac{x_L - 64}{2.6} = -1.96 \quad \text{and} \quad \frac{x_U - 64}{2.6} = 1.96$$

$$\Rightarrow x_L = 64 - 1.96(2.6) = 64 - 5.10 = 58.90$$

$$\Rightarrow x_U = 64 + 1.96(2.6) = 64 + 5.10 = 69.10$$

e. Find x_L and x_U such that $P(x_L < x < x_U) = .99$.

First, we need to find z_0 such that $P(-z_0 < z < z_0) = .99$. Since we know that the center of the z distribution is 0, half of the area or $.99/2 = .495$ will be between $-z_0$ and 0 and half will be between 0 and z_0.

We look up .495 in the body of Table IV, Appendix A to find $z_0 = 2.575$.

We know

$$P(x_L < x < x_U) = P\left(\frac{x_L - 64}{2.6} < z < \frac{x_U - 64}{2.6}\right) = P(-2.575 < z < 2.575) = .99$$

Thus,

$$\frac{x_L - 64}{2.6} = -2.575 \quad \text{and} \quad \frac{x_U - 64}{2.6} = 2.575$$

$$\Rightarrow x_L = 64 - 2.575(2.6) = 64 - 6.70 = 57.30$$

$$\Rightarrow x_U = 64 + 2.575(2.6) = 64 + 6.70 = 70.70$$

5.45   a.   Using Table IV, Appendix A, with $\mu = 9.06$ and $\sigma = 2.11$,

$$P(x < 6) = \left(z < \frac{6 - 9.06}{2.11} \right) = P(z < -1.45) = .5 - P(-1.45 \le z \le 0)$$
$$= .5 - .4265 = .0735$$

  b. $P(8 < x < 10) = P\left(\frac{8 - 9.06}{2.11} < z < \frac{10 - 9.06}{2.11} \right) = P(-.50 < z < .45)$

$$= P(-.50 < z < 0) + P(0 < z < .45) = .1915 + .1736 = .3651$$

  c. $P(x < x_0) = .2000$.

   If $P(x < x_0) = .2000$, then $P\left(z < \frac{x_0 - 9.06}{2.11} \right) = P(z < z_0) = .2000$

   If $P(z < z_0)$, then $P(z_0 < z < 0) = .3000$. Looking up .3000 in Table IV, the z-score is .84. Since $z_0 < 0$, $z_0 = -.84$. Now, we must convert z_0 back to an x score.

$$z_0 = \frac{x_0 - 9.06}{2.11} \Rightarrow -.84 = \frac{z_0 - 9.06}{2.11} \Rightarrow x_0 = 2.11(-.84) + 9.06 = 7.29$$

5.47   Using Table IV, Appendix A, with $\mu = 50$ and $\sigma = 12$,

$$P(x > d) = .30 \Rightarrow P\left(z > \frac{d - 50}{12} \right) = P(z > z_0) = .30$$

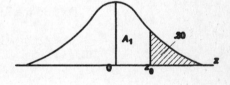

$A_1 = .5 - .30 = .2000$
Looking up the area .2000 in Table IV gives $z_0 = .52$.

$$z_0 = .52 = \frac{d - 50}{12} \Rightarrow d = 12(.52) + 50 = 56.24 \text{ cm}$$

5.49   a.   If z is a standard normal random variable,

   $Q_L = z_L$ is the value of the standard normal distribution which has 25% of the data to the left and 75% to the right.

   Find z_L such that $P(z < z_L) = .25$

   $A_1 = .50 - .25 = .25$.

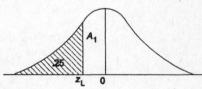

   Look up the area $A_1 = .25$ in the body of Table IV of Appendix A; $z_L = -.67$ (taking the closest value). If interpolation is used, $-.675$ would be obtained.

$Q_U = z_U$ is the value of the standard normal distribution which has 75% of the data to the left and 25% to the right.

Find z_U such that $P(z < z_U) = .75$

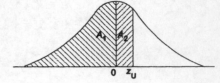

$$A_1 + A_2 = P(z \le 0) + P(0 \le z \le z_U)$$
$$= .5 + P(0 \le z \le z_U)$$
$$= .75$$

Therefore, $P(0 \le z \le z_U) = .25$.

Look up the area .25 in the body of Table IV of Appendix A; $z_U = .67$ (taking the closest value).

b.   Recall that the inner fences of a box plot are located $1.5(Q_U - Q_L)$ outside the hinges (Q_L and Q_U).

To find the lower inner fence,

$$Q_L - 1.5(Q_U - Q_L) = -.67 - 1.5(.67 - (-.67))$$
$$= -.67 - 1.5(1.34)$$
$$= -2.68 \ (-2.70 \text{ if } z_L = -.675 \text{ and } z_U = +.675)$$

The upper inner fence is:

$$Q_U + 1.5(Q_U - Q_L) = .67 + 1.5(.67 - (-.67))$$
$$= .67 + 1.5(1.34)$$
$$= 2.68 \ (+2.70 \text{ if } z_L = -.675 \text{ and } z_U = +.675)$$

c.   Recall that the outer fences of a box plot are located $3(Q_U - Q_L)$ outside the hinges (Q_L and Q_U).

To find the lower outer fence,

$$Q_L - 3(Q_U - Q_L) = -.67 - 3(.67 - (.67))$$
$$= -.67 - 3(1.34)$$
$$= -4.69 \ (-4.725 \text{ if } z_L = -.675 \text{ and } z_U = +.675)$$

The upper outer fence is:

$$Q_U + 3(Q_U - Q_L) = .67 + 3(.67 - (-.67))$$
$$= .67 + 3(1.34)$$
$$= 4.69 \ (4.725 \text{ if } z_L = -,675 \text{ and } z_U = +.675)$$

d. $P(z < -2.68) + P(z > 2.68)$
$\quad = 2P(z > 2.68)$
$\quad = 2(.5000 - .4963)$
$\qquad$ (Table IV, Appendix A)
$\quad = 2(.0037) = .0074$

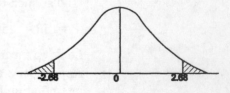

(or $2(.5000 - .4965) = .0070$ if -2.70 and 2.70 are used)

$P(z < -4.69) + P(z > 4.69)$
$\quad = 2P(z > 4.69)$
$\quad \approx 2(.5000 - .5000) \approx 0$

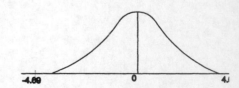

e.   In a normal probability distribution, the probability of an observation being beyond the inner fences is only .0074 and the probability of an observation being beyond the outer fences is approximately zero. Since the probability is so small, there should not be any observations beyond the inner and outer fences. Therefore, they are probably outliers.

5.51   Several statistical techniques are based on the assumption that the population is approximately normally distributed. Thus, it is important to determine whether the sample data come from a normal population so that the techniques can be properly applied.

5.53   a.   The proportion of measurements that one would expect to fall in the interval $\mu \pm \sigma$ is about .68.

        b.   The proportion of measurements that one would expect to fall in the interval $\mu \pm 2\sigma$ is about .95.

        c.   The proportion of measurements that one would expect to fall in the interval $\mu \pm 3\sigma$ is about 1.00.

5.55   If the data are normally distributed, then the normal probability plot should be an approximate straight line. Of the three plots, only plot c implies that the data are normally distributed. The data points in plot c form an approximately straight line. In both plots a and b, the plots of the data points do not form a straight line.

5.57   a.   Using MINITAB, the stem-and-leaf display is:

Stem-and-Leaf Display: Data

```
Stem-and-leaf of Data   N  = 40
Leaf Unit = 10

   14    0   00012223333344
  (10)   0   5556666889
   16    1   001334
   10    1   5689
    6    2   01
    4    2
    4    3   000
    1    3
    1    4   0
```

The distribution from the stem-and-leaf display looks like it is skewed to the right. Thus, it does not appear that the data are normally distributed.

b. Using MINITAB, the descriptive statistics are:

Descriptive Statistics: Data

```
Variable   N  N*   Mean  SE Mean  StDev  Minimum    Q1  Median     Q3  Maximum
Data      40   0  105.1     15.1   95.8     1.00  37.5    65.0  152.8    401.0
```

The lower quartile is $Q_L = 37.5$ and the upper quartile is $Q_U = 152.8$.

c. The interquartile range is IQR $= Q_U - Q_L = 152.8 - 37.5 = 115.3$. From the printout, $s = 95.8$. IQR $/ s = 115.3 / 95.8 = 1.204$. If the data are approximately normal, then this ratio should be 1.3. Since the actual ratio is 1.204 which is close to 1.3, it appears that the data could be approximately normal.

d. Using MINITAB, the normal probability plot is:

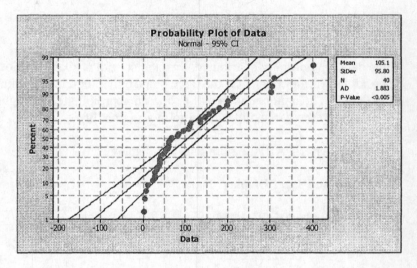

Since the data points do not fall along the straight line, the data do not appear to be normally distributed.

5.59 a. IQR $= Q_U - Q_L = 53 - 47 = 6$.

b. From the printout, $s = 5.275$.

c. If the data are approximately normal, then IQR $/ s \approx 1.3$. For this data, IQR $/ s = 6 / 5.275 = 1.14$. This is somewhat close to 1.3, so the data may be normal.

d.   Using MINITAB, a relative frequency histogram of the data is:

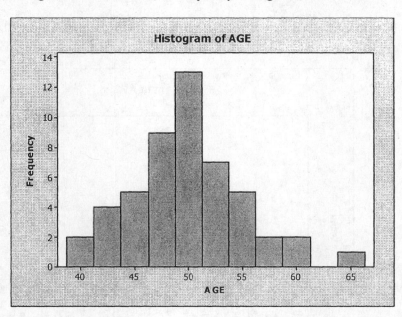

From the histogram, the data appear to be close to mound-shaped, so the data may be normally distributed.

5.61   The interquartile range is IQR = $Q_U - Q_L$ = 6 − 1 = 5. IQR / s = 5 / 6.09 = .82. This is much less than the 1.3 that we would expect if the data were normal. It appears that these data are not approximately normal.   In addition, $\bar{x}$ = 4.71 and s = 6.09. Since the number of updates cannot be normal, the data must be skewed to the right since one standard deviation below the mean would be a negative number which is impossible.

5.63   a.   From the graph, the histogram of the difference between the elevation estimates is too peaked. This means that there are too many observations in the very center of the distribution and not enough observations in the tails for the distribution to be normal.

b.   The interval μ ± 2σ would be .28 ± 2(1.6) => .28 ± 3.2 => (-2.92, 3.48). From the graph, almost all the observations are between are between -2 and 2. Thus, the interval (-2.92, 3.48) will contain more that 95% of the 400 elevation differences.

5.65   American League

To determine if the distribution of batting averages is approximately normal, we will run through the tests. Using MINITAB, a histogram of the data with a normal curve included is:

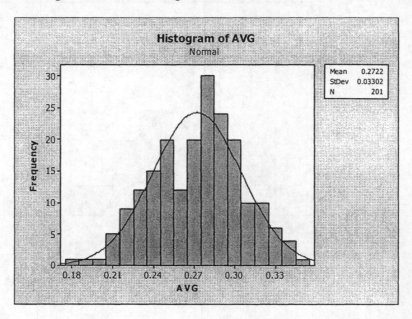

From the graph, the data appear to be approximately mound-shaped. The data may be normal.

Using MINITAB, the descriptive statistics are:

Descriptive Statistics: AVG

Variable	N	Mean	StDev	Minimum	Q1	Median	Q3	Maximum
AVG	201	0.27223	0.03302	0.18400	0.24650	0.27600	0.29500	0.34700

The interval $\bar{x} \pm s \Rightarrow .27223 \pm .03302 \Rightarrow (.23921, .30525)$ contains 133 of the 201 observations.   The proportion is 133 / 201 = .662. This is very close to the .68 from the Empirical Rule.

The interval $\bar{x} \pm 2s \Rightarrow .27223 \pm 2(.03302) \Rightarrow .27223 \pm .06604 \Rightarrow (.20619, .33827)$ contains 193 of the 201 observations. The proportion is 193 / 201 = .960. This is very close to the .95 from the   Empirical Rule.

The interval $\bar{x} \pm 3s \Rightarrow .27223 \pm 3(.03302) \Rightarrow .27223 \pm .09906 \Rightarrow (.17317, .37129)$ contains 201 of the 201 observations. The proportion is 201 / 201 = 1.00. This is very close to the .997 from the   Empirical Rule. Thus, it appears that the data may be normal.

The lower quartile is $Q_L = .2465$ and the upper quartile is $Q_U = .295$. The interquartile range is $IQR = Q_U - Q_L = .295 - .2456 = .0494$. From the printout, $s = .03302$. $IQR / s = .0494 / .03302   = 1.496$. This is a fairly close to the 1.3 that we would expect if the data were normal. Thus,   there is evidence that the data may be normal.

Using MINITAB, the normal probability plot is:

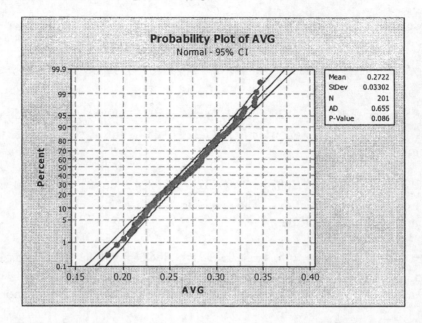

The data are very close to a straight line. Thus, it appears that the data may be normal.

From all of the indicators, it appears that the batting averages from the American League come from an approximate normal distribution.

National League

To determine if the distribution of batting averages is approximately normal, we will run through the tests. Using MINITAB, a histogram of the data with a normal curve included is:

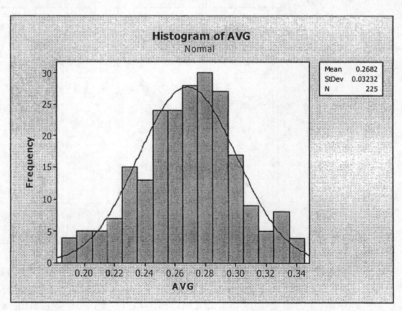

From the graph, the data appear to be approximately mound-shaped. The data may be normal.

Using MINITAB, the descriptive statistics are:

Descriptive Statistics: AVG

Variable	N	Mean	StDev	Minimum	Q1	Median	Q3	Maximum
AVG	225	0.26824	0.03232	0.19000	0.24700	0.27000	0.29050	0.34400

The interval $\bar{x} \pm s \Rightarrow .26824 \pm (.03232) \Rightarrow (.23592, .30056)$ contains 154 of the 225 observations. The proportion is $154 / 225 = .684$. This is very close to the .68 from the Empirical Rule.

The interval $\bar{x} \pm 2s \Rightarrow .26824 \pm 2(.03232) \Rightarrow .26824 \pm .06464 \Rightarrow (.20360, .33288)$ contains 211 of the 225 observations. The proportion is $211 / 225 = .938$. This is very close to the .95 from the   Empirical Rule.

The interval $\bar{x} \pm 3s \Rightarrow .26824 \pm 3(.03232) \Rightarrow .26824 \pm .09696 \Rightarrow (.17128, .36520)$ contains 225 of the 225 observations. The proportion is $225 / 225 = 1.00$. This is very close to the .997 from the   Empirical Rule. Thus, it appears that the data may be normal.

The lower quartile is $Q_L = .247$ and the upper quartile is $Q_U = .2905$. The interquartile range is $IQR = Q_U - Q_L = .2905 - .247 = .0435$. From the printout, s = .03232. IQR / s = .0435 / .03232   = 1.346. This is a very close to the 1.3 that we would expect if the data were normal. Thus,   there is evidence that the data may be normal.

Using MINITAB, the normal probability plot is:

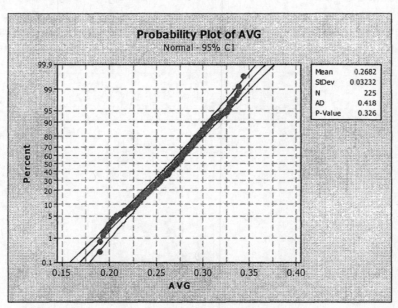

The data are very close to a straight line. Thus, it appears that the data may be normal.

From all of the indicators, it appears that the batting averages from the National League come from an approximate normal distribution.

5.67   We will look at the 4 methods for determining if the data are normal. First, we will look at a histogram of the data. Using MINITAB, the histogram of the driver's head injury ratings is:

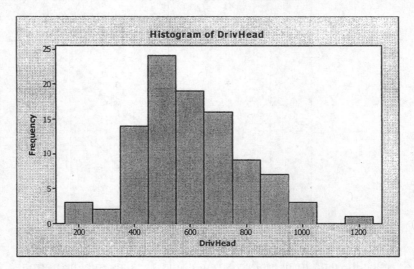

From the histogram, the data appear to be somewhat mound-shaped, but also somewhat skewed to the right. This indicates that the data may be normal.

Next, we look at the intervals $\bar{x} \pm s$, $\bar{x} \pm 2s$, $\bar{x} \pm 3s$. If the proportions of observations falling in each interval are approximately .68, .95, and 1.00, then the data are approximately normal.

Using MINITAB, the summary statistics are:

Descriptive Statistics: DrivHead

```
Variable    N   N*   Mean   SE Mean   StDev   Minimum     Q1   Median      Q3   Maximum
DrivHead   98    0  603.7      18.7   185.4     216.0  475.0    605.0   724.3    1240.0
```

$\bar{x} \pm s \Rightarrow 603.7 \pm 185.4 \Rightarrow (418.3,\ 789.1)$ 68 of the 98 values fall in this interval. The proportion is .69. This is very close to the .68 we would expect if the data were normal.

$\bar{x} \pm 2s \Rightarrow 603.7 \pm 2(185.4) \Rightarrow 603.7 \pm 370.8 \Rightarrow (232.9,\ 974.5)$ 96 of the 98 values fall in this interval. The proportion is .98. This is a fair amount larger than the .95 we would expect if the data were normal.

$\bar{x} \pm 3s \Rightarrow 603.7 \pm 3(185.4) \Rightarrow 603.7 \pm 556.2 \Rightarrow (47.5,\ 1,159.9)$ 97 of the 98 values fall in this interval. The proportion is .99. This is fairly close to the 1.00 we would expect if the data were normal.

From this method, it appears that the data may be normal.

Next, we look at the ratio of the IQR to s. IQR $= Q_U - Q_L = 724.3 - 475 = 249.3$.

$\dfrac{\text{IQR}}{s} = \dfrac{249.3}{185.4} = 1.3$ This is equal to the 1.3 we would expect if the data were normal. This method indicates the data may be normal.

Finally, using MINITAB, the normal probability plot is:

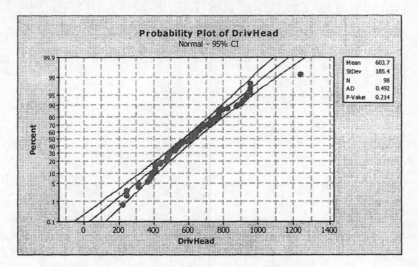

Since the data form a fairly straight line, the data may be normal.

From the 4 different methods, all indications are that the driver's head injury rating data are normal.

5.69   The lowest score possible is 0. The sample mean is 7.62 and the standard deviation is 8.91. Since the lower limit of 0 is less than one standard deviation below the mean, this implies that the data are not normal, but skewed to the right.

5.71   The binomial probability distribution is a discrete distribution. The random variable can take on only a limited number of values. The normal distribution is a continuous distribution. The random variable can take on an infinite number of values. To get a better estimate of probabilities for the binomial probability distribution using the normal distribution, we use the continuity correction factor.

5.73   a.   In order to approximate the binomial distribution with the normal distribution, the interval $\mu \pm 3\sigma \Rightarrow np \pm 3\sqrt{npq}$ should lie in the range 0 to n.

When $n = 25$ and $p = .4$,
$$np \pm 3\sqrt{npq} \Rightarrow 25(.4) \pm 3\sqrt{25(.4)(1-.4)}$$
$$\Rightarrow 10 \pm 3\sqrt{6} \Rightarrow 10 \pm 7.3485 \Rightarrow (2.6515, 17.3485)$$

Since the interval calculated does lie in the range 0 to 25, we can use the normal approximation.

b. $\mu = np = 25(.4) = 10$
$\sigma^2 = npq = 25(.4)(.6) = 6$

c. $P(x \geq 9) = 1 - P(x \leq 8) = 1 - .274 = .726$ (Table II, Appendix A)

d. $P(x \geq 9) \approx P\left(z \geq \dfrac{(9-.5)-10}{\sqrt{6}}\right)$

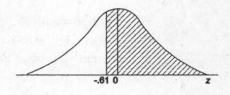

$= P(z \geq -.61)$

$= .5000 + .2291 = .7291$

(Using Table IV in Appendix A.)

5.75 x is a binomial random variable with $n = 100$ and $p = .4$.

$$\mu \pm 3\sigma \Rightarrow np \pm 3\sqrt{npq} \Rightarrow 100(.4) \pm 3\sqrt{100(.4)(1-.4)}$$

$$\Rightarrow 40 \pm 3(4.8990) \Rightarrow (25.303, 54.697)$$

Since the interval lies in the range 0 to 100, we can use the normal approximation to approximate the probabilities.

a. $P(x \leq 35) \approx P\left(z \leq \dfrac{(35+.5)-40}{4.899}\right)$

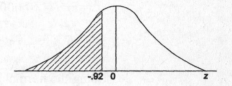

$= P(z \leq -.92)$

$= .5000 - .3212 = .1788$

(Using Table IV in Appendix A.)

b. $P(40 \leq x \leq 50)$

$\approx P\left(\dfrac{(40-.5)-40}{4.899} \leq z \leq \dfrac{(50+.5)-40}{4.899}\right)$

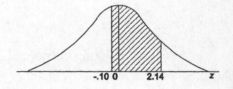

$= P(-.10 \leq z \leq 2.14)$

$= P(-.10 \leq z \leq 0) + P(0 \leq z \leq 2.14)$

$= .0398 + .4838 = .5236$

(Using Table IV in Appendix A.)

c. $P(x \geq 38) \approx P\left(z \geq \dfrac{(38-.5)-40}{4.899}\right)$

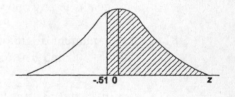

$= P(z \geq -.51)$

$= .5000 + .1950 = .6950$

(Using Table IV in Appendix A.)

5.77   a.   Let x = number of bottles that contain tap water in 65 trials. Then x is a binomial random variable with $n = 65$ and $p = .25$.

$E(x) = \mu = np = 65(.25) = 16.25$

b. $\sigma = \sqrt{npq} = \sqrt{65(.25)(.75)} = \sqrt{12.1875} = 3.4911$

c. $z = \dfrac{x - \mu}{\sigma} = \dfrac{20 - 16.25}{3.4911} = 1.07$

d. $\mu \pm 3\sigma \Rightarrow 16.25 \pm 3(3.4911) \Rightarrow 16.25 \pm 10.4733 \Rightarrow (5.7767, \ 26.7233)$ Since the interval lies in the range 0 to 65, we can use the normal approximation to approximate the binomial probability.

$$P(x \geq 20) = P\left(z \geq \frac{(20-.5)-16.25}{3.4911} \right) = P(z \geq .93) = .5 - .3238 = .1762$$

(Using Table IV, Appendix A)

5.79 a. $n = 10,000$ and $p = .15$

$E(x) = \mu = np = 10,000(.15) = 1,500$

$\sigma^2 = npq = 10,000(.15)(.85) = 1,275$

b. $\sigma = \sqrt{1,275} = 35.707$

Using the normal approximation to the binomial,

$$P(x > 1,600) \approx P\left(z > \frac{(1,600+.5)-1,500}{35.707} \right) = P(z > 2.81) = .5 - .4975 = .0025$$
(from Table IV, Appendix A)

c. Using the normal approximation to the binomial,

$$P(x > 6,500) \approx P\left(z > \frac{(6,500+.5)-1,500}{35.707} \right) = P(z > 140.04) \approx .5 - .5 = 0$$
(from Table IV, Appendix A)

Since this probability is essentially 0, we would not expect to see more than 6,500 deaths in any one year.

5.81 Let x = number of guppies that survive in 300 trials. Then x is a binomial random variable with $n = 300$ and $p = .6$.

$\mu \pm 3\sigma \Rightarrow np \pm 3\sqrt{npq} \Rightarrow 300(.6) \pm 3\sqrt{300(.6)(1-.6)} \Rightarrow 180 \pm 3\sqrt{72} \Rightarrow 180 \pm 25.456$
$\Rightarrow (154.544, \ 205.456)$

Since the interval lies in the range 0 to 300, we can use the normal approximation to approximate the binomial probability.

$$P(x < 100) = P\left(z < \frac{(100-.5)-180}{\sqrt{72}} \right) = P(z < -9.49) \approx .5 - .5 = 0$$
(Using Table IV, Appendix A)

5.83   a.   For $n = 100$ and $p = .01$:

$$\mu \pm 3\sigma \Rightarrow np \pm 3\sqrt{npq} \Rightarrow 100(.01) \pm 3\sqrt{100(.01)(.99)}$$

$$\Rightarrow 1 \pm 3(.995) \Rightarrow 1 \pm 2.985 \Rightarrow (-1.985, 3.985)$$

Since the interval does not lie in the range 0 to 100, we cannot use the normal approximation to approximate the probabilities.

   b.   For $n = 100$ and $p = .5$:

$$\mu \pm 3\sigma \Rightarrow np \pm 3\sqrt{npq} \Rightarrow 100(.5) \pm 3\sqrt{100(.5)(.5)}$$

$$\Rightarrow 50 \pm 3(5) \Rightarrow 50 \pm 15 \Rightarrow (35, 65)$$

Since the interval lies in the range 0 to 100, we can use the normal approximation to approximate the probabilities.

   c.   For $n = 100$ and $p = .9$:

$$\mu \pm 3\sigma \Rightarrow np \pm 3\sqrt{npq} \Rightarrow 100(.9) \pm 3\sqrt{100(.9)(.1)}$$

$$\Rightarrow 90 \pm 3(3) \Rightarrow 90 \pm 9 \Rightarrow (81, 99)$$

Since the interval lies in the range 0 to 100, we can use the normal approximation to approximate the probabilities.

5.85   Let x = number of beech trees damaged by fungi in 200 trees. Then x is a binomial random variable with $n = 200$ and $p = .25$.

$$\mu = np = 200(.25) = 50$$

$$\sigma = \sqrt{npq} = \sqrt{200(.25)(.75)} = \sqrt{37.5} = 6.124$$

$$P(x > 100) \approx P\left(z > \frac{(100 + .5) - 50}{6.124}\right) = P(z > 8.25) \approx 0$$

Since this probability is so small, it is almost impossible for more than half of the trees to have fungi damage.

5.87   a.   If 80% of the passengers pass through without their luggage being inspected, then 20% will be detained for luggage inspection. The expected number of passengers detained will be:

$$E(x) = np = 1,500(.2) = 300$$

   b.   For $n = 4,000$, $E(x) = np = 4,000(.2) = 800$

   c. $P(x > 600) \approx P\left(z > \dfrac{(600 + .5) - 800}{\sqrt{4000(.2)(.8)}}\right) = P(z > -7.89) = .5 + .5 = 1.0$

5.89   The amount of time or distance between occurrences of random events can be described by the exponential probability distribution. This distribution has the property that the mean is equal to the standard deviation.

5.91   For $\theta = 3, f(x) = \dfrac{1}{3} e^{-x/3}$. Using Table V, Appendix A:

x	$f(x)$
0	.333
1	.239
2	.171
3	.123
4	.088
5	.063
6	.045

For $\theta = 1, f(y) = e^{-y}$. Using Table V, Appendix A:

x	$f(x)$
0	1.000
1	.368
2	.135
3	.050
4	.018
5	.007
6	.002

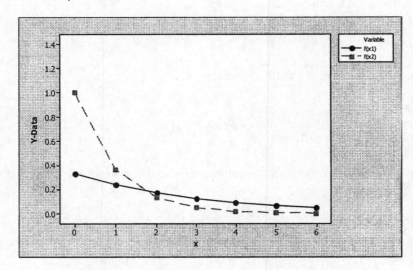

5.93 $P(x \geq a) = e^{-a/\theta} = e^{-a/1}$. Using Table V, Appendix A:

a. $P(x > 1) = e^{-1/1} = e^{-1} = .367879$

b. $P(x \leq 3) = 1 - P(x > 3) = 1 - e^{-3/1} = 1 - e^{-3} = 1 - .049787 = .950213$

c. $P(x > 1.5) = e^{-1.5/1} = e^{-1.5} = .223130$

d. $P(x \leq 5) = 1 - P(x > 5) = 1 - e^{-5/1} = 1 - e^{-5} = 1 - .006738 = .993262$

5.95 $\mu = \theta = 1$, $\sigma = \theta = 1$

$\mu \pm 2\sigma \Rightarrow 1 \pm 2(1) \Rightarrow 1 \pm 2 \Rightarrow (-1, 3)$

$P(-1 < x < 3) = P(0 < x < 3) = 1 - P(x \geq 3) = 1 - e^{-3/1}$
$= 1 - e^{-3} = 1 - .049787 = .950213$
(from Table V, Appendix A)

5.97   a. $P(x > 2) = e^{-2/2.5} = e^{-.8} = .449329$ (using Table V, Appendix A)

b. $P(x < 5) = 1 - P(x \geq 5) = 1 - e^{-5/2.5} = 1 - e^{-2} = 1 - .135335 = .864665$
(Using Table V, Appendix A)

5.99   Let x = breast height diameter of the western hemlock. Then x has an exponential distribution with a mean of 30 cm.

$P(x > 25) = e^{-25/30} = e^{-.833333} = .434598$

5.101   a.   Let x_1 = repair time for machine 1. Then x_1 has an exponential distribution with $\mu_1 = 1$ hour.

$P(x_1 > 1) = e^{-1/1} = e^{-1} = .367879$ (using Table V, Appendix A)

b.   Let x_2 = repair time for machine 2. Then x_2 has an exponential distribution with $\mu_2 = 2$ hours.

$P(x_2 > 1) = e^{-1/2} = e^{-.5} = .606531$ (using Table V, Appendix A)

c.   Let x_3 = repair time for machine 3. Then x_3 has an exponential distribution with $\mu_3 = .5$ hours.

$P(x_3 > 1) = e^{-1/.5} = e^{-2} = .135335$ (using Table V, Appendix A)

Since the mean repair time for machine 4 is the same as for machine 3, $P(x_4 > 1)$ $= P(x_3 > 1) = .135335$.

d.   The only way that the repair time for the entire system will not exceed 1 hour is if all four machines are repaired in less than 1 hour. Thus, the probability that the repair time for the entire system exceeds 1 hour is:

P(Repair time entire system exceeds 1 hour)
$= 1 - P((x_1 \leq 1) \cap (x_2 \leq 1) \cap (x_3 \leq 1) \cap (x_4 \leq 1))$
$= 1 - P(x_1 \leq 1)P(x_2 \leq 1)P(x_3 \leq 1)P(x_4 \leq 1)$
$= 1 - (1 - .367879)(1 - .606531)(1 - .135335)(1 - .135335)$
$= 1 - (.632121)(.393469)(.864665)(.864665) = 1 - .185954 = .814046$

5.103 a. For $\mu = 17 = \theta$. To graph the distribution, we will pick several values of x and find the value of $f(x)$, where x = time between arrivals of the smaller craft at the pier.

$$f(x) = \frac{1}{\theta}e^{-x/\theta} = \frac{1}{17}e^{-x/17}$$

$$f(1) = \frac{1}{17}e^{-1/17} = .0555$$

$$f(3) = \frac{1}{17}e^{-3/17} = .0493$$

$$f(5) = \frac{1}{17}e^{-5/17} = .0438$$

$$f(7) = \frac{1}{17}e^{-7/17} = .0390$$

$$f(10) = \frac{1}{17}e^{-10/17} = .0327$$

$$f(15) = \frac{1}{17}e^{-15/17} = .0243$$

$$f(20) = \frac{1}{17}e^{-20/17} = .0181$$

$$f(25) = \frac{1}{17}e^{-25/17} = .0135$$

The graph is:

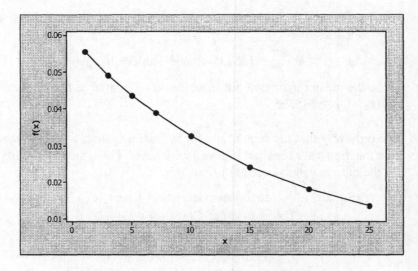

b. We want to find the probability that the time between arrivals is less than 15 minutes.

$$P(x < 15) = 1 - P(x \geq 15) = 1 - e^{-15/17} = 1 - .4138 = .5862$$

5.105 a. For $\theta = 250$, $P(x > a) = e^{-a/250}$

For $a = 300$ and $b = 200$, show $P(x > a + b) \geq P(x > a)P(x > b)$

$P(x > 300 + 200) = P(x > 500) = e^{-500/250} = e^{-2} = .1353$

$P(x > 300)\, P(x > 200) = e^{-300/250}\, e^{-200/250} = e^{-1.2}\, e^{-.8} = .3012(.4493) = .1353$

Since $P(x > 300 + 200) = P(x > 300)\, P(x > 200)$, then
$P(x > 300 + 200) \geq P(x > 300)\, P(x > 200)$

Also, show $P(x > 300 + 200) \leq P(x > 300)\, P(x > 200)$. Since we already showed that
$P(x > 300 + 200) = P(x > 300)\, P(x > 200)$,
then $P(x > 300 + 200) \leq P(x > 300)\, P(x > 200)$.

b. Let $a = 50$ and $b = 100$. Show $P(x > a + b) \geq P(x > a)\, P(x > b)$

$P(x > 50 + 100) = P(x > 150) = e^{-150/250} = e^{-.6} = .5488$

$P(x > 50)\, P(x > 100) = e^{-50/250}\, e^{-100/250} = e^{-.2}e^{-.4} = .8187(.6703) = .5488$

Since $P(x > 50 + 100) = P(x > 50)\, P(x > 100)$, then
$P(x > 50 + 100) \geq P(x > 50)\, P(x > 100)$

Also, show $P(x > 50 + 100) \leq P(x > 50)\, P(x > 100)$. Since we already showed that
$P(x > 50 + 100) = P(x > 50)\, P(x > 100)$,
then $P(x > 50 + 100) \leq P(x > 50)\, P(x > 100)$.

c. Show $P(x > a + b) \geq P(x > a)\, P(x > b)$
$P(x > a + b) = e^{-(a+b)/250} = e^{-a/250}\, e^{-b/250} = P(x > a)\, P(x > b)$

5.107 a. For the probability density function, $f(x) = \dfrac{e^{-x/7}}{7}$, $x > 0$, x is an exponential random variable.

b. For the probability density function, $f(x) = \dfrac{1}{20}$, $5 < x < 25$, x is a uniform random variable.

c. For the probability function, $f(x) = \dfrac{e^{-.5[(x-10)/5]^2}}{5\sqrt{2\pi}}$, x is a normal random variable.

5.109 a. $P(z \leq -2.1) = A_1 + A_2$
$= .5 + .4821$
$= .9821$

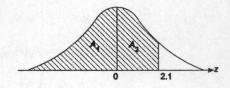

b. $P(z \geq 2.1) = A_2 = .5 - A_1$
$= .5 - .4821$
$= .0179$

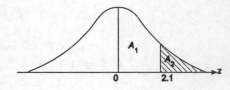

c. $P(z \geq -1.65) = A_1 + A_2$
$= .4505 + .5000$
$= .9505$

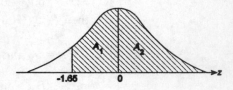

d. $P(-2.13 \leq z \leq -.41)$
$= P(-2.13 \leq z \leq 0) - P(-.41 \leq z \leq 0)$
$= .4834 - .1591$
$= .3243$

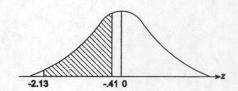

e. $P(-1.45 \leq z \leq 2.15) = A_1 + A_2$
$= .4265 + .4842$
$= .9107$

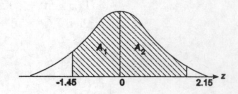

f. $P(z \leq -1.43) = A_1 = .5 - A_2$
$= .5000 - .4236$
$= .0764$

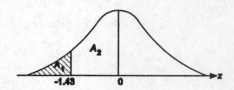

5.111 a. $P(x \leq 75) = P\left(z \leq \dfrac{75-70}{10}\right)$

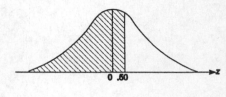

$= P(z \leq .50)$
$= .5 + .1915 = .6915$

b. $P(x \geq 90) = P\left(z \geq \dfrac{90-70}{10}\right)$

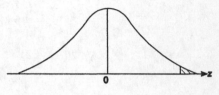

$= P(z \geq 2.00)$
$= .5000 - .4772 = .0228$

c. $P(60 \leq x \leq 75) = P\left(\dfrac{60-70}{10} \leq z \leq \dfrac{75-70}{10}\right)$

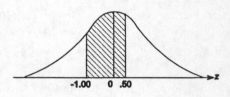

$= P(-1.00 \leq z \leq .50)$
$= .3413 + .1915 = .5328$

d. This is the probability of the complement of the event in part **a**. Therefore:

$P(x > 75) = 1 - P(x \leq 75) = 1 - .6915 = .3085$

e. $P(x = 75) = 0$ (True for any continuous random variable.)

f. $P(x \leq 95) = P\left(z \leq \dfrac{95-70}{10}\right)$

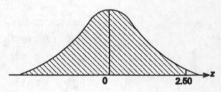

$= P(z \leq 2.50)$
$= .5000 + .4938 = .9938$

5.113 $\mu = np = 100(.5) = 50, \ \sigma = \sqrt{npq} = \sqrt{100(.5)(.5)} = 5$

a. $P(x \leq 48) = P\left(z \leq \dfrac{(48+.5)-50}{5}\right)$

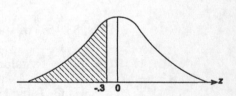

$= P(z \leq -.30)$
$= .5 - .1179 = .3821$

b. $P(50 \leq x \leq 65)$

$= P\left(\dfrac{(50-.5)-50}{5} \leq z \leq \dfrac{(65+.5)-50}{5}\right)$

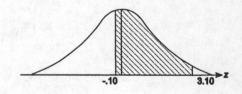

$= P(-.10 \leq z \leq 3.10)$
$= .0398 + .5000 = .5398$

c. $P(x \geq 70) = P\left(z \geq \dfrac{(70-.5)-50}{5}\right)$

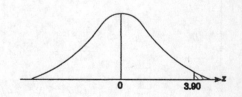

$= (z \geq 3.90)$
$= .5 - .5 = 0$

d. $P(55 \le x \le 58)$

$$= P\left(\frac{(55-.5)-50}{5} \le z \le \frac{(58+.5)-50}{5}\right)$$

$= P(.90 \le z \le 1.70)$
$= P(0 \le z \le 1.70) - P(0 \le z \le .90)$
$= .4554 - .3159 = .1395$

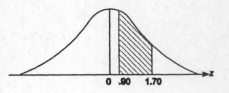

e. $P(x = 62)$

$$= P\left(\frac{(62-.5)-50}{5} \le z \le \frac{(62+.5)-50}{5}\right)$$

$= P(2.30 \le z \le 2.50)$
$= P(0 \le z \le 2.50) - (0 \le z \le 2.30)$
$= .4938 - .4893 = .0045$

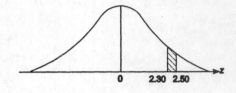

f. $P(x \le 49 \text{ or } x \ge 72)$

$$= P\left(z \le \frac{(49+.5)-50}{5}\right) + P\left(z \ge \frac{(72-.5)-50}{5}\right)$$

$= P(z \le -.10) + P(z \ge 4.30)$
$= (.5 - .0398) + (.5 - .5) = .4602$

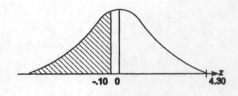

5.115 Let x = score for first time students on mathematics achievement test. Then x is a normal random variable with $\mu = 77$ and $\sigma = 7.3$.

$$P(x \ge 70) = P\left(z \ge \frac{70-77}{7.3}\right)$$

$= P(z \ge -.96)$
$= .3315 + .5000 = .8315$

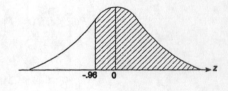

Thus, 83.15% of students will pass the test the first time.

5.117 a. Let x = alkalinity level of water specimens collected from the Han River.

Using Table IV, Appendix A,

$$P(x > 45) = P\left(z > \frac{45-50}{3.2}\right) = P(z > -1.56) = .5 + .4406 = .9406.$$

b. Using Table IV, Appendix A,

$$P(x < 55) = P\left(z < \frac{55-50}{3.2}\right) = P(z < 1.56) = .5 + .4406 = .9406.$$

c. Using Table IV, Appendix A,

$$P(51 < x < 52) = P\left(\frac{51-50}{3.2} < z < \frac{52-50}{3.2}\right) = P(.31 < z < .63) = .2357 - .1217 = .1140.$$

5.119   Let x = number of African-Americans with sickle-cell anemia. Then x is a binomial random variable with $n = 1,000$ and $p = .08$.

$$\mu = np = 1,000(.08) = 80 \quad \text{and} \quad \sigma = \sqrt{npq} = \sqrt{1,000(.08)(.92)} = 8.579$$

a. $P(x > 175) = P\left(z > \dfrac{(175 + .5) - 80}{8.579}\right) = P(z > 11.13) \approx .5 - .5 = 0$

b. $P(x < 140) = P\left(z < \dfrac{(140 - .5) - 80}{8.579}\right) = P(z < 6.94) = .5 + .5 = 1.0$

5.121   For the uniform distribution,

$$f(x) = \begin{cases} \dfrac{1}{d-c} = \dfrac{1}{90-75} = \dfrac{1}{15} & 75 \le x \le 90 \\ 0 & \text{otherwise} \end{cases}$$

a. $P(x > 80) = (90 - 80)\left(\dfrac{1}{15}\right) = .667$

b. $P(80 < x < 85) = (85 - 80)\left(\dfrac{1}{15}\right) = .333$

c. $E(x) = \mu = \dfrac{c+d}{2} = \dfrac{75+90}{2} = 82.5°F$

5.123   a.   Using Table IV, Appendix A, with $\mu = 208$ and $\sigma = 25$,

$$P(x < 200) = P\left(z < \dfrac{200 - 208}{25}\right) = P(z < -.32) = .5 - .1255 = .3745$$

b.   Let y = number of patients out of three who have cholesterol levels below 200. Assuming that the patients are independent, then y is a binomial random variable with $n = 3$ and $p = .3745$.

$$P(y \ge 1) = 1 - P(y = 0) = 1 - \binom{3}{0}.3745^0 .6255^3 = 1 - .2447 = .7553$$

5.125 $c = 0, d = 4$

$$f(x) = \begin{cases} \dfrac{1}{d-c} = \dfrac{1}{4-0} = \dfrac{1}{4} & 0 \leq x \leq 4 \\ 0 & \text{otherwise} \end{cases}$$

$$\mu = \frac{c+d}{2} = \frac{0+4}{2} = 2, \ \sigma = \frac{d-c}{\sqrt{12}} = \frac{4-0}{\sqrt{12}} = 1.155$$

$$15 \text{ seconds} = \frac{15}{60} = .25 \text{ minutes}$$

It takes .25 minutes to go from floor 2 to floor 1. Thus, we must find the probability that the waiting time is less than $1.5 - .25 = 1.25$.

$$P(x < 1.25) = (1.25 - 0)\frac{1}{4} = 1.25\left(\frac{1}{4}\right) = .3125$$

5.127   Let x = velocity of a galaxy located within the galaxy cluster A2142. Then x is a normal random variable with mean $\mu = 27{,}117$ km/s and standard deviation $\sigma = 1{,}280$ km/s.

$$P(x \leq 24{,}350) = P\left(z \leq \frac{24{,}350 - 27{,}117}{1{,}280}\right) = P(z \leq -2.16) = .5 - .4846 = .0154$$

Since the probability of observing a galaxy from the galaxy cluster A2142 with a velocity of 24,350 km/s or slower is so small ($p = .0154$), it would be very unlikely that the galaxy observed came from the galaxy cluster A2142.

5.129   Let x = number of smokers who will enter into a treatment program in 200 trials. Then x is a binomial random variable with $n = 200$ and $p = .05$.

$$\mu \pm 3\sigma \Rightarrow np \pm 3\sqrt{npq} \Rightarrow 200(.05) \pm 3\sqrt{200(.05)(1-.05)} \Rightarrow 10 \pm 3\sqrt{9.5} \Rightarrow 10 \pm 9.247$$
$$\Rightarrow (.753, \ 19.247)$$

Since the interval lies in the range 0 to 200, we can use the normal approximation to approximate the binomial probability.

$$P(x \geq 11) = P\left(z \geq \frac{(11-.5)-10}{\sqrt{9.5}}\right) = P(z \geq .16) = .5 - .0636 = .4364$$

(Using Table IV, Appendix A)

5.131   From Table IV, Appendix A, with $\mu = 6.3$ and $\sigma = .6$

a. $P(x < 5) = P\left(z < \dfrac{5 - 6.3}{.6}\right) = P(z < -2.17) = .5 - .4850 = .0150$

Thus, the percentage of days when the oxygen content is undesirable is 1.5%.

b.   We would expect the oxygen contents to fall within 2 standard deviations of the mean, or

$$\mu \pm 2\sigma \Rightarrow 6.3 \pm 2(.6) \Rightarrow 6.3 \pm 1.2 \Rightarrow (5.1, 7.5)$$

5.133 a. (i) $P(0 < z < 1.2) \approx z(4.4 - z)/10 = 1.2(4.4 - 1.2)/10 = .384$
          (ii) $P(0 < z < 2.5) \approx .49$
          (iii) $P(z > .8) = .5 - P(0 < z < .8) \approx .5 - .8(4.4 - .8)/10 = .5 - .288 = .212$
          (iv) $P(z < 1.0) = .5 + P(0 < z < 1.0) = .5 + 1(4.4 - 1)/10 = .5 + .34 = .84$

       b. Using Table IV, Appendix A:
          (i) $P(0 < z < 1.2) = .3849$
          (ii) $P(0 < z < 2.5) = .4938$
          (iii) $P(z > .8) = .5 - P(0 < z < .8) = .5 - .2881 = .2119$
          (iv) $P(z < 1.0) = .5 + P(0 < z < 1.0) = .5 + .3413 = .8413$

       c. For each part, the absolute error is:
          (i) Error $= |.384 - .3869| = .0029$
          (ii) Error $= |.49 - .4938| = .0038$
          (iii) Error $= |.212 - .2119| = .0001$
          (iv) Error $= |.84 - .8413| = .0013$

          In all cases, the absolute error is less than .0052.

5.135 Since we know for a mound-shaped distribution that approximately 95% of the data lies
       within two standard deviations of the mean, by symmetry approximately 2.5% would fall
       below 2 standard deviations below the mean and approximately 2.5% would fall more than 2
       standard deviations above the mean. Therefore, the amount of time allotted should be two
       standard deviations above the mean. This is:

       $$\mu + 2\sigma = 40 + 2(6) = 52 \text{ minutes}$$

5.137 From Table IV, Appendix A, and $\sigma = .4$:

       $$P(x > 6) = .01$$

       $$P(x > 6) = P\left(z > \frac{6 - \mu}{.4}\right) = P(z > z_0) = .01$$

       $A_1 = .5 - .01 = .4900$
       From Table IV, $z_0 = 2.33$

       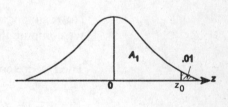

       $$2.33 = \frac{6 - \mu}{.4} \Rightarrow \mu = 6 - 2.33(.4) = 5.068$$

5.139   Using Table IV, Appendix A, with $\mu = .27$ and $\sigma = .04$,

$$P(x \le .14) = P\left(z \le \frac{.14 - .27}{.04}\right) = P(z < -3.25) \approx .5 - .5 = .0000$$

Since the probability of observing a reading of .14 or smaller is so small ($p \approx .0000$) if the pressure is .1 MPa, it would be extremely unlikely that this reading was obtained at a pressure of .1 MPa.

5.141   a.   Class I (very bright) consists of those with IQs above the 95th percentile. Thus, there would be .05 of the people in Class I.

Class II (bright) consists of those with IQs between the 75th and 95th percentile. Thus, there would be $.95 - .75 = .20$ of the people in Class II.

Class III (normal) consists of those with IQs between the 25th and 75th percentile. Thus, there would be $.75 - .25 = .50$ of the people in Class III.

Class IV (dull) consists of those with IQs between the 5th and 25th percentile. Thus, there would be $.25 - .05 = .20$ of the people in Class IV.

Class V (very dull) consists of those with IQs below the 5th percentile. Thus, there would be .05 of the people in Class V.

b.   Suppose we pick a pair of percentiles and compute the corresponding z-scores. Suppose we compute the z-scores for the 50th and 55th percentiles.

The z-score corresponding to the 50th percentile is $z = 0$. The z-score corresponding to the 55th percentile is:

$$P(z < z_0) = .55 \Rightarrow z_0 = .13$$

Now, suppose we pick another pair of percentiles that are also 5 points apart. Suppose we compute the z-scores for the 94th and 99th percentiles.

The z-score corresponding to the 94th percentile is:

$$P(z < z_0) = .94 \Rightarrow z_0 = 1.555$$

The z-score corresponding to the 99th percentile is:

$$P(z < z_0) = .99 \Rightarrow z_0 = 2.33$$

The z-scores corresponding to the 50th and 55th percentiles are 0 and .13. The difference is $.13 - 0 = .13$. The z-scores corresponding to the 94th and 99th percentiles are 1.555 and 2.33. The difference is $2.33 - 1.555 = .775$. Even though the differences in the percentiles are the same, the differences in the z-scores are much different. The information in the z-scores is more informative.

c. If the distribution of the IQs is skewed to the right, then the tail to the right is much longer than the tail to the left. If the distribution of the IQs is skewed to the left, then the tail to the left is much longer than the tail to the right. However, the proportions in the 5 cognitive classes would not differ, regardless of the shape of the distribution.

Class I (very bright) consists of those with IQs above the 95th percentile. Thus, there would be .05 of the people in Class I.

Class II (bright) consists of those with IQs between the 75th and 95th percentile. Thus, there would be .95 − .75 = .20 of the people in Class II.

Class III (normal) consists of those with IQs between the 25th and 75th percentile. Thus, there would be .75 − .25 = .50 of the people in Class III.

Class IV (dull) consists of those with IQs between the 5th and 25th percentile. Thus, there would be .25 − .05 = .20 of the people in Class IV.

Class V (very dull) consists of those with IQs below the 5th percentile. Thus, there would be .05 of the people in Class V.

Sampling Distributions

6.1   A population parameter is a numerical descriptive measure of a population. Because it is based on the observations in a population, its value is almost always unknown.

A sample statistic is a numerical descriptive measure of a sample. It is calculated from the observations in the sample.

6.3   a–b. The different samples of $n = 2$ with replacement and their means are:

Possible Samples	$\bar{x}$	Possible Samples	$\bar{x}$
0, 0	0	4, 0	2
0, 2	1	4, 2	3
0, 4	2	4, 4	4
0, 6	3	4, 6	5
2, 0	1	6, 0	3
2, 2	2	6, 2	4
2, 4	3	6, 4	5
2, 6	4	6, 6	6

c.   Since each sample is equally likely, the probability of any 1 being selected is

$$\frac{1}{4}\left(\frac{1}{4}\right) = \frac{1}{16}$$

d. $P(\bar{x} = 0) = \dfrac{1}{16}$

$P(\bar{x} = 1) = \dfrac{1}{16} + \dfrac{1}{16} = \dfrac{2}{16}$

$P(\bar{x} = 2) = \dfrac{1}{16} + \dfrac{1}{16} + \dfrac{1}{16} = \dfrac{3}{16}$

$P(\bar{x} = 3) = \dfrac{1}{16} + \dfrac{1}{16} + \dfrac{1}{16} + \dfrac{1}{16} = \dfrac{4}{16}$

$P(\bar{x} = 4) = \dfrac{1}{16} + \dfrac{1}{16} + \dfrac{1}{16} = \dfrac{3}{16}$

$P(\bar{x} = 5) = \dfrac{1}{16} + \dfrac{1}{16} = \dfrac{2}{16}$

$P(\bar{x} = 6) = \dfrac{1}{16}$

$\bar{x}$	$p(\bar{x})$
0	1/16
1	2/16
2	3/16
3	4/16
4	3/16
5	2/16
6	1/16

e.

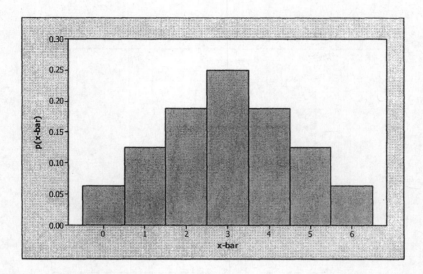

6.5   If the observations are independent of each other, then

$P(1, 1) = p(1)p(1) = .2(.2) = .04$
$P(1, 2) = p(1)p(2) = .2(.3) = .06$
$P(1, 3) = p(1)p(3) = .2(.2) = .04$
    etc.

a.

Possible Samples	$\bar{x}$	$p(\bar{x})$	Possible Samples	$\bar{x}$	$p(\bar{x})$
1, 1	1	.04	3, 4	3.5	.04
1, 2	1.5	.06	3, 5	4	.02
1, 3	2	.04	4, 1	2.5	.04
1, 4	2.5	.04	4, 2	3	.06
1, 5	3	.02	4, 3	3.5	.04
2, 1	1.5	.06	4, 4	4	.04
2, 2	2	.09	4, 5	4.5	.02
2, 3	2.5	.06	5, 1	3	.02
2, 4	3	.06	5, 2	3.5	.03
2, 5	3.5	.03	5, 3	4	.02
3, 1	2	.04	5, 4	4.5	.02
3, 2	2.5	.06	5, 5	5	.01
3, 3	3	.04			

Summing the probabilities, the probability distribution of $\bar{x}$ is:

$\bar{x}$	$p(\bar{x})$
1	.04
1.5	.12
2	.17
2.5	.20
3	.20
3.5	.14
4	.08
4.5	.04
5	.01

b.

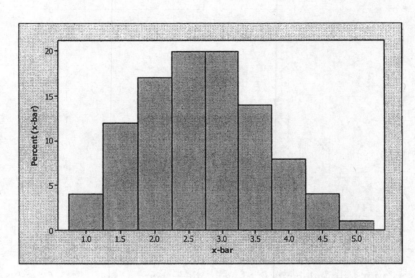

c. $P(\bar{x} \geq 4.5) = .04 + .01 = .05$

d. No. The probability of observing $\bar{x}$ = 4.5 or larger is small (.05).

6.7 a. For a sample of size $n = 2$, the sample mean and sample median are exactly the same. Thus, the sampling distribution of the sample median is the same as that for the sample mean (see Exercise 6.5a).

b. The probability histogram for the sample median is identical to that for the sample mean (see Exercise 6.5b).

6.11 If the sampling distribution of a sample statistic has a mean equal to the population parameter the statistic is intended to estimate, the statistic is said to be an unbiased estimate of the parameter. If the mean of the sampling distribution is not equal to the parameter, the statistic is said to be a biased estimate of the parameter.

6.13 The properties of an ideal estimator are unbiased and minimum variance. Ideally, when we are estimating a parameter, we would like to use an unbiased estimator. In addition, of all unbiased estimators, we would like to choose the estimator with the minimum variance.

6.15 a. $\mu = \sum xp(x) = 2\left(\dfrac{1}{3}\right) + 4\left(\dfrac{1}{3}\right) + 9\left(\dfrac{1}{3}\right) = \dfrac{15}{3} = 5$

b. The possible samples of size $n = 3$, the sample means, and the probabilities are:

Possible Samples	$\bar{x}$	$p(\bar{x})$	m	Possible Samples	$\bar{x}$	$p(\bar{x})$	m
2, 2, 2	2	1/27	2	4, 4, 4	4	1/27	4
2, 2, 4	8/3	1/27	2	4, 4, 9	17/3	1/27	4
2, 2, 9	13/3	1/27	2	4, 9, 2	5	1/27	4
2, 4, 2	8/3	1/27	2	4, 9, 4	17/3	1/27	4
2, 4, 4	10/3	1/27	4	4, 9, 9	22/3	1/27	9
2, 4, 9	5	1/27	4	9, 2, 2	13/3	1/27	2
2, 9, 2	13/3	1/27	2	9, 2, 4	5	1/27	4
2, 9, 4	5	1/27	4	9, 2, 9	20/3	1/27	9
2, 9, 9	20/3	1/27	9	9, 4, 2	5	1/27	4
4, 2, 2	8/3	1/27	2	9, 4, 4	17/3	1/27	4
4, 2, 4	10/3	1/27	4	9, 4, 9	22/3	1/27	9
4, 2, 9	5	1/27	4	9, 9, 2	20/3	1/27	9
4, 4, 2	10/3	1/27	4	9, 9, 4	22/3	1/27	9
				9, 9, 9	9	1/27	9

The sampling distribution of $\bar{x}$ is:

$\bar{x}$	$p(\bar{x})$
2	1/27
8/3	3/27
10/3	3/27
4	1/27
13/3	3/27
5	6/27
17/3	3/27
20/3	3/27
22/3	3/27
9	1/27
	27/27

$$E(\bar{x}) = \sum \bar{x}p(\bar{x}) = 2\left(\frac{1}{27}\right) + \frac{8}{3}\left(\frac{3}{27}\right) + \frac{10}{3}\left(\frac{3}{27}\right) + 4\left(\frac{1}{27}\right) + \frac{13}{3}\left(\frac{3}{27}\right)$$

$$+ 5\left(\frac{6}{27}\right) + \frac{17}{3}\left(\frac{3}{27}\right) + \frac{20}{3}\left(\frac{3}{27}\right) + \frac{22}{3}\left(\frac{3}{27}\right) + 9\left(\frac{1}{27}\right)$$

$$= \frac{2}{27} + \frac{8}{27} + \frac{10}{27} + \frac{4}{27} + \frac{13}{27} + \frac{30}{27} + \frac{17}{27} + \frac{20}{27} + \frac{22}{27} + \frac{9}{27}$$

$$= \frac{135}{27} = 5$$

Since $\mu = 5$ in part **a**, and $E(\bar{x}) = \mu = 5$, $\bar{x}$ is an unbiased estimator of μ.

c.	The median was calculated for each sample and is shown in the table in part **b**. The sampling distribution of m is:

m	$p(m)$
2	7/27
4	13/27
9	7/27
	27/27

$$E(m) = \sum mp(m) = 2\left(\frac{7}{27}\right) + 4\left(\frac{13}{27}\right) + 9\left(\frac{7}{27}\right) = \frac{14}{27} + \frac{52}{27} + \frac{63}{27} = \frac{129}{27} = 4.778$$

The $E(m) = 4.778 \neq \mu = 5$. Thus, m is a biased estimator of μ.

d.	Use the sample mean, $\bar{x}$. It is an unbiased estimator.

6.19	a.	Refer to the solution to Exercise 6.9. The values of s^2 and the corresponding probabilities are listed below:

$$s^2 = \frac{\sum (x^2) - \frac{\left(\sum x\right)^2}{n}}{n-1}$$

For sample 1, 1, $s^2 = \dfrac{2 - \frac{2^2}{2}}{1} = 0$

For sample 1, 2, $s^2 = \dfrac{5 - \frac{3^2}{2}}{1} = .5$

The rest of the values are calculated and shown:

s^2	$p(s^2)$	s^2	$p(s^2)$
0.0	.04	0.5	.04
0.5	.06	2.0	.02
2.0	.04	4.5	.04
4.5	.04	2.0	.06
8.0	.02	0.5	.04
0.5	.06	0.0	.04
0.0	.09	0.5	.02
0.5	.06	8.0	.02
2.0	.06	4.5	.03
4.5	.03	2.0	.02
2.0	.04	0.5	.02
0.5	.06	0.0	.01
0.0	.04		

The sampling distribution of s^2 is:

s^2	$p(s^2)$
0.0	.22
0.5	.36
2.0	.24
4.5	.14
8.0	.04

b. $\sigma^2 = \sum (x - \mu)^2 p(x) = (1 - 2.7)^2(.2) + (2 - 2.7)^2(.3) + (3 - 2.7)^2(.2)$

$$+ (4 - 2.7)^2(.2) + (5 - 2.7)^2(.1)$$

$$= 1.61$$

c. $E(s^2) = \sum s^2 p(s^2) = 0(.22) + .5(.36) + 2(.24) + 4.5(.14) + 8(.04) = 1.61$

d. The sampling distribution of s is listed below, where $s = \sqrt{s^2}$:

s	$p(s)$
0.000	.22
0.707	.36
1.414	.24
2.121	.14
2.828	.04

e. $E(s) = \sum s p(s) = 0(.22) + .707(.36) + 1.414(.24) + 2.121(.14) + 2.828(.04)$

$$= 1.00394$$

Since $E(s) = 1.00394$ is not equal to $\sigma = \sqrt{\sigma^2} = \sqrt{1.61} = 1.269$, s is a biased estimator of σ.

6.21 The symbol $\mu_{\bar{x}}$ represents the mean of the sampling distribution of $\bar{x}$. The symbol $\sigma_{\bar{x}}$ represents the standard deviation of the sampling distribution of $\bar{x}$.

6.23 The standard deviation of the sampling distribution of $\bar{x}$, $\sigma_{\bar{x}}$, is the standard deviation of the population from which the sample is selected divided by $\sqrt{n}$.

6.25 The Central Limit Theorem states: Consider a random sample of n observations selected from a population (*any* population) with mean μ and standard deviation σ. Then, when n is sufficiently large, the sampling distribution of $\bar{x}$ will be approximately a normal distribution with mean $\mu_{\bar{x}} = \mu$ and standard deviation $\sigma_{\bar{x}} = \sigma / \sqrt{n}$. The larger the sample size, the better will be the normal approximation to the sampling distribution of $\bar{x}$.

6.27 a. $\mu_{\bar{x}} = \mu = 100, \sigma_{\bar{x}} = \dfrac{\sigma}{\sqrt{n}} = \dfrac{\sqrt{100}}{\sqrt{4}} = 5$

b. $\mu_{\bar{x}} = \mu = 100, \sigma_{\bar{x}} = \dfrac{\sigma}{\sqrt{n}} = \dfrac{\sqrt{100}}{\sqrt{25}} = 2$

c. $\mu_{\bar{x}} = \mu = 100, \sigma_{\bar{x}} = \dfrac{\sigma}{\sqrt{n}} = \dfrac{\sqrt{100}}{\sqrt{100}} = 1$

d. $\mu_{\bar{x}} = \mu = 100, \sigma_{\bar{x}} = \dfrac{\sigma}{\sqrt{n}} = \dfrac{\sqrt{100}}{\sqrt{50}} = 1.414$

e. $\mu_{\bar{x}} = \mu = 100, \sigma_{\bar{x}} = \dfrac{\sigma}{\sqrt{n}} = \dfrac{\sqrt{100}}{\sqrt{500}} = .447$

f. $\mu_{\bar{x}} = \mu = 100, \sigma_{\bar{x}} = \dfrac{\sigma}{\sqrt{n}} = \dfrac{\sqrt{100}}{\sqrt{1000}} = .316$

6.29 a. $\mu = \sum xp(x) = 1(.1) + 2(.4) + 3(.4) + 8(.1) = 2.9$

$\sigma^2 = \sum (x - \mu)^2 \, p(x) = (1 - 2.9)^2(.1) + (2 - 2.9)^2(.4) + (3 - 2.9)^2(.4) + (8 - 2.9)^2(.1)$
$$= .361 + .324 + .004 + 2.601 = 3.29$$
$\sigma = \sqrt{3.29} = 1.814$

b. The possible samples, values of $\bar{x}$, and associated probabilities are listed:

Possible Samples	$\bar{x}$	$p(\bar{x})$	Possible Samples	$\bar{x}$	$p(\bar{x})$
1, 1	1	.01	3, 1	2	.04
1, 2	1.5	.04	3, 2	2.5	.16
1, 3	2	.04	3, 3	3	.16
1, 8	4.5	.01	3, 8	5.5	.04
2, 1	1.5	.04	8, 1	4.5	.01
2, 2	2	.16	8, 2	5	.04
2, 3	2.5	.16	8, 3	5.5	.04
2, 8	5	.04	8, 8	8	.01

$P(1, 1) = p(1)p(1) = .1(.1) = .01$
$P(1, 2) = p(1)p(2) = .1(.4) = .04$
$P(1, 3) = p(1)p(3) = .1(.4) = .04$
  etc.

The sampling distribution of $\bar{x}$ is:

$\bar{x}$	$p(\bar{x})$
1	.01
1.5	.08
2	.24
2.5	.32
3	.16
4.5	.02
5	.08
5.5	.08
8	.01
	1.00

c. $\mu_{\bar{x}} = E(\bar{x}) = \sum \bar{x}p(\bar{x}) = 1(.01) + 1.5(.08) + 2(.24) + 2.5(.32) + 3(.16) + 4.5(.02)$
$$+ 5(.08) + 5.5(.08) + 8(.01)$$
$$= 2.9 = \mu$$

$\sigma_{\bar{x}}^2 = \sum (\bar{x} - \mu_{\bar{x}})^2 \, p(\bar{x}) = (1 - 2.9)^2(.01) + (1.5 - 2.9)^2(.08) + (2 - 2.9)^2(.24)$
$$+ (2.5 - 2.9)^2(.32) + (3 - 2.9)^2(.16) + (4.5 - 2.9)^2(.02)$$
$$+ (5 - 2.9)^2(.08) + (5.5 - 2.9)^2(.08) + (8 - 2.9)^2(.01)$$
$$= .0361 + .1568 + .1944 + .0512 + .0016 + .0512 + .3528$$
$$+ .5408 + .2601$$
$$= 1.645$$

$\sigma_{\bar{x}} = \sqrt{1.645} = 1.283$

$\sigma_{\bar{x}} = \sigma/\sqrt{n} = 1.814/\sqrt{2} = 1.283$

6.31   a. $\mu_{\bar{x}} = \mu = 30$, $\sigma_{\bar{x}} = \dfrac{\sigma}{\sqrt{n}} = \dfrac{16}{\sqrt{100}} = 1.6$

b.   By the Central Limit Theorem, the sampling distribution of $\bar{x}$ is approximately normal.

c. $P(\bar{x} \geq 28) = P\left(z \geq \dfrac{28 - 30}{1.6} \right) = P(z \geq -1.25) = .5 + .3944 = .8944$

d. $P(22.1 \leq \bar{x} \leq 26.8) = P\left(\dfrac{22.1 - 30}{1.6} \leq z \leq \dfrac{26.8 - 30}{1.6} \right) = P(-4.94 \leq z \leq -2)$
$$= .5 - .4772 = .0228$$

e. $P(\bar{x} \leq 28.2) = P\left(z \leq \dfrac{28.2 - 30}{1.6} \right) = P(z \leq -1.13) = .5 - .3708 = .1292$

f. $P(\bar{x} \geq 27.0) = P\left(z \geq \dfrac{27.0 - 30}{1.6} \right) = P(z \geq -1.88) = .5 + .4699 = .9699$

6.35   a.   By the Central Limit Theorem, the sampling distribution of $\bar{x}$ is approximately normal. The mean of the $\bar{x}$ distribution is $\mu_{\bar{x}} = \mu = 320$ and the standard deviation of the

$\bar{x}$ distribution is $\sigma_{\bar{x}} = \dfrac{\sigma}{\sqrt{n}} = \dfrac{100}{\sqrt{100}} = \dfrac{100}{10} = 10$.

   b. $P(300 < \bar{x} < 310) = P\left(\dfrac{300 - 320}{10} < z < \dfrac{310 - 320}{10} \right)$

$= P(-2 < z < -1) = .4772 - .3413 = .1359$

(Using Table IV, Appendix A)

   c. $P(\bar{x} > 360) = P\left(z > \dfrac{360 - 320}{10} \right) = P(z > 4) \approx .5 - .5 = 0$ (Using Table IV, Appendix A)

6.37   a. $\mu_{\bar{x}}$ is the mean of the sampling distribution of $\bar{x}$. $\mu_{\bar{x}} = \mu = 79$.

   b. $\sigma_{\bar{x}}$ is the standard deviation of the sampling distribution of $\bar{x}$. $\sigma_{\bar{x}} = \dfrac{\sigma}{\sqrt{n}} = \dfrac{23}{\sqrt{100}} = 2.3$

   c.   By the Central Limit Theorem, the sampling distribution of $\bar{x}$ is approximately normal.

   d. $z = \dfrac{\bar{x} - \mu_{\bar{x}}}{\sigma_{\bar{x}}} = \dfrac{80 - 79}{2.3} = 0.43$

   e. $P(\bar{x} > 80) = P(z > 0.43) = .5 - .1664 = .3336$ (Using Table IV, Appendix A.)

6.39   a. $E(\bar{x}) = \mu_{\bar{x}} = \mu = .10$ and $V(\bar{x}) = \sigma_{\bar{x}}^2 = \dfrac{\sigma^2}{n} = \dfrac{.10^2}{50} = .0002$

   b.   By the Central Limit Theorem, the sampling distribution of $\bar{x}$ is approximately normal if the sample size is sufficiently large. In this case, n = 50 is considered large.

   c. $P(\bar{x} > .13) = P\left(z > \dfrac{.13 - .10}{\sqrt{.0002}} \right) = P(z > 2.12) = .5 - .4830 = .0170$

6.41   a.   By the Central Limit Theorem, the sampling distribution of $\bar{x}$ is approximately normal if the sample size is sufficiently large. In this case, n = 50 is considered large.

$\mu_{\bar{x}} = \mu = .53$, $\sigma_{\bar{x}} = \dfrac{\sigma}{\sqrt{n}} = \dfrac{.193}{\sqrt{50}} = .0273$

   b. $P(\bar{x} > .58) = P\left(z > \dfrac{.58 - .53}{.0273} \right) = P(z > 1.83) = .5 - .4664 = .0336$

(Using Table IV, Appendix A)

c. $P(\bar{x} > .59 \mid \mu = .58) = P\left(z > \dfrac{.59 - .58}{.0273}\right) = P(z > .37) = .5 - .1443 = .3557$

(Using Table IV, Appendix A) The above is the probability *after tensioning and loading*.

$P(\bar{x} > .59 \mid \mu = .53) = P\left(z > \dfrac{.59 - .53}{.0273}\right) = P(z > 2.20) = .5 - .4772 = .0228$

(Using Table IV, Appendix A) The above is the probability *before tensioning*.

Since the probability *after tensioning* is not small, this would not be an unusual event. Since the probability before tensioning is very small, this would be an unusual event. The sample measurements were probably obtained after tensioning and loading.

6.43 a. By the Central Limit Theorem, the sampling distribution of $\bar{x}$ is approximately normal if the sample size is sufficiently large. In this case, n = 326 is considered large.

$\mu_{\bar{x}} = \mu = 6, \quad \sigma_{\bar{x}} = \dfrac{\sigma}{\sqrt{n}} = \dfrac{10}{\sqrt{326}} = .5538$

$P(\bar{x} > 7.5) = P\left(z > \dfrac{7.5 - 6}{.5538}\right) = P(z > 2.71) = .5 - .4966 = .0034$

b. $P(\bar{x} > 300) = P\left(z > \dfrac{300 - 6}{.5538}\right) = P(z > 530.88) \approx .5 - .5 = 0$

Since it is essentially impossible to get a sample mean of 300 if the population mean was 6, we would conclude that the mean PFOA in people who live near DuPont's Teflon-making facility is not 6 ppb but something much larger.

6.45 **Handrubbing**

$\mu_{\bar{x}} = \mu = 35, \quad \sigma_{\bar{x}} = \dfrac{\sigma}{\sqrt{n}} = \dfrac{59}{\sqrt{50}} = 8.344$

$P(\bar{x} < 30) = P\left(z < \dfrac{30 - 35}{8.344}\right) = P(z < -.60) = .5 - .2257 = .2743$

(Using Table IV, Appendix A.)

Handwashing

$$\mu_{\bar{x}} = \mu = 69, \quad \sigma_{\bar{x}} = \frac{\sigma}{\sqrt{n}} = \frac{106}{\sqrt{50}} = 14.991$$

$$P(\bar{x} < 30) = P\left(z < \frac{30-69}{14.991}\right) = P(z < -2.60) = .5 - .4953 = .0047$$

(Using Table IV, Appendix A.)

This sample of workers probably came from the handrubbing group. The probability of seeing a mean of 30 or smaller for this group is .2743, while the probability of observing a mean of 30 or less for the handwashers is .0047. This would be an extremely rare event if the workers were handwashers because the probability is so small. It would not be a very unusual event for the handrubbers because the probability is not that small.

6.47   The statement "The sample mean, $\bar{x}$, will always be equal to μ_x" is false. The sample mean is used to estimate the population mean, but the probability that it is equal to the population mean is 0.

6.49   The statement "The standard error of $\bar{x}$ will always be small than σ" is true. The standard error of $\bar{x}$ is $\sigma_{\bar{x}} = \dfrac{\sigma}{\sqrt{n}}$. Thus, the standard error of $\bar{x}$ will always be smaller than σ.

6.51   a.   "The sampling distribution of the sample statistic A" is the probability distribution of the variable A.

       b.   "A" is an unbiased estimator of α because the mean of the sampling distribution of A is α.

       c.   If both A and B are unbiased estimators of α, then the statistic whose standard deviation is smaller is a better estimator of α.

       d.   No. The Central Limit Theorem applies only to the sample mean. If A is the sample mean, $\bar{x}$, and n is sufficiently large, then the Central Limit Theorem will apply. However, both A and B cannot be sample means. Thus, we cannot apply the Central Limit Theorem to both A and B.

6.53   By the Central Limit Theorem, the sampling distribution of $\bar{x}$ is approximately normal.

$$\mu_{\bar{x}} = \mu = 19.6, \quad \sigma_{\bar{x}} = \frac{3.2}{\sqrt{68}} = .388$$

       a. $P(\bar{x} \leq 19.6) = P\left(z \leq \dfrac{19.6-19.6}{.388}\right) = P(z \leq 0) = .5$ (using Table IV, Appendix A)

       b. $P(\bar{x} \leq 19) = P\left(z \leq \dfrac{19-19.6}{.388}\right) = P(z \leq -1.55) = .5 - .4394 = .0606$

(using Table IV, Appendix A)

c. $P(\bar{x} \geq 20.1) = P\left(z \geq \dfrac{20.1-19.6}{.388}\right) = P(z \geq 1.29) = .5 - .4015 = .0985$

(using Table IV, Appendix A)

d. $P(19.2 < \bar{x} < 20.6) = P\left(\dfrac{19.2-19.6}{.388} < z < \dfrac{20.6-19.6}{.388}\right)$

$$= P(-1.03 < z < 2.58) = .3485 + .4951 = .8436$$

(using Table IV, Appendix A)

6.57 Given: $\mu = 100$ and $\sigma = 10$

n	1	5	10	20	30	40	50
$\dfrac{\sigma}{\sqrt{n}}$	10	4.472	3.162	2.236	1.826	1.581	1.414

The graph of $\sigma/\sqrt{n}$ against n is given here:

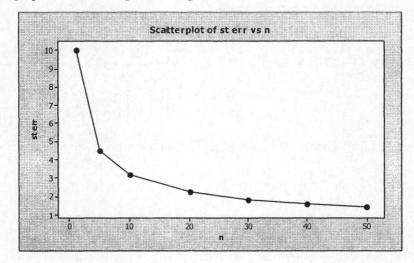

6.59 a. $\mu_{\bar{x}} = \mu = 89.34; \ \sigma_{\bar{x}} = \dfrac{\sigma}{\sqrt{n}} = \dfrac{7.74}{\sqrt{35}} = 1.3083$

b.

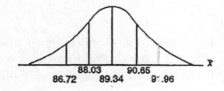

c. $P(\bar{x} > 88) = P\left(z > \dfrac{88-89.34}{1.3083}\right) = P(z > -1.02) = .5 + .3461 = .8461$

(using Table IV, Appendix A)

d. $P(\bar{x} < 87) = P\left(z < \dfrac{87-89.34}{1.3083}\right) = P(z < -1.79) = .5 - .4633 = .0367$

(using Table IV, Appendix A)

6.61   a. $\mu_{\bar{x}} = \mu = 1.3$, $\sigma_{\bar{x}} = \sigma/\sqrt{n} = 1.7/\sqrt{50} = .240$

        b.   Yes. By the Central Limit Theorem, the distribution of $\bar{x}$ is approximately normal for n sufficiently large. For this problem, $n = 50$ is sufficiently large.

        c. $P(\bar{x} < 1) = P\left(z < \dfrac{1-1.3}{.24}\right) = P(z < -1.25) = .5 - .3944 = .1056$

              (using Table IV, Appendix A)

        d. $P(\bar{x} > 1.9) = P\left(z > \dfrac{1.9-1.3}{.24}\right) = P(z > 2.5) = .5 - .4938 = .0062$

              (using Table IV, Appendix A)

6.63   From Exercise 6.62, $\sigma = .001$. We must assume the Central Limit theorem applies (n is only 25). Thus, the distribution of $\bar{x}$ is approximately normal with $\mu_{\bar{x}} = \mu = .501$ and $\sigma_{\bar{x}} = \sigma/\sqrt{n} = .001/\sqrt{25} = .0002$. Using Table IV, Appendix A,

$$P(\bar{x} < .4994) + P(\bar{x} > .5006) = P\left(z < \frac{.4994 - .501}{.0002}\right) + P\left(z < \frac{.5006 - .501}{.0002}\right)$$
$$= P(z < -8) + P(z > -2)$$
$$= (.5 - .5) + (.5 + .4772) = .9772$$

6.65   By the Central Limit Theorem, the sampling distribution of $\bar{x}$ is approximately normal.

$$\mu_{\bar{x}} = \mu = 75; \quad \sigma_{\bar{x}} = \frac{\sigma}{\sqrt{n}} = \frac{10}{\sqrt{36}} = 1.6667$$

        a. $P(\bar{x} > 79) = P\left(z > \dfrac{79-75}{1.6667}\right) = P(z > 2.4) = .5 - .4918 = .008$

              (using Table IV, Appendix A)

        b.   We must assume the sample size of $n = 36$ is sufficiently large so that the Central Limit Theorem applies.

6.67   a.   We must assume that the distribution of surface roughness scores is normally distributed because the sample size is only $n = 20$. Thus, the Central Limit Theorem will not be valid. If the distribution being sampled from is normal, then the sampling distribution of $\bar{x}$ is normal.

$$\mu_{\bar{x}} = \mu = 1.8, \qquad \sigma_{\bar{x}} = \frac{\sigma}{\sqrt{n}} = \frac{.5}{\sqrt{20}} = .1118$$

$$P(\bar{x} > 1.85) = P\left(z > \frac{1.85 - 1.8}{.1118}\right) = P(z > .45) = .5 - .1736 = .3264$$

b. $\bar{x} = \dfrac{\sum x}{n} = \dfrac{37.62}{20} = 1.881$

c. $P(\bar{x} > 1.881) = P\left(z > \dfrac{1.881 - 1.8}{.1118} \right) = P(z > .72) = .5 - .2642 = .2358$

Since the probability of getting a sample mean of 1.881 or anything more unusual is not small (probability is .2358), the assumptions made in part **a** appear to be valid.

6.69 Even though the number of flaws per piece of siding has a Poisson distribution, the Central Limit Theorem implies that the distribution of the sample mean will be approximately normal with $\mu_{\bar{x}} = \mu = 2.5$ and $\sigma_{\bar{x}} = \dfrac{\sigma}{\sqrt{n}} = \dfrac{\sqrt{2.5}}{\sqrt{35}} = .2673$. Therefore,

$$P(\bar{x} > 2.1) + P\left(z > \dfrac{2.1 - 2.5}{\sqrt{2.5}/\sqrt{35}} \right) = P(z > -1.50) = .5 + .4332 = .9332$$

(using Table IV, Appendix A)

6.71 a. By the Central Limit Theorem, the distribution of $\bar{x}$ is approximately normal, with $\mu_{\bar{x}} = \mu = 157$ and $\sigma_{\bar{x}} = \sigma/\sqrt{n} = 3/\sqrt{40} = .474$.

The sample mean is 1.3 psi below 157 or $\bar{x} = 157 - 1.3 = 155.7$

$$P(\bar{x} \le 155.7) = P\left(z \le \dfrac{155.7 - 157}{.474} \right) = P(z \le -2.74) = .5 - .4969 = .0031$$

(using Table IV, Appendix A)

If the claim is true, it is very unlikely (probability = .0031) to observe a sample mean 1.3 psi below 157 psi. Thus, the actual population mean is probably not 157 but something lower.

b. $P(\bar{x} \le 155.7) = P\left(z \le \dfrac{155.7 - 156}{.474} \right) = P(z \le -.63) = .5 - .2357 = .2643$

(using Table IV, Appendix A)

The observed sample is more likely if $\mu = 156$ rather than 157.

$$P(\bar{x} \le 155.7) = P\left(z \le \dfrac{155.7 - 158}{.474} \right) = P(z \le -4.85) = .5 - .5 = 0$$

The observed sample is less likely if $\mu = 158$ rather than 157.

c. $\sigma_{\bar{x}} = \sigma/\sqrt{n} = 2/\sqrt{40} = .316 \qquad \mu_{\bar{x}} = 157$

$$P(\bar{x} \le 155.7) = P\left(z \le \frac{155.7 - 157}{.316}\right) = P(z \le -4.11) = .5 - .5 = 0$$

(using Table IV, Appendix A)

The observed sample is less likely if $\sigma = 2$ rather than 3.

$\sigma_{\bar{x}} = \sigma/\sqrt{n} = 6/\sqrt{40} = .949 \qquad \mu_{\bar{x}} = 157$

$$P(\bar{x} \le 155.7) = P\left(z \le \frac{155.7 - 157}{.949}\right) = P(z \le -1.37) = .5 - .4147 = .0853$$

(using Table IV, Appendix A)

The observed sample is more likely if $\sigma = 6$ rather than 3.

Inferences Based on a Single Sample: Estimation with Confidence Intervals

7.1   The unknown population parameter (e.g., mean or proportion) that we are interested in estimating is called the target parameter.

7.3   An interval estimator estimates μ with a range of values, while a point estimator estimates μ with a single point.

7.5   Yes. As long as the sample size is sufficiently large, the Central Limit Theorem says the distribution of $\bar{x}$ is approximately normal regardless of the original distribution.

7.7   a.   For $\alpha = .10$, $\alpha/2 = .10/2 = .05$. $z_{\alpha/2} = z_{.05}$ is the z-score with .05 of the area to the right of it. The area between 0 and $z_{.05}$ is $.5 - .05 = .4500$. Using Table IV, Appendix A, $z_{.05} = 1.645$.

         b.   For $\alpha = .01$, $\alpha/2 = .01/2 = .005$. $z_{\alpha/2} = z_{.005}$ is the z-score with .005 of the area to the right of it. The area between 0 and $z_{.005}$ is $.5 - .005 = .4950$. Using Table IV, Appendix A, $z_{.005} = 2.58$.

         c.   For $\alpha = .05$, $\alpha/2 = .05/2 = 025$. $z_{\alpha/2} = z_{.025}$ is the z-score with .025 of the area to the right of it. The area between 0 and $z_{.025}$ is $.5 - .025 = .4750$. Using Table IV, Appendix A, $z_{.025} = 1.96$.

         d.   For $\alpha = .20$, $\alpha/2 = .20/2 = .10$. $z_{\alpha/2} = z_{.10}$ is the z-score with .10 of the area to the right of it. The area between 0 and $z_{.10}$ is $.5 - .10 = .4000$. Using Table IV, Appendix A, $z_{.10} = 1.28$.

7.9   a.   For confidence coefficient .95, $\alpha = .05$ and $\alpha/2 = .05/2 = .025$. From Table IV, Appendix A, $z_{.025} = 1.96$. The confidence interval is:

$$\bar{x} \pm z_{.025}\frac{s}{\sqrt{n}} \Rightarrow 28 \pm 1.96\frac{\sqrt{12}}{\sqrt{75}} \Rightarrow 28 \pm .784 \Rightarrow (27.216, 28.784)$$

         b. $\bar{x} \pm z_{.025}\frac{s}{\sqrt{n}} \Rightarrow 102 \pm 1.96\frac{\sqrt{22}}{\sqrt{200}} \Rightarrow 102 \pm .65 \Rightarrow (101.35, 102.65)$

         c. $\bar{x} \pm z_{.025}\frac{s}{\sqrt{n}} \Rightarrow 15 \pm 1.96\frac{.3}{\sqrt{100}} \Rightarrow 15 \pm .0588 \Rightarrow (14.9412, 15.0588)$

         d. $\bar{x} \pm z_{.025}\frac{s}{\sqrt{n}} \Rightarrow 4.05 \pm 1.96\frac{.83}{\sqrt{100}} \Rightarrow 4.05 \pm .163 \Rightarrow (3.887, 4.213)$

         e.   No. Since the sample size in each part was large (n ranged from 75 to 200), the Central Limit Theorem indicates that the sampling distribution of $\bar{x}$ is approximately normal.

7.11   a.   For confidence coefficient .95, $\alpha = .05$ and $\alpha/2 = .05/2 = .025$. From Table IV, Appendix A, $z_{.025} = 1.96$. The confidence interval is:

$$\bar{x} \pm z_{\alpha/2} \frac{s}{\sqrt{n}} \Rightarrow 83.2 \pm 1.96 \frac{6.4}{\sqrt{100}} \Rightarrow 83.2 \pm 1.25 \Rightarrow (81.95, 84.45)$$

        b.   The confidence coefficient of .95 means that in repeated sampling, 95% of all confidence intervals constructed will include μ.

        c.   For confidence coefficient .99, $\alpha = .01$ and $\alpha/2 = .01/2 = .005$. From Table IV, Appendix A, $z_{.005} = 2.58$. The confidence interval is:

$$\bar{x} \pm z_{\alpha/2} \frac{s}{\sqrt{n}} \Rightarrow 83.2 \pm 2.58 \frac{6.4}{\sqrt{100}} \Rightarrow 83.2 \pm 1.65 \Rightarrow (81.55, 84.85)$$

        d.   As the confidence coefficient increases, the width of the confidence interval also increases.

        e.   Yes. Since the sample size is 100, the Central Limit Theorem applies. This ensures the distribution of $\bar{x}$ is normal, regardless of the original distribution.

7.13   a.   The point estimate for the average number of latex gloves used per week by all healthcare workers with latex allergy is $\bar{x} = 19.3$.

        b.   For confidence coefficient .95, $\alpha = .05$ and $\alpha/2 = .05/2 = .025$. From Table IV, Appendix A, $z_{.025} = 1.96$. The 95% confidence interval is:

$$\bar{x} \pm z_{.025} \sigma_{\bar{x}} \Rightarrow \bar{x} \pm 1.96 \frac{\sigma}{\sqrt{n}} \Rightarrow 19.3 \pm 1.96 \frac{11.9}{\sqrt{46}} \Rightarrow 19.3 \pm 3.44 \Rightarrow (15.86, \ 22.74)$$

        c.   We are 95% confident that the average number of latex gloves used pr week by all healthcare workers with a latex allergy is between 15.86 and 22.74.

        d.   We must assume that we have a random sample from the target population and that the sample size is sufficiently large.

7.15   a.   Some preliminary calculations are:

$$\bar{x} = \frac{\sum x}{n} = \frac{181.56}{504} = .3602$$

$$s^2 = \frac{\sum x^2 - \frac{\left(\sum x\right)^2}{n}}{n-1} = \frac{74.6546 - \frac{181.56^2}{504}}{504-1} = \frac{9.24977143}{503} = .018389$$

$$s = \sqrt{.018389} = .13561$$

For confidence coefficient .90, $\alpha = .10$ and $\alpha/2 = .10/2 = .05$. From Table IV, Appendix A, $z_{.05} = 1.645$. The 90% confidence interval is:

$$\bar{x} \pm z_{.05}\sigma_{\bar{x}} \Rightarrow \bar{x} \pm 1.645\frac{\sigma}{\sqrt{n}} \Rightarrow .3602 \pm 1.645\frac{.13561}{\sqrt{504}}$$

$$\Rightarrow .3602 \pm .00994 \Rightarrow (.35026, \ .37014)$$

b.   We are 90% confident that the true mean visible albedo value of all Canadian Arctic ice ponds is between .35026 and .37014.

In repeated sampling 90% of all intervals constructed in the same manner will contain the true mean and 10% will not.

c.   **First-year Ice:**

Some preliminary calculations are:

$$\bar{x} = \frac{\sum x}{n} = \frac{26.63}{88} = .3026$$

$$s^2 = \frac{\sum x^2 - \frac{\left(\sum x\right)^2}{n}}{n-1} = \frac{10.6923 - \frac{26.63^2}{88}}{88-1} = \frac{2.633698864}{87} = .03027$$

$$s = \sqrt{.03027} = .17399$$

The 90% confidence interval is:

$$\bar{x} \pm z_{.05}\sigma_{\bar{x}} \Rightarrow \bar{x} \pm 1.645\frac{\sigma}{\sqrt{n}} \Rightarrow .3026 \pm 1.645\frac{.17399}{\sqrt{88}}$$

$$\Rightarrow .3026 \pm .03051 \Rightarrow (.27209, \ .33311)$$

We are 90% confident that the true mean visible albedo value of all First-year Canadian Arctic ice ponds is between .35026 and .37014.

Landfast Ice:

Some preliminary calculations are:

$$\bar{x} = \frac{\sum x}{n} = \frac{71.04}{196} = .3624$$

$$s^2 = \frac{\sum x^2 - \frac{\left(\sum x\right)^2}{n}}{n-1} = \frac{30.161 - \frac{71.04^2}{196}}{196-1} = \frac{4.41262449}{195} = .02263$$

$$s = \sqrt{.02263} = .15043$$

The 90% confidence interval is:

$$\bar{x} \pm z_{.05}\sigma_{\bar{x}} \Rightarrow \bar{x} \pm 1.645\frac{\sigma}{\sqrt{n}} \Rightarrow .3624 \pm 1.645\frac{.15043}{\sqrt{196}}$$

$$\Rightarrow .3624 \pm .01768 \Rightarrow (.34472, \ .38008)$$

We are 90% confident that the true mean visible albedo value of all Landfast Canadian Arctic ice ponds is between .34472 and .38008.

Multi-year Ice:

Some preliminary calculations are:

$$\bar{x} = \frac{\sum x}{n} = \frac{83.89}{220} = .3813$$

$$s^2 = \frac{\sum x^2 - \frac{(\sum x)^2}{n}}{n-1} = \frac{33.8013 - \frac{83.89^2}{220}}{220-1} = \frac{1.81251773}{219} = .00827$$

$$s = \sqrt{.00827} = .09097$$

The 90% confidence interval is:

$$\bar{x} \pm z_{.05}\sigma_{\bar{x}} \Rightarrow \bar{x} \pm 1.645\frac{\sigma}{\sqrt{n}} \Rightarrow .3813 \pm 1.645\frac{.09097}{\sqrt{220}}$$

$$\Rightarrow .3813 \pm .01009 \Rightarrow (.37121, \ .39139)$$

We are 90% confident that the true mean visible albedo value of all Multi-year Canadian Arctic ice ponds is between .37121 and .39139.

7.17 a. The parameter of interest is the mean effect size for all psychological studies of personality and aggressive behavior.

b. No, the distribution does not look normal. The data appear to be skewed to the right. The shape of this distribution is not of interest because the sample size is large, n = 109. By the Central Limit Theorem, the sampling distribution of $\bar{x}$ is normal regardless of the original distribution.

c. From the printout, the 95% confidence interval for μ is (0.4786, 0.8167). We are 95% confident that the true mean effect size is between 0.4786 and 0.8167.

d. Yes. Since 0 is not contained in the 95% confidence interval, it is not a likely value for the true mean effect size. Since all the values in the 95% confident interval are above 0, the researchers are justified in concluding the true effect size mean is greater than 0 or that those who score high on a personality test are more aggressive than those who score low.

7.19   a.   For confidence coefficient .99, $\alpha = .01$ and $\alpha/2 = .01/2 = .005$. From Table IV, Appendix A, $z_{.005} = 2.58$. The confidence interval is:

$$\bar{x} \pm z_{\alpha/2}\frac{s}{\sqrt{n}} \Rightarrow 1.13 \pm 2.58\frac{2.21}{\sqrt{72}} \Rightarrow 1.13 \pm .672 \Rightarrow (.458,\ 1.802)$$

We are 99% confident that the true mean number of pecks made by chickens pecking at blue string is between .458 and 1.802.

b.   Yes, there is evidence that chickens are more apt to peck at white string. The mean number of pecks for white string is 7.5. Since 7.5 is not in the 99% confidence interval for the mean number of pecks at blue string, it is not a likely value for the true mean for blue string.

7.21   a.   For confidence coefficient .95, $\alpha = .05$ and $\alpha/2 = .05/2 = .025$. From Table IV, Appendix A, $z_{.025} = 1.96$. The confidence interval is:

$$\bar{x} \pm z_{\alpha/2}\frac{s}{\sqrt{n}} \Rightarrow 19 \pm 1.96\frac{65}{\sqrt{265}} \Rightarrow 19 \pm 7.826 \Rightarrow (11.174,\ 26.826)$$

b.   For confidence coefficient .95, $\alpha = .05$ and $\alpha/2 = .05/2 = .025$. From Table IV, Appendix A, $z_{.025} = 1.96$. The confidence interval is:

$$\bar{x} \pm z_{\alpha/2}\frac{s}{\sqrt{n}} \Rightarrow 7 \pm 1.96\frac{49}{\sqrt{265}} \Rightarrow 7 \pm 5.90 \Rightarrow (1.10,\ 12.90)$$

c.   The SAT-Mathematics test would be more likely to have a mean change of 15 because 15 is in the 95% confidence interval for the mean change in SAT-Mathematics. Since 15 is in the confidence interval, it is a likely value. The value 15 is not in the 95% confidence interval for the SAT-Verbal. Thus, it is not a likely value.

7.23   a.   For confidence coefficient .95, $\alpha = .05$ and $\alpha/2 = .05/2 = .025$. From Table IV, Appendix A, $z_{.025} = 1.96$.

For Males, the 95% confidence interval is:

$$\bar{x} \pm z_{.025}\sigma_{\bar{x}} \Rightarrow \bar{x} \pm 1.96\frac{\sigma}{\sqrt{n}} \Rightarrow 16.79 \pm 1.96\frac{13.57}{\sqrt{128}} \Rightarrow 16.79 \pm 2.35 \Rightarrow (14.44,\ 19.14)$$

For Females, the 95% confidence interval is:

$$\bar{x} \pm z_{.025}\sigma_{\bar{x}} \Rightarrow \bar{x} \pm 1.96\frac{\sigma}{\sqrt{n}} \Rightarrow 10.79 \pm 1.96\frac{11.53}{\sqrt{184}} \Rightarrow 10.79 \pm 1.67 \Rightarrow (9.12,\ 12.46)$$

b. Since the two intervals are independent, the probability that at least one of the 2 confidence intervals will not contain the population mean is equal to 1 minus the probability that neither of the 2 confidence intervals will not contain the population mean. Thus the probability is:

P(at least one interval will not contain mean)

$= 1 - P$(neither interval will not contain the mean)

$= 1 - P$(both will contain the mean)$= 1 - .95(.95) = 1 - .9025 = .0975$.

c. The 95% confidence interval for the males is (14.44, 19.14). The 95% confidence interval for the females is (9.12, 12.46). Since all of the values in the male interval are larger than all the values in the female interval, we can infer that the males consume the most alcohol, on average, per week.

7.25 The two problems (and corresponding solutions) with using a small sample to estimate μ are:

1. The shape of the sampling distribution of the sample mean $\bar{x}$ now depends on the shape of the population that is sampled. The Central Limit Theorem no longer applies since the sample size is small.

2. The population standard deviation σ is almost always unknown. Although it is still true that $\sigma_{\bar{x}} = \sigma / \sqrt{n}$, the sample standard deviation s may provide a poor approximation for σ when the sample size is small.

7.27 a. If x is normally distributed, the sampling distribution of is normal, regardless of the sample size.

b. If nothing is known about the distribution of x, the sampling distribution of is approximately normal if n is sufficiently large. If n is not large, the distribution of is unknown if the distribution of x is not known.

7.29 a. $P(t \geq t_0) = .025$ where df $= 10$
$t_0 = 2.228$

b. $P(t \geq t_0) = .01$ where df $= 17$
$t_0 = 2.567$

c. $P(t \leq t_0) = .005$ where df $= 6$

Because of symmetry, the statement can be rewritten

$P(t \geq -t_0) = .005$ where df $= 6$
$t_0 = -3.707$

d. $P(t \leq t_0) = .05$ where df $= 13$
$t_0 = -1.771$

7.31   First, we must compute $\bar{x}$ and s.

$$\bar{x} = \frac{\sum x}{n} = \frac{30}{6} = 5$$

$$s^2 = \frac{\sum x^2 - \frac{\left(\sum x\right)^2}{n}}{n-1} = \frac{176 - \frac{(30)^2}{6}}{6-1} = \frac{26}{5} = 5.2$$

$$s = \sqrt{5.2} = 2.2804$$

a.   For confidence coefficient .90, $\alpha = 1 - .90 = .10$ and $\alpha/2 = .10/2 = .05$. From Table VI, Appendix A, with $df = n - 1 = 6 - 1 = 5$, $t_{.05} = 2.015$. The 90% confidence interval is:

$$\bar{x} \pm t_{.05}\frac{s}{\sqrt{n}} \Rightarrow 5 \pm 2.015\frac{2.2804}{\sqrt{6}} \Rightarrow 5 \pm 1.876 \Rightarrow (3.124, 6.876)$$

b.   For confidence coefficient .95, $\alpha = 1 - .95 = .05$ and $\alpha/2 = .05/2 = .025$. From Table VI, Appendix A, with $df = n - 1 = 6 - 1 = 5$, $t_{.025} = 2.571$. The 95% confidence interval is:

$$\bar{x} \pm t_{.025}\frac{s}{\sqrt{n}} \Rightarrow 5 \pm 2.571\frac{2.2804}{\sqrt{6}} \Rightarrow 5 \pm 2.394 \Rightarrow (2.606, 7.394)$$

c.   For confidence coefficient .99, $\alpha = 1 - .99 = .01$ and $\alpha/2 = .01/2 = .005$. From Table VI, Appendix A, with $df = n - 1 = 6 - 1 = 5$, $t_{.005} = 4.032$. The 99% confidence interval is:

$$\bar{x} \pm t_{.005}\frac{s}{\sqrt{n}} \Rightarrow 5 \pm 4.032\frac{2.2804}{\sqrt{6}} \Rightarrow 5 \pm 3.754 \Rightarrow (1.246, 8.754)$$

d.   a)   For confidence coefficient .90, $\alpha = 1 - .90 = .10$ and $\alpha/2 = .10/2 = .05$. From Table VI, Appendix A, with $df = n - 1 = 25 - 1 = 24$, $t_{.05} = 1.711$. The 90% confidence interval is:

$$\bar{x} \pm t_{.05}\frac{s}{\sqrt{n}} \Rightarrow 5 \pm 1.711\frac{2.2804}{\sqrt{25}} \Rightarrow 5 \pm .780 \Rightarrow (4.220, 5.780)$$

b)   For confidence coefficient .95, $\alpha = 1 - .95 = .05$ and $\alpha/2 = .05/2 = .025$. From Table VI, Appendix A, with $df = n - 1 = 25 - 1 = 24$, $t_{.025} = 2.064$. The 95% confidence interval is:

$$\bar{x} \pm t_{.025}\frac{s}{\sqrt{n}} \Rightarrow 5 \pm 2.064\frac{2.2804}{\sqrt{25}} \Rightarrow 5 \pm .941 \Rightarrow (4.059, 5.941)$$

c)   For confidence coefficient .99, $\alpha = 1 - .99 = .01$ and $\alpha/2 = .01/2 = .005$. From Table VI, Appendix A, with $df = n - 1 = 25 - 1 = 24$, $t_{.005} = 2.797$. The 99% confidence interval is:

$$\bar{x} \pm t_{.005}\frac{s}{\sqrt{n}} \Rightarrow 5 \pm 2.797\frac{2.2804}{\sqrt{25}} \Rightarrow 5 \pm 1.276 \Rightarrow (3.724, 6.276)$$

Increasing the sample size decreases the width of the confidence interval.

7.33   a. $\bar{x} = \dfrac{\sum x}{n} = \dfrac{39}{21} = 1.857$

$$s^2 = \frac{\sum x^2 - \dfrac{\left(\sum x\right)^2}{n}}{n-1} = \frac{101 - \dfrac{39^2}{21}}{21-1} = \frac{28.57143}{20} = 1.4286$$

$s = \sqrt{1.4286} = 1.1952$

   b.   Using MINITAB, a histogram of the data is:

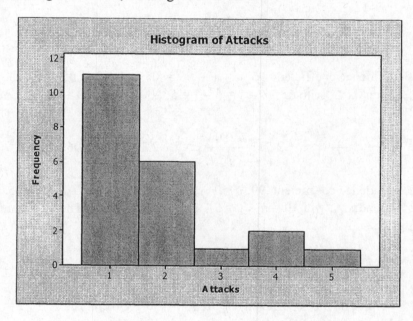

Although this is a histogram of the sample, it should reflect the distribution of the population fairly well. The population distribution appears to be skewed to the right. However, the distribution is fairly mound-shaped. Since the sample size of n = 21 is somewhat close to 30, the sampling distribution of $\bar{x}$ will be approximately normal.

   c.   For confidence coefficient .90, $\alpha = .10$ and $\alpha/2 = .10/2 = .05$. From Table VI, Appendix A, with df $= n - 1 = 21 - 1 = 20$, $t_{.05} = 1.725$. The 90% confidence interval is:

$$\bar{x} \pm t_{.05,20}\frac{s}{\sqrt{n}} \Rightarrow 1.857 \pm 1.725\frac{1.1952}{\sqrt{21}} \Rightarrow 1.857 \pm .450 \Rightarrow (1.407,\ 2.307)$$

   d.   We are 90% confident that the true mean number of individual suicide bombings and/or attacks per incident is between 1.407 and 2.307.

   e.   With repeated sampling, 90% of all confidence intervals constructed in this manner will contain the true mean and 10% will not.

7.35   a.   The point estimate for the mean amount of cesium in lichen specimens collected in Alaska is $\bar{x} = .009027$.

b.   For confidence coefficient .95, $\alpha = .05$ and $\alpha/2 = .05/2 = .025$. From Table VI, Appendix A, with df $= n - 1 = 9 - 1 = 8$, $t_{.025} = 2.306$.

c.   The 95% confidence interval is:

$$\bar{x} \pm t_{.025}\frac{s}{\sqrt{n}} \Rightarrow .009027 \pm 2.306\frac{.004854}{\sqrt{9}} \Rightarrow .009027 \pm .003731 \Rightarrow (.005296, \ .012758)$$

d.   This interval is the same as that found on the printout.

e.   We are 95% confident that the mean amount of cesium in lichen specimens collected in Alaska is between .005296 and .012758 microcuries per milliliter.

7.37   Some preliminary calculations:

$$\bar{x} = \frac{\sum x}{n} = \frac{6.44}{6} = 1.073$$

$$s^2 = \frac{\sum x^2 - \frac{\left(\sum x\right)^2}{n}}{n-1} = \frac{7.1804 - \frac{6.44^2}{6}}{6-1} = .0536$$

$$s = \sqrt{.0536} = .2316$$

a.   For confidence coefficient .95, $\alpha = .05$ and $\alpha/2 = .05/2 = .025$. From Table VI, Appendix A, with df $= n - 1 = 6 - 1 = 5$, $t_{.025} = 2.571$. The confidence interval is:

$$\bar{x} \pm t_{\alpha/2}\frac{s}{\sqrt{n}} \Rightarrow 1.073 \pm 2.571\frac{.2316}{\sqrt{6}} \Rightarrow 1.073 \pm .243 \Rightarrow (.830, \ 1.316)$$

We are 95% confident that the true average decay rate of fine particles produced from oven cooking or toasting is between .830 and 1.316.

b.   The phrase "95% confident" means that in repeated sampling, 95% of all confidence intervals constructed will contain the true mean.

c.   In order for the inference above to be valid, the distribution of decay rates must be normally distributed.

7.39   a.   **Both Untreated**: For confidence coefficient .90, $\alpha = .10$ and $\alpha/2 = .10/2 = .05$. From Table VI, Appendix A, with df $= n - 1 = 29 - 1 = 28$, $t_{.05} = 1.701$. The 90% confidence interval is:

$$\bar{x} \pm t_{.05,28} \frac{s}{\sqrt{n}} \Rightarrow 20.9 \pm 1.701 \frac{3.34}{\sqrt{29}} \Rightarrow 20.9 \pm 1.055 \Rightarrow (19.845, \ 21.955)$$

Male Treated: For confidence coefficient .90, $\alpha = .10$ and $\alpha/2 = .10/2 = .05$. From Table VI, Appendix A, with df $= n - 1 = 23 - 1 = 22$, $t_{.05} = 1.717$. The 90% confidence interval is:

$$\bar{x} \pm t_{.05,22} \frac{s}{\sqrt{n}} \Rightarrow 20.3 \pm 1.717 \frac{3.50}{\sqrt{23}} \Rightarrow 20.3 \pm 1.253 \Rightarrow (19.047, \ 21.553)$$

Female Treated: For confidence coefficient .90, $\alpha = .10$ and $\alpha/2 = .10/2 = .05$. From Table VI, Appendix A, with df $= n - 1 = 18 - 1 = 17$, $t_{.05} = 1.740$. The 90% confidence interval is:

$$\bar{x} \pm t_{.05,17} \frac{s}{\sqrt{n}} \Rightarrow 22.9 \pm 1.740 \frac{4.37}{\sqrt{18}} \Rightarrow 22.9 \pm 1.792 \Rightarrow (21.108, \ 24.692)$$

Both Treated: For confidence coefficient .90, $\alpha = .10$ and $\alpha/2 = .10/2 = .05$. From Table VI, Appendix A, with df $= n - 1 = 21 - 1 = 20$, $t_{.05} = 1.725$. The 90% confidence interval is:

$$\bar{x} \pm t_{.05,20} \frac{s}{\sqrt{n}} \Rightarrow 18.6 \pm 1.725 \frac{2.11}{\sqrt{21}} \Rightarrow 18.6 \pm 0.794 \Rightarrow (17.806, \ 19.394)$$

   b.   The female/male pair that appears to produce the highest mean number of eggs is the **Female treated** pair. The 90% confidence interval for this pair is the highest. This interval does overlap with the **Both treated** pair and the **Males treated** pair, but just barely.

7.41   a.   Some preliminary calculations are:

$$\bar{x} = \frac{\sum x}{n} = \frac{160.9}{22} = 7.314$$

$$s^2 = \frac{\sum x^2 - \dfrac{\left(\sum x\right)^2}{n}}{n-1} = \frac{1{,}389.1 - \dfrac{160.9^2}{22}}{22-1} = 10.1112$$

$$s = \sqrt{10.1112} = 3.180$$

For confidence coefficient .95, $\alpha = .05$ and $\alpha/2 = .05/2 = .025$. From Table VI, Appendix A with df $= n - 1 = 22 - 1 = 21$, $t_{.025} = 2.080$. The 95% confidence interval is:

$$\bar{x} \pm t_{\alpha/2} \frac{s}{\sqrt{n}} \Rightarrow 7.314 \pm 2.080 \frac{3.180}{\sqrt{22}} \Rightarrow 7.314 \pm 1.410 \Rightarrow (5.904, \ 8.724)$$

b. We are 95% confident that the mean PMI for all human brain specimens obtained at autopsy is between 5.904 and 8.724.

c. We must assume that the population of all PMI's is normally distributed. From the dot plot in Exercise 2.25, the distribution does not appear to be normal.

d. "95% confidence" means that if repeated samples of size n were selected from the population and 95% confidence intervals formed for μ, 95% of the intervals formed will contain the true mean and 5% will not.

7.43 a. For confidence coefficient .95, $\alpha = .05$, and $\alpha/2 = .05/2 = .025$. From Table VI, Appendix A, with df $= n - 1 = 15 - 1 = 14$, $t_{.025} = 2.145$. The confidence interval is:

$$\bar{x} \pm t_{.025}\frac{s}{\sqrt{n}} \Rightarrow 37.3 \pm 2.145\frac{13.9}{\sqrt{15}} \Rightarrow 37.3 \pm 7.70 \Rightarrow (29.60, 45.00)$$

b. We are 95% confident that the true mean adrenocorticotropin level in sleepers one-hour prior to anticipated waking is between 29.60 and 45.00

c. Suppose we assume that the true mean adrenocorticotropin level of sleepers three hours before anticipated wake-up time is 25.5. Since the 95% confidence interval for the true mean adrenocorticotropin level in sleepers one-hour prior to anticipated waking (29.60, 45.00) does not contain 25.5, there is evidence to indicate that the true mean adrenocorticotropin level of sleepers three hours before anticipated wake-up time is different from that of sleepers one-hour prior to anticipated waking.

7.45 An unbiased estimator is one in which the mean of the sampling distribution is the parameter of interest, i.e., $E(\hat{p}) = p$.

7.47 a. The sample size is large enough if both $n\hat{p} \geq 15$ and $n\hat{q} \geq 15$.

$n\hat{p} = 196(.64) = 125.44$ and $n\hat{q} = 196(.36) = 70.56$. Since both of these numbers are greater than or equal to 15, the sample size is sufficiently large to conclude the normal approximation is reasonable.

b. For confidence coefficient .95, $\alpha = .05$ and $\alpha/2 = .025$. From Table IV, Appendix A, $z_{.025} = 1.96$. The confidence interval is:

$$\hat{p} \pm z_{.025}\sqrt{\frac{\hat{p}\hat{q}}{n}} \Rightarrow .64 \pm 1.96\sqrt{\frac{.64(.36)}{196}} \Rightarrow .64 \pm .067 \Rightarrow (.573, .707)$$

c. We are 95% confident the true value of p is between .372 and .468.

d. "95% confidence" means that if repeated samples of size 196 were selected from the population and 95% confidence intervals formed, 95% of all confidence intervals will contain the true value of p.

7.49   The sample size is large enough if both $n\hat{p} \geq 15$ and $n\hat{q} \geq 15$.

   a. $n\hat{p} = 500(.05) = 25$ and $n\hat{q} = 500(.95) = 475$. Since both of these numbers are greater than or equal to 15, the sample size is sufficiently large to conclude the normal approximation is reasonable.

   b. $n\hat{p} = 100(.05) = 5$ and $n\hat{q} = 100(.95) = 95$. Since the first number is not greater than 15, the sample size is not sufficiently large to conclude the normal approximation is reasonable.

   c. $n\hat{p} = 10(.5) = 5$ and $n\hat{q} = 10(.5) = 5$. Since neither of these numbers is greater than or equal to 15, the sample size is not sufficiently large to conclude the normal approximation is reasonable.

   d. $n\hat{p} = 10(.3) = 3$ and $n\hat{q} = 10(.7) = 7$. Since neither of these numbers is greater than or equal to 15, the sample size is not sufficiently large to conclude the normal approximation is reasonable.

7.51   a.   The population of interest is the set of gun ownerships (Yes or No) of all adults in the U.S.

   b.   The parameter of interest is the true percentage of all adults in the U.S. who own at least one gun.

   c.   The estimate of the population proportion is $\hat{p} = .26$. The estimate of the population percentage is .26(100%) = 26%.

   d.   The sample size is large enough if both $n\hat{p} \geq 15$ and $n\hat{q} \geq 15$.

   $n\hat{p} = 2,770(.26) = 720.2$ and $n\hat{q} = 2,770(.74) = 2,049.8$. Since both of these numbers are greater than or equal to 15, the sample size is sufficiently large to conclude the normal approximation is reasonable.

   For confidence coefficient .99, $\alpha = .01$ and $\alpha/2 = .01/2 = .005$. From Table IV, Appendix A, $z_{.005} = 2.575$. The 99% confidence interval is:

   $$\hat{p} \pm z_{.005}\sqrt{\frac{pq}{n}} \Rightarrow \hat{p} \pm 2.575\sqrt{\frac{\hat{p}\hat{q}}{n}} \Rightarrow .26 \pm 2.575\sqrt{\frac{.26(.74)}{2,770}} \Rightarrow .26 \pm .021 \Rightarrow (.239, .281)$$

   The 99% confidence interval for the true percentage is (23.9%, 28.1%).

   e.   We are 99% confident that the true percentage of adults in the U.S who own at least one gun is between 23.9% and 28.1%.

   f.   "99% confidence" means that in repeated sampling 99% of all confidence intervals constructed in this manner will contain the true percentage.

7.53   a.   The sample size is large enough if both $n\hat{p} \geq 15$ and $n\hat{q} \geq 15$.

$n\hat{p} = 1,000(.63) = 630$ and $n\hat{q} = 1,000(.37) = 370$. Since both of the numbers are greater than or equal to 15, the sample size is sufficiently large to conclude the normal approximation is reasonable.

For confidence coefficient .95, $\alpha = .05$ and $\alpha/2 = .05/2 = .025$. From Table IV, Appendix A, $z_{.025} = 1.96$. The 95% confidence interval is:

$$\hat{p} \pm z_{.025} \sqrt{\frac{pq}{n}} \Rightarrow \hat{p} \pm 1.96 \sqrt{\frac{\hat{p}\hat{q}}{n}} \Rightarrow .63 \pm 1.96 \sqrt{\frac{.63(.37)}{1,000}}$$

$$\Rightarrow .63 \pm .030 \Rightarrow (.600, \ .660)$$

b.   Yes, we would be surprised. Since .70 is not in the 95% confidence interval, it is not a likely value for the true value of the population proportion of adults who would choose to sleep when home sick.

7.55   First, we compute $\hat{p}$: $\hat{p} = \dfrac{x}{n} = \dfrac{183}{837} = .219$

The sample size is large enough if both $n\hat{p} \geq 15$ and $n\hat{q} \geq 15$.

$n\hat{p} = 837(.219) = 183.303$ and $n\hat{q} = 837(.781) = 653.697$. Since both of the numbers are greater than or equal to 15, the sample size is sufficiently large to conclude the normal approximation is reasonable.

For confidence coefficient .90, $\alpha = .10$ and $\alpha/2 = .10/2 = .05$. From Table IV, Appendix A, $z_{.05} = 1.645$. The confidence interval is:

$$\hat{p} \pm z_{.05} \sqrt{\frac{pq}{n}} \approx \hat{p} \pm 1.645 \sqrt{\frac{\hat{p}\hat{q}}{n}} \Rightarrow .219 \pm 1.645 \sqrt{\frac{.219(.781)}{837}} \Rightarrow .219 \pm .024 \Rightarrow (.195, \ .243)$$

7.57   a.   First, we compute $\hat{p}$. There were 1333 useable responses. Of these, 450 chose "Bible is the actual word of God and is to be taken literally".

$$\hat{p} = \frac{x}{n} = \frac{450}{1333} = .338$$

The sample size is large enough if both $n\hat{p} \geq 15$ and $n\hat{q} \geq 15$.

$n\hat{p} = 1,333(.338) = 450.6$ and $n\hat{q} = 1,333(.662) = 882.4$. Since both of the numbers are greater than or equal to 15, the sample size is sufficiently large to conclude the normal approximation is reasonable.

For confidence coefficient .95, $\alpha = .05$ and $\alpha/2 = .05/2 = .025$. From Table IV, Appendix A, $z_{.025} = 1.96$. The 95% confidence interval is:

$$\hat{p} \pm z_{.025}\sqrt{\frac{pq}{n}} \Rightarrow \hat{p} \pm 1.96\sqrt{\frac{\hat{p}\hat{q}}{n}} \Rightarrow .338 \pm 1.96\sqrt{\frac{.338(.662)}{1,333}}$$

$$\Rightarrow .338 \pm .025 \Rightarrow (.313, \ .363)$$

b. We are 95% confident that the true proportion of all Americans who believe that the "Bible is the actual word of God and is to be taken literally" is between .313 and .363.

7.59 a. First we must compute $\hat{p}$: $\hat{p} = \dfrac{x}{n} = \dfrac{97}{185} = .524$

The sample size is large enough if both $n\hat{p} \geq 15$ and $n\hat{q} \geq 15$.

$n\hat{p} = 185(.524) = 96.9$ and $n\hat{q} = 185(.476) = 88.1$. Since both of the numbers are greater than or equal to 15, the sample size is sufficiently large to conclude the normal approximation is reasonable.

For confidence coefficient .99, $\alpha = .01$ and $\alpha/2 = .01/2 = .005$. From Table IV, Appendix A, $z_{.005} = 2.58$. The confidence interval is:

$$\hat{p} \pm z_{.005}\sqrt{\frac{pq}{n}} \approx \hat{p} \pm 2.58\sqrt{\frac{\hat{p}\hat{q}}{n}} \Rightarrow .524 \pm 2.58\sqrt{\frac{.524(.476)}{185}} \Rightarrow .524 \pm .095$$

$$\Rightarrow (.429, .619)$$

b. We are 99% confident that the true proportion of bottlenose dolphin signature whistles that are Type "a" whistles is between .429 and .619.

7.61 The point estimate for the proportion of health care workers with latex allergy who suspect that he/she has the allergy is $\hat{p} = \dfrac{x}{n} = \dfrac{36}{83} = .434$.

The sample size is large enough if both $n\hat{p} \geq 15$ and $n\hat{q} \geq 15$.

$n\hat{p} = 83(.434) = 36$ and $n\hat{q} = 83(.566) = 47$. Since both of the numbers are greater than or equal to 15, the sample size is sufficiently large to conclude the normal approximation is reasonable.

Since no confidence limit is given, we will use 95%. For confidence coefficient .95, $\alpha = .05$ and $\alpha/2 = .05/2 = .025$. From Table IV, Appendix A, $z_{.025} = 1.96$. The 95% confidence interval is:

$$\hat{p} \pm z_{.025}\sigma_{\hat{p}} \Rightarrow \hat{p} \pm 1.96\sqrt{\frac{\hat{p}\hat{q}}{n}} \Rightarrow .434 \pm 1.96\sqrt{\frac{.434(.566)}{83}} \Rightarrow .434 \pm .107 \Rightarrow (.327, \ .541)$$

We are 95% confident that the proportion of health care workers with latex allergy who suspect that he/she has the allergy is between .327 and .541.

7.63   The statement "For a specified sampling error SE, increasing the confidence level $(1 - \alpha)$ will lead to a larger n when determining sample size" is True.

7.65   To compute the necessary sample size, use

$$n = \frac{\left(z_{\alpha/2}\right)^2 \sigma^2}{(SE)^2} \text{ where } \alpha = 1 - .95 = .05 \text{ and } \alpha/2 = .05/2 = .025$$

From Table IV, Appendix A, $z_{.025} = 1.96$. Thus,

$$n = \frac{(1.96)^2(5.4)}{.2^2} = 518.616 \approx 519$$

You would need to take 519 samples.

7.67   a.   Range = 39 − 31 = 8. $\sigma \approx \dfrac{\text{Range}}{4} = \dfrac{8}{4} = 2$

For confidence coefficient .90, $\alpha = .10$ and $\alpha/2 = .05$. From Table IV, Appendix A, $z_{.05} = 1.645$.

The sample size is $n = \dfrac{z_{\alpha/2}^2 \sigma^2}{(SE)^2} = \dfrac{1.645^2(2^2)}{.15^2} = 481.07 \approx 482$

b. $\sigma \approx \dfrac{\text{Range}}{6} = \dfrac{8}{6} = 1.333$

The sample size is $n = \dfrac{z_{\alpha/2}^2 \sigma^2}{(SE)^2} = \dfrac{1.645^2(1.333^2)}{.15^2} = 213.7 \approx 214$

7.69   For confidence coefficient .90, $\alpha = .10$ and $\alpha/2 = .05$. From Table IV, Appendix A, $z_{.05} = 1.645$.

We know $\hat{p}$ is in the middle of the interval, so $\hat{p} = \dfrac{.54 + .26}{2} = .4$

The confidence interval is $\hat{p} \pm z_{.05}\sqrt{\dfrac{\hat{p}\hat{q}}{n}} \Rightarrow .4 \pm 1.645\sqrt{\dfrac{.4(.6)}{n}}$

We know $.4 - 1.645\sqrt{\dfrac{.4(.6)}{n}} = .26$

$$\Rightarrow .4 - \frac{.8059}{\sqrt{n}} = .26$$

$$\Rightarrow .4 - .26 = \frac{.8059}{\sqrt{n}} \Rightarrow \sqrt{n} = \frac{.8059}{.14} = 5.756$$

$$\Rightarrow n = 5.756^2 = 33.1 \approx 34$$

7.71   a.   The width of a confidence interval is $2(SE) = 2z_{\alpha/2}\dfrac{\sigma}{\sqrt{n}}$

For confidence coefficient .95, $\alpha = 1 - .95 = .05$ and $\alpha/2 = .05/2 = .025$. From Table IV, Appendix A, $z_{.025} = 1.96$.

For $n = 16$,
$$W = 2z_{\alpha/2}\frac{\sigma}{\sqrt{n}} = 2(1.96)\frac{1}{\sqrt{16}} = 0.98$$

For $n = 25$,
$$W = 2z_{\alpha/2}\frac{\sigma}{\sqrt{n}} = 2(1.96)\frac{1}{\sqrt{25}} = 0.784$$

For $n = 49$,
$$W = 2z_{\alpha/2}\frac{\sigma}{\sqrt{n}} = 2(1.96)\frac{1}{\sqrt{49}} = 0.56$$

For $n = 100$,
$$W = 2z_{\alpha/2}\frac{\sigma}{\sqrt{n}} = 2(1.96)\frac{1}{\sqrt{100}} = 0.392$$

For $n = 400$,
$$W = 2z_{\alpha/2}\frac{\sigma}{\sqrt{n}} = 2(1.96)\frac{1}{\sqrt{400}} = 0.196$$

b.

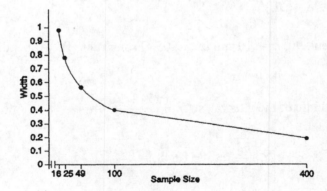

7.73   a.   The sample size may not be large enough. The sample size is large enough if both $n\hat{p} \geq 15$ and $n\hat{q} \geq 15$.

$$\hat{p} = \frac{x}{n} = \frac{52}{60} = .867$$

$n\hat{p} = 60(.867) = 52$ and $n\hat{q} = 60(.133) = 8$. Since the second number is not greater than or equal to 15, the sample size is not sufficiently large to conclude the normal approximation is reasonable.

b. For confidence coefficient .90, $\alpha = .10$ and $\alpha/2 = .10/2 = .05$. From Table IV, Appendix A, $z_{.05} = 1.645$. We will use the sample proportion to estimate the population proportion:

$$\hat{p} = \frac{x}{n} = \frac{52}{60} = .867$$

$$n = \frac{z_{\alpha/2}^2 pq}{(SE)^2} = \frac{1.645^2(.867)(.133)}{.05^2} = 124.8 \approx 125$$

7.75 For confidence coefficient .90, $\alpha = .10$ and $\alpha/2 = .10/2 = .05$. From Table IV, Appendix A, $z_{.05} = 1.645$.

$$n = \frac{z_{\alpha/2}^2 \sigma^2}{(SE)^2} = \frac{1.645^2 10.9^2}{4^2} = 20.09 \approx 21$$

7.77 For confidence coefficient .99, $\alpha = .01$ and $\alpha/2 = .01/2 = .005$. From Table IV, Appendix A, $z_{.005} = 2.58$. From the previous estimate, we will use $\hat{p} = .333$ to estimate p.

$$n = \frac{z_{\alpha/2}^2 pq}{(SE)^2} = \frac{2.58^2(.333)(.667)}{.01^2} = 14,784.5966 \approx 14,785$$

7.79 From Exercise 7.37, $s = .2316$. For confidence coefficient .95, $\alpha = .05$ and $\alpha/2 = .05/2 = .025$. From Table IV, Appendix A, $z_{.025} = 1.96$.

$$n = \frac{z_{\alpha/2}^2 \alpha^2}{(SE)^2} = \frac{1.96^2(.2316)^2}{.04^2} = 128.786 \approx 129$$

7.81 For confidence coefficient .90, $\alpha = .10$ and $\alpha/2 = .05$. From Table IV, Appendix A, $z_{.05} = 1.645$.

The sample size is $n = \dfrac{(z_{\alpha/2})^2 \sigma^2}{(SE)^2} = \dfrac{(1.645)^2(10^2)}{1^2} = 270.6 \approx 271$

7.83 a. The target parameter for the average score on the SAT is μ.

b. The target parameter for the mean time waiting at a supermarket checkout lane is μ.

c. The target parameter for the proportion of voters in favor of legalizing marijuana is p.

b. The target parameter for the percentage of NFL players who have ever made the Pro Bowl is p.

e. The target parameter for the dropout rate of American college students is p.

7.85   a.   For a small sample from a normal distribution with unknown standard deviation, we use the t statistic. For confidence coefficient .95, $\alpha = 1 - .95 = .05$ and $\alpha/2 = .05/2 = .025$. From Table VI, Appendix A, with df $= n - 1 = 21 - 1 = 20$, $t_{.025} = 2.086$.

b.   For a large sample from a distribution with an unknown standard deviation, we can estimate the population standard deviation with s and use the z statistic. For confidence coefficient .95, $\alpha = 1 - .95 = .05$ and $\alpha/2 = .05/2 = .025$. From Table IV, Appendix A, $z_{.025} = 1.96$.

c.   For a small sample from a normal distribution with known standard deviation, we use the z statistic. For confidence coefficient .95, $\alpha = 1 - .95 = .05$ and $\alpha/2 = .05/2 = .025$. From Table IV, Appendix A, $z_{.025} = 1.96$.

d.   For a large sample from a distribution about which nothing is known, we can estimate the population standard deviation with s and use the z statistic. For confidence coefficient .95, $\alpha = 1 - .95 = .05$ and $\alpha/2 = .05/2 = .025$. From Table IV, Appendix A, $z_{.025} = 1.96$.

e.   For a small sample from a distribution about which nothing is known, we can use neither z nor t.

7.87   a.   The point estimate for the proportion of measurements in the population with characteristic A is $\hat{p} = \dfrac{x}{n} = \dfrac{227}{400} = .5675$.

The sample size is large enough if both $n\hat{p} \geq 15$ and $n\hat{q} \geq 15$.

$n\hat{p} = 400(.5675) = 227$ and $n\hat{q} = 400(.4325) = 173$. Since both of the numbers are greater than or equal to 15, the sample size is sufficiently large to conclude the normal approximation is reasonable.

For confidence coefficient .95, $\alpha = .05$ and $\alpha/2 = .05/2 = .025$. From Table IV, Appendix A, $z_{.025} = 1.96$. The 95% confidence interval is:

$$\hat{p} \pm z_{.025}\sigma_{\hat{p}} \Rightarrow \hat{p} \pm 1.96\sqrt{\frac{\hat{p}\hat{q}}{n}} \Rightarrow .5675 \pm 1.96\sqrt{\frac{.5675(.4325)}{400}}$$
$$\Rightarrow .5675 \pm .0486 \Rightarrow (.5189, \ .6161)$$

We are 95% confident that the proportion of measurements in the population with characteristic A is between .5189 and .6161.

b.   We will use $\hat{p} = .5675$ to estimate p.

For confidence coefficient .95, $\alpha = .05$ and $\alpha/2 = .05/2 = .025$. From Table IV, Appendix A, $z_{.025} = 1.96$.

The sample size is $n = \dfrac{(z_{.025})^2 pq}{(SE)^2} = \dfrac{1.96^2(.5675)(.4325)}{.02^2} = 2,357.2 \approx 2,358$.

7.89   The parameters of interest for the problems are:

    (1)   The question requires a categorical response. One parameter of interest might be the proportion, p, of all Americans over 18 years of age who think their health is generally very good or excellent.

    (2)   A parameter of interest might be the mean number of days, μ, in the previous 30 days that all Americans over 18 years of age felt that their physical health was not good because of injury or illness.

    (3)   A parameter of interest might be the mean number of days, μ, in the previous 30 days that all Americans over 18 years of age felt that their mental health was not good because of stress, depression, or problems with emotions.

    (4)   A parameter of interest might be the mean number of days, μ, in the previous 30 days that all Americans over 18 years of age felt that their physical or mental health prevented them from performing their usual activities.

7.91   a.   The point estimate of p, the true driver phone cell use rate, is $\hat{p} = \dfrac{x}{n} = \dfrac{35}{1,165} = .030$.

    b.   The sample size is large enough if both $n\hat{p} \geq 15$ and $n\hat{q} \geq 15$.

    $n\hat{p} = 1,165(.030) = 35$ and $n\hat{q} = 1,165(.970) = 1130$. Since both of the numbers are greater than or equal to 15, the sample size is sufficiently large to conclude the normal approximation is reasonable.

    For confidence coefficient .95, $\alpha = .05$ and $\alpha/2 = .05/2 = .025$. From Table IV, Appendix A, $z_{.025} = 1.96$. The 95% confidence interval is:

$$\hat{p} \pm z_{.025}\sigma_{\hat{p}} \Rightarrow \hat{p} \pm 1.96\sqrt{\frac{\hat{p}\hat{q}}{n}} \Rightarrow .030 \pm 1.96\sqrt{\frac{.030(.970)}{1,165}} \Rightarrow .030 \pm .010 \Rightarrow (.020,\ .040)$$

    c.   We are 95% confident that the true driver phone cell use rate is between .020 and .040.

7.93   a.   The confidence level desired by the researchers is .90.

    b.   The sampling error desired by the researchers is $SE = .05$.

    c.   Since no value was given for p, we will use $\hat{p} = \dfrac{x}{n} = \dfrac{64}{106} = .604$.

    For confidence coefficient .90, $\alpha = .10$ and $\alpha/2 = .10/2 = .05$. From Table IV, Appendix A, $z_{.05} = 1.645$.

    The sample size is $n = \dfrac{(z_{.05})^2\, pq}{(SE)^2} = \dfrac{1.645^2(.604)(.396)}{.05^2} = 258.9 \approx 259$.

7.95 Using MINITAB, the descriptive statistics are:

Descriptive Statistics: SeedWt

Variable	N	N*	Mean	SE Mean	StDev	Minimum	Q1	Median	Q3	Maximum
SeedWt	16	0	1.373	0.258	1.034	0.201	0.475	1.347	2.009	3.751

For confidence coefficient .99, $\alpha = .01$ and $\alpha/2 = .01/2 = .005$. From Table VI, Appendix A, with df $= n - 1 = 16 - 1 = 15$, $t_{.005} = 2.947$.

The 99% confidence interval is:

$$\bar{x} \pm t_{0.05} \frac{s}{\sqrt{n}} \Rightarrow 1.373 \pm 2.947 \frac{1.034}{\sqrt{16}} \Rightarrow 1.373 \pm .762 \Rightarrow (.611, \ 2.135)$$

We are 99% confident that the average weight of dry seeds in the crops of spinifex pigeons inhabiting the Western Australian desert is between .611 and 2.135 grams.

7.97 The 95% confidence interval is (.045, .091). We are 95% confident that the true proportion of people who were/are homeless is between .045 and .091.

7.99 a. First, we compute $\hat{p}$: $\hat{p} = \frac{x}{n} = \frac{665}{1007} = .660$

The sample size is large enough if both $n\hat{p} \geq 15$ and $n\hat{q} \geq 15$.

$n\hat{p} = 1,007(.660) = 665$ and $n\hat{q} = 1,007(.340) = 342$. Since both of the numbers are greater than or equal to 15, the sample size is sufficiently large to conclude the normal approximation is reasonable.

For confidence coefficient .95, $\alpha = .05$ and $\alpha/2 = .05/2 = .025$. From Table IV, Appendix A, $z_{.025} = 1.96$. The confidence interval is:

$$\hat{p} \pm z_{.025} \sqrt{\frac{pq}{n}} \approx \hat{p} \pm 1.96 \sqrt{\frac{\hat{p}\hat{q}}{n}} \Rightarrow .660 \pm 1.96 \sqrt{\frac{.660(.340)}{1007}} \Rightarrow .660 \pm .029 \Rightarrow (.631, \ .689)$$

We are 95% confident that the proportion of U.S. workers who take their lunch to work is between .631 and .689.

b. First, we compute $\hat{p}$: $\hat{p} = \frac{x}{n} = \frac{200}{665} = .301$

The sample size is large enough if both $n\hat{p} \geq 15$ and $n\hat{q} \geq 15$.

$n\hat{p} = 665(.301) = 200$ and $n\hat{q} = 665(.699) = 465$. Since both of the numbers are greater than or equal to 15, the sample size is sufficiently large to conclude the normal approximation is reasonable.

For confidence coefficient .95, $\alpha = .05$ and $\alpha/2 = .05/2 = .025$. From Table IV, Appendix A, $z_{.025} = 1.96$. The confidence interval is:

$$\hat{p} \pm z_{.025}\sqrt{\frac{pq}{n}} \approx \hat{p} \pm 1.96\sqrt{\frac{\hat{p}\hat{q}}{n}} \Rightarrow .301 \pm 1.96\sqrt{\frac{.301(.699)}{665}} \Rightarrow .301 \pm .035 \Rightarrow (.266, .336)$$

We are 95% confident that the proportion of U.S. workers who take their lunch to work who brown-bag their lunch is between .266 and .336.

7.101   a.   For confidence coefficient .99, $\alpha = .01$ and $\alpha/2 = .01/2 = .005$. From Table IV, Appendix A, $z_{.005} = 2.58$. The confidence interval is:

$$\bar{x} \pm z_{.005}\frac{s}{\sqrt{n}} \Rightarrow .044 \pm 2.58\frac{.884}{\sqrt{197}} \Rightarrow .044 \pm .162 \Rightarrow (-.118, .206)$$

We are 99% confident that the mean inbreeding coefficient for this species of wasps is between $-.118$ and .206.

b.   A coefficient of 0 indicates that the wasp has no tendency to inbreed. Since 0 is in the 99% confidence interval, it is a likely value for the true mean inbreeding coefficient. Thus, there is no evidence to indicate that this species of wasps has a tendency to inbreed.

7.103   a.   Using MINITAB, the descriptive statistics are:

Descriptive Statistics: CPU Time

Variable	N	N*	Mean	SE Mean	StDev	Minimum	Q1	Median	Q3	Maximum
CPU Time	52	0	0.812	0.209	1.505	0.0360	0.136	0.275	0.595	8.788

For confidence coefficient .95, $\alpha = .05$ and $\alpha/2 = .05/2 = .025$. From Table IV, Appendix A, $z_{.025} = 1.96$. The 95% confidence interval is:

$$\bar{x} \pm z_{.025}\sigma_{\bar{x}} \Rightarrow \bar{x} \pm 1.96\frac{\sigma}{\sqrt{n}} \Rightarrow .812 \pm 1.96\frac{1.505}{\sqrt{52}} \Rightarrow .812 \pm .409 \Rightarrow (.403, \ 1.221)$$

We are 95% confident that the mean solution time for the hybrid algorithm is between .403 and 1.221.

b.   The sample size needed would be $n = \dfrac{(z_{.025})^2\sigma^2}{(SE)^2} = \dfrac{1.96^2(1.505)^2}{.25^2} = 139.2 \approx 140$.

7.105   For all parts to this problem, we will use a 95% confidence interval. For confidence coefficient, .95, $\alpha = .05$ and $\alpha/2 = .05/2 = .025$. From Table IV, Appendix A, $z_{.025} = 1.96$.

a.   Some preliminary calculations:

$\hat{p} = x/n = 14/37 = .378$

The sample size is large enough if both $n\hat{p} \geq 15$ and $n\hat{q} \geq 15$.

$n\hat{p} = 37(.378) = 14$ and $n\hat{q} = 37(.622) = 23$. Since the first number is not greater than or equal to 15, the sample size is not sufficiently large to conclude the normal approximation is reasonable.

Since the normal approximation is not reasonable, we will use the Wilson adjusted confidence interval.

$$\tilde{p} = \frac{x+2}{n+4} = \frac{14+2}{37+4} = \frac{16}{41} = .390$$

The Wilson adjusted 95% confidence interval is:

$$\tilde{p} \pm z_{.025}\sqrt{\frac{\tilde{p}(1-\tilde{p})}{n+4}} \Rightarrow .390 \pm 1.96\sqrt{\frac{.390(.610)}{37+4}} \Rightarrow .390 \pm .149 \Rightarrow (.241, .539)$$

We are 95% confident that the true proportion of suicides at the jail that are committed by inmates charged with murder/manslaughter is between .241 and .539.

b.   Some preliminary calculations:

$\hat{p} = x/n = 26/37 = .703$

The sample size is large enough if both $n\hat{p} \geq 15$ and $n\hat{q} \geq 15$.

$n\hat{p} = 37(.703) = 26$ and $n\hat{q} = 37(.297) = 11$. Since the second number is not greater than or equal to 15, the sample size is not sufficiently large to conclude the normal approximation is reasonable.

Since the normal approximation is not reasonable, we will use the Wilson adjusted confidence interval.

$$\tilde{p} = \frac{x+2}{n+4} = \frac{26+2}{37+4} = \frac{28}{41} = .683$$

The Wilson adjusted 95% confidence interval is:

$$\tilde{p} \pm z_{.025}\sqrt{\frac{\tilde{p}(1-\tilde{p})}{n+4}} \Rightarrow .683 \pm 1.96\sqrt{\frac{.683(.317)}{37+4}} \Rightarrow .683 \pm .142 \Rightarrow (.541, .825)$$

We are 95% confident that the true proportion of suicides at the jail that are committed at night is between .541 and .825.

c. Some preliminary calculations are:

$$\bar{x} = \frac{\sum x}{n} = \frac{1532}{37} = 41.405$$

$$s^2 = \frac{\sum x^2 - \frac{\left(\sum x\right)^2}{n}}{n-1} = \frac{223,606 - \frac{(1532)^2}{37}}{37-1} = 4,449.2477$$

$$s = \sqrt{s^2} = \sqrt{4,449.2477} = 66.703$$

The confidence interval is:

$$\bar{x} \pm z_{.05}\frac{s}{\sqrt{n}} \Rightarrow 41.405 \pm 1.96\frac{66.703}{\sqrt{37}} \Rightarrow 41.405 \pm 21.493 \Rightarrow (19.912, 62.898)$$

We are 95% confident that the true average length of time an inmate is in jail before committing suicide is between 19.912 and 62.898 days.

d. Some preliminary calculations:

$$\hat{p} = x/n = 14/37 = .378$$

The sample size is large enough if both $n\hat{p} \geq 15$ and $n\hat{q} \geq 15$.

$n\hat{p} = 37(.378) = 14$ and $n\hat{q} = 37(.622) = 23$. Since the first number is not greater than or equal to 15, the sample size is not sufficiently large to conclude the normal approximation is reasonable.

Since the normal approximation is not reasonable, we will use the Wilson adjusted confidence interval.

$$\tilde{p} = \frac{x+2}{n+4} = \frac{14+2}{37+4} = \frac{16}{41} = .390$$

The Wilson adjusted 95% confidence interval is:

$$\tilde{p} \pm z_{.025}\sqrt{\frac{\tilde{p}(1-\tilde{p})}{n+4}} \Rightarrow .390 \pm 1.96\sqrt{\frac{.390(.610)}{37+4}} \Rightarrow .390 \pm .149 \Rightarrow (.241, .539)$$

We are 95% confident that the true percentage of suicides at the jail that are committed by white inmates is between 24.1% and 53.9%.

7.107   a.   The point estimate for the fraction of the entire market who refuse to purchase bars is:

$$\hat{p} = \frac{x}{n} = \frac{23}{244} = .094$$

   b.   The sample size is large enough if both $n\hat{p} \geq 15$ and $n\hat{q} \geq 15$.

   $n\hat{p} = 244(.094) = 23$ and $n\hat{q} = 244(.906) = 221$. Since both of the numbers are greater than or equal to 15, the sample size is sufficiently large to conclude the normal approximation is reasonable.

   c.   For confidence coefficient .95, $\alpha = 1 - .95 = .05$ and $\alpha/2 = .05/2 = .025$. From Table IV, Appendix A, $z_{.025} = 1.96$. The confidence interval is:

$$\hat{p} \pm z_{.025}\sqrt{\frac{\hat{p}\hat{q}}{n}} \Rightarrow .094 \pm 1.96\sqrt{\frac{.094(.906)}{244}} \Rightarrow .094 \pm .037 \Rightarrow (.057, .131)$$

   d.   The best estimate of the true fraction of the entire market who refuse to purchase bars six months after the poisoning is .094. We are 95% confident the true fraction of the entire market who refuse to purchase bars six months after the poisoning is between .057 and .131.

7.109   a.   For confidence coefficient .99, $\alpha = .01$ and $\alpha/2 = .01/2 = .005$. From Table VI, Appendix A, with df $= n - 1 = 3 - 1 = 2$, $t_{.005} = 9.925$. The confidence interval is:

$$\bar{x} \pm t_{.005}\frac{s}{\sqrt{n}} \Rightarrow 49.3 \pm 9.925\frac{1.5}{\sqrt{3}} \Rightarrow 49.3 \pm 8.60 \Rightarrow (40.70, 57.90)$$

   b.   We are 99% confident that the mean percentage of B(a)p removed from all soil specimens using the poison is between 40.70% and 57.90%.

   c.   We must assume that the distribution of the percentages of B(a)p removed from all soil specimens using the poison is normal.

   d.   For confidence coefficient .99, $\alpha = .01$ and $\alpha/2 = .01/2 = .005$. From Table IV, Appendix A, $z_{.005} = 2.58$.

$$n = \frac{z_{\alpha/2}^2\sigma^2}{(SE)^2} = \frac{1.96^2(1.5)^2}{.5^2} = 34.57 \approx 35$$

7.111   a.   From Chebyshev's Rule, we know that at least $1 - \dfrac{1}{k^2}$ of the observations will fall within k standard deviations of the mean. We need to find k so that $1 - \dfrac{1}{k^2} = .60$.

$$1 - \frac{1}{k^2} = .60 \Rightarrow .40 = \frac{1}{k^2} \Rightarrow k^2 = \frac{1}{.40} = 2.5 \Rightarrow k = \sqrt{2.5} = 1.58$$

$$s = \frac{80^{th} \text{ percentile} - 60^{th} \text{ percentile}}{k} = \frac{73,000 - 35,100}{1.58} = 23,987.34$$

For confidence coefficient .98, $\alpha = .02$ and $\alpha/2 = .02/2 = .01$. From Table IV, Appendix A, $z_{.01} = 2.33$. The sample size required is:

$$n = \frac{(z_{.01})^2 \sigma^2}{(SE)^2} = \frac{2.33^2 (23,987.34)^2}{2,000^2} = 780.9 \approx 781$$

b.   See part **a**.

c.   We must assume that any sample selected will be random.

7.113   a.   As long as the sample is random (and thus representative), a reliable estimate of the mean weight of all the scallops can be obtained.

b.   The government is using only the sample mean to make a decision. Rather than using a point estimate, they should probably use a confidence interval to estimate the true mean weight of the scallops.

c.   We will form a 95% confidence interval for the mean weight of the scallops. Using MINITAB, the descriptive statistics are:

Descriptive Statistics: Weight

Variable	N	N*	Mean	SE Mean	StDev	Minimum	Q1	Median	Q3
Weight	18	0	0.9317	0.0178	0.0753	0.8400	0.8800	0.9100	9800

Variable	Maximum
Weight	1.1400

For confidence coefficient .95, $\alpha = .05$ and $\alpha/2 = .05/2 = .025$. From Table VI, Appendix A, with df $= n - 1 = 18 - 1 = 17$, $t_{.025} = 2.110$.

The 95% confidence interval is:

$$\bar{x} \pm t_{.025} \frac{s}{\sqrt{n}} \Rightarrow .9317 \pm 2.110 \frac{.0753}{\sqrt{18}} \Rightarrow .9317 \pm .0374 \Rightarrow (.8943, \ .9691)$$

We are 95% confident that the true mean weight of the scallops is between .8943 and .9691. Recall that the weights have been scaled so that a mean weight of 1 corresponds to 1/36 of a pound. Since the above confidence interval does not include 1, we have sufficient evidence to indicate that the minimum weight restriction was violated.

Inferences Based on a Single Sample: Tests of Hypotheses

Chapter 8

8.1 The null hypothesis is the "status quo" hypothesis, while the alternative hypothesis is the research hypothesis.

8.3 The "level of significance" of a test is α. This is the probability that the test statistic will fall in the rejection region when the null hypothesis is true.

8.5 The four possible results are:

1. Rejecting the null hypothesis when it is true. This would be a Type I error.
2. Accepting the null hypothesis when it is true. This would be a correct decision.
3. Rejecting the null hypothesis when it is false. This would be a correct decision.
4. Accepting the null hypothesis when it is false. This would be a Type II error.

8.7 When you reject the null hypothesis in favor of the alternative hypothesis, this does not prove the alternative hypothesis is correct. We are $100(1 - \alpha)\%$ confident that there is sufficient evidence to conclude that the alternative hypothesis is correct.

If we were to repeatedly draw samples from the population and perform the test each time, approximately $100(1 - \alpha)\%$ of the tests performed would yield the correct decision.

8.9 Let p = proportion of liars correctly detected by the new thermal imaging camera. To determine if the camera can correctly detect liars 75% of the time, we test:

H_0: $p = .75$
H_a: $p \neq .75$

8.11 a. To determine if the percentage of senior women who use herbal therapies to prevent or treat health problems is different from 45%, we test:

H_0: $p = .45$
H_a: $p \neq .45$

b. To determine if the average number of herbal products used in a year by senior women who use herbal therapies is different from 2.5, we test:

H_0: $\mu = 2.5$
H_a: $\mu \neq 2.5$

8.13 Let p = error rate of the DNA-reading device. To determine if the error rate of the DNA-reading device is less than 5%, we test:

H_0: $p = .05$
H_a: $p < .05$

8.15 a. To determine if the average level of mercury uptake in wading birds in the Everglades today is less than 15 parts per million, we test:

$H_0: \mu = 15$
$H_a: \mu < 15$

b. A Type I error is rejecting H_0 when H_0 is true. In terms of this problem, we would be concluding that the average level of mercury uptake in wading birds in the Everglades today is less than 15 parts per million, when in fact, the average level of mercury uptake in wading birds in the Everglades today is equal to 15 parts per million.

c. A Type II error is accepting H_0 when H_0 is false. In terms of this problem, we would be concluding that the average level of mercury uptake in wading birds in the Everglades today is equal to 15 parts per million, when in fact, the average level of mercury uptake in wading birds in the Everglades today is less than 15 parts per million.

8.17 a. The null hypothesis is:
H_0: No intrusion occurs

b. The alternative hypothesis is:
H_a: Intrusion occurs

c. α = Probability of a Type I error = Probability of rejecting H_o when it is true = P(system provides warning when no intrusion occurs) = 1 / 1000 = .001.

β = Probability of a Type II error = Probability of accepting H_o when it is false = P(system does not provide warning when an intrusion occurs) = 500 / 1000 = .5.

8.19 There are 2 conditions required for a valid large-sample hypothesis test for μ. They are:

1. A random sample is selected from a target population.
2. The sample size n is large, i.e., $n \geq 30$. (Due to the Central Limit Theorem, this condition guarantees that the test statistic will be approximately normal regardless of the shape of the underlying probability distribution of the population.)

8.21 a. Probability of Type I error
$= P(z > 1.96) = .5 - .4750 = .0250$
(From Table IV, Appendix A)

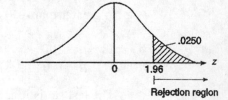

b. Probability of Type I error
$= P(z > 1.645) = .5 - .4500 = .05$
(From Table IV, Appendix A)

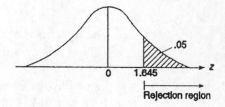

c. Probability of Type I error
$= P(z > 2.58) = .5 - .4951 = .0049$
(From Table IV, Appendix A)

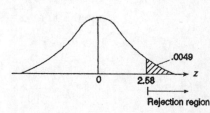

Inferences Based on a Single Sample: Tests of Hypotheses

d. Probability of Type I error
$= P(z < -1.28) = .5 - .3997 = .1003$
(From Table IV, Appendix A)

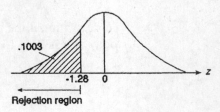

e. Probability of Type I error
$= P(z < -1.645) + P(z > 1.645)$
$= .5 - .4500 + .5 - .4500 = .05 + .05 = .10$
(From Table IV, Appendix A)

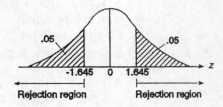

f. Probability of Type I error
$= P(z < -2.58) + P(z > 2.58)$
$= .5 - .4951 + .5 - .4951 = .0049 + .0049$
$= .0098$ (From Table IV, Appendix A)

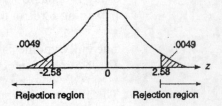

8.23 a. H_0: $\mu = 100$
H_a: $\mu > 100$

The test statistic is $z = \dfrac{\bar{x} - \mu_0}{\sigma_{\bar{x}}} = \dfrac{\bar{x} - \mu_0}{\sigma / \sqrt{n}} = \dfrac{110 - 100}{60 / \sqrt{100}} = 1.67$

The rejection region requires $\alpha = .05$ in the upper tail of the z distribution. From Table IV, Appendix A, $z_{.05} = 1.645$. The rejection region is $z > 1.645$.

Since the observed value of the test statistic falls in the rejection region, ($z = 1.67 > 1.645$), H_0 is rejected. There is sufficient evidence to indicate the true population mean is greater than 100 at $\alpha = .05$.

b. H_0: $\mu = 100$
H_a: $\mu \neq 100$

The test statistic is $z = \dfrac{\bar{x} - \mu_0}{\sigma_{\bar{x}}} = \dfrac{110 - 100}{60 / \sqrt{100}} = 1.67$

The rejection region requires $\alpha/2 = .05/2 = .025$ in each tail of the z distribution. From Table IV, Appendix A, $z_{.025} = 1.96$. The rejection region is $z < -1.96$ or $z > 1.96$.

Since the observed value of the test statistic does not fall in the rejection region, ($z = 1.67 \not> 1.96$), H_0 is not rejected. There is insufficient evidence to indicate μ does not equal 100 at $\alpha = .05$.

c. In part **a**, we rejected H_0 and concluded the mean was greater than 100. In part **b**, we did not reject H_0. There was insufficient evidence to conclude the mean was different from 100. Because the alternative hypothesis in part **a** is more specific than the one in **b**, it is easier to reject H_0.

8.25 a. A Type I error would be to conclude that the mean response for the population of all New York City public school children is not 3 when, in fact, the mean response is equal to 3.

A Type II error would be to conclude that the mean response for the population of all New York City public school children is equal to 3 when, in fact, the mean response is not equal to 3.

b. To determine if the mean response for the population of all New York City public school children is different from 3, we test:

H_0: $\mu = 3$
H_a: $\mu \neq 3$

The test statistic is $z = \dfrac{\bar{x} - \mu_o}{\sigma_{\bar{x}}} = \dfrac{2.15 - 3}{\dfrac{1.05}{\sqrt{11,160}}} = -85.52$

The rejection region requires $\alpha/2 = .05/2 = .025$ in each tail of the z distribution. From Table IV, Appendix A, $z_{.025} = 1.96$. The rejection region is $z < -1.96$ or $z > 1.96$.

Since the observed value of the test statistic falls in the rejection region ($z = -85.52 < -1.96$), H_0 is rejected. There is sufficient evidence to indicate the mean response for the population of all New York City public school children different from 3 at $\alpha = .05$.

c. To determine if the mean response for the population of all New York City public school children is different from 3, we test:

H_0: $\mu = 3$
H_a: $\mu \neq 3$

The test statistic is $z = \dfrac{\bar{x} - \mu_o}{\sigma_{\bar{x}}} = \dfrac{2.15 - 3}{\dfrac{1.05}{\sqrt{11,160}}} = -85.52$

The rejection region requires $\alpha/2 = .10/2 = .05$ in each tail of the z distribution. From Table IV, Appendix A, $z_{.05} = 1.645$. The rejection region is $z < -1.645$ or $z > 1.645$.

Since the observed value of the test statistic falls in the rejection region ($z = -85.52 < -1.645$), H_0 is rejected. There is sufficient evidence to indicate the mean response for the population of all New York City public school children different from 3 at $\alpha = .10$.

8.27   To determine if the true mean heart rate during laughter exceeds 71 beats/minute, we test:

H_0: $\mu = 71$
H_a: $\mu > 71$

The test statistic is $z = \dfrac{\bar{x} - \mu_o}{\sigma_{\bar{x}}} = \dfrac{73.5 - 71}{\dfrac{6}{\sqrt{90}}} = 3.95$

The rejection region requires $\alpha = .05$ in the upper tail of the z distribution. From Table IV, Appendix A, $z_{.05} = 1.645$. The rejection region is $z > 1.645$.

Since the observed value of the test statistic falls in the rejection region ($z = 3.95 > 1.645$). H_0 is rejected. There is sufficient evidence to indicate the true mean heart rate during laughter exceeds 71 beats/minute at $\alpha = .05$.

8.29   a.   To determine if the true mean PTSD score of all World War II avaiator POWs is less than 16, we test:

H_0: $\mu = 16$
H_a: $\mu < 16$

b.   The test statistic is $z = z = \dfrac{\bar{x} - \mu_o}{\sigma_{\bar{x}}} = \dfrac{9.00 - 16}{9.32 / \sqrt{33}} = -4.31$

The rejection region requires $\alpha = .10$ in the lower tail of the z distribution. From Table IV, Appendix A, $z_{.10} = 1.28$. The rejection region is $z < -1.28$.

Since the observed value of the test statistic falls in the rejection region ($z = -4.31 < -1.28$), H_0 is rejected. There is sufficient evidence to indicate that the true mean PTSD score of all World War II aviator POWs is less than 16 at $\alpha = .10$.

c.   Only 33 of the total 239 World War II aviator POWs volunteered to participate in the study. These POWs were self-selected, not random. Usually those who self-select themselves tend to have strong feelings about the subject. They may not be representative of the entire group.

8.31   a.   Using MINITAB, the descriptive statistics are:

Descriptive Statistics: Bones

Variable	N	N*	Mean	SE Mean	StDev	Minimum	Q1	Median	Q3	Maximum
Bones	41	.0	9.258	0.188	1.204	6.230	8.615	9.200	9.935	12.000

To determine if the population mean ratio of all bones of this particular species differs from 8.5, we test:

H_0: $\mu = 8.5$
H_a: $\mu \neq 8.5$

The test statistic is $z = \dfrac{\bar{x} - \mu_0}{\sigma_{\bar{x}}} = \dfrac{9.258 - 8.5}{1.204 / \sqrt{41}} = 4.03$

The rejection region requires $\alpha/2 = .01/2 = .005$ in each tail of the z distribution. From Table IV, Appendix A, $z_{.005} = 2.58$. The rejection region is $z < -2.58$ or $z > 2.58$.

Since the observed value of the test statistic falls in the rejection region ($z = 4.03 > 2.58$), H_0 is rejected. There is sufficient evidence to indicate that the true population mean ratio of all bones of this particular species differs from 8.5 at $\alpha = .01$.

b. The practical implications of the test in part **a** is that the species from which these particular bones came from is probably not species A.

8.33   a. Since the hypothesized value of μ_M ($60,000) falls in the 95% confidence interval, it is not an unusual value. There is no evidence to reject it. There is no evidence to indicate that the mean salary of all males with post-graduate degrees differs from $60,000.

b. To determine if the mean salary of all males with post-graduate degrees differs from $60,000, we test:

H_0: $\mu_M = 60,000$
H_a: $\mu_M \neq 60,000$

The test statistic is $z = \dfrac{\bar{x} - \mu_0}{\sigma_{\bar{x}}} = \dfrac{61,340 - 60,000}{2,185} = 0.61$

The rejection region requires $\alpha/2 = .05/2 = .025$ in each tail of the z distribution. From Table IV, Appendix A, $z_{.025} = 1.96$. The rejection region is $z < -1.96$ or $z > 1.96$.

Since the observed value of the test statistic does not fall in the rejection region ($z = 0.61 \not> 1.96$), H_0 is not rejected. There is insufficient evidence to indicate the mean salary of all males with post-graduate degrees differs from $60,000 at $\alpha = .05$.

c. The inferences in parts **a** and **b** agree because the same values are used for both. The z value used in the 95% confidence interval is the same z used for the rejection region. The value of $\bar{x}$ and $s_{\bar{x}}$ are the same in both the confidence interval and the test statistic.

d. Since the hypothesized value of μ_F ($33,000) falls in the 95% confidence interval, it is not an unusual value. There is no evidence to reject it. There is no evidence to indicate that the mean salary of all females with post-graduate degrees differs from $33,000.

e. To determine if the mean salary of all females with post-graduate degrees differs from $33,000, we test:
H_0: $\mu_M = 33,000$
H_a: $\mu_M \neq 33,000$

The test statistic is $z = \dfrac{\bar{x} - \mu_0}{\sigma_{\bar{x}}} = \dfrac{32,227 - 33,000}{932} = -0.83$

The rejection region requires $\alpha/2 = .05/2 = .025$ in each tail of the z distribution. From Table IV, Appendix A, $z_{.025} = 1.96$. The rejection region is $z < -1.96$ or $z > 1.96$.

Since the observed value of the test statistic does not fall in the rejection region ($z = -0.83 \not< -1.96$), H_0 is not rejected. There is insufficient evidence to indicate the mean salary of all females with post-graduate degrees differs from \$33,000 at $\alpha = .05$.

f. The inferences in parts **d** and **e** agree because the same values are used for both. The z value used in the 95% confidence interval is the same z used for the rejection region. The value of $\bar{x}$ and $s_{\bar{x}}$ are the same in both the confidence interval and the test statistic.

8.35 a. To determine if the mean social interaction score of all Connecticut mental health patients differs from 3, we test:

H_0: $\mu = 3$
H_a: $\mu \neq 3$

The test statistic is $z = \dfrac{\bar{x} - \mu_0}{\sigma_{\bar{x}}} = \dfrac{2.95 - 3}{1.10/\sqrt{6,681}} = -3.72$

The rejection region requires $\alpha/2 = .01/2 = .005$ in each tail of the z distribution. From Table IV, Appendix A, $z_{.005} = 2.58$. The rejection region is $z < -2.58$ or $z > 2.58$.

Since the observed value of the test statistic falls in the rejection region ($z = -3.72 < -2.58$), H_0 is rejected. There is sufficient evidence to indicate that the mean social interaction score of all Connecticut mental health patients differs from 3 at $\alpha = .01$.

b. From the test in part a, we found that the mean social interaction score was statistically different from 3. However, the sample mean score was 2.95. Practically speaking, 2.95 is very similar to 3.0. The very large sample size, $n = 6,681$, makes it very easy to find statistical significance, even when no practical significance exists.

c. Because the variable of interest is measured on a 5-point scale, it is very unlikely that the population of the ratings will be normal. However, because the sample size was extremely large, ($n = 6,681$), the Central Limit Theorem will apply. Thus, the distribution of $\bar{x}$ will be normal, regardless of the distribution of x. Thus, the analysis used above is appropriate.

8.37 In general, small p-values support the alternative hypothesis, H_a. The p-value is the probability of observing your test statistic or anything more unusual, given H_0 is true. If the p-value is small, it would be very unusual to observe your test statistic, given H_0 is true. This would indicate that H_0 is probably false and H_a is probably true.

8.39   a.   Since the p-value = .10 is greater than α = .05, H_0 is not rejected.

       b.   Since the p-value = .05 is less than α = .10, H_0 is rejected.

       c.   Since the p-value = .001 is less than α = .01, H_0 is rejected.

       d.   Since the p-value = .05 is greater than α = .025, H_0 is not rejected.

       e.   Since the p-value = .45 is greater than α = .10, H_0 is not rejected.

8.41 p-value = $P(z \geq 2.17) = .5 - P(0 < z < 2.17) = .5 - .4850 = .0150$
                                                        (using Table IV, Appendix A)

8.43 $z = \dfrac{\bar{x} - \mu_0}{\sigma_{\bar{x}}} = \dfrac{49.4 - 50}{4.1/\sqrt{100}} = -1.46$

       p-value = $P(z \geq -1.46) = .5 + .4279 = .9279$

       There is no evidence to reject H_0 for $\alpha \leq .10$.

8.45   a.   The p-value reported by SPSS is for a two-tailed test. Thus, $P(z \leq -1.63) + P(z \geq 1.63)$ = .1032. For this one-tailed test, the p-value = $P(z \leq -1.63) = .1032/2 = .0516$.

            Since the p-value = .0516 > α = .05, H_0 is not rejected. There is insufficient evidence to indicate $\mu < 75$ at α = .05.

       b.   For this one-tailed test, the p-value = $P(z \leq 1.63)$. Since $P(z \leq -1.63) = .1032/2 = .0516$, $P(z \leq 1.63) = 1 - .0516 = .9484$.

            Since the p-value = .9484 > α = .10, H_0 is not rejected. There is insufficient evidence to indicate $\mu < 75$ at α = .10.

       c.   For this one-tailed test, the p-value = $P(z \geq 1.63) = .1032/2 = .0516$.

            Since the p-value = .0516 < α = .10, H_0 is rejected. There is sufficient evidence to indicate $\mu > 75$ at α = .10.

       d.   For this two-tailed test, the p-value = .1032.

            Since the p-value = .1032 > α = .01, H_0 is not rejected. There is insufficient evidence to indicate $\mu \neq 75$ at α = .01.

8.47   a.   From Exercise 8.26, $z = -.40$. The p-value is $p = P(z \leq -.40) = .5 - .1554 = .3446$ (using Table IV, Appendix A).

       b.   The p-value is $p = .3446$. Since this is greater than α = .01, H_0 is not rejected. There is insufficient evidence to indicate the mean number of latex gloves used per week by hospital employees diagnosed with a latex allergy from exposure to the powder on latex gloves is less than 20 at α = .01.

8.49　From the printout, the p-value $< .0001$. Since the p-value $< .0001 < \alpha = .01$, H_0 is rejected. There is sufficient evidence to indicate that the true population mean ratio of all bones of this particular species differs from 8.5 at $\alpha = .01$.

8.51　a.　The test statistic is $z = \dfrac{\bar{x} - \mu_0}{\sigma_{\bar{x}}} = \dfrac{10.2 - 0}{31.3/\sqrt{50}} = 2.3$

　　　b.　To determine if the mean level of feminization differs from 0%, we test:

$$H_0: \ \mu = 0$$
$$H_a: \ \mu \neq 0$$

Since the alternative hypothesis contains $\neq$, this is a two-tailed test. The p-value is $p = P(z \leq -2.3) + P(z \geq 2.3) = .5 - P(-2.3 < z < 0) + .5 - P(0 < z < 2.3) = .5 - .4893 + .5 - .4893 = .0214$.

　　　b.　The test statistic is $z = \dfrac{\bar{x} - \mu_0}{\sigma_{\bar{x}}} = \dfrac{15 - 0}{25.1/\sqrt{50}} = 4.22$.

Since the alternative hypothesis contains $\neq$, this is a two-tailed test. The p-value is $p = P(z \leq -4.22) + P(z \geq 4.22) = .5 - P(-4.22 < z < 0) + .5 - P(0 < z < 4.22) \approx .5 - .5 + .5 - .5 = 0$.

8.53　a.　$z = \dfrac{\bar{x} - \mu_0}{\sigma_{\bar{x}}} = \dfrac{52.3 - 51}{7.1/\sqrt{50}} = 1.29$

$$p\text{-value} = P(z \leq -1.29) + P(z \geq 1.29) = (.5 - .4015) + (.5 - .4015)$$
$$= .0985 + .0985 = .1970$$

　　　b.　$z = \dfrac{\bar{x} - \mu_0}{\sigma_{\bar{x}}} = \dfrac{52.3 - 51}{7.1/\sqrt{50}} = 1.29$

$$p\text{-value} = P(z \geq 1.29) = (.5 - .4015) = .0985$$

　　　c.　$z = \dfrac{\bar{x} - \mu_0}{\sigma_{\bar{x}}} = \dfrac{52.3 - 51}{10.4/\sqrt{50}} = 0.88$

$$p\text{-value} = P(z \leq -0.88) + P(z \geq 0.88) = (.5 - .3106) + (.5 - .3106)$$
$$= .1894 + .1894 = .3788$$

　　　d.　For part **a**, any value of α greater than .1970 would lead to the rejection of the null hypothesis.

For part **b**, any value of α greater than .0985 would lead to the rejection of the null hypothesis.

For part **c**, any value of α greater than .3788 would lead to the rejection of the null hypothesis.

e.   For p-value of .01 and a two-tailed test, we need to find a z-value so that $.01/2 = .005$ is to the right of it. From Table IV, Appendix A, $z_{.005} = 2.58$.

$$z = \frac{\bar{x} - \mu_o}{\sigma_{\bar{x}}} \Rightarrow 2.58 = \frac{52.3 - 51}{s/\sqrt{50}} \Rightarrow 2.58 \frac{s}{\sqrt{50}} = 1.3 \Rightarrow s = \frac{1.3(\sqrt{50})}{2.58} \Rightarrow s = 3.56$$

For any value of s less than or equal to 3.56, the p-value will be less than or equal to .01 for a two-tailed test.

For p-value of .01 and a one-tailed test, we need to find a z-value so that .01 is to the right of it. From Table IV, Appendix A, $z_{.01} = 2.33$.

$$z = \frac{\bar{x} - \mu_0}{\sigma_{\bar{x}}} \Rightarrow 2.33 = \frac{52.3 - 51}{s/\sqrt{50}} \Rightarrow 2.33 \frac{s}{\sqrt{50}} = 1.3 \Rightarrow s = \frac{1.3(\sqrt{50})}{2.33} \Rightarrow s = 3.95$$

For any value of s less than or equal to 3.95, the p-value will be less than or equal to .01 for a one-tailed test.

8.55   We should use the t distribution in testing a hypothesis about a population mean if the sample size is small, the population being sampled from is normal, and the variance of the population is unknown.

8.57 $\alpha = P(\text{Type I error}) = P(\text{Reject } H_0 \text{ when } H_0 \text{ is true})$

a. $\alpha = P(t > 1.440)$ where df $= 6$
      $= .10$ \qquad\qquad Table VI, Appendix A

b. $\alpha = P(t < -1.782)$ where df $= 12$
      $= P(t > 1.782)$
      $= .05$ \qquad\qquad Table VI, Appendix A

c. $\alpha = P(t < -2.060 \text{ or } t > 2.060)$ where df $= 25$
      $= 2P(t > 2.060)$
      $= 2(.025)$
      $= .05$ \qquad\qquad Table VI, Appendix A

8.59   a. H_0: $\mu = 6$
             H_a: $\mu < 6$

The test statistic is $t = \dfrac{\bar{x} - \mu_0}{s/\sqrt{n}} = \dfrac{4.8 - 6}{1.3/\sqrt{5}} = -2.064$.

The necessary assumption is that the population is normal.

The rejection region requires $\alpha = .05$ in the lower tail of the t distribution with df $= n - 1 = 5 - 1 = 4$. From Table VI, Appendix A, $t_{.05} = 2.132$. The rejection region is $t < -2.132$.

Since the observed value of the test statistic does not fall in the rejection region ($t = -2.064 \Leftrightarrow -2.132$), H_0 is not rejected. There is insufficient evidence to indicate the mean is less than 6 at $\alpha = .05$.

b. H_0: $\mu = 6$
    H_a: $\mu \neq 6$

The test statistic is $t = -2.064$ (from **a**).

The assumption is the same as in **a**.

The rejection region requires $\alpha/2 = .05/2 = .025$ in each tail of the t distribution with df $= n - 1 = 5 - 1 = 4$. From Table VI, Appendix A, $t_{.025} = 2.776$. The rejection region is $t < -2.776$ or $t > 2.776$.

Since the observed value of the test statistic does not fall in the rejection region ($t = -2.064 \not< -2.776$), H_0 is not rejected. There is insufficient evidence to indicate the mean is different from 6 at $\alpha = .05$.

c. For part **a**, the p-value $= P(t \leq -2.064)$.

From Table VI, with df $= 4$, $.05 < P(t \leq -2.064) < .10$ or $.05 < p\text{-value} < .10$.

For part **b**, the p-value $= P(t \leq -2.064) + P(t \geq 2.064)$.

From Table VI, with df $= 4$, $2(.05) < p\text{-value} < 2(.10)$ or $.10 < p\text{-value} < .20$.

8.61 a. We must assume that a random sample was drawn from a normal population.

b. The hypotheses are:

   H_0: $\mu = 1000$
   H_a: $\mu > 1000$

The test statistic is $t = 1.89$.

The p-value is .038.

There is evidence to reject H_0 for $\alpha > .038$. There is evidence to indicate the mean is greater than 1000 for $\alpha > .038$.

c. The hypotheses are:

   H_0: $\mu = 1000$
   H_a: $\mu \neq 1000$

The test statistic is $t = 1.89$.

The p-value is $2(.038) = .076$.
There is no evidence to reject H_0 for $\alpha = .05$. There is insufficient evidence to indicate the mean is different than 1000 for $\alpha = .05$.

There is evidence to reject H_0 for $\alpha > .076$. There is evidence to indicate the mean is different than 1000 for $\alpha > .076$.

8.63   a.   To determine if the mean breaking strength of the new bonding adhesive is less than
              5.70 Mpa, we test:

              H_0: $\mu = 5.70$
              H_a: $\mu < 5.70$

       b.   The rejection region requires $\alpha = .01$ in the lower tail of the t distribution with df $= n - 1$
              $= 10 - 1 = 9$. From Table VI, Appendix A, $t_{.01} = 2.821$. The rejection region is $t < -2.821$.

       c.   The test statistic is $t = \dfrac{\bar{x} - \mu_0}{\dfrac{s}{\sqrt{n}}} = \dfrac{5.07 - 5.70}{\dfrac{.46}{\sqrt{10}}} = -4.33$.

       d.   Since the observed value of the test statistic falls in the rejection region
              ($t = -4.33 < -2.821$), H_o is rejected. There is sufficient evidence to indicate that the
              mean breaking strength of the new bonding adhesive is less than 5.70 Mpa at $\alpha = .01$.

       e.   The conditions required for this test is a random sample from the target population and
              the population from which the sample is selected is approximately normal.

8.65   To determine if the mean Dental Anxiety Scale score for college students differs from 11, we
       test:

       H_0: $\mu = 11$
       H_a: $\mu \neq 11$

       The test statistic is $t = \dfrac{\bar{x} - \mu_0}{s/\sqrt{n}} = \dfrac{10.7 - 11}{3.6/\sqrt{27}} = -.43$

       The rejection region requires $\alpha/2 = .05/2 = .025$ in each tail of the t distribution with df $=$
       $n - 1 = 27 - 1 = 26$. From Table VI, Appendix A, $t_{.025} = 2.056$. The rejection region is
       $t < -2.056$ or $t > 2.056$.

       Since the observed value of the test statistic does not fall in the rejection region ($t = -.43 \not<$
       -2.056), H_0 is not rejected. There is insufficient evidence to indicate the mean Dental Anxiety
       Scale score for college students differs from 11 at $\alpha = .05$.

8.67   To determine if the mean amount of cesium in lichen specimens differs from .003, we test:

       H_0: $\mu = .003$
       H_a: $\mu \neq .003$

       From the printout, the test statistic is $t = 3.725$ and the p-value is $p = .0058$.

       Since the p-value is less than α ($p = .0058 < \alpha = .10$), H_0 is rejected. There is sufficient
       evidence to indicate the mean amount of cesium in lichen specimens differs from .003 at
       $\alpha = .10$.

8.69   From Exercise 7.40, $\bar{x} = 358.45$ and $s = 117.8172$. To determine if the mean skidding distance is less than 425 meters, we test:

H_0: $\mu = 425$
H_a: $\mu < 425$

The test statistic is $t = \dfrac{\bar{x} - \mu_o}{s/\sqrt{n}} = \dfrac{358.45 - 425}{\dfrac{117.8172}{\sqrt{20}}} = -2.53$

The rejection region requires $\alpha = .10$ in the lower tail of the t-distribution with df $= n - 1$ $= 20 - 1 = 19$. From Table VI, Appendix A, $t_{.10} = 1.328$. The rejection region is $t < -1.328$.

Since the observed value of the test statistic falls in the rejection region ($t = -2.53 < -1.328$), H_0 is rejected. There is sufficient evidence to indicate the mean skidding distance is less than 425 meters at $\alpha = .10$. Thus, there is enough evidence to refute the claim.

8.71   To determine if the true mean PST score of all patients with mild to moderate traumatic brain injury exceeds the "normal" value of 40, we test:

H_0: $\mu = 40$
H_a: $\mu > 40$

The test statistic is $t = \dfrac{\bar{x} - \mu_0}{s/\sqrt{n}} = \dfrac{48.43 - 48}{20.76/\sqrt{23}} = 1.95$

The rejection region requires $\alpha = .05$ in the upper tail of the t distribution with df $= n - 1$ $= 23 - 1 = 22$. From Table VI, Appendix A, $t_{.05} = 1.717$. The rejection region is $t > 1.717$.

Since the observed value of the test statistic falls in the rejection region ($t = 1.95 > 1.717$), H_0 is rejected. There is sufficient evidence to indicate that the true mean PST score of all patients with mild to moderate traumatic brain injury exceeds the "normal" value of 40 at $\alpha = .05$.

8.73   Typically, qualitative data are associated with making inferences about a population proportion.

8.75   The sample size is large enough if both np_o and nq_o are greater than or equal to 15.

a. $np_o = 500(.05) = 25$ and $nq_o = 500(1 - .05) = 500(.95) = 475$. Since both of these values are greater than 15, the sample size is large enough to use the normal approximation.

b. $np_o = 100(.99) = 99$ and $nq_o = 100(1 - .99) = 100(.01) = 1$. Since the second number is less than 15, the sample size is not large enough to use the normal approximation.

c. $np_o = 50(.2) = 10$ and $nq_o = 50(1 - .2) = 50(.8) = 40$. Since the first number is less than 15, the sample size is not large enough to use the normal approximation.

d. $np_o = 20(.2) = 4$ and $nq_o = 20(1 - .2) = 20(.8) = 16$. Since the first number is less than 15, the sample size is not large enough to use the normal approximation.

e. $np_o = 10(.4) = 4$ and $nq_o = 10(1 - .4) = 10(.6) = 6$. Since both numbers are less than 15, the sample size is not large enough to use the normal approximation.

8.77   a. $z = \dfrac{\hat{p} - p_0}{\sqrt{\dfrac{p_0 q_0}{n}}} = \dfrac{.84 - .9}{\sqrt{\dfrac{.9(.1)}{100}}} = -2.00$

   b.   The denominator in Exercise 8.76 is $\sqrt{\dfrac{.75(.25)}{100}} = .0433$ as compared to $\sqrt{\dfrac{.9(.1)}{100}} = .03$ in part **a**. Since the denominator in this problem is smaller, the absolute value of z is larger.

   c.   The rejection region requires $\alpha = .05$ in the lower tail of the z distribution. From Table IV, Appendix A, $z_{.05} = 1.645$. The rejection region is $z < -1.645$.

   Since the observed value of the test statistic falls in the rejection region ($z = -2.00 < -1.645$), H_0 is rejected. There is sufficient evidence to indicate the population proportion is less than .9 at $\alpha = .05$.

   d.   The p-value $= P(z \leq -2.00) = .5 - .4772 = .0228$ (from Table IV, Appendix A).

8.79   From Exercise 7.50, $n = 50$ and since p is the proportion of consumers who do not like the snack food, will be:

$$\hat{p} = \frac{\text{Number of 0's in sample}}{n} = \frac{29}{50} = .58$$

In order for the inference to be valid, the sample size must be large enough. The sample size is large enough if both np_o and nq_o are greater than or equal to 15.

$np_o = 50(.5) = 25$ and $nq_o = 50(1 - .5) = 50(.5) = 25$. Since both of these values are greater than 15, the sample size is large enough to use the normal approximation.

   a. H_0: $p = .5$
        H_a: $p > .5$

   The test statistic is $z = \dfrac{\hat{p} - p_0}{\sigma_{\hat{p}}} = \dfrac{\hat{p} - p_0}{\sqrt{\dfrac{p_0 q_0}{n}}} = \dfrac{.58 - .5}{\sqrt{\dfrac{.5(1 - .5)}{50}}} = 1.13$

   The rejection region requires $\alpha = .10$ in the upper tail of the z distribution. From Table IV, Appendix A, $z_{.10} = 1.28$. The rejection region is $z > 1.28$.

   Since the observed value of the test statistic does not fall in the rejection region ($z = 1.13 \not> 1.28$), H_0 is not rejected. There is insufficient evidence to indicate the proportion of customers who do not like the snack food is greater than .5 at $\alpha = .10$.

   b. p-value $= P(z \geq 1.13) = .5 - .3708 = .1292$

8.81 a. The population parameter of interest is p, the proportion of items that had the wrong price when scanned through the electronic checkout scanner at California Wal-Mart stores.

b. To determine if the true proportion of items scanned at California Wal-Mart stores exceeds the 2% NIST standard, we test:

H_0: p = .02
H_a: p > .02

c. The test statistic is $z = \dfrac{\hat{p} - p_o}{\sqrt{\dfrac{p_o q_o}{n}}} = \dfrac{.083 - .02}{\sqrt{\dfrac{.02(.98)}{1000}}} = 14.23$

The rejection region requires $\alpha = .05$ in the upper tail of the z-distribution. From Table IV, Appendix A, $z_{.05} = 1.645$. The rejection region is $z > 1.645$.

d. Since the observed value of the test statistic falls in the rejection region ($z = 14.23 > 1.645$), H_o is rejected. There is sufficient evidence to indicate the true proportion of items scanned at California Wal-Mart stores exceeds the 2% NIST standard at $\alpha = .05$.

e. In order for the inference to be valid, the sample size must be large enough. The sample size is large enough if both np_o and nq_o are greater than or equal to 15.

$np_o = 1000(.02) = 20$ and $nq_o = 1000(1 - .02) = 1000(.98) = 980$. Since both of these values are greater than 15, the sample size is large enough to use the normal approximation.

8.83 Some preliminary calculations:

$\hat{p} = \dfrac{x}{n} = \dfrac{401}{835} = .48$

In order for the inference to be valid, the sample size must be large enough. The sample size is large enough if both np_o and nq_o are greater than or equal to 15.

$np_o = 835(.45) = 375.75$ and $nq_o = 835(1 - .45) = 835(.55) = 459.25$. Since both of these values are greater than 15, the sample size is large enough to use the normal approximation.

To determine if more than 45% of male youths are raised in a single-parent family, we test:

H_0: p = .45
H_a: p > .45

The test statistic is $z = \dfrac{\hat{p} - p_0}{\sqrt{\dfrac{p_0 q_0}{n}}} = \dfrac{.48 - .45}{\sqrt{\dfrac{.45(.55)}{835}}} = 1.74$

The rejection region requires $\alpha = .05$ in the upper tail of the z distribution. From Table IV, Appendix A, $z_{.05} = 1.645$. The rejection region is $z > 1.645$.

Since the observed value of the test statistic falls in the rejection region ($z = 1.74 > 1.645$), H_0 is rejected. There is sufficient evidence that more than 45% of male youths are raised in a single-parent family at $\alpha = .05$.

8.85   The point estimate for p is $\hat{p} = \dfrac{x}{n} = \dfrac{71 + 68}{678} = \dfrac{139}{678} = .205$

In order for the inference to be valid, the sample size must be large enough. The sample size is large enough if both np_o and nq_o are greater than or equal to 15.

$np_o = 678(.25) = 169.5$ and $nq_o = 678(1 - .25) = 678(.75) = 508.5$. Since both of these values are greater than 15, the sample size is large enough to use the normal approximation.

To determine if the true percentage of appealed civil cases that are actually reversed is less than 25%, we test:

H_0: $p = .25$
H_a: $p < .25$

The test statistic is $z = \dfrac{\hat{p} - p_0}{\sqrt{\dfrac{p_0 q_0}{n}}} = \dfrac{.205 - .25}{\sqrt{\dfrac{.25(.75)}{678}}} = -2.71$

The rejection region requires $\alpha = .01$ in the lower tail of the z distribution. From Table IV, Appendix A, $z_{.01} = 2.33$. The rejection region is $z < -2.33$.

Since the observed value of the test statistic falls in the rejection region ($z = -2.71 < -2.33$), H_0 is rejected. There is sufficient evidence to indicate the true percentage of appealed civil cases that are actually reversed is less than 25% at $\alpha = .01$.

8.87   First, we calculate the estimate for p: $\hat{p} = \dfrac{x}{n} = \dfrac{37}{148} = .25$

In order for the inference to be valid, the sample size must be large enough. The sample size is large enough if both np_o and nq_o are greater than or equal to 15.

$np_o = 148(.20) = 29.6$ and $nq_o = 148(1 - .20) = 148(.80) = 118.4$. Since both of these values are greater than 15, the sample size is large enough to use the normal approximation.

To determine if more than 20% of all freshman college students believe in the Big Bang Theory, we test:

H_0: $p = .20$
H_a: $p > .20$

The test statistic is $z = \dfrac{\hat{p} - p_o}{\sqrt{\dfrac{p_o q_o}{n}}} = \dfrac{.25 - .20}{\sqrt{\dfrac{.20(.80)}{148}}} = 1.52$

Since no value of α was given, we will choose $\alpha = .05$. The rejection region requires $\alpha = .05$ in the upper tail of the z-distribution. From Table IV, Appendix A, $z_{.05} = 1.645$. The rejection region is $z > 1.645$.

Since the observed value of the test statistic does not fall in the rejection region ($z = 1.52 \not> 1.645$), H_0 is not rejected. There is insufficient evidence to indicate that more than 20% of all freshman college students believe in the Big Bang Theory at $\alpha = .05$.

Thus, we would not be willing to state that more than 20% of all freshman college students believe in the Big Bang Theory. We are 95% confident of this decision.

8.89   a.   The point estimate for p is $\hat{p} = \dfrac{x}{n} = \dfrac{24}{33} = .727$

In order for the inference to be valid, the sample size must be large enough. The sample size is large enough if both np_o and nq_o are greater than or equal to 15.

$np_o = 33(.60) = 19.8$ and $nq_o = 33(1 - .60) = 33(.40) = 13.2$. Since the second value is not greater than 15, the sample size is may not be large enough to use the normal approximation.

To determine if the cream will improve the skin of more than 60% of middle-aged women, we test:

H_0: $p = .60$
H_a: $p > .60$

The test statistic is $z = \dfrac{\hat{p} - p_0}{\sqrt{\dfrac{p_0 q_0}{n}}} = \dfrac{.727 - .60}{\sqrt{\dfrac{.60(.40)}{33}}} = 1.49$

The rejection region requires $\alpha = .05$ in the upper tail of the z distribution. From Table IV, Appendix A, $z_{.05} = 1.645$. The rejection region is $z > 1.645$.

Since the observed value of the test statistic does not fall in the rejection region ($z = 1.49 \not> 1.645$), H_0 is not rejected. There is insufficient evidence to indicate the cream will improve the skin of more than 60% of middle-aged women at $\alpha = .05$.

b.   The p-value of the test is:

$p = P(z \geq 1.49) = .5 - .4319 = .0681$

Since the p-value is not less than α ($p = .0681 \not< .05$), H_0 is not rejected. There is insufficient evidence to indicate the cream will improve the skin of more than 60% of middle-aged women at $\alpha = .05$.

8.91   Let p = proportion of patients taking the pill who reported an improved condition.

In order for the inference to be valid, the sample size must be large enough. The sample size is large enough if both np_0 and nq_0 are greater than or equal to 15.

$np_0 = 7000(.50) = 3500$ and $nq_0 = 7000(1 - .50) = 7000(.50) = 3500$. Since both of these values are greater than 15, the sample size is large enough to use the normal approximation.

To determine if there really is a placebo effect at the clinic, we test:

H_0: $p = .5$
H_a: $p > .5$

The test statistic is $z = \dfrac{\hat{p} - p_0}{\sqrt{\dfrac{p_0 q_0}{n}}} = \dfrac{.7 - .5}{\sqrt{\dfrac{.5(.5)}{7000}}} = 33.47$

The rejection region requires $\alpha = .05$ in the upper tail of the z distribution. From Table IV, Appendix A, $z_{.05} = 1.645$. The rejection region is $z > 1.645$.

Since the observed value of the test statistic falls in the rejection region ($z = 33.47 > 1.645$), H_0 is rejected. There is sufficient evidence to indicate that there really is a placebo effect at the clinic at $\alpha = .05$.

8.93   a.   The power of a test is equal to $1 - \beta$. As β increases, the power decreases.

8.95   a.

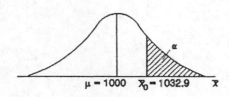

b. $z = \dfrac{\bar{x}_0 - \mu_0}{\sigma_{\bar{x}}} \Rightarrow \bar{x}_0 = \mu_0 + z_\alpha \sigma_{\bar{x}} = \mu_0 + z_\alpha \dfrac{\sigma}{\sqrt{n}}$ where $z_\alpha = z_{.05} = 1.645$ from Table IV, Appendix A.

Thus, $\bar{x}_0 = 1000 + 1.645 \dfrac{120}{\sqrt{36}} = 1032.9$.

c.

d. $\beta = P(\bar{x}_0 < 1032.9 \text{ when } \mu = 1020) = P\left(z < \dfrac{1032.9 - 1020}{120/\sqrt{36}}\right) = P(z < .65)$

$$= .5 + .2422 = .7422$$

e. Power $= 1 - \beta = 1 - .7422 = .2578$

8.97 a. The sampling distribution of $\bar{x}$ will be approximately normal (by the Central Limit Theorem) with $\mu_{\bar{x}} = \mu = 50$ and $\sigma_{\bar{x}} = \dfrac{\sigma}{\sqrt{n}} = \dfrac{20}{\sqrt{64}} = 2.5$.

b. The sampling distribution of $\bar{x}$ will be approximately normal (by the Central Limit Theorem) with $\mu_{\bar{x}} = \mu = 45$ and $\sigma_{\bar{x}} = \dfrac{\sigma}{\sqrt{n}} = \dfrac{20}{\sqrt{64}} = 2.5$.

c. First, find $\bar{x}_0$ so that $P(\bar{x} < \bar{x}_0) = .10$.

$$P(\bar{x} < \bar{x}_0) = P\left(z < \dfrac{\bar{x}_0 - 50}{20/\sqrt{64}}\right) = P\left(z < \dfrac{\bar{x}_0 - 50}{2.5}\right) = P(z < z_0) = .10$$

From Table IV, Appendix A, $z_{.10} = -1.28$

$$z_0 = \dfrac{\bar{x}_0 - 50}{2.5} \Rightarrow -1.28 = \dfrac{\bar{x}_0 - 50}{2.5} \Rightarrow \bar{x}_0 = 46.8$$

Now, find $\beta = P(\bar{x} > 46.8 \text{ when } \mu = 45) = P\left(z > \dfrac{46.8 - 45}{20/\sqrt{64}}\right) = P(z > .72)$

$$= .5 - .2642 = .2358$$

d. Power $= 1 - \beta = 1 - .2358 = .7642$

8.99 a. The sampling distribution of $\bar{x}$ will be approximately normal (by the Central Limit Theorem) with $\mu_{\bar{x}} = \mu = 10$ and $\sigma_{\bar{x}} = \dfrac{\sigma}{\sqrt{n}} = \dfrac{1.00}{\sqrt{100}} = 0.1$.

b. The sampling distribution of $\bar{x}$ will be approximately normal (CLT) with $\mu_{\bar{x}} = \mu = 9.9$ and $\sigma_{\bar{x}} = \dfrac{\sigma}{\sqrt{n}} = \dfrac{1.00}{\sqrt{100}} = 0.1$.

c. First, find $\bar{x}_{0,L}$ and $\bar{x}_{0,U}$ such that $P(\bar{x} < \bar{x}_{0,L}) = P(\bar{x} > \bar{x}_{0,U}) = .025$

$$P(\bar{x} < \bar{x}_{0,L}) = P\left(z < \frac{\bar{x}_{0,L} - 10}{.1}\right) = P(z < z_{0,L}) = .025$$

From Table IV, Appendix A, $z_{0,L} = -1.96$

Thus, $z_{0,L} = \dfrac{\bar{x}_{0,L} - 10}{.1} \Rightarrow \bar{x}_{0,L} = -1.96(.1) + 10 = 9.804$

$$P(\bar{x} > \bar{x}_{0,U}) = P\left(z > \frac{\bar{x}_{0,U} - 10}{.1}\right) = P(z > z_{0,U}) = .025$$

From Table IV, Appendix A, $z_{0,U} = 1.96$

Thus, $z_{0,U} = \dfrac{\bar{x}_{0,U} - 10}{.1} \Rightarrow \bar{x}_{0,U} = 1.96(.1) + 10 = 10.196$

Now, find $\beta = P(9.804 < \bar{x} < 10.196$ when $\mu = 9.9)$

$$= P\left(\frac{9.804 - 9.9}{1/\sqrt{100}} < z < \frac{10.196 - 9.9}{1/\sqrt{100}}\right) = P(-.96 < z < 2.96)$$

$$= .3315 + .4985 = .8300$$

d. $\beta = P(9.804 < \bar{x} < 10.196$ when $\mu = 10.1)$

$$= P\left(\frac{9.804 - 10.1}{1/\sqrt{100}} < z < \frac{10.196 - 10.1}{1/\sqrt{100}}\right) = P(-2.96 < z < .96)$$

$$= .4985 + .3315 = .8300$$

8.101 a. First, find $\bar{x}_0$ such that $P(\bar{x} < \bar{x}_0) = .10$

$$P(\bar{x} < \bar{x}_0) = P\left(z < \frac{\bar{x}_0 - 16}{9.32/\sqrt{100}}\right) = P(z < z_0) = .10$$

From Table IV, Appendix A, $z_{.10} = -1.28$

Thus, $z_0 = \dfrac{\bar{x}_0 - 16}{9.32/\sqrt{100}} \Rightarrow \bar{x}_0 = -1.28(1.622) + 16 = 14.81$

For $\mu = 15.5$, the power is:

$$\text{Power} = P(\bar{x} < 14.81) = P\left(z < \frac{14.81 - 15.5}{9.32/\sqrt{100}}\right) = P(z < -.74) = .5 - .2704$$

$$= .2296$$

For $\mu = 15.0$, the power is:

$$\text{Power} = P(\bar{x} < 14.81) = P\left(z < \frac{14.81 - 15.0}{9.32/\sqrt{100}}\right) = P(z < -.20) = .5 - .0793$$
$$= .4207$$

For $\mu = 14.5$, the power is:
$$\text{Power} = P(\bar{x} < 14.81) = P\left(z < \frac{14.81 - 14.5}{9.32/\sqrt{100}}\right) = P(z < .33) = .5 - .1293$$
$$= .6293$$

For $\mu = 14.0$, the power is:
$$\text{Power} = P(\bar{x} < 14.81) = P\left(z < \frac{14.81 - 14.0}{9.32/\sqrt{100}}\right) = P(z < .87) = .5 - .3078$$
$$= .8078$$

For $\mu = 13.5$, the power is:
$$\text{Power} = P(\bar{x} < 14.81) = P\left(z < \frac{14.81 - 13.5}{9.32/\sqrt{100}}\right) = P(z < 1.41) = .5 + .4207$$
$$= .9207$$

b.   The plot of the power is:

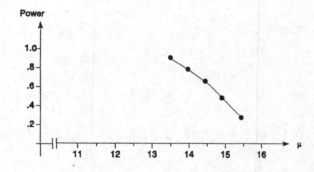

8.103   First, we must compute the value of $\bar{x}$ corresponding to the border of the rejection region. From Exercise 8.32,

H_0: $\mu = 10,000$
H_a: $\mu > 10,000$

The rejection region is $z > 1.645$ with $\alpha = .05$. The border value for the one-sided test is:

$$\bar{x}_0 = \mu_0 + z_\alpha \sigma_{\bar{x}} = \mu_0 + z_\alpha \frac{\sigma}{\sqrt{n}} = 10,000 + 1.645\frac{1600}{\sqrt{67}} = 10,000 + 321.55 = 10,321.55$$

Next, we convert this border value to a z-score using the alternative distribution with mean $\mu_a = 10{,}500$. The z-score is:

$$z = \frac{\bar{x}_0 - \mu_a}{\sigma_{\bar{x}}} = \frac{10{,}321.55 - 10{,}500}{\frac{1{,}600}{\sqrt{67}}} = -.91$$

The probability of making a Type II error is:

$$\beta = P(\bar{x} \le 10{,}321.55 \mid \mu_a = 10{,}500) = P(z \le -.91) = .5 - .3186 = .1814$$

8.105   The conditions required for a valid test for σ^2 are:

1.   A random sample is selected from the target population.
2.   The population from which the sample is selected has a distribution that is approximately normal.

8.107   The statement "When the sample size n is large, no assumptions about the population are necessary to test the population variance σ^2" is false. The population from which the sample is drawn must be approximately normal.

8.109   a.   df $= n - 1 = 16 - 1 = 15$; reject H_0 if $\chi^2 < 6.26214$ or $\chi^2 > 27.4884$

   b.   df $= n - 1 = 23 - 1 = 22$; reject H_0 if $\chi^2 > 40.2894$

   c.   df $= n - 1 = 15 - 1 = 14$; reject H_0 if $\chi^2 > 21.0642$

   d.   df $= n - 1 = 13 - 1 = 12$; reject H_0 if $\chi^2 < 3.57056$

   e.   df $= n - 1 = 7 - 1 = 6$; reject H_0 if $\chi^2 < 1.63539$ or $\chi^2 > 12.5916$

   f.   df $= n - 1 = 25 - 1 = 24$; reject H_0 if $\chi^2 < 13.8484$

8.111   a. H_0: $\sigma^2 = 1$
        H_a: $\sigma^2 > 1$
        The test statistic is $\chi^2 = \dfrac{(100-1)(1.84)}{1} = 182.16$

   The rejection region requires $\alpha = .05$ in the upper tail of the χ^2 distribution with df $= n - 1 = 100 - 1 = 99$. From Table VII, Appendix A, $\chi^2_{.05} \approx 124.324$. The rejection region is $\chi^2 > 124.324$.

   Since the observed value of the test statistic falls in the rejection region ($\chi^2 = 182.16 > 124.324$), H_0 is rejected. There is sufficient evidence to indicate the variance is larger than 1 at $\alpha = .05$.

   b.   In part **b** of Exercise 8.110, the test statistic was $\chi^2 = 11.04$. In Exercise 8.108, we did not reject H_0, but we did in this Exercise. As the sample size increases, it becomes easier to reject H_0.

8.113   a.   The rejection region requires $\alpha/2 = .01/2 = .005$ in each tail of the χ^2 distribution with df $= n - 1 = 46 - 1 = 45$. From Table VII, Appendix A, $\chi^2_{.005} \approx 66.7659$ and $\chi^2_{.995} \approx 20.7065$. The rejection region is $\chi^2 < 20.7065$ or $\chi^2 > 66.7659$.

   b.   The test statistic is $\chi^2 = \dfrac{(n-1)s^2}{\sigma_o^2} = \dfrac{(46-1)11.9^2}{100} = 63.7245$

   c.   Since the observed value of the test statistic does not fall in the rejection region ($\chi^2 = 63.7245 \not> 66.7659$ and $\chi^2 = 63.7245 \not< 20.7065$), H_0 is not rejected. There is insufficient evidence to indicate the variance is different from 100 at $\alpha = .01$.

8.115   To determine if the variance of birthweights of babies delivered by cocaine-dependent women is less than 200,000, we test:

H_0: $\sigma^2 = 200,000$
H_a: $\sigma^2 < 200,000$

The test statistic is $\chi^2 = \dfrac{(n-1)s^2}{\sigma_0^2} = \dfrac{(16-1)410^2}{200,000} = 12.6075$

The rejection region requires $\alpha = .01$ in the lower tail of the χ^2 distribution with df $= n - 1 = 16 - 1 = 15$. From Table VII, Appendix A, $\chi^2_{.99} = 5.22935$. The rejection region is $\chi^2 < 5.22935$.

Since the observed value of the test statistic does not fall in the rejection region ($\chi^2 = 12.6075 \not< 5.22935$), H_0 is not rejected. There is insufficient evidence to indicate that the variance of birthweights of babies delivered by cocaine-dependent women is less than 200,000 at $\alpha = .01$.

8.117   To determine if the true standard deviation of the point-spread errors exceed 15 (variance exceeds 225), we test:

H_0: $\sigma^2 = 225$
H_a: $\sigma^2 > 225$

The test statistic is $\chi^2 = \dfrac{(n-1)s^2}{\sigma_0^2} = \dfrac{(240-1)13.3^2}{225} = 187.896$

The rejection region requires $\alpha = .05$ in the upper tail of the χ^2 distribution with df $= n - 1 = 240 - 1 = 239$. The maximum value of df in Table VII is 100. Thus, we cannot find the rejection region using Table VII. Using a statistical package, the p-value associated with $\chi^2 = 187.896$ is .9938.

Since the p-value is so large, there is no evidence to reject H_0. There is insufficient evidence to indicate that the true standard deviation of the point-spread errors exceed 15 for any reasonable value of α.

(Since the observed variance (or standard deviation) is less than the hypothesized value of the variance (or standard deviation) under H_0, there is no way H_0 will be rejected for any reasonable value of α.)

8.119   To determine if the standard deviation of the percent recovery using the new method differs from 15, we test:

H_0: $\sigma^2 = 15^2 = 225$
H_a: $\sigma^2 \neq 225$

The statistic is $\chi^2 = \dfrac{(n-1)s^2}{\sigma_0^2} = \dfrac{(7-1)9^2}{225} = 2.16$

The rejection region requires $\alpha/2 = .10/2 = .05$ in each tail of the χ^2 distribution with df $= n - 1 = 7 - 1 = 6$. From Table VII, Appendix A, $\chi^2_{.05} = 12.5916$ and $\chi^2_{.95} = 1.63539$. The rejection region is $\chi^2 < 1.63539$ or $\chi^2 > 12.5916$.

Since the observed value of the test statistic does not fall in the rejection region ($\chi^2 = 2.16 \not< 1.63539$ and $\chi^2 = 2.16 \not> 12.5916$), H_0 is not rejected. There is insufficient evidence to indicate that the standard deviation of the percent recovery using the new method differs from 15 (or the variance differs from 225) at $\alpha = .10$.

8.121   a.   Some preliminary calculations:

$$s^2 = \frac{\sum x^2 - \dfrac{\left(\sum x\right)^2}{n}}{n-1} = \frac{29345.78 - \dfrac{(342.6)^2}{4}}{4-1} = .6967$$

To determine if the variation in the CO_2 concentration measurements using the LRM method differs from 1, we test:

H_0: $\sigma^2 = 1$
H_a: $\sigma^2 \neq 1$

The test statistic is $\chi^2 = \dfrac{(n-1)s^2}{\sigma_0^2} = \dfrac{(4-1).6967^2}{1} = 2.09$

Since no α was given we will use $\alpha = .05$. The rejection region requires $\alpha/2 = .05/2 = .025$ in each tail of the χ^2 distribution with df $= n - 1 = 4 - 1 = 3$. From Table VII, Appendix A, $\chi^2_{.025} = 9.34840$ and $\chi^2_{.975} = 0.215795$. The rejection region is $\chi^2 < 0.215795$ or $\chi^2 > 9.34840$.

Since the observed value of the test statistic does not fall in the rejection region ($\chi^2 = 2.09 \not< 0.215795$ and $\chi^2 = 2.09 \not> 9.34840$), H_0 is not rejected. There is insufficient evidence to indicate that the variation in the CO_2 concentration measurements using the LRM method differs from 1 at $\alpha = .05$.

b.   We must assume that the population of CO_2 amounts are normally distributed.

8.123 The smaller the p-value associated with a test of hypothesis, the stronger the support for the **alternative** hypothesis. The p-value is the probability of observing your test statistic or anything more unusual, given the null hypothesis is true. If this value is small, it would be very unusual to observe this test statistic if the null hypothesis were true. Thus, it would indicate the alternative hypothesis is true.

8.125 The elements of the test of hypothesis that should be specified prior to analyzing the data are: null hypothesis, alternative hypothesis, and significance level.

8.127 The larger the p-value associated with a test of hypothesis, the stronger the support for the **null** hypothesis. The p-value is the probability of observing your test statistic or anything more unusual, given the null hypothesis is true. If this value is large, it would not be very unusual to observe this test statistic if the null hypothesis were true. Thus, it would lend support that the null hypothesis is true.

8.129 a. H_a: $p = .35$
H_a: $p < .35$

The test statistic is $z = \dfrac{\hat{p} - p_0}{\sqrt{\dfrac{p_0 q_0}{n}}} = \dfrac{.29 - .35}{\sqrt{\dfrac{.35(.65)}{200}}} = -1.78$

The rejection region requires $\alpha = .05$ in the lower tail of the z distribution. From Table IV, Appendix A, $z_{.05} = 1.645$. The rejection region is $z < -1.645$.

Since the observed value of the test statistic falls in the rejection region ($z = -1.78 < -1.645$), H_0 is rejected. There is sufficient evidence to indicate $p < .35$ at $\alpha = .05$.

b. H_0: $p = .35$
H_a: $p \neq .35$

The test statistic is $z = -1.78$ (from **a**).

The rejection region requires $\alpha/2 = .05/2 = .025$ in each tail of the z distribution. From Table IV, Appendix A, $t_{.025} = 1.96$. The rejection region is $z < -1.96$ or $z > 1.96$.

Since the observed value of the test statistic does not fall in the rejection region ($z = -1.78 \not< -1.96$), H_0 is not rejected. There is insufficient evidence to indicate p is different from .35 at $\alpha = .05$.

c. For confidence coefficient .95, $\alpha = .05$ and $\alpha/2 = .025$. From Table IV, Appendix A, $z_{.025} = 1.96$. The confidence interval is:

$$\hat{p} \pm z_{.025} \sqrt{\dfrac{\hat{p}\hat{q}}{n}} \Rightarrow .29 \pm 1.96 \sqrt{\dfrac{.29(.71)}{200}} \Rightarrow .29 \pm .063 \Rightarrow (.227, .353)$$

d.   For confidence coefficient .99, $\alpha = .01$ and $\alpha/2 = .005$. From Table IV, Appendix A, $z_{.005} = 2.58$. The confidence interval is:

$$\hat{p} \pm z_{.005}\sqrt{\frac{\hat{p}\hat{q}}{n}} \Rightarrow .29 \pm 2.58\sqrt{\frac{.29(.71)}{200}} \Rightarrow .29 \pm .083 \Rightarrow (.207, .373)$$

e. $n = \dfrac{(z_{\alpha/2})^2 pq}{B^2} = \dfrac{2.58^2(.29)(.71)}{.05^2} = 548.2 \approx 549$

(We will use $\hat{p} = .29$ to estimate the value of p.)

8.131 a. H_0: $\sigma^2 = 30$
           H_a: $\sigma^2 > 30$

The test statistic is $\chi^2 = \dfrac{(n-1)s^2}{\sigma_0^2} = \dfrac{(41-1)(6.9)^2}{30} = 63.48$

The rejection region requires $\alpha = .05$ in the upper tail of the χ^2 distribution with $df = n - 1 = 40$. From Table VII, Appendix A, $\chi^2_{.05} = 55.7585$. The rejection region is $\chi^2 > 55.7585$.

Since the observed value of the test statistic falls in the rejection region ($\chi^2 = 63.48 > 55.7585$), H_0 is rejected. There is sufficient evidence to indicate the variance is larger than 30 at $\alpha = .05$.

b. H_0: $\sigma^2 = 30$
      H_a: $\sigma^2 \neq 30$

The test statistic is $\chi^2 = 63.48$ (from part **a**).

The rejection region requires $\alpha/2 = .05/2 = .025$ in each tail of the χ^2 distribution with $df = n - 1 = 40$. From Table VII, Appendix A, $\chi^2_{.025} = 59.3417$ and $\chi^2_{.975} = 24.4331$. The rejection region is $\chi^2 < 24.4331$ or $\chi^2 > 59.3417$.

Since the observed value of the test statistic falls in the rejection region ($\chi^2 = 63.48 > 59.3417$), H_0 is rejected. There is sufficient evidence to indicate the variance is not 30 at $\alpha = .05$.

8.133 a.   Since the company must give proof the drug is safe, the null hypothesis would be the drug is unsafe. The alternative hypothesis would be the drug is safe.

b.   A Type I error would be concluding the drug is safe when it is not safe. A Type II error would be concluding the drug is not safe when it is. α is the probability of concluding the drug is safe when it is not. β is the probability of concluding the drug is not safe when it is.

c.   In this problem, it would be more important for α to be small. We would want the probability of concluding the drug is safe when it is not to be as small as possible.

8.135 a. To determine if the true mean score of all sleep-deprived subjects is less than 80, we test:

H_0: $\mu = 80$
H_a: $\mu < 80$

The test statistic is $t = \dfrac{\bar{x}_0 - \mu_0}{s/\sqrt{n}} = \dfrac{63 - 80}{17/\sqrt{12}} = -3.46$

The rejection region requires $\alpha = .05$ in the lower tail of the t distribution with df $= n - 1 = 12 - 1 = 11$. From Table VI, Appendix A, $t_{.05} = 1.796$. The rejection region is $t < -1.796$.

Since the observed value of the test statistic falls in the rejection region ($t = -3.46 < -1.796$), H_0 is rejected. There is sufficient evidence to indicate that the true mean score of all sleep-deprived subjects is less than 80 at $\alpha = .05$.

b. We must assume that the overall test scores of sleep-deprived students are normally distributed.

8.137 a. Let p = proportion of the subjects who felt that the masculinization of face shape decreased attractiveness of the male face. If there is no preference for the unaltered or morphed male face, then $p = .5$.

For this problem, $\hat{p} = x/n = 58/67 = .866$.

In order for the inference to be valid, the sample size must be large enough. The sample size is large enough if both np_0 and nq_0 are greater than or equal to 15.

$np_0 = 67(.5) = 33.5$ and $nq_0 = 67(1 - .5) = 67(.5) = 33.5$. Since both of these values are greater than 15, the sample size is large enough to use the normal approximation.

To determine if the subjects showed a preference for either the unaltered or morphed face, we test:

H_0: $p = .5$
H_a: $p \neq .5$

b. The test statistic is $z = \dfrac{\hat{p} - p_0}{\sqrt{\dfrac{p_0 q_0}{n}}} = \dfrac{.866 - .50}{\sqrt{\dfrac{.50(.50)}{67}}} = 5.99$

c. By definition, the p-value is the probability of observing our test statistic or anything more unusual, given H_0 is true. For this problem, $p = P(z \leq -5.99) + P(z \geq 5.99)$
$= .5 - P(-5.99 < z < 0) + .5 - P(0 < z < 5.99) \approx .5 - .5 + .5 - .5 = 0$. This corresponds to what is listed in the Exercise.

d.   The rejection region requires $\alpha/2 = .01/2 = .005$ in each tail of the z distribution. From Table IV, Appendix A, $z_{.005} = 2.575$. The rejection region is $z < -2.575$ or $z > 2.575$.

Since the observed value of the test statistic falls in the rejection region ($z = 5.99 > 2.575$), H_0 is rejected. There is sufficient evidence to indicate that the subjects showed a preference for either the unaltered or morphed face at $\alpha = .01$.

8.139   To determine if the mean alkalinity level of water in the tributary exceeds 50 mpl, we test:

H_0: $\mu = 50$
H_a: $\mu > 50$

The test statistic is $z = \dfrac{\bar{x} - \mu_0}{\sigma_{\bar{x}}} = \dfrac{67.8 - 50}{14.4/\sqrt{100}} = 12.36$

The rejection region requires $\alpha = .01$ in the upper tail of the z distribution. From Table IV, Appendix A, $z_{.01} = 2.33$. The rejection region is $z > 2.33$.

Since the observed value of the test statistic falls in the rejection region ($z = 12.36 > 2.33$), H_0 is rejected. There is sufficient evidence to indicate that the mean alkalinity level of water in the tributary exceeds 50 mpl at $\alpha = .01$.

8.141   Using MINITAB, the descriptive statistics are:

Descriptive Statistics: Weight

Variable	N	N*	Mean	SE Mean	StDev	Minimum	Q1	Median	Q3	Maximum
Weight	16	0	1.373	0.258	1.034	0.201	0.475	1.347	2.009	3.751

To determine whether the mean weight in the crops of all spinifex pigeons differs from 1 gram, we test:

H_0: $\mu = 1$
H_a: $\mu \neq 1$

The test statistic is $t = \dfrac{\bar{x} - \mu_0}{s/\sqrt{n}} = \dfrac{1.373 - 1}{1.034/\sqrt{16}} = 1.44$

The rejection region requires $\alpha/2 = .05/2 = .025$ in each tail of the t distribution with df = $n - 1 = 16 - 1 = 15$. From Table VI, Appendix A, $t_{.025} = 2.131$. The rejection region is $t < -2.131$ and $t > 2.131$.

Since the observed value of the test statistic does not fall in the rejection region ($t = 1.44 \not> 2.131$), H_0 is not rejected. There is insufficient evidence to indicate the mean weight in the crops of all spinifex pigeons differs from 1 gram at $\alpha = .05$.

We must assume that the population being sampled from is normal. The stem-and-leaf display of the data is:

Stem-and-Leaf Display: Weight

```
Stem-and-leaf of Weight  N  = 16
Leaf Unit = 0.10

   4    0   2234
   7    0   588
  (4)   1   3344
   5    1   6
   4    2   1
   3    2   59
   1    3
   1    3   7
```

The data do not appear to be mound-shaped. Thus, the test above may not be valid.

8.143   a.   To determine if the true mean inbreeding coefficient μ for this species of wasp exceeds 0, we test:

H_0: $\mu = 0$
H_a: $\mu > 0$

The test statistic is $z = \dfrac{\bar{x} - \mu_0}{\sigma_{\bar{x}}} = \dfrac{.044 - 0}{.884/\sqrt{197}} = 0.70$

The rejection region requires $\alpha = .05$ in the upper tail of the z distribution. From Table IV, Appendix A, $z_{.05} = 1.645$. The rejection region is $z > 1.645$.

Since the observed value of the test statistic does not fall in the rejection region ($z = 0.70 \not> 1.645$), H_0 is not rejected. There is insufficient evidence to indicate that the true mean inbreeding coefficient μ for this species of wasp exceeds 0 at $\alpha = .05$.

   b.   This result agrees with that of Exercise 7.101. The confidence interval in Exercise 7.101 was $(-.118, .206)$. Since this interval contains 0, there is no evidence to indicate that the mean is different from 0. This is the same conclusion that was reached in part **a**.

8.145   a.   To determine if the production process should be halted, we test:

H_0: $\mu = 3$
H_a: $\mu > 3$

where $\mu =$ mean amount of PCB in the effluent.

The test statistic is $z = \dfrac{\bar{x} - \mu_0}{\sigma_{\bar{x}}} = \dfrac{3.1 - 3}{.5/\sqrt{50}} = 1.41$

The rejection region requires $\alpha = .01$ in the upper tail of the z distribution. From Table IV, Appendix A, $z_{.01} = 2.33$. The rejection region is $z > 2.33$.

Since the observed value of the test statistic does not fall in the rejection region, ($z = 1.41 \not> 2.33$), H_0 is not rejected. There is insufficient evidence to indicate the mean amount of PCB in the effluent is more than 3 parts per million at $\alpha = .01$. Do not halt the manufacturing process.

b.   As plant manager, I do not want to shut down the plant unnecessarily. Therefore, I want $\alpha = P(\text{shut down plant when } \mu = 3)$ to be small.

8.147   a.   No, it increases the risk of falsely rejecting H_0, i.e., closing the plant unnecessarily.

b.   First, find $\bar{x}_0$ such that $P(\bar{x} > \bar{x}_0) = P(z > z_0) = .05$.

From Table IV, Appendix A, $z_0 = 1.645$

$$z = \frac{\bar{x} - \mu_0}{\sigma/\sqrt{n}} \Rightarrow 1.645 = \frac{\bar{x}_0 - 3}{.5/\sqrt{50}} \Rightarrow \bar{x}_0 = 3.116$$

Then, compute

$$\beta = P(\bar{x}_0 \leq 3.116 \text{ when } \mu = 3.1) = P\left(z \leq \frac{3.116 - 3.1}{.5/\sqrt{50}}\right) = P(z \leq .23)$$

$$= .5 + .0910 = .5910$$

Power $= 1 - \beta = 1 - .5910 = .4090$

c.   The power of the test increases as α increases.

8.149   The point estimate for p is $\hat{p} = \dfrac{x}{n} = \dfrac{85}{124} = .685$

In order for the inference to be valid, the sample size must be large enough. The sample size is large enough if both np_0 and nq_0 are greater than or equal to 15.

$np_0 = 124(.167) = 20.708$ and $nq_0 = 124(1 - .167) = 124(.833) = 103.292$. Since both of these values are greater than 15, the sample size is large enough to use the normal approximation.

To determine if the students will tend to choose the three-grill display so that Grill #2 is a compromise between a more and a less desirable grill ($p > .167$), we test:

H_0: $p = .167$
H_a: $p > .167$

The test statistic is $z = \dfrac{\hat{p} - p_0}{\sqrt{\dfrac{p_0 q_0}{n}}} = \dfrac{.685 - .167}{\sqrt{\dfrac{.167(.822)}{124}}} = 15.46$

The rejection region requires $\alpha = .05$ in the upper tail of the z distribution. From Table IV, Appendix A, $z_{.05} = 1.645$. The rejection region is $z > 1.645$.

Since the observed value of the test statistic falls in the rejection region ($z = 15.46 > 1.645$), H_0 is rejected. There is sufficient evidence to indicate the students will tend to choose the three-grill display so that Grill #2 is a compromise between a more and a less desirable grill ($p > .167$) at $\alpha = .05$.

8.151   a.   A Type I error is rejecting the null hypothesis when it is true. In this problem, we would be concluding that the individual is a liar when, in fact, the individual is telling the truth.

         A Type II error is accepting the null hypothesis when it is false. In this problem, we would be concluding that the individual is telling the truth when, in fact, the individual is a liar.

     b.   The probability of a Type I error would be the probability of concluding the individual is a liar when he/she is telling the truth. From the problem, it stated that the polygraph would indicate that of 500 individuals who were telling the truth, 185 would be liars. Thus, an estimate of the probability of a Type I error would be $185/500 = .37$.

         The probability of a Type II error would be the probability of concluding the individual is telling the truth when he/she is a liar. From the problem, it stated that the polygraph would indicate that of 500 individuals who were liars, 120 would be telling the truth. Thus, an estimate of the probability of a Type II error would be $120/500 = .24$.

8.153   a.   Using MINITAB, the descriptive statistics are:

Descriptive Statistics: PCB

```
Variable    N  N*    Mean  SE Mean  StDev      Minimum     Q1  Median      Q3
PCB        48   0   9.292    0.304  2.103  0.000000000  9.000   9.000  10.000

Variable  Maximum
PCB        13.000
```

To determine if the data can refute the manufacturer's claim, we test:

H_0: $\mu = 10$
H_a: $\mu < 10$

The test statistic is $z = \dfrac{\bar{x} - \mu_0}{\sigma_{\bar{x}}} = \dfrac{9.292 - 10}{2.103/\sqrt{48}} = -2.33$

The rejection region requires $\alpha = .05$ in the lower tail of the z distribution. From Table IV, Appendix A, $z_{.05} = 1.645$. The rejection region is $z < -1.645$.

Since the observed value of the test statistic falls in the rejection region ($z = -2.33 < -1.645$), H_0 is rejected. There is sufficient evidence to indicate that the true average number of inspections is less than 10 at $\alpha = .05$.

We agree with the potential buyer who doubts the manufacturer's claim.

b. First, we must compute the value of $\bar{x}$ corresponding to the border of the rejection region. The rejection region is $z < -1.645$ with $\alpha = .05$. The border value for the one-sided test is:

$$\bar{x}_0 = \mu_0 - z_\alpha \sigma_{\bar{x}} = \mu_0 - z_\alpha \frac{\sigma}{\sqrt{n}} = 10 - 1.645 \frac{1.2}{\sqrt{48}} = 10 - .285 = 9.715$$

Next, we convert this border value to a z-score using the alternative distribution with mean $\mu_a = 9.5$. The z-score is:

$$z = \frac{\bar{x}_0 - \mu_a}{\sigma_{\bar{x}}} = \frac{9.715 - 9.5}{\frac{1.2}{\sqrt{48}}} = 1.24$$

The probability of making a Type II error is:

$$\beta = P(\bar{x} \geq 9.715 \mid \mu_a = 9.5) = P(z \geq 1.24) = .5 - .3925 = .1075$$

8.155   a. We want to show that more than 17,000 of the 18,200 signatures are valid. Thus, we want to show that the proportion of valid signatures is greater than $17{,}000/18{,}200 = .934$.

From the Exercise, $\hat{p} = 98/100 = .98$.

In order for the inference to be valid, the sample size must be large enough. The sample size is large enough if both np_0 and nq_0 are greater than or equal to 15.

$np_0 = 100(.934) = 93.4$ and $nq_0 = 100(1 - .934) = 100(.066) = 6.6$. Since the second value is not greater than 15, the sample size may not be large enough to use the normal approximation.

To determine if more than .934 of the signatures are valid, we test:

H_0: $p = .934$
H_a: $p > .934$

The test statistic is $z = \dfrac{\hat{p} - p_0}{\sqrt{\dfrac{p_0 q_0}{n}}} = \dfrac{.98 - .934}{\sqrt{\dfrac{.934(.066)}{100}}} = 1.85$

Since no α is given in the problem, we will choose $\alpha = .05$. The rejection region requires $\alpha = .05$ in the upper tail of the z distribution. From Table IV, Appendix A, $z_{.05} = 1.645$. The rejection region is $z > 1.645$.

Since the observed value of the test statistic falls in the rejection region ($z = 1.85 > 1.645$), H_0 is rejected. There is sufficient evidence to indicate that the true proportion of valid signatures is greater than .934 at $\alpha = .05$. This indicates that more than 17,000 of the 18,200 signatures are valid.

b. We want to show that more than 16,000 of the 18,200 signatures are valid. Thus, we want to show that the proportion of valid signatures is greater than $16,000/18,200 = .879$.

Again, from the Exercise, $\hat{p} = 98/100 = .98$.

In order for the inference to be valid, the sample size must be large enough. The sample size is large enough if both np_o and nq_o are greater than or equal to 15.

$np_o = 100(.879) = 87.9$ and $nq_o = 100(1 - .879) = 100(.121) = 12.1$. Since the second value is not greater than 15, the sample size may not be large enough to use the normal approximation.

To determine if more than .879 of the signatures are valid, we test:

H_0: $p = .879$
H_a: $p > .879$

The test statistic is $z = \dfrac{\hat{p} - p_0}{\sqrt{\dfrac{p_0 q_0}{n}}} = \dfrac{.98 - .879}{\sqrt{\dfrac{.879(.121)}{100}}} = 3.09$

Since no α is given in the problem, we will choose $\alpha = .05$. The rejection region requires $\alpha = .05$ in the upper tail of the z distribution. From Table IV, Appendix A, $z_{.05} = 1.645$. The rejection region is $z > 1.645$.

Since the observed value of the test statistic falls in the rejection region ($z = 3.09 > 1.645$), H_0 is rejected. There is sufficient evidence to indicate that the true proportion of valid signatures is greater than .879 at $\alpha = .05$. This indicates that more than 16,000 of the 18,200 signatures are valid.

Inferences Based on Two Samples: Confidence Intervals and Tests of Hypotheses

9.1   When the samples are large, the sampling distribution of $(\bar{x}_1 - \bar{x}_2)$ is approximately normal by the Central Limit Theorem. The mean of the sampling distribution is $(\mu_1 - \mu_2)$ and the

standard deviation is $\sqrt{\dfrac{\sigma_1^2}{n_1} + \dfrac{\sigma_2^2}{n_2}}$.

9.3   a.   No. Both populations must be normal.

        b.   No. Both populations variances must be equal.

        c.   No. Both populations must be normal.

        d.   Yes.

        e.   No. Both populations must be normal.

9.5   The confidence interval for $(\mu_1 - \mu_2)$ is (–10, –4). The correct inference is **b** – $(\mu_1 < \mu_2)$.
Since all the values in the interval are negative, there is evidence that $(\mu_1 < \mu_2)$.

9.7   a. $\mu_{\bar{x}_1} = \mu_1 = 14$

            $\sigma_{\bar{x}_1} = \dfrac{\sigma_1}{\sqrt{n_1}} = \dfrac{4}{\sqrt{100}} = .4$

        b. $\mu_{\bar{x}_2} = \mu_2 = 10$

            $\sigma_{\bar{x}_2} = \dfrac{\sigma_2}{\sqrt{n_2}} = \dfrac{3}{\sqrt{100}} = .3$

        c. $\mu_{\bar{x}_1 - \bar{x}_2} = \mu_1 - \mu_2 = 14 - 10 = 4$

            $\sigma_{\bar{x}_1 - \bar{x}_2} = \sqrt{\dfrac{\sigma_1^2}{n_1} + \dfrac{\sigma_2^2}{n_2}} = \sqrt{\dfrac{4^2}{100} + \dfrac{3^2}{100}} = \sqrt{\dfrac{25}{100}} = .5$

        d.   Since $n_1 \geq 30$ and $n_2 \geq 30$, the sampling distribution of $\bar{x}_1 - \bar{x}_2$ is approximately normal by the Central Limit Theorem.

9.9   Some preliminary calculations are:

$$\bar{x}_1 = \frac{\sum x_1}{n_1} = \frac{11.8}{5} = 2.36 \qquad s_1^2 = \frac{\sum x_1^2 - \dfrac{\left(\sum x_1\right)^2}{n_1}}{n_1 - 1} = \frac{30.78 - \dfrac{(11.8)^2}{5}}{5 - 1} = .733$$

$$\bar{x}_2 = \frac{\sum x_2}{n_2} = \frac{14.4}{4} = 3.6 \qquad s_2^2 = \frac{\sum x_2^2 - \dfrac{\left(\sum x_2\right)^2}{n_2}}{n_2 - 1} = \frac{53.1 - \dfrac{(14.4)^2}{4}}{4 - 1} = .42$$

a. $$s_p^2 = \frac{(n_1 - 1)s_1^2 + (n_2 - 1)s_2^2}{n_1 + n_2 - 2} = \frac{(5-1).773 + (4-1).42}{5 + 4 - 2} = \frac{4.192}{7} = .5989$$

b. H_0: $\mu_1 - \mu_2 = 0$
      H_a: $\mu_1 - \mu_2 < 0$

The test statistic is $t = \dfrac{(\bar{x}_1 - \bar{x}_2) - D_0}{\sqrt{s_p^2\left(\dfrac{1}{n_1} + \dfrac{1}{n_2}\right)}} = \dfrac{(2.36 - 3.6) - 0}{\sqrt{.5989\left(\dfrac{1}{5} + \dfrac{1}{4}\right)}} = \dfrac{-1.24}{.5191} = -2.39$

The rejection region requires $\alpha = .10$ in the lower tail of the t distribution with df = $n_1 + n_2 - 2 = 5 + 4 - 2 = 7$. From Table VI, Appendix A, $t_{.10} = 1.415$. The rejection region is $t < -1.415$.

Since the test statistic falls in the rejection region ($t = -2.39 < -1.415$), H_0 is rejected. There is sufficient evidence to indicate that $\mu_2 > \mu_1$ at $\alpha = .10$.

c.   A small sample confidence interval is needed because $n_1 = 5 < 30$ and $n_2 = 4 < 30$.

For confidence coefficient .90, $\alpha = .10$ and $\alpha/2 = .05$. From Table VI, Appendix A, with df = $n_1 + n_2 - 2 = 5 + 4 - 2 = 7$, $t_{.05} = 1.895$. The 90% confidence interval for $(\mu_1 - \mu_2)$ is:

$$(\bar{x}_1 - \bar{x}_2) \pm t_{.05}\sqrt{s_p^2\left(\frac{1}{n_1} + \frac{1}{n_2}\right)} \Rightarrow (2.36 - 3.6) \pm 1.895\sqrt{.5989\left(\frac{1}{5} + \frac{1}{4}\right)}$$

$$\Rightarrow -1.24 \pm .98 \Rightarrow (-2.22, -0.26)$$

d.   The confidence interval in part **c** provides more information about $(\mu_1 - \mu_2)$ than the test of hypothesis in part **b**. The test in part **b** only tells us that μ_2 is greater than μ_1. However, the confidence interval estimates what the difference is between μ_1 and μ_2.

9.11   a.   The *p*-value is $p = .115$.

Since the *p*-value is not small, there is no evidence to reject H_0. There is no evidence to indicate that the population means are different for $\alpha \le .10$.

b. The *p*-value would be half of the *p*-value in part **a**. The *p*-value = 1/2(.1150) = .0575.

There is no evidence to reject H_0 for $\alpha = .05$. There is no evidence to indicate that the mean for population 1 is less than the mean for population 2 for $\alpha = .05$.

There is evidence to reject H_0 for $\alpha > .0575$. There is evidence to indicate that the mean for population 1 is less than the mean for population 2 for $\alpha > .0575$.

9.13 a. Let μ_1 = mean recall of those who received only visual aspects of the ad and μ_2 = mean recall of those who received an audiovisual presentation. To determine if the mean recall of ad information by those who receive an audiovisual presentation is different from those who receive only the visual aspects of the ad, we test:

H_0: $\mu_1 = \mu_2$
H_a: $\mu_1 \neq \mu_2$

b. Some preliminary calculations are:

$$s_p^2 = \frac{(n_1 - 1)s_1^2 + (n_2 - 1)s_2^2}{n_1 + n_2 - 2} = \frac{(20 - 1)1.98^2 + (20 - 1)2.13^2}{20 + 20 - 2} = 4.22865$$

The test statistic is $t = \dfrac{\bar{x}_1 - \bar{x}_2}{\sqrt{s_p^2\left(\dfrac{1}{n_1} + \dfrac{1}{n_2}\right)}} = \dfrac{3.70 - 3.30}{\sqrt{4.22865\left(\dfrac{1}{20} + \dfrac{1}{20}\right)}} = .615$

c. The rejection region requires $\alpha/2 = .10/2 = .05$ in each tail of the *t* distribution with df $= n_1 + n_2 - 2 = 20 + 20 - 2 = 38$. From Table VI, Appendix A, $t_{.05} \approx 1.684$. The rejection region is $t > 1.684$ or $t < -1.684$.

d. Since the observed value of the test statistic does not fall in the rejection region ($t = .615 \not> 1.684$), H_0 is not rejected. There is insufficient evidence to indicate the mean recall of ad information by those who receive an audiovisual presentation is different from those who receive only the visual aspects of the ad at $\alpha = .10$.

Although we cannot prove the researchers' theory is correct (Accept H_o), there is evidence that supports the theory.

e. A *p*-value of .62 is the probability of observing our test statistic or anything more unusual. For this problem, the *p*-value = $P(t > .615) + P(t < -.615) = .62$. Since the *p*-value is so large, H_o will not be rejected for any reasonable value of α.

f. In order for the inference to be valid, the following conditions must be met:

1. Both populations being sampled from are normal
2. Both population variances must be the same
3. The samples must be random and independent.

9.15 a.   Let μ_1 = mean number of books read by students who earned an "A" grade and μ_2 = mean number of books read by students who earned a "B or C" grade. The parameter of interest for this problem is $\mu_1 - \mu_2$, the difference in the mean number of books read between the students who received an "A" grade and those who received a "B or C" grade.

b.   Using MINITAB, the descriptive statistics are:

Descriptive Statistics: No. Books

```
Variable    Grade2   N   N*    Mean   SE Mean   StDev   Minimum      Q1   Median      Q3
No. Books    A       8   0    37.00      3.08    8.70     24.00   31.00    36.50   41.50
             B-C     6   0    24.50      3.48    8.53     16.00   19.00    21.50   31.00

Variable    Grade2   Maximum
No. Books    A         53.00
             B-C       40.00
```

Some preliminary calculations are:

$$s_p^2 = \frac{(n_1 - 1)s_1^2 + (n_2 - 1)s_2^2}{n_1 + n_2 - 2} = \frac{(8-1)8.7^2 + (6-1)8.53^2}{8 + 6 - 2} = 74.470$$

For confidence coefficient .95, $\alpha = .05$ and $\alpha/2 = .05/2 = .025$. From Table VI, Appendix A, with df $= n_1 + n_2 - 2 = 8 + 6 - 2 = 12$, $t_{.025} = 2.179$. The confidence interval is:

$$(\overline{x}_1 - \overline{x}_2) \pm t_{.025}\sqrt{s_p^2\left(\frac{1}{n_1} + \frac{1}{n_2}\right)} \Rightarrow (37 - 24.5) \pm 2.179\sqrt{74.47\left(\frac{1}{8} + \frac{1}{6}\right)}$$

$$\Rightarrow 12.5 \pm 10.155 \Rightarrow (2.345, \ 22.655)$$

c.   We are 95% confident that the true difference in the mean number of books read between the students who received an "A" grade and those who received a "B or C" grade is between 2.345 and 22.655.

d.   In Exercise 2.31, the stem-and-leaf display indicated that the students who earned A's tended to read more books than those who earned B's or C's. This agrees with the confidence interval in part b. Because all the values in the confidence interval are positive, the mean number of books read by the students earning A's is larger than the mean number read by students earning B's or C's.

9.17   a.   Let μ_1 = mean height of Australian boys who repeated a grade and μ_2 = mean height of Australian boys who never repeated a grade.
To determine if the average height of Australian boys who repeated a grade is less than the average height of boys who never repeated, we test:

H_0: $\mu_1 = \mu_2$
H_a: $\mu_1 < \mu_2$

   b.   The test statistic is $z = \dfrac{\bar{x}_1 - \bar{x}_2}{\sqrt{\dfrac{s_1^2}{n_1} + \dfrac{s_2^2}{n_2}}} = \dfrac{-.04 - .30}{\sqrt{\dfrac{1.17^2}{86} + \dfrac{.97^2}{1346}}} = -2.64$

The rejection region requires $\alpha = .05$ in the lower tail of the z distribution. From Table IV, Appendix A, $z_{.05} = 1.645$. The rejection region is $z < -1.645$.

Since the observed value of the test statistic falls in the rejection region ($z = -2.64 < -1.645$), H_0 is rejected. There is sufficient evidence to indicate that the average height of Australian boys who repeated a grade is less than the average height of boys who never repeated at $\alpha = .05$.

   c.   Let μ_1 = mean height of Australian girls who repeated a grade and μ_2 = mean height of Australian girls who never repeated a grade.

To determine if the average height of Australian girls who repeated a grade is less than the average height of girls who never repeated, we test:

H_0: $\mu_1 = \mu_2$
H_a: $\mu_1 < \mu_2$

The test statistic is $z = \dfrac{\bar{x}_1 - \bar{x}_2}{\sqrt{\dfrac{s_1^2}{n_1} + \dfrac{s_2^2}{n_2}}} = \dfrac{.26 - .22}{\sqrt{\dfrac{.94^2}{43} + \dfrac{1.04^2}{1366}}} = .27$

The rejection region requires $\alpha = .05$ in the lower tail of the z distribution. From Table IV, Appendix A, $z_{.05} = 1.645$. The rejection region is $z < -1.645$.

Since the observed value of the test statistic does not fall in the rejection region ($z = .27 \not< -1.645$), H_0 is not rejected. There is insufficient evidence to indicate that the average height of Australian girls who repeated a grade is less than the average height of girls who never repeated at $\alpha = .05$.

9.19   Let μ_1 = mean FNE scores for bulimic students and μ_2 = mean FNE score for normal students.

Some preliminary calculations are:

$$\bar{x}_1 = \frac{\sum x_1}{n_1} = \frac{196}{11} = 17.82$$

$$s_1^2 = \frac{\sum x_1^2 - \frac{\left(\sum x_1\right)^2}{n_1}}{n_1 - 1} = \frac{3,734 - \frac{196^2}{11}}{11 - 1} = 24.1636$$

$$s_1 = \sqrt{s_1^2} = \sqrt{24.1636} = 4.916$$

$$\bar{x}_2 = \frac{\sum x_2}{n_2} = \frac{198}{14} = 14.14$$

$$s_2^2 = \frac{\sum x_2^2 - \frac{\left(\sum x_2\right)^2}{n_2}}{n_2 - 1} = \frac{3,164 - \frac{198^2}{14}}{14 - 1} = 27.9780$$

$$s_2 = \sqrt{s_2^2} = \sqrt{27.9780} = 5.289$$

$$s_p^2 = \frac{(n_1 - 1)s_1^2 + (n_2 - 1)s_2^2}{n_1 + n_2 - 2} = \frac{(11-1)24.1636 + (14-1)27.9780}{11 + 14 - 2} = 26.3196$$

a.   For confidence coefficient .95, $\alpha = .05$ and $\alpha/2 = .05/2 = .025$. From Table VI, Appendix A, with df $= n_1 + n_2 - 2 = 11 + 14 - 2 = 23$, $t_{.025} = 2.069$. The confidence interval is:

$$\left(\bar{x}_1 - \bar{x}_2\right) \pm t_{.025}\sqrt{s_p^2\left(\frac{1}{n_1} + \frac{1}{n_2}\right)} \Rightarrow (17.82 - 14.14) \pm 2.069\sqrt{26.3196\left(\frac{1}{11} + \frac{1}{14}\right)}$$

$$\Rightarrow 3.68 \pm 4.277 \Rightarrow (-.597, \ 7.957)$$

We are 95% confident that the difference in mean FNE scores for bulimic and normal students is between −.597 and 7.957,

b.   We must assume that the distribution of FNE scores for the bulimic students and the distribution of the FNE scores for the normal students are normally distributed. We must also assume that the variances of the two populations are equal.

Both sample distributions look somewhat mound-shaped and the sample variances are fairly close in value. Thus, both assumptions appear to be reasonably satisfied.

9.21   Let μ_1 = mean total rating for group 1 (support favored position) and μ_2 = mean total rating for group 2 (weaken opposing position).

Some preliminary calculations are:

$$s_p^2 = \frac{(n_1 - 1)s_1^2 + (n_2 - 1)s_2^2}{n_1 + n_2 - 2} = \frac{(26-1)12.5^2 + (26-1)12.2^2}{26 + 26 - 2} = \frac{7,627.25}{50} = 152.545$$

To determine if the difference in the mean total ratings for the two groups differ, we test:

H_0: $\mu_1 = \mu_2$
H_a: $\mu_1 \neq \mu_2$

The test statistic is $t = \dfrac{(\bar{x}_1 - \bar{x}_2) - 0}{\sqrt{s_p^2\left(\dfrac{1}{n_1} + \dfrac{1}{n_2}\right)}} = \dfrac{(28.6 - 24.9) - 0}{\sqrt{152.545\left(\dfrac{1}{26} + \dfrac{1}{26}\right)}} = 1.08$

The rejection region requires $\alpha/2 = .05/2 = .025$ in each tail of the t distribution. From Table VI, Appendix A, with df $= n_1 + n_2 - 2 = 26 + 26 - 2 = 50$, $t_{.025} \approx 2.010$. The rejection region is $t < -2.010$ or $z > 2.010$.

Since the observed value of the test statistic does not fall in the rejection region ($t = 1.08 \not> 2.010$ and $t = 1.08 \not< -2.010$), H_0 is not rejected. There is insufficient evidence to indicate there is a difference in the mean total ratings for the two groups at $\alpha = .05$.

9.23   Using MINITAB, the descriptive statistics are:

Descriptive Statistics: Species

Variable	Region	N	N*	Mean	SE Mean	StDev	Minimum	Q1	Median
Species	Dry Steppe	5	0	14.00	9.53	21.31	3.00	3.00	5.00
	Gobi Desert	6	0	11.83	7.44	18.21	4.00	4.00	4.50

Variable	Region	Q3	Maximum
Species	Dry Steppe	29.50	52.00
	Gobi Desert	16.00	49.00

Let μ_1 = average number of ant species found in the Dry Steppe and μ_2 = average number of ant species found in the Gobi Desert.

$$s_p^2 = \frac{(n_1 - 1)s_1^2 + (n_2 - 1)s_2^2}{n_1 + n_2 - 2} = \frac{(5-1)21.31^2 + (6-1)18.21^2}{5 + 6 - 2} = 386.0539$$

To determine if there is a difference in the average number of ant species found at sites in the two regions, we test:

H_0: $\mu_1 = \mu_2$
H_a: $\mu_1 \neq \mu_2$

The test statistic is $t = \dfrac{(\bar{x}_1 - \bar{x}_2) - D_o}{\sqrt{s_p^2\left(\dfrac{1}{n_1} + \dfrac{1}{n_2}\right)}} = \dfrac{(14 - 11.83) - 0}{\sqrt{386.0539\left(\dfrac{1}{5} + \dfrac{1}{6}\right)}} = 0.18$

The rejection region requires $\alpha/2 = .05/2 = .025$ in each tail of the t-distribution with df $= n_1 + n_2 - 2 = 5 + 6 - 2 = 9$. From Table VI, Appendix B, $t_{.025} = 2.262$. The rejection region is $t < -2.262$ or $t > 2.262$.

Since the observed value of the test statistic does not fall in the rejection region ($t = 0.18 \not> 2.262$), H_0 is not rejected. There is insufficient evidence to indicate a difference in the average number of ant species found at sites in the two regions at $\alpha = .05$.

9.25   a.   Let μ_1 = mean MFS score for the violent event group and μ_2 = mean MFS score for the avoided-violent event group. The target parameter of interest is $\mu_1 - \mu_2$ = difference in the mean MFS score between the 2 groups.

        b.   We are given the sample mean scores for the two groups. However, in order to determine if there is a difference in the 2 population means, we also need to know the population standard deviations for the 2 groups or the sample standard deviations for the 2 groups. This information was not provided.

        c.   The p-value of the test is

$$p = P(z \leq -1.21) + P(z \geq 1.21) = (.5 - .3869) + (.5 - .3869) = .1131 + .1131 = .2262$$

        d.   Since the observed p-value is greater than α ($p = .2262 > \alpha = .10$), H_0 is not rejected. There is insufficient evidence to indicate a difference in the mean MFS score between the violent event group and the avoided-violent event group at $\alpha = .10$.

9.27   Let μ_1 = mean DIQ score for SLI children and μ_2 = mean DIQ score for YND children.

      Some preliminary calculations are:

$$\bar{x}_1 = \frac{\sum x_1}{n_1} = \frac{936}{10} = 93.6$$

$$s_1^2 = \frac{\sum x_1^2 - \dfrac{\left(\sum x_1\right)^2}{n_1}}{n_1 - 1} = \frac{88352 - \dfrac{(936)^2}{10}}{10 - 1} = 82.4889$$

$$\bar{x}_2 = \frac{\sum x_2}{n_2} = \frac{953}{10} = 95.3$$

$$s_2^2 = \frac{\sum x_2^2 - \frac{\left(\sum x_2\right)^2}{n_2}}{n_1 - 1} = \frac{91329 - \frac{(953)^2}{10}}{10 - 1} = 56.4556$$

$$s_p^2 = \frac{(n_1 - 1)s_1^2 + (n_2 - 1)^2 s_2^2}{n_1 + n_2 - 2} = \frac{(10-1)82.4889 + (10-1)56.4556}{10 + 10 - 2} = 69.4723$$

To determine if the mean DIQ scores differ for the two groups, we test:

H_0: $\mu_1 - \mu_2 = 0$
H_a: $\mu_1 - \mu_2 \neq 0$

The test statistic is $t = \dfrac{\bar{x}_1 - \bar{x}_2 - 0}{\sqrt{s_p^2\left(\dfrac{1}{n_1} + \dfrac{1}{n_2}\right)}} = \dfrac{(93.6 - 95.3) - 0}{\sqrt{69.4723\left(\dfrac{1}{10} + \dfrac{1}{10}\right)}} = -0.46$

The rejection region requires $\alpha/2 = .10/2 = .05$ in each tail of the t distribution with df $= n_1 + n_2 - 2 = 10 + 10 - 2 = 18$. From Table VI, Appendix A, $t_{.05} = 1.734$. The rejection region is $t < -1.734$ or $t > 1.734$.

Since the observed value of the test statistic does not fall in the rejection region ($t = -.46 \not< -1.734$), H_0 is not rejected. There is insufficient evidence to indicate that the mean DIQ scores differ for the two groups at $\alpha = .10$.

9.29   a.   From the information given, we have no idea what the standard deviations are. Whether the population means are different or not depends on how variable the data are.

      b.   Let μ_1 = mean pain intensity rating for blacks and μ_2 = mean pain intensity rating for whites.

        To determine if blacks, on average, have a higher pain intensity rating than whites, we test:
H_0: $\mu_1 = \mu_2$
H_a: $\mu_1 > \mu_2$

        The rejection region requires $\alpha = .05$ in the upper tail of the z distribution. From Table IV, Appendix A, $z_{.05} = 1.645$. Thus, in order to reject H_0, the test statistic would have to be greater than 1.645.

        Substituting known values into the test statistic, we can solve for s_p.

$$z = \frac{\bar{x}_1 - \bar{x}_2}{\sqrt{s_p^2 \left(\dfrac{1}{n_1} + \dfrac{1}{n_2} \right)}} > 1.645$$

$$\Rightarrow \frac{8.2 - 6.9}{\sqrt{s_p^2 \left(\dfrac{1}{55} + \dfrac{1}{159} \right)}} > 1.645$$

$$\Rightarrow 1.3 > 1.645(s_p)\sqrt{\left(\frac{1}{55} + \frac{1}{159} \right)}$$

$$\Rightarrow \frac{1.3}{1.645(.1564)} > s_p$$

$$\Rightarrow 5.053 > s_p$$

Thus, if both s_1 and s_2 were less than 5.053, then we would conclude that blacks, on average would have a higher pain intensity than whites. (It is possible that either s_1 or s_2 could be greater than 5.053 and we would still reject H_0. As long as $s_p < 5.053$, we would reject H_0.)

c. From part **b**, we know we would reject H_0 if $s_p < 5.053$. Thus, we would not reject H_0 if $s_p \geq 5.053$. This could happen if both s_1 and s_2 are greater than 5.053. (It is possible that either s_1 or s_2 could be less than 5.053 and we would still reject H_0. As long as $s_p \geq 5.053$, we would reject H_0.)

9.31 In a paired difference experiment, the observations should be paired before collecting the data.

9.33 a. The rejection region requires $\alpha = .05$ in the upper tail of the t distribution with df = $n_d - 1 = 10 - 1 = 9$. From Table VI, Appendix A, $t_{.05} = 1.833$. The rejection region is $t > 1.833$.

b. From Table VI, with df = $n_d - 1 = 20 - 1 = 19$, $t_{.10} = 1.328$. The rejection region is $t > 1.328$.

c. From Table VI, with df = $n_d - 1 = 5 - 1 = 4$, $t_{.025} = 2.776$. The rejection region is $t > 2.776$.

d. From Table VI, with df = $n_d - 1 = 9 - 1 = 8$, $t_{.01} = 2.896$. The rejection region is $t > 2.896$.

9.35   a.

Pair	Difference
1	3
2	2
3	2
4	4
5	0
6	1

$$\bar{x}_d = \frac{\sum x_d}{n_d} = \frac{12}{6} = 2$$

$$s_d^2 = \frac{\sum x_d^2 - \frac{\left(\sum x_d\right)^2}{n_d}}{n_d - 1} = \frac{\left(34 - \frac{(12)^2}{6}\right)}{5} = 2$$

b. $\mu_d = \mu_1 - \mu_2$

c.   For confidence coefficient .95, $\alpha = .05$ and $\alpha/2 = .025$. From Table VI, Appendix A, with df $= n_d - 1 = 6 - 1 = 5$, $t_{.025} = 2.571$. The confidence interval is:

$$\bar{x}_d \pm t_{\alpha/2}\frac{s_d}{\sqrt{n_d}} \Rightarrow 2 \pm 2.571\frac{\sqrt{2}}{\sqrt{6}} \Rightarrow 2 \pm 1.484 \Rightarrow (.516, 3.484)$$

d. H_0: $\mu_d = 0$
      H_a: $\mu_d \neq 0$

The test statistic is $t = \dfrac{\bar{x}_d - 0}{s_d / \sqrt{n_d}} = \dfrac{2}{\sqrt{2}/\sqrt{6}} = 3.46$

The rejection region requires $\alpha/2 = .05/2 = .025$ in each tail of the t distribution with df $= n_d - 1 = 6 - 1 = 5$. From Table VI, Appendix A, $t_{.025} = 2.571$. The rejection region is $t < -2.571$ or $t > 2.571$.

Since the observed value of the test statistic falls in the rejection region ($t = 3.46 > 2.571$), H_0 is rejected. There is sufficient evidence to indicate that the mean difference is different from 0 at $\alpha = .05$.

9.37   a. H_0: $\mu_d = 10$
              H_a: $\mu_d \neq 10$

The test statistic is $z = \dfrac{\bar{x}_d - 0}{s_d / \sqrt{n_d}} = \dfrac{11.7 - 10}{6 / \sqrt{40}} = 1.79$

The rejection region requires $\alpha/2 = .05/2 = .025$ in each tail of the z distribution. From Table IV, Appendix A, $z_{.025} = 1.96$. The rejection region is $z < -1.96$ or $z > 1.96$. Since the observed value of the test statistic does not fall in the rejection region ($z = 1.79 \not> 1.96$), H_0 is not rejected. There is insufficient evidence to indicate the difference in the population means is different from 10 at $\alpha = .05$.

b. The p-value is

$$p = P(z \leq -1.79) + P(z \geq 1.79) = (.5 - .4633) + (.5 - .4633) = .0367 + .0367 = .0734$$

9.39 For confidence coefficient .95, $\alpha = .05$ and $\alpha/2 = .05/2 = .025$. From Table VI, Appendix A, with df $= n_d - 1 = 26 - 1 = 25$, $t_{.025} = 2.060$. The 95% confidence interval is:

$$\bar{x}_d \pm t_{.025} \frac{s_d}{\sqrt{n_d}} \Rightarrow 10.5 \pm 2.060 \frac{7.6}{\sqrt{26}} \Rightarrow 10.5 \pm 3.07 \Rightarrow (7.43, \ 13.57)$$

We are 95% confident that the difference in the mean anxiety levels between the before and after the visit is between 7.43 and 13.57. Since all of these values are positive, we can conclude that mean anxiety level after the visit is significantly less than the anxiety level before the visit.

9.41 a. Let μ_1 = mean BOLD score for the first session and μ_2 = mean BOLD score when the placebo was applied. The target parameter is $\mu_d = \mu_1 - \mu_2$, the difference in the mean BOLD scores between the first session and the session involving the placebo.

b. A paired difference design was used to collect the data. Each experimental person had two observations measured on it.

c. To determine if there was a placebo effect, we test:

H_o: $\mu_d = 0$
H_a: $\mu_d > 0$

d. The test statistic is $t = \dfrac{\bar{x}_d - 0}{s_d / \sqrt{n_d}} = \dfrac{0.21 - 0}{.47 / \sqrt{24}} = 2.19$

Since no α value was given, we will use $\alpha = .05$. The rejection region requires $\alpha = .05$ in the upper tail of the t distribution with df $= n_d - 1 = 24 - 1 = 23$. From Table VI, Appendix A, $t_{.025} = 1.714$. The rejection region is $t > 1.714$.

Since the observed value of the test statistic falls in the rejection region ($t = 2.19 > 1.714$), H_0 is rejected. There is sufficient evidence to indicate that there is a placebo effect at $\alpha = .05$.

e. Since the p-value is less than α ($p = .02 < \alpha = .05$), H_0 is rejected. There is sufficient evidence to indicate that there is a placebo effect at $\alpha = .05$.

9.43   a.   Let μ_1 = mean reading response times for the tongue-twister list and μ_2 = mean reading response times for the control list. Then $\mu_d = \mu_1 - \mu_2$.

To compare the mean reading response times for the tongue-twister and control lists, we test:

H_0: $\mu_d = 0$
H_a: $\mu_d \neq 0$

b.   The test statistic is $z = \dfrac{\bar{x}_d - 0}{s_c / \sqrt{n_d}} = \dfrac{.25 - 0}{.78 / \sqrt{42}} = 2.08$

The p-value is $P(z \leq -2.08) + P(z \geq 2.08) = (.5000 - .4812) + (.5000 - .4812) = .0188 + .0188 = .0376$. (Table IV, Appendix A.)

c.   Since the observed p-value is less than α ($p = .0376 < .05$), H_0 is rejected. There is sufficient evidence to conclude that the mean reading response times differ for the tongue-twister and control lists at $\alpha = .05$.

9.45   a.   Since all the data were collected from twins, the observations from each pair are not independent of each other. Thus, the data were collected as paired data. If the data are collected as paired data, they must be analyzed as paired data.

b.   Let μ_1 = mean level of overall leisure activity for the control group and μ_2 = mean level of overall leisure activity for the demented group. The target parameter is $\mu_d = \mu_1 - \mu_2$, the difference in the mean level of overall leisure activity for the control group and the demented group.

Using MINITAB, the descriptive statistics are:

Descriptive Statistics: Diff

Variable	N	N*	Mean	SE Mean	StDev	Minimum	Q1	Median	Q3
Diff	107	0	1.953	0.975	10.082	-28.000	-4.000	2.000	9.000

Variable	Maximum
Diff	29.000

We will use a 95% confidence interval to estimate the difference in the mean level of leisure activity between the control group and the demented group. For confidence coefficient .95, $\alpha = .05$ and $\alpha/2 = .05/2 = .025$. From Table IV, appendix A, $z_{.025} = 1.96$. The 95% confidence interval is:

$$\bar{x}_d \pm z_{.025} \frac{s_d}{\sqrt{n_d}} \Rightarrow 1.953 \pm 1.96 \frac{10.082}{\sqrt{107}} \Rightarrow 1.953 \pm 1.910 \Rightarrow (0.043, \ 3.863)$$

We are 95% confident that the difference in the mean level of leisure activity between the control group and the demented group is between 0.043 and 3.863. Since 0 is not contained in this interval, there is evidence of a difference in the mean level of leisure activity between the control group and the demented group at $\alpha = .05$. Since all the values are positive, there is evidence that the mean level of leisure activity for the control group is larger than that for the demented group.

9.47   a.   Let μ_1 = mean standardized growth of genes in the full-dark condition and μ_2 = mean standardized growth of genes in the transient light condition. Then $\mu_d = \mu_1 - \mu_2$ = difference in mean standardized growth between genes in full-dark condition and genes in transient light condition.

Some preliminary calculations are:

Gene ID	Full-Dark	Transient Light	Difference
SLR2067	-0.00562	1.40989	-1.41551
SLR1986	-0.68372	1.83097	-2.51469
SLR3383	-0.25468	-0.79794	0.54326
SLR0928	-0.18712	-1.20901	1.02189
SLR0335	-0.20620	1.71404	-1.92024
SLR1459	-0.53477	2.14156	-2.67633
SLR1326	-0.06291	1.03623	-1.09914
SLR1329	-0.85178	-0.21490	-0.63688
SLR1327	0.63588	1.42608	-0.79020
SLR1325	-0.69866	1.93104	-2.62970

$$\bar{x}_d = \frac{\sum x_d}{n_d} = \frac{-12.11754}{10} = -1.211754$$

$$s_d^2 = \frac{\sum x_d^2 - \frac{\left(\sum x_d\right)^2}{n_d}}{n_d - 1} = \frac{29.67025188 - \frac{(-12.11754)^2}{10}}{10 - 1} = \frac{14.98677431}{9} = 1.665197146$$

$$s = \sqrt{1.665197146} = 1.29042518$$

To determine if there is a difference in mean standardized growth of genes in the full-dark condition and genes in the transient light condition, we test:

H_0: $\mu_d = 0$
H_a: $\mu_d \neq 0$

The test statistic is $t = \dfrac{\bar{x}_d - 0}{s_d / \sqrt{n_d}} = \dfrac{-1.211754 - 0}{1.29042518 / \sqrt{10}} = -2.97$.

The rejection region requires $\alpha/2 = .01/2 = .005$ in each tail of the t distribution with df $= n_d - 1 = 10 - 1 = 9$. From Table VI, Appendix A, $t_{.005} = 3.250$. The rejection region is $t < -3.250$ or $t > 3.250$.

Since the observed value of the test statistic does not fall in the rejection region ($t = -2.97 \not< -3.250$), H_0 is not rejected. There is insufficient evidence to indicate a difference in mean standardized growth of genes in the full-dark condition and genes in the transient light condition at $\alpha = .01$.

b. Using MINITAB, the mean difference in standardized growth of the 103 genes in the full-dark condition and the transient light condition is:

Descriptive Statistics: FD-TL

```
Variable    N    Mean  StDev  Minimum      Q1  Median     Q3  Maximum
FD-TL     103  -0.420  1.422   -2.765  -1.463  -0.296  0.803    2.543
```

The sample mean difference is -.420. The test in part **a** did not detect this difference.

c. Let μ_3 = mean standardized growth of genes in the transient dark condition.
Then $\mu_d = \mu_1 - \mu_3$ = difference in mean standardized growth between genes in full-dark condition and genes in transient dark condition.

Some preliminary calculations are:

Gene ID	Full-Dark	Transient Dark	Difference
SLR2067	-0.00562	-1.28569	1.28007
SLR1986	-0.68372	-0.68723	0.00351
SLR3383	-0.25468	-0.39719	0.14251
SLR0928	-0.18712	-1.18618	0.99906
SLR0335	-0.20620	-0.73029	0.52409
SLR1459	-0.53477	-0.33174	-0.20303
SLR1326	-0.06291	0.30392	-0.36683
SLR1329	-0.85178	0.44545	-1.29723
SLR1327	0.63588	-0.13664	0.77252
SLR1325	-0.69866	-0.24820	-0.45046

$$\bar{x}_d = \frac{\sum x_d}{n_d} = \frac{1.40421}{10} = 0.140421$$

$$s_d^2 = \frac{\sum x_d^2 - \frac{\left(\sum x_d\right)^2}{n_d}}{n_d - 1} = \frac{5.589984302 - \frac{(1.40421)^2}{10}}{10 - 1} = \frac{5.39280373}{9} = 0.599200414$$

$$s = \sqrt{0.599200414} = .774080366$$

To determine if there is a difference in mean standardized growth of genes in the full-dark condition and genes in the transient dark condition, we test:

H_0: $\mu_d = 0$
H_a: $\mu_d \neq 0$

The test statistic is $t = \dfrac{\bar{x}_d - 0}{s_d / \sqrt{n_d}} = \dfrac{.140421 - 0}{.774080366 / \sqrt{10}} = 0.57$.

From part **a**, the rejection region is $t < -3.250$ or $t > 3.250$.

Since the observed value of the test statistic does not fall in the rejection region ($t = 0.57 \not> 3.250$), H_0 is not rejected. There is insufficient evidence to indicate a difference in mean standardized growth of genes in the full-dark condition and genes in the transient dark condition at $\alpha = .01$.

Using MINITAB, the mean difference in standardized growth of the 103 genes in the full dark condition and the transient dark condition is:

Descriptive Statistics: FD-TD

```
Variable     N     Mean    StDev  Minimum      Q1   Median      Q3  Maximum
FD-TD      103  -0.2274   0.9239  -2.8516  -0.8544  -0.1704  0.2644   2.3998
```

The sample mean difference is $-.2274$. The test above did not detect this difference.

d. $\mu_d = \mu_2 - \mu_3 =$ difference in mean standardized growth between genes in transient light condition and genes in transient dark condition.

Some preliminary calculations are:

Gene ID	Transient Light	Transient Dark	Difference
SLR2067	1.40989	−1.28569	2.69558
SLR1986	1.83097	−0.68723	2.51820
SLR3383	−0.79794	−0.39719	−0.40075
SLR0928	−1.20901	−1.18618	−0.02283
SLR0335	1.71404	−0.73029	2.44433
SLR1459	2.14156	−0.33174	2.47330
SLR1326	1.03623	0.30392	0.73231
SLR1329	−0.21490	0.44545	−0.66035
SLR1327	1.42608	−0.13664	1.56272
SLR1325	1.93104	−0.24820	2.17924

$$\bar{x}_d = \frac{\sum x_d}{n_d} = \frac{13.52175}{10} = 1.352175$$

$$s_d^2 = \frac{\sum x_d^2 - \frac{\left(\sum x_d\right)^2}{n_d}}{n_d - 1} = \frac{34.02408743 - \frac{(13.52175)^2}{10}}{10 - 1} = \frac{15.74031512}{9} = 1.748923902$$

$$s = \sqrt{1.748923902} = 1.322468866$$

To determine if there is a difference in mean standardized growth of genes in the transient light condition and genes in the transient dark condition, we test:

H_0: $\mu_d = 0$
H_a: $\mu_d \neq 0$

The test statistic is $t = \dfrac{\overline{x}_d - 0}{s_d / \sqrt{n_d}} = \dfrac{1.352175 - 0}{1.322468866 / \sqrt{10}} = 3.23$.

From part **a**, the rejection region is t < -3.250 or t > 3.250.

Since the observed value of the test statistic does not fall in the rejection region ($t = 3.23 \not> 3.250$), H_0 is not rejected. There is insufficient evidence to indicate a difference in mean standardized growth of genes in the transient light condition and genes in the transient dark condition at $\alpha = .01$.

Using MINITAB, the mean difference in standardized growth of the 103 genes in the full-dark condition and the transient dark condition is:

Descriptive Statistics: TL-TD

Variable	N	Mean	StDev	Minimum	Q1	Median	Q3	Maximum
TL-TD	103	0.192	1.499	-3.036	-1.166	0.149	1.164	2.799

The sample mean difference is .192. The test above did not detect this difference.

9.49 Let μ_1 = mean density of the wine measured with the hydrometer and μ_2 = mean density of the wine measured with the hydrostatic balance. The target parameter is $\mu_d = \mu_1 - \mu_2$, the difference in the mean density of the wine measured with the hydrometer and the hydrostatic balance.

Using MINITAB, the descriptive statistics for the difference data are:

Descriptive Statistics: Diff

Variable	N	N*	Mean	SE Mean	StDev	Minimum	Q1
Diff	40	0	-0.000523	0.000204	0.001291	-0.004480	-0.001078

Variable	Median	Q3	Maximum
Diff	-0.000165	0.000317	0.001580

We will use a 95% confidence interval to estimate the difference in the mean density of the wine measured with the hydrometer and the hydrostatic balance. For confidence coefficient .95, $\alpha = .05$ and $\alpha/2 = .05/2 = .025$. From Table IV, appendix A, $z_{.025} = 1.96$. The 95% confidence interval is:

$$\overline{x}_d \pm z_{.025}\frac{s_d}{\sqrt{n_d}} \Rightarrow -.000523 \pm 1.96\frac{.001292}{\sqrt{40}} \Rightarrow -.000523 \pm .000400 \Rightarrow (-.000923, \ -.000123)$$

We are 95% confident that the difference in the mean density of the wine measured with the hydrometer and the hydrostatic balance is between -.000923 and -.000123. Thus, we are 95% confident that the difference in the means ranges from .000123 and .000923 in absolute value. Since this entire confidence interval is less than .002, we can conclude that the difference in the mean scores does not exceed .002. Thus, we would recommend that the winery switch to the hydrostatic balance.

9.51   The conditions required for a large-sample inference about $(p_1 - p_2)$ are:

1. The two samples are randomly selected in an independent manner from the two target populations.
2. The sample sizes, n_1 and n_2, are both large so that the sampling distribution of $(\hat{p}_1 - \hat{p}_2)$ will be approximately normal. (This condition will be satisfied if all of the following are true: $n_1 \hat{p}_1 \geq 15$, $n_1 \hat{q}_1 \geq 15$ and $n_2 \hat{p}_2 \geq 15$, $n_2 \hat{q}_2 \geq 15$.)

9.53   The sample sizes are large enough if $n_1 \hat{p}_1 \geq 15$, $n_1 \hat{q}_1 \geq 15$ and $n_2 \hat{p}_2 \geq 15$, $n_2 \hat{q}_2 \geq 15$.

a. $n_1 \hat{p}_1 = 10(.5) = 5 \not\geq 15$, $n_1 \hat{q}_1 = 10(.5) = 5 \not\geq 15$

$n_2 \hat{p}_2 = 12(.5) = 6 \not\geq 15$, $n_2 \hat{q}_2 = 12(.5) = 6 \not\geq 15$

Since none of the products are greater than or equal to 15, the sample sizes are not large enough to assume normality.

b. $n_1 \hat{p}_1 = 10(.1) = 1 \not\geq 15$, $n_1 \hat{q}_1 = 10(.9) = 9 \not\geq 15$

$n_2 \hat{p}_2 = 12(.08) = .96 \not\geq 15$, $n_2 \hat{q}_2 = 12(.92) = 11.04 \not\geq 15$

Since none of the products are greater than or equal to 15, the sample sizes are not large enough to assume normality.

c. $n_1 \hat{p}_1 = 30(.2) = 6 \not\geq 15$, $n_1 \hat{q}_1 = 30(.8) = 24 > 15$

$n_2 \hat{p}_2 = 30(.3) = 9 \not\geq 15$, $n_2 \hat{q}_2 = 30(.7) = 21 > 15$

Since two of the products are not greater than or equal to 15, the sample sizes are not large enough to assume normality.

d. $n_1 \hat{p}_1 = 100(.05) = 5 \not\geq 15$, $n_1 \hat{q}_1 = 100(.95) = 95 > 15$

$n_2 \hat{p}_2 = 200(.09) = 18 > 15$, $n_2 \hat{q}_2 = 200(.91) = 182 > 15$

Since one of the products is not greater than or equal to 15, the sample sizes are not large enough to assume normality.

e. $n_1 \hat{p}_1 = 100(.95) = 95 > 15$, $n_1 \hat{q}_1 = 100(.05) = 5 \not\geq 15$

$n_2 \hat{p}_2 = 200(.91) = 182 > 15$, $n_2 \hat{q}_2 = 200(.09) = 18 > 15$

Since one of the products is not greater than or equal to 15, the sample sizes are not large enough to assume normality.

9.55   For confidence coefficient .95, $\alpha = 1 - .95 = .05$ and $\alpha/2 = .05/2 = .025$. From Table IV, Appendix A, $z_{.025} = 1.96$. The 95% confidence interval for $p_1 - p_2$ is approximately:

a. $(\hat{p}_1 - \hat{p}_2) \pm z_{\alpha/2} \sqrt{\dfrac{\hat{p}_1 \hat{q}_1}{n_1} + \dfrac{\hat{p}_2 \hat{q}_2}{n_2}}$

$\Rightarrow (.65 - .58) \pm 1.96 \sqrt{\dfrac{.65(1-.65)}{400} + \dfrac{.58(1-.58)}{400}} \Rightarrow .07 \pm .067 \Rightarrow (.003, .137)$

b. $(\hat{p}_1 - \hat{p}_2) \pm z_{\alpha/2} \sqrt{\dfrac{\hat{p}_1 \hat{q}_1}{n_1} + \dfrac{\hat{p}_2 \hat{q}_2}{n_2}}$

$\Rightarrow (.31 - .25) \pm 1.96 \sqrt{\dfrac{.31(1-.31)}{180} + \dfrac{.25(1-.25)}{250}} \Rightarrow .06 \pm .086 \Rightarrow (-026, .146)$

c. $(\hat{p}_1 - \hat{p}_2) \pm z_{\alpha/2} \sqrt{\dfrac{\hat{p}_1 \hat{q}_1}{n_1} + \dfrac{\hat{p}_2 \hat{q}_2}{n_2}}$

$\Rightarrow (.46 - .61) \pm 1.96 \sqrt{\dfrac{.46(1-.46)}{100} + \dfrac{.61(1-.61)}{120}} \Rightarrow -.15 \pm .131 \Rightarrow (-.281, -.019)$

9.57 H_0: $(p_1 - p_2) = .1$
H_a: $(p_1 - p_2) > .1$

Since D_0 is not equal to 0, the test statistic is:

$$z = \frac{(\hat{p}_1 - \hat{p}_2) - D_0}{\sqrt{\dfrac{p_1 q_1}{n_1} + \dfrac{p_2 q_2}{n_2}}} = \frac{(\hat{p}_1 - \hat{p}_2) - D_0}{\sqrt{\dfrac{\hat{p}_1 \hat{q}_1}{n_1} + \dfrac{\hat{p}_2 \hat{q}_2}{n_2}}} = \frac{(.4 - .2) - .1}{\sqrt{\dfrac{.4(1-.4)}{50} + \dfrac{.2(1-.2)}{60}}} = \frac{.1}{.0864} = 1.16$$

The rejection region requires $\alpha = .05$ in the upper tail of the z distribution. From Table IV, Appendix A, $z_{.05} = 1.645$. The rejection region is $z > 1.645$.

Since the observed value of the test statistic does not fall in the rejection region ($z = 1.16 \not> 1.645$), H_0 is not rejected. There is insufficient evidence to show $(p_1 - p_2) > .1$ at $\alpha = .05$.

9.59   a. $\hat{p}_1 = \dfrac{x_1}{n_1} = \dfrac{746}{1358} = .549$

b. $\hat{p}_2 = \dfrac{x_2}{n_2} = \dfrac{967}{1379} = .701$

c.   For confidence coefficient .90, $\alpha = .10$ and $\alpha/2 = .10/2 = .05$. From Table IV, Appendix A, $z_{.05} = 1.645$. The 90% confidence interval is:

$$(\hat{p}_1 - \hat{p}_2) \pm z_{.05} \sqrt{\dfrac{\hat{p}_1 \hat{q}}{n_1} + \dfrac{\hat{p}_2 \hat{q}_2}{n_2}} \Rightarrow (.549 - .701) \pm 1.645 \sqrt{\dfrac{.549(.451)}{1358} + \dfrac{.701(.299)}{1379}}$$

$$\Rightarrow -.152 \pm .030 \Rightarrow (-.182, \ -.122)$$

d. We are 90% confident that the true value of the difference in the 2 population proportions is between $-.182$ and $-.122$.

e. Since the 90% confidence interval contains only negative numbers, the proportion of girls who do not bully is larger than that for boys. Thus, the proportion of Dutch boys who bully is significantly greater than the proportion of Dutch girls who bully.

9.61 a. The observed proportion of St. John's wort patients who were in remission is:

$$\hat{p}_1 = \frac{x_1}{n_1} = \frac{14}{98} = .143$$

b. The observed proportion of placebo patients who were in remission is:

$$\hat{p}_2 = \frac{x_2}{n_2} = \frac{5}{102} = .049$$

c. Some preliminary calculations are:

$$\hat{p} = \frac{x_1 + x_2}{n_1 + n_2} = \frac{14+5}{98+102} = \frac{19}{200} = .095 \qquad \hat{q} = 1 - \hat{p} = 1 - .095 = .905$$

To determine if the proportion of St. John wort patients in remission exceeds the proportion of placebo patients in remission, we test:

H_0: $p_1 = p_2$
H_a: $p_1 > p_2$

The test statistic is $z = \dfrac{\hat{p}_1 - \hat{p}_2}{\sqrt{\hat{p}\hat{q}\left(\dfrac{1}{n_1}+\dfrac{1}{n_2}\right)}} = \dfrac{.143 - .049}{\sqrt{.095(.905)\left(\dfrac{1}{98}+\dfrac{1}{102}\right)}} = 2.27$

The rejection region requires $\alpha = .01$ in the upper tail of the z distribution. From Table IV, Appendix A, $z_{.01} = 2.33$. The rejection region is $z > 2.33$.

Since the observed value of the test statistic does not fall in the rejection region ($z = 2.27 \not> 2.33$), H_0 is not rejected. There is insufficient evidence to indicate the proportion of St. John wort patients in remission exceeds the proportion of placebo patients in remission at $\alpha = .01$.

d. The hypotheses and test statistic are the same as in part c.

The rejection region requires $\alpha = .10$ in the upper tail of the z distribution. From Table IV, Appendix A, $z_{.10} = 1.28$. The rejection region is $z > 1.28$.

Since the observed value of the test statistic falls in the rejection region ($z = 2.27 > 1.28$), H_0 is rejected. There is sufficient evidence to indicate the proportion of St. John wort patients in remission exceeds the proportion of placebo patients in remission $\alpha = .10$.

e. By changing the value of α in this problem, the conclusion changes. When α is small ($\alpha = .01$), we did not reject H_0. When α increases ($\alpha = .10$), we rejected H_0. As α increases, the chance of making a Type I error increase, but the chance of a Type II error decrease.

9.63 a. Let p_1 = proportion of African-American drivers who were searched and p_2 = proportion of White drivers who were searched. The parameter of interest is $p_1 - p_2$, or the difference in the proportions of drivers searched between the two ethnic groups.

Some preliminary calculations are:

$$\hat{p}_1 = \frac{x_1}{n_1} = \frac{12,016}{61,688} = .195 \qquad \hat{p}_2 = \frac{x_2}{n_2} = \frac{5,312}{106,892} = .050$$

$$\hat{p} = \frac{x_1 + x_2}{n_1 + n_2} = \frac{12,016 + 5,312}{61,688 + 106,892} = .103$$

To determine if there is a difference in the proportions of African-American and White drivers who are searched by the LA police, we test:

$H_0: p_1 - p_2 = 0$
$H_a: p_1 - p_2 \neq 0$

The test statistic is $z = \dfrac{(\hat{p}_1 - \hat{p}_2) - 0}{\sqrt{\hat{p}\hat{q}\left(\dfrac{1}{n_1} + \dfrac{1}{n_2}\right)}} = \dfrac{(.195 - .050) - 0}{\sqrt{.103(.897)\left(\dfrac{1}{61,688} + \dfrac{1}{106,892}\right)}} = 94.35$

The rejection region requires $\alpha/2 = .05/2 = .025$ in each tail of the z distribution. From Table IV, Appendix A, $z_{.025} = 1.96$. The rejection region is $z < -1.96$ or $z > 1.96$.

Since the observed value of the test statistic falls in the rejection region ($z = 94.35 > 1.96$), H_0 is rejected. There is sufficient evidence to indicate a difference in the proportions of African-American and White drivers who are searched by the LA police at $\alpha = .05$.

b. Let p_1 = "hit rate" of African-American drivers and p_2 = "hit rate" of White drivers. The parameter of interest is $p_1 - p_2$, or the difference in the "hit rates" for the two ethnic groups.

Some preliminary calculations are:

$$\hat{p}_1 = \frac{x_1}{n_1} = \frac{5,134}{12,016} = .427 \qquad \hat{p}_2 = \frac{x_2}{n_2} = \frac{3,006}{5,312} = .566$$

For confidence coefficient .95, $\alpha = .05$ and $\alpha/2 = .05/2 = .025$. From Table IV, Appendix A, $z_{.025} = 1.96$. The 95% confidence interval is:

$$(.427 - .566) \pm 1.96\sqrt{\frac{.427(.573)}{12,016} + \frac{.566(.434)}{5,312}} \Rightarrow -.139 \pm .016 \Rightarrow (-.155, \ -.123)$$

We are 95% confident that the difference in "hit rates" between African-American and White drivers searched by the LA police is between −.155 and −.123. Since both end points of the interval are negative, there is evidence that the "hit rate" for African-American drives is less than that for White drivers.

9.65   Let p_1 = proportion of rice weevils found dead after 4 days of exposure to nitrogen gas and p_2 = proportion of rice weevils found dead after 3.5 days of exposure to nitrogen gas. The parameter of interest is $p_1 - p_2$, or the difference in the proportions of rice weevils found dead between the 2 different exposures to nitrogen gas.

Some preliminary calculations are:

$$\hat{p}_1 = \frac{x_1}{n_1} = \frac{31,386}{31,421} = .999 \qquad\qquad \hat{p}_2 = \frac{x_2}{n_2} = \frac{23,516}{23,676} = .993$$

$$\hat{p} = \frac{x_1 + x_2}{n_1 + n_2} = \frac{31,386 + 23,516}{31,421 + 23,676} = .996$$

To determine if there is a difference in the proportions of rice weevils found dead between the 2 exposure times, we test:

H_o: $p_1 - p_2 = 0$
H_a: $p_1 - p_2 \neq 0$

The test statistic is $z = \dfrac{(\hat{p}_1 - \hat{p}_2) - 0}{\sqrt{\hat{p}\hat{q}\left(\dfrac{1}{n_1} + \dfrac{1}{n_2}\right)}} = \dfrac{(.999 - .993) - 0}{\sqrt{.996(.004)\left(\dfrac{1}{31,421} + \dfrac{1}{23,676}\right)}} = 11.05$

The rejection region requires $\alpha/2 = .10/2 = .05$ in each tail of the z distribution. From Table IV, Appendix A, $z_{.05} = 1.645$. The rejection region is $z < -1.645$ or $z > 1.645$.

Since the observed value of the test statistic falls in the rejection region ($z = 11.05 > 1.645$), H_o is rejected. There is sufficient evidence to indicate a difference in the proportions of rice weevils found dead between the 2 exposure times at $\alpha = .10$.

9.67   Let p_1 = recall rate of the 60 to 72 year-old seniors and p_2 = recall rate of the 73 to 83 year-old seniors.

Some preliminary calculations are:

$$\hat{p}_1 = \frac{x_1}{n_1} = \frac{31}{40} = .775 \qquad \hat{p}_2 = \frac{x_2}{n_2} = \frac{22}{40} = .55 \qquad \hat{p} = \frac{x_1 + x_2}{n_1 + n_2} = \frac{31 + 22}{40 + 40} = .6625$$

To determine if the recall rates of the two groups differ, we test:

H_0: $p_1 = p_2$
H_a: $p_1 \neq p_2$

The test statistic is $z = \dfrac{(\hat{p}_1 - \hat{p}_2) - 0}{\sqrt{\hat{p}\hat{q}\left(\dfrac{1}{n_1} + \dfrac{1}{n_2}\right)}} = \dfrac{(.775 - .55) - 0}{\sqrt{.6625(.3375)\left(\dfrac{1}{40} + \dfrac{1}{40}\right)}} = 2.13$

The rejection region requires $\alpha/2 = .05/2 = .025$ in each tail of the z distribution. From Table IV, Appendix A, $z_{.025} = 1.96$. The rejection region is $z < -1.96$ or $z > 1.96$.

Since the observed value of the test statistic falls in the rejection region ($z = 2.13 > 1.96$), H_0 is rejected. There is sufficient evidence to indicate the recall rates of the two groups differ at $\alpha = .05$.

9.69   Let p_1 = proportion of volunteers who figured out the third rule in the group that slept and p_2 = proportion of volunteers who figured out the third rule in the group that stayed awake all night. The parameter of interest is $p_1 - p_2$, or the difference in the proportions of volunteers who figured out the third rule between the 2 groups.

Some preliminary calculations are:

$$\hat{p}_1 = \frac{x_1}{n_1} = \frac{39}{50} = .78 \qquad\qquad \hat{p}_2 = \frac{x_2}{n_2} = \frac{15}{100} = .30$$

For confidence coefficient .90, $\alpha = .10$ and $\alpha/2 = .10/2 = .05$. From Table IV, Appendix A, $z_{.05} = 1.645$. The 90% confidence interval is:

$$(.78 - .30) \pm 1.645\sqrt{\frac{.78(.22)}{50} + \frac{.30(.70)}{50}} \Rightarrow .48 \pm .144 \Rightarrow (.336, \ .624)$$

We are 90% confident that the difference in the proportions of volunteers who figured out the third rule between the 2 groups is between .336 and .624. Since both end points are greater than 0, there is evidence that the proportion of those who slept who figured out the third rule is greater than the proportion of those who stayed awake all night who figured out the third rule.

9.71   a.   Let p_1 = true proportion of women who acknowledge having food cravings and p_2 = true proportion of men who acknowledge having food cravings.

To determine if the true proportion of women who acknowledge having food cravings exceeds the corresponding proportion of men, we test:

H_0: $p_1 = p_2$
H_a: $p_1 > p_2$

The test statistic would be $z = \dfrac{(\hat{p}_1 - \hat{p}_2) - 0}{\sqrt{\hat{p}\hat{q}\left(\dfrac{1}{n_1} + \dfrac{1}{n_2}\right)}}$

The rejection region requires $\alpha = .01$ in the upper tail of the z distribution. From Table IV, Appendix A, $z_{.01} = 2.33$. The rejection region is $z > 2.33$.

We need to find $n_1 = n_2$ so that the test statistic falls in the rejection region or $z > 2.33$. We know that $\hat{p}_1 = .97$ and $\hat{p}_2 = .67$. Assuming $n_1 = n_2 = n$,

$$\hat{p} = \frac{\hat{p}_1 + \hat{p}_2}{2} = \frac{.97 + .67}{2} = .82$$

$$z = \frac{(\hat{p}_1 - \hat{p}_2) - 0}{\sqrt{\hat{p}\hat{q}\left(\frac{1}{n_1} + \frac{1}{n_2}\right)}} = \frac{(.97 - .67) - 0}{\sqrt{.82(.18)\left(\frac{1}{n} + \frac{1}{n}\right)}} = 2.33$$

$$\Rightarrow .30 = 2.33\sqrt{.82(.18)\left(\frac{1}{n} + \frac{1}{n}\right)}$$

$$\Rightarrow .30 = 2.33\sqrt{.1476\left(\frac{2}{n}\right)} \Rightarrow .30 = 2.33\sqrt{\frac{.2952}{n}}$$

$$\Rightarrow .30^2 n = 2.33^2(.2952) \Rightarrow n = \frac{2.33^2(.2952)}{.3^2} = 17.807 \approx 18$$

We would need at least 18 observations from each group.

    b.   This study involved Students at McMaster University students. It is very dangerous to generalize the results of this study to the general adult population of North America. The sample of students used may not be representative of the population of interest.

9.73   We can estimate the values of p_1 and p_2 based on prior samples, by using an educated guess of the values, or most conservatively, estimating p_1 and p_2 with .5.

9.75 $n_1 = n_2 = \dfrac{(z_{\alpha/2})^2(\sigma_1^2 + \sigma_2^2)}{(SE)^2}$

For confidence coefficient .95, $\alpha = 1 - .95 = .05$ and $\alpha/2 = .05/2 = .025$. From Table IV, Appendix A, $z_{.025} = 1.96$.

$$n_1 = n_2 = \frac{1.96^2(15 + 15)}{2.2^2} = 23.8 \approx 24$$

9.77   a.   For confidence coefficient .99, $\alpha = 1 - .99 = .01$ and $\alpha/2 = .01/2 = .005$. From Table IV, Appendix A, $z_{.005} = 2.58$.

$$n_1 = n_2 = \frac{(z_{\alpha/2})^2(p_1 q_1 + p_2 q_2)}{(SE)^2} = \frac{2.58^2(.4(1 - .4) + .7(1 - .7))}{.01^2}$$

$$= \frac{2.99538}{.0001} = 29,953.8 \approx 29,954$$

b. For confidence coefficient .90, $\alpha = 1 - .90 = .10$ and $\alpha/2 = .10/2 = .05$. From Table IV, Appendix A, $z_{.05} = 1.645$. Since we have no prior information about the proportions, we use $p_1 = p_2 = .5$ to get a conservative estimate. For a width of .05, the standard error is .5.

$$n_1 = n_2 = \frac{(z_{\alpha/2})^2 (p_1 q_1 + p_2 q_2)}{(SE)^2} = \frac{(1.645)^2 (.5(1-.5) + .5(1-.5))}{.025^2} = 2164.82 \approx 2165$$

c. From part **b**, $z_{.05} = 1.645$.

$$n_1 = n_2 = \frac{(z_{\alpha/2})^2 (p_1 q_1 + p_2 q_2)}{(SE)^2} = \frac{1.645^2 (.2(1-.2) + .3(1-.3))}{.03^2}$$

$$= \frac{1.00123}{.0009} = 1112.48 \approx 1113$$

9.79 For confidence coefficient .95, $\alpha = .05$ and $\alpha/2 = .05/2 = .025$. From Table IV, Appendix A, $z_{.025} = 1.96$.

$$n_1 = n_2 = \frac{(z_{\alpha/2})^2 (\sigma_B^2 + \sigma_N^2)}{(SE)} = \frac{1.96^2 (25 + 25)}{2^2} = 48.02 \approx 49$$

9.81 For confidence coefficient .90, $\alpha = .10$ and $\alpha/2 = .10/2 = .05$. From Table IV, Appendix A, $z_{.05} = 1.645$.

$$n_1 = n_2 = \frac{(z_{\alpha/2})^2 (p_1 q_1 + p_2 q_2)}{(SE)^2} = \frac{1.645^2 (.4(.6) + .6(.4))}{.1^2} = 129.889 \approx 130$$

9.83 For confidence coefficient .95, $\alpha = .05$ and $\alpha/2 = .025$. From Table IV, Appendix A, $z_{.025} = 1.96$.

$$n_1 = n_2 = \frac{(z_{\alpha/2})^2 (\sigma_1^2 + \sigma_2^2)}{(SE)^2} = \frac{1.96^2 (15^2 + 15^2)}{1^2} = 1728.72 \approx 1729$$

9.85 For confidence coefficient .9, $\alpha = 1 - .9 = .1$ and $\alpha/2 = .1/2 = .05$. From Table IV, Appendix A, $z_{.05} = 1.645$.

$$n_1 = n_2 = \frac{(z_{\alpha/2})^2 (\sigma_1^2 + \sigma_2^2)}{(SE)^2} = \frac{1.645^2 (5^2 + 5^2)}{1^2} = 135.3 \Rightarrow 136$$

9.87 The conditions required for valid inferences about σ_1^2 / σ_2^2 are:

1. The two sampled populations are normally distributed.
2. The samples are randomly and independently selected from their respective populations.

9.89 The statement "H_0: $\sigma_1^2 = \sigma_2^2$ is equivalent to H_0: $\sigma_1^2 / \sigma_2^2 = 0$" is false. The correct statement is:

H_0: $\sigma_1^2 = \sigma_2^2$ is equivalent to H_0: $\sigma_1^2 / \sigma_2^2 = 1$.

9.91 a. With $\nu_1 = 2$ and $\nu_2 = 30$,
  $P(F \geq 4.18) = .025$ (Table X, Appendix A)
  b. With $\nu_1 = 24$ and $\nu_2 = 10$,
  $P(F \geq 1.94) = .10$ (Table VIII, Appendix A)

  Thus, $P(F < 1.94) = 1 - P(F \geq 1.94) = 1 - .10 = .90$

  c. With $\nu_1 = 9$ and $\nu_2 = 1$,
  $P(F \geq 6022) = .01$ (Table XI, Appendix A)

  Thus, $P(F < 6022) = 1 - P(F \geq 6022) = 1 - .01 = .99$

  d. With $\nu_1 = 30$ and $\nu_2 = 30$,
  $P(F \geq 1.84) = .05$ (Table IX, Appendix A)

9.93 The test statistic for each of these is:

$$F = \frac{\text{Larger sample variance}}{\text{Smaller sample variance}}$$

  a. The rejection region requires $\alpha/2 = .20/2 = .10$ in the upper tail of the F distribution. If $s_1^2 > s_2^2$, numerator df = $\nu_1 = 8$ and denominator df = $\nu_2 = 40$. From Table VIII, Appendix A, $F_{.10} = 1.83$. The rejection region is $F > 1.83$. If $s_1^2 < s_2^2$, numerator df = $\nu_2 = 40$ and denominator df = $\nu_1 = 8$. From Table VIII, Appendix A, $F_{.10} = 2.36$. The rejection region is $F > 2.36$.

  b. The rejection region requires $\alpha/2 = .10/2 = .05$ in the upper tail of the F distribution. If $s_1^2 > s_2^2$, numerator df = $\nu_1 = 8$ and denominator df = $\nu_2 = 40$. From Table IX, Appendix A, $F_{.05} = 2.18$. The rejection region is $F > 2.18$. If $s_1^2 < s_2^2$, numerator df = $\nu_2 = 40$ and denominator df = $\nu_1 = 8$. From Table IX, Appendix A, $F_{.05} = 3.04$. The rejection region is $F > 3.04$.

  c. The rejection region requires $\alpha/2 = .05/2 = .025$ in the upper tail of the F distribution. If $s_1^2 > s_2^2$, numerator df = $\nu_1 = 8$ and denominator df = $\nu_2 = 40$. From Table X, Appendix A, $F_{.025} = 2.53$. The rejection region is $F > 2.53$. If $s_1^2 < s_2^2$, numerator df = $\nu_2 = 40$ and denominator df = $\nu_1 = 8$. From Table X, Appendix A, $F_{.025} = 3.84$. The rejection region is $F > 3.84$.

d. The rejection region requires $\alpha/2 = .02/2 = .01$ in the upper tail of the F distribution. If $s_1^2 > s_2^2$, numerator df $= v_1 = 8$ and denominator df $= v_2 = 40$. From Table XI, Appendix A, $F_{.01} = 2.99$. The rejection region is $F > 2.99$. If $s_1^2 < s_2^2$, numerator df $= v_2 = 40$ and denominator df $= v_1 = 8$. From Table XI, Appendix A, $F_{.01} = 5.12$. The rejection region is $F > 5.12$.

9.95 a. H_0: $\sigma_1^2 = \sigma_2^2$

H_a: $\sigma_1^2 \neq \sigma_2^2$

The test statistic is $F = \dfrac{\text{Larger sample variance}}{\text{Smaller sample variance}} = \dfrac{s_2^2}{s_1^2} = \dfrac{9.85}{2.87} = 3.43$

The rejection region requires $\alpha/2 = .05/2 = .025$ in the upper tail of the F distribution with numerator df $v_1 = n_2 - 1 = 25 - 1 = 24$ and denominator df $v_2 = n_1 - 1 = 16 - 1 = 15$. From Table X, Appendix A, $F_{.025} = 2.70$. The rejection region is $F > 2.70$.

Since the observed value of the test statistic falls in the rejection region ($F = 3.43 > 2.70$), H_0 is rejected. There is sufficient evidence to indicate $\sigma_1^2 \neq \sigma_2^2$ at $\alpha = .05$.

b. The p-value for the test is $p = 2P(F \geq 3.43)$. Using Table XI, Appendix A, with df $v_1 = n_2 - 1 = 25 - 1 = 24$ and df $v_2 = n_1 - 1 = 16 - 1 = 15$, $P(F > 3.29) = .01$. Thus, the p-value is $2P(F \geq 3.43) < 2(.01) = .02$ or $p < .02$.

9.97 a. To determine whether σ_M^2 is less than σ_F^2, we test:

H_0: $\sigma_M^2 = \sigma_F^2$

H_a: $\sigma_M^2 < \sigma_F^2$

b. The test statistic is $F = \dfrac{s_F^2}{s_M^2} = \dfrac{6.94^2}{6.73^2} = 1.063$.

c. The rejection region requires $\alpha = .10$ in the upper tail of the F distribution with $v_1 = n_F - 1 = 114 - 1 = 113$ and $v_2 = n_M - 1 = 127 - 1 = 126$. From Table VIII, Appendix A, $F_{.10} \approx 1.26$. The rejection region is $F > 1.26$.

d. The p-value is $p = P(F \geq 1.063)$. Using Table VIII with $v_1 = n_F - 1 = 114 - 1 = 113$ and $v_2 = n_M - 1 = 127 - 1 = 126$, $P(F > 1.26) = .10$. Thus, the p-value is $p = P(F \geq 1.063) > .10$.

e. Since the observed value of the test statistic does not fall in the rejection region ($F = 1.063 \not> 1.26$), H_0 is not rejected. There is insufficient evidence to indicate σ_M^2 is less than σ_F^2 at $\alpha = .10$.

f.   The conditions necessary for the test results to be valid are:

   1.   The two sampled populations are normally distributed.
   2.   The samples are randomly and independently selected from their respective populations.

9.99   To determine if the variances of the 2 groups are different, we test:

H_0: $\sigma_1^2 = \sigma_2^2$
H_a: $\sigma_1^2 \neq \sigma_2^2$

The test statistic is $F = \dfrac{\text{Larger sample variance}}{\text{Smaller sample variance}} = \dfrac{s_1^2}{s_2^2} = \dfrac{12.5^2}{12.2^2} = 1.050$

The rejection region requires $\alpha/2 = .05/2 = .025$ in the upper tail of the F distribution with $\nu_1 = n_1 - 1 = 26 - 1 = 25$ and $\nu_2 = n_2 - 1 = 26 - 1 = 25$. From Table X, Appendix A, $F_{.025} \approx 2.24$. The rejection region is $F > 2.24$.

Since the observed value of the test statistic does not fall in the rejection region ($F = 1.050 \not> 2.24$), H_0 is not rejected. There is insufficient evidence to indicate a difference in the population variances between the two groups at $\alpha = .05$. Thus, the assumption appears to be valid.

9.101   Let σ_1^2 = variance of the mathematics achievement test scores for males and σ_2^2 = variance of the mathematics achievement test scores for females. To determine if the test scores are more variable for the males than the females, we test:

H_0: $\sigma_1^2 = \sigma_2^2$
H_a: $\sigma_1^2 > \sigma_2^2$

The test statistic is $F = \dfrac{\text{Larger sample variance}}{\text{Smaller sample variance}} = \dfrac{s_1^2}{s_2^2} = \dfrac{12.96^2}{11.85^2} = 1.196$

The rejection region requires $\alpha = .01$ in the upper tail of the F distribution with numerator df $\nu_1 = n_1 - 1 = 1764 - 1 = 1763$ and denominator df $\nu_2 = n_2 - 1 = 1739 - 1 = 1738$. From Table XI, Appendix A, $F_{.01} \approx 1.00$. The rejection region is $F > 1.00$.

Since the observed value of the test statistic falls in the rejection region ($F = 1.196 > 1.00$), H_0 is rejected. There is sufficient evidence to indicate the test scores are more variable for the males than the females at $\alpha = .01$.

9.103 a.   Using MINITAB, some descriptive statistics are:

Descriptive Statistics: Novice, Exper

Variable	N	N*	Mean	SE Mean	StDev	Minimum	Q1	Median	Q3	Maximum
Novice	12	0	32.83	2.49	8.64	20.00	26.75	32.00	39.00	48.00
Exper	12	0	20.58	1.66	5.74	10.00	17.25	19.50	24.75	31.00

Let σ_1^2 = variance of inspection errors of novice inspectors and σ_2^2 = variance of inspection errors of experienced inspectors.

To determine if the variance of inspection errors of experienced inspectors is lower than at for novice inspectors, we test:

H_0: $\sigma_1^2 = \sigma_2^2$

H_a: $\sigma_1^2 > \sigma_2^2$

The test statistic is $F = \dfrac{\text{Larger sample variance}}{\text{Smaller sample variance}} = \dfrac{s_1^2}{s_2^2} = \dfrac{8.64^2}{5.74^2} = 2.266$

The rejection region requires $\alpha = .05$ in the upper tail of the F distribution with $v_1 = n_1 - 1 = 12 - 1 = 11$ and $v_2 = n_2 - 1 = 12 - 1 = 11$. Using Table IX, Appendix A, $F_{.05} \approx 2.82$. The rejection region is $F > 2.82$.

Since the observed value of the test statistic does not fall in the rejection region $(F = 2.266 \not> 2.82)$, H_0 is not rejected. There is insufficient evidence to indicate the variance of inspection errors of experienced inspectors is lower than that for novice inspectors at $\alpha = .05$.

b.   The p-value is $p = P(F \geq 2.266)$. Using Table VIII with $v_1 = n_1 - 1 = 12 - 1 = 11$ and $v_2 = n_2 - 1 = 12 - 1 = 11$, $P(F > 2.82) = .10$. Thus, the p-value is:

$p = P(F \geq 2.266) > .10$.

9.105   Some preliminary calculations:

$$s_1^2 = \frac{\sum x_1^2 - \dfrac{\left(\sum x_1\right)^2}{n_1}}{n_1 - 1} = \frac{6620.96 - \dfrac{227^2}{9}}{9-1} = 110.6778$$

$$s_2^2 = \frac{\sum x_2^2 - \dfrac{\left(\sum x_2\right)^2}{n_2}}{n_2 - 1} = \frac{596.68 - \dfrac{66.4^2}{9}}{9-1} = 13.3494$$

a. Let σ_1^2 = variance of the order-to-delivery times for the Persian Gulf and σ_2^2 = variance of the order-to-delivery times for Bosnia. To determine if the variances of the order-to-delivery times for the Persian Gulf and Bosnia differ, we test:

$H_0: \sigma_1^2 = \sigma_2^2$

$H_a: \sigma_1^2 \neq \sigma_2^2$

The test statistic is $F = \dfrac{\text{Larger sample variance}}{\text{Smaller sample variance}} = \dfrac{s_1^2}{s_2^2} = \dfrac{110.6778}{13.3494} = 8.29$

The rejection region requires $\alpha/2 = .05/2 = .025$ in the upper tail of the F distribution with numerator df $v_1 = n_1 - 1 = 9 - 1 = 8$ denominator df $v_2 = n_2 - 1 = 9 - 1 = 8$. From Table X, Appendix A, $F_{.025} = 4.03$. The rejection region is $F > 4.03$.

Since the observed value of the test statistic falls in the rejection region ($F = 8.29 > 4.03$), H_0 is rejected. There is sufficient evidence to indicate the variances of the order-to-delivery times for Persian Gulf and Bosnia differ at $\alpha = .05$.

b. No. Since both sample sizes are less than 30, we must use the two-sample *t*-test to test for differences in means. However, one of the assumptions for the two-sample *t*-test is that the variances of the two populations being sampled from are equal. Since we rejected H_0 in part a, there is evidence to indicate the variances of the two populations are not the same.

9.107 a. To compare average SAT scores of males and females, the target parameter is $\mu_1 - \mu_2$.

b. To compare the difference between mean waiting times at two supermarket checkout lines, the target parameter is $\mu_1 - \mu_2$.

c. To compare proportions of Democrats and Republicans who favor legalization of marijuana, the target parameter is $p_1 - p_2$.

d. To compare variation in salaries of NBA players picked in the 1st round and the 2nd round, the target parameter is $\sigma_1^2 \big/ \sigma_2^2$.

e. To compare the difference in dropout rates of college student-athletes and regular students, the target parameter is $p_1 - p_2$.

9.109 a. $s_p^2 = \dfrac{(n_1 - 1)s_1^2 + (n_1 - 1)s_2^2}{n_1 + n_2 - 2} = \dfrac{11(74.2) + 13(60.5)}{12 + 14 - 2} = 66.7792$

$H_0: \mu_1 - \mu_2 = 0$

$H_a: \mu_1 - \mu_2 > 0$

The test statistic is $t = \dfrac{(\bar{x}_1 - \bar{x}_2) - 0}{\sqrt{s_p^2\left(\dfrac{1}{n_1} + \dfrac{1}{n_2}\right)}} = \dfrac{(17.8 - 15.3) - 0}{\sqrt{66.7792\left(\dfrac{1}{12} + \dfrac{1}{14}\right)}} = .78$

The rejection region requires $\alpha = .05$ in the upper tail of the t distribution with df $= n_1 + n_2 - 2 = 12 + 14 - 2 = 24$. From Table VI, Appendix A, for df $= 24$, $t_{.05} = 1.711$. The rejection region is $t > 1.711$.

Since the observed value of the test statistic does not fall in the rejection region ($t = 0.78 \not> 1.711$), H_0 is not rejected. There is insufficient evidence to indicate that $\mu_1 > \mu_2$ at $\alpha = .05$.

b. For confidence coefficient .99, $\alpha = .01$ and $\alpha/2 = .01/2 = .005$. From Table VI, Appendix A, with df $= n_1 + n_2 - 2 = 12 + 14 - 2 = 24$, $t_{.005} = 2.797$. The confidence interval is:

$$(\bar{x}_1 - \bar{x}_2) \pm t_{.005}\sqrt{s_p^2\left(\frac{1}{n_1} + \frac{1}{n_2}\right)} \Rightarrow (17.8 - 15.3) \pm 2.797\sqrt{66.7792\left(\frac{1}{12} + \frac{1}{14}\right)}$$

$$\Rightarrow 2.50 \pm 8.99 \Rightarrow (-6.49, 11.49)$$

c. For confidence coefficient .99, $\alpha = .01$ and $\alpha/2 = .01/2 = .005$. From Table IV, Appendix A, $z_{.005} = 2.58$.

$$n_1 = n_2 = \frac{(z_{\alpha/2})^2(\sigma_1^2 + \sigma_2^2)}{(SE)^2} = \frac{(2.58)^2(74.2 + 60.5)}{2^2} = 224.15 \approx 225$$

9.111 a. For confidence coefficient .90, $\alpha = .10$ and $\alpha/2 = .05$. From Table IV, Appendix A, $z_{.05} = 1.645$. The confidence interval is:

$$(\bar{x}_1 - \bar{x}_2) \pm z_{.05}\sqrt{\frac{s_1^2}{n_1} + \frac{s_2^2}{n_2}} \Rightarrow (12.2 - 8.3) \pm 1.645\sqrt{\frac{2.1}{135} + \frac{3.0}{148}}$$

$$\Rightarrow 3.90 \pm .31 \Rightarrow (3.59, 4.21)$$

b. H_0: $\mu_1 - \mu_2 = 0$
    H_a: $\mu_1 - \mu_2 \neq 0$

The test statistic is $z = \dfrac{(\bar{x}_1 - \bar{x}_2) - 0}{\sqrt{\dfrac{s_1^2}{n_1} + \dfrac{s_2^2}{n_2}}} = \dfrac{(12.2 - 8.3) - 0}{\sqrt{\dfrac{2.1}{135} + \dfrac{3.0}{148}}} = 20.60$

The rejection region requires $\alpha/2 = .01/2 = .005$ in each tail of the z distribution. From Table IV, Appendix A, $z_{.005} = 2.58$. The rejection region is $z < -2.58$ or $z > 2.58$.

Since the observed value of the test statistic falls in the rejection region ($20.60 > 2.58$), H_0 is rejected. There is sufficient evidence to indicate that $\mu_1 \neq \mu_2$ at $\alpha = .01$.

c. For confidence coefficient .90, $\alpha = .10$ and $\alpha/2 = .05$. From Table IV, Appendix A, $z_{.05} = 1.645$.

$$n_1 = n_2 = \frac{(z_{\alpha/2})^2(\sigma_1^2 + \sigma_2^2)}{(SE)^2} = \frac{(1.645)^2(2.1 + 3.0)}{.2^2} = 345.02 \approx 346$$

9.113   From the printout, the 95% confidence interval for the difference in mean seabird densities of oiled and unoiled transects is (−2.93, 2.49). We are 95% confident that the true difference in mean seabird densities of oiled and unoiled transects is between −2.93 and 2.49. Since 0 is contained in the interval, there is no evidence to indicate that the mean seabird densities are different for the oiled and unoiled transects.

9.115   a.   The variable measured for this experiment is the time needed to match the picture with the sentence.

       b.   The experimental units are the climbers.

       c.   Since the same climbers were timed at the base camp and at a camp 5 miles above sea level, the data should be analyzed as a paired experiment.

9.117   a.   Let μ_1 = mean trait for senior year in high school and μ_2 = mean trait during sophomore year of college. Also, let $\mu_d = \mu_1 - \mu_2$. To determine if there is a decrease in the mean self-concept of females between the senior year in high school and the sophomore year in college, we test:

$$H_0: \; \mu_d = 0$$
$$H_a: \; \mu_d > 0$$

       b.   Since the same female students were used in each study, the data should be analyzed as paired differences. The samples are not independent.

       c.   Since the number of pairs is $n_d = 133$, no assumptions about the population of differences is necessary. We still assume that a random sample of differences is selected.

       d.   Leadership: The p-value > .05. There is no evidence to reject H_0 for $\alpha = .05$. There is no evidence to indicate that there is a decrease in mean self-concept of females on leadership between the senior year in high school and the sophomore year in college for $\alpha = .05$.

Popularity: The p-value < .05. There is evidence to reject H_0 for $\alpha = .05$. There is evidence to indicate that there is a decrease in mean self-concept of females on popularity between the senior year in high school and the sophomore year in college for $\alpha = .05$.

Intellectual self-confidence: The p-value < .05. There is evidence to reject H_0 for $\alpha = .05$. There is evidence to indicate that there is a decrease in mean self-concept of females on intellectual self-concept between the senior year in high school and the sophomore year in college for $\alpha = .05$.

9.119   a. $\hat{p}_1 = \dfrac{x_1}{n_1} = \dfrac{29}{189} = .153$

       b. $\hat{p}_2 = \dfrac{x_2}{n_2} = \dfrac{32}{149} = .215$

c. For confidence coefficient .90, $\alpha = .10$ and $\alpha/2 = .10/2 = .05$. From Table IV, Appendix A, $z_{.05} = 1.645$. The 90% confidence interval is:

$$(\hat{p}_1 - \hat{p}_2) \pm z_{.05} \sqrt{\frac{\hat{p}_1 \hat{q}_1}{n_1} + \frac{\hat{p}_2 \hat{q}_2}{n_2}} \Rightarrow (.153 - .215) \pm 1.645 \sqrt{\frac{.153(.847)}{189} + \frac{.215(.785)}{149}}$$

$$\Rightarrow -.062 \pm .070 \Rightarrow (-.132, \quad .008)$$

d. We are 90% confident that the difference in the true proportion of super-experienced bidders who fall prey to the winner's curse and the true proportion of less-experienced bidders who fall prey to the winner's curse is between −0.132 and 0.008.

9.121 a. Let μ_1 = mean score on the moderate questions for Group A (had 15 difficult questions first) and μ_2 = mean score on the moderate questions for Group B (had 15 easy questions first).

To determine if the students with the easy questions first (Group B) scored better than the students with the difficult questions (Group A), we test:

$H_0:$ $\mu_1 - \mu_2 = 0$
$H_a:$ $\mu_1 - \mu_2 < 0$

b. The p-value = .3248. Since this p-value is not small, H_0 would not be rejected for any reasonable α. There is insufficient evidence to indicate the students with the easy questions first (Group B) scored better than the students with the difficult questions (Group A) at $\alpha \leq .10$.

c. We must assume that the samples were independent.

9.123 a. First, some preliminary calculations:

Patient	Pretreatment	Testosterone Levels After 6 months of Depro-Provera	$D = P - A$
1	849	96	753
2	903	41	862
3	890	31	859
4	1,092	124	968
5	362	46	316
6	900	53	847
7	1,006	113	893
8	672	174	498

$$\bar{x}_d = \frac{\sum x_d}{n_d} = \frac{5,996}{8} = 749.5$$

$$s_d^2 = \frac{\sum x_d^2 - \frac{\left(\sum x_d\right)^2}{n_d}}{n_d - 1} = \frac{4,847,676 - \frac{5,996^2}{8}}{8 - 1} = 50,524.857$$

For confidence coefficient .99, $\alpha = .01$ and $\alpha/2 = .01/2 = .005$. From Table VI, Appendix A, with df $= n_d - 1 = 8 - 1 = 7$, $t_{.005} = 3.499$. The confidence interval is:

$$\bar{x}_d \pm t_{.005} \frac{s_d}{\sqrt{n_d}} \Rightarrow 749.5 \pm 3.499 \frac{\sqrt{50{,}524.857}}{\sqrt{8}} \Rightarrow 749.5 \pm 278.068$$

$$\Rightarrow (471.432, 1027.568)$$

We are 99% confident that the true mean difference between the pretreatment and after-treatment testosterone levels of young male TBI patients is between 471.432 and 1,027.568.

b. Since 0 is not in the confidence interval, there is evidence that there is a difference in the testosterone levels pre and post treatment. Thus, there is evidence that Depo-Provera is effective in reducing testosterone levels in young male TBI patients.

9.125 a. Since we want to compare the means of two independent groups, we would use a two sample test of hypothesis for independent samples. Since the sample sizes are small (21 for each), we would use a small sample test.

b. We must assume that both populations being sampled from are approximately normally distributed and that the population variances of the two groups are equal.

c. No. We need to know the common variance of the two groups or an estimate of it.

d. Since the p-value was so large ($p = .79$), H_0 would not be rejected for any reasonable value of α. There is insufficient evidence to indicate a difference in the mean scores for the two groups on the question dealing with clean/sterile gloves at $\alpha \leq .10$.

e. Since the p-value was so small ($p = .02$), H_0 is rejected for any $\alpha > .02$. There is sufficient evidence to indicate a difference in the mean scores for the two groups on the question dealing with a stethoscope at $\alpha > .02$.

9.127 Some preliminary calculations:

	Panic Attacks per Week				Panic Attacks per Week		
Patient	Placebo	Inositol	D = Pl – In	Patient	Placebo	Inositol	D = Pl – In
1	0	0	0	12	0	2	–2
2	2	1	1	13	3	1	2
3	0	3	–3	14	3	1	2
4	1	2	–1	15	3	4	–1
5	0	0	0	16	4	2	2
6	10	5	5	17	6	4	2
7	2	0	2	18	15	21	–6
8	6	4	2	19	28	8	20
9	1	1	0	20	30	0	30
10	1	0	1	21	13	0	13
11	1	3	–2				

$$\bar{x}_d = \frac{\sum x_d}{n_d} = \frac{67}{21} = 3.19$$

$$s_d^2 = \frac{\sum x_d^2 - \frac{\left(\sum x_d\right)^2}{n_d}}{n_d - 1} = \frac{\left(1{,}575 - \frac{67^2}{21}\right)}{21 - 1} = 68.0619$$

To determine if there is a difference in the mean number of panic attacks between the placebo and the drug Inositol, we test:

H_0: $\mu_d = 0$
H_a: $\mu_d \neq 0$

The test statistic is $t = \dfrac{\bar{x}_d - 0}{s_d / \sqrt{n_d}} = \dfrac{3.19 - 0}{\sqrt{68.0619} / \sqrt{21}} = 1.77$

Since no α was given, we will use $\alpha = .05$. The rejection region requires $\alpha/2 = .05/2 = .025$ in each tail of the t distribution with df $= n_d - 1 = 21 - 1 = 20$. From Table VI, Appendix A, $t_{.025} = 2.086$. The rejection region is $t < -2.086$ or $t > 2.086$.

Since the observed value of the test statistic does not fall in the rejection region ($t = 1.77 \not> 2.086$), H_0 is not rejected. There is insufficient evidence to indicate that there is a difference in the mean number of panic attacks between the placebo and the drug Inositol at $\alpha = .05$. There is no evidence that the drug is effective.

9.129   a.   Let p_1 = proportion of female students who switched due to loss of interest in SME and p_2 = proportion of male students who switched due to lack of interest in SME.

Some preliminary calculations are:

$$\hat{p}_1 = \frac{x_1}{n_1} = \frac{74}{172} = .43; \quad \hat{p}_2 = \frac{x_2}{n_2} = \frac{72}{163} = .44; \quad \hat{p} = \frac{x_1 + x_2}{n_1 + n_2} = \frac{74 + 72}{172 + 163} = .436$$

To determine if the proportion of female students who switch due to lack of interest in SME differs from the proportion of males who switch due to a lack of interest, we test:

H_0: $p_1 - p_2 = 0$
H_a: $p_1 - p_2 \neq 0$

The test statistic is $z = \dfrac{(\hat{p}_1 - \hat{p}_2) - 0}{\sqrt{\hat{p}\hat{q}\left(\dfrac{1}{n_1} + \dfrac{1}{n_2}\right)}} = \dfrac{(.43 - .44) - 0}{\sqrt{.436(.564)\left(\dfrac{1}{172} + \dfrac{1}{163}\right)}} = -0.18$

The rejection region requires $\alpha/2 = .10/2 = .05$ in each tail of the z distribution. From Table IV, Appendix A, $z_{.05} = 1.645$. The rejection region is $z < -1.645$ or $z > 1.645$.

Since the observed value of the test statistic does not fall in the rejection region ($z = -0.18 \not< -1.645$), H_0 is not rejected. There is insufficient evidence to indicate the proportion of female students who switch due to lack of interest in SME differs from the proportion of males who switch due to a lack of interest in SME at $\alpha = .10$.

b. Let p_1 = proportion of female students who switched due to low grades in SME and p_2 = proportion of male students who switched due to low grades in SME.

Some preliminary calculations are:

$$\hat{p}_1 = \frac{x_1}{n_1} = \frac{33}{172} = .19; \quad \hat{p}_2 = \frac{x_2}{n_2} = \frac{44}{163} = .27$$

For confidence coefficient .90, $\alpha = .10$ and $\alpha/2 = .10/2 = .05$. From Table IV, Appendix A, $z_{.05} = 1.645$. The confidence interval is:

$$(\hat{p}_1 - \hat{p}_2) \pm z_{.05} \sqrt{\frac{\hat{p}_1 \hat{q}_1}{n_1} + \frac{\hat{p}_2 \hat{q}_2}{n_2}} \Rightarrow (.19 - .27) \pm 1.645 \sqrt{\frac{.19(.81)}{172} + \frac{.27(.73)}{163}}$$

$$\Rightarrow -.08 \pm .075 \Rightarrow (-.155, -.005)$$

We are 90% confident that the difference between the proportions of female and male switchers who lost confidence due to low grades in SME is between $-.155$ and $-.005$. Since the interval does not include 0, there is evidence to indicate the proportion of female switchers due to low grades is less than the proportion of male switchers due to low grades.

9.131 Let μ_1 = mean rating for music majors and μ_2 = mean rating of nonmusic majors.

For confidence coefficient .95, $\alpha = .05$ and $\alpha/2 = .05/2 = .025$. From Table IV, Appendix A, $z_{.025} = 1.96$. The confidence interval is:

$$(\bar{x}_1 - \bar{x}_2) \pm z_{.025} \sqrt{\frac{s_1^2}{n_1} + \frac{s_2^2}{n_2}} \Rightarrow (4.26 - 4.59) \pm 1.96 \sqrt{\frac{.81^2}{100} + \frac{.78^2}{100}}$$

$$\Rightarrow -.33 \pm .220 \Rightarrow (-.550, -.110)$$

We are 95% confident that the difference in the mean rating between music majors and nonmusic majors is between $-.55$ and $-.11$.

9.133 a. Some preliminary calculations are:

$$s_1^2 = \frac{\sum x_1^2 - \frac{\left(\sum x_1\right)^2}{n_1}}{n_1 - 1} = \frac{4209 - \frac{(145)^2}{5}}{5 - 1} = 1$$

$$s_2^2 = \frac{\sum x_2^2 - \frac{\left(\sum x_2\right)^2}{n_2}}{n_2 - 1} = \frac{4540 - \frac{(150)^2}{5}}{5 - 1} = 10$$

Let σ_1^2 = variance of measurements for instrument 1 and σ_2^2 = variance of measurements for instrument 2.

To determine if the variances of the measurements for the two instruments differ, we test:

$$H_0: \sigma_1^2 = \sigma_2^2$$
$$H_a: \sigma_1^2 \neq \sigma_2^2$$

The test statistic is $F = \dfrac{\text{Larger sample variance}}{\text{Smaller sample variance}} = \dfrac{s_2^2}{s_1^2} = \dfrac{10}{1} = 10$

Since no α was given, we will use $\alpha = .05$. The rejection region requires $\alpha/2 = .05/2 = .025$ in the upper tail of the F distribution with numerator df $v_1 = n_2 - 1 = 5 - 1 = 4$ and denominator df $v_2 = n_1 - 1 = 5 - 1 = 4$. From Table X, Appendix A, $F_{.025} = 9.60$. The rejection region is $F > 9.60$.

Since the observed value of the test statistic falls in the rejection reigon ($F = 10 > 9.60$), H_0 is rejected. There is sufficient evidence to indicate the variances of the measurements for the two instruments differ at $\alpha = .05$.

b. We must assume that the samples are randomly and independently selected from populations that are normally distributed.

9.135 Let p_1 = proportion of patients receiving Zyban who were not smoking one year later and p_2 = proportion of patients not receiving Zyban who were not smoking one year later.

Some preliminary calculations are:

$$\hat{p}_1 = \frac{x_1}{n_1} = \frac{71}{309} = .230 \qquad\qquad \hat{p}_2 = \frac{x_2}{n_2} = \frac{37}{306} = .121$$

$$\hat{p} = \frac{x_1 + x_2}{n_1 + n_2} = \frac{71 + 37}{309 + 306} = .176 \qquad \hat{q} = 1 = \hat{p} = 1 - .176 = .824$$

To determine if the antidepressant drug Zyban helped cigarette smokers kick their habit, we test:

$$H_0: p_1 - p_2 = 0$$
$$H_a: p_1 - p_2 > 0$$

The test statistic is $z = \dfrac{(\hat{p}_1 - \hat{p}_2) - 0}{\sqrt{\hat{p}\hat{q}\left(\dfrac{1}{n_1} + \dfrac{1}{n_2}\right)}} = \dfrac{(.230 - .121) - 0}{\sqrt{.176(.824)\left(\dfrac{1}{309} + \dfrac{1}{306}\right)}} = 3.55$

The rejection region requires $\alpha = .05$ in the upper tail of the z distribution. From Table IV, Appendix A, $z_{.05} = 1.645$. The rejection region is $z > 1.645$.

Since the observed value of the test statistic falls in the rejection region ($z = 3.55 > 1.645$), H_0 is rejected. There is sufficient evidence to indicate that the antidepressant drub Zyban helped cigarette smokers kick their habit at $\alpha = .05$.

9.137 Let μ_1 = mean number of timeout incidents for Option II students and μ_2 = mean number of timeout incidents for Option III students. To determine if the mean number of timeout incidents for students in Option III is greater than the mean number of timeout incidents for students in Option II, we test:

H_0: $\mu_1 = \mu_2$
H_0: $\mu_1 < \mu_2$

The test statistic is $z = \dfrac{(\bar{x}_1 - \bar{x}_2) - 0}{\sqrt{\dfrac{s_1^2}{n_1} + \dfrac{s_2^2}{n_2}}} = \dfrac{(78.67 - 102.87) - 0}{\sqrt{\dfrac{59.08^2}{100} + \dfrac{69.33^2}{55}}} = -2.19$

Since no value of α was given, we will use $\alpha = .05$. The rejection region requires $\alpha = .05$ in the lower tail of the z distribution. From Table IV, Appendix A, $z_{.05} = 1.645$. The rejection region is $z < -1.645$.

Since the observed value of the test statistic falls in the rejection region ($z = -2.19 < -1.645$), H_0 is rejected. There is sufficient evidence to indicate the mean number of timeout incidents for students in Option III is greater than the mean number of timeout incidents for students in Option II at $\alpha = .05$.

9.139 Using MINITAB, the descriptive statistics are:

Descriptive Statistics: CREATIVE, INFO, DECPERS, SKILLS, TASKID, AGE, EDYRS

Variable	GROUP	N	N*	Mean	SE Mean	StDev	Minimum	Q1	Median	Q3	Maximum
CREATIVE	NOSPILL	67	0	4.4478	0.0648	0.5304	3.0000	4.0000	4.0000	5.0000	5.0000
	SPILLOV	47	0	5.2553	0.0643	0.4408	5.0000	5.0000	5.0000	6.0000	6.0000
INFO	NOSPILL	67	0	4.657	0.272	2.226	1.000	3.000	5.000	7.000	7.000
	SPILLOV	47	0	5.213	0.251	1.719	1.000	4.000	5.000	7.000	7.000
DECPERS	NOSPILL	67	0	2.746	0.240	1.964	1.000	1.000	2.000	5.000	7.000
	SPILLOV	47	0	3.319	0.295	2.023	1.000	1.000	3.000	5.000	7.000
SKILLS	NOSPILL	67	0	4.821	0.134	1.100	2.000	4.000	5.000	5.000	7.000
	SPILLOV	47	0	5.851	0.169	1.161	2.000	5.000	6.000	7.000	7.000
TASKID	NOSPILL	67	0	4.239	0.141	1.156	1.000	3.000	4.000	5.000	7.000
	SPILLOV	47	0	4.809	0.296	2.028	1.000	3.000	5.000	7.000	7.000
AGE	NOSPILL	67	0	45.343	0.974	7.971	21.000	41.000	45.000	50.000	63.000
	SPILLOV	47	0	46.298	0.841	5.767	30.000	44.000	47.000	51.000	54.000
EDYRS	NOSPILL	67	0	13.224	0.160	1.312	12.000	12.000	13.000	14.000	18.000
	SPILLOV	47	0	13.085	0.155	1.060	12.000	12.000	13.000	14.000	16.000

Let μ_1 = mean characteristic for the workers with positive spillover and μ_2 = mean characteristic for the workers with no spillover. To determine if there is a difference in the mean characteristic between the positive spillover group and the no spillover group, we test:

H_0: $\mu_1 = \mu_2$
H_a: $\mu_1 \neq \mu_2$

The test statistic is $z = \dfrac{\bar{x}_1 - \bar{x}_2}{\sqrt{\left(\dfrac{s_1^2}{n_1} + \dfrac{s_2^2}{n_2}\right)}}$

Since no α was given, we will run all tests at $\alpha = .05$. The rejection region requires $\alpha/2 = .05/2 = .025$ in each tail of the z distribution. From Table IV, Appendix A, $z_{.025} = 1.96$. The rejection region is $z < -1.96$ or $z > 1.96$.

Use of Creative Ideas:

The test statistic is $z = \dfrac{\bar{x}_1 - \bar{x}_2}{\sqrt{\left(\dfrac{s_1^2}{n_1} + \dfrac{s_2^2}{n_2}\right)}} = \dfrac{5.2553 - 4.4478}{\sqrt{\dfrac{.4408^2}{47} + \dfrac{.5304^2}{67}}} = 8.85$

Since the observed value of the test statistic falls in the rejection region ($z = 8.85 > 1.96$), H_0 is rejected. There is sufficient evidence to indicate a difference in the mean use of creative ideas between the positive spillover group and the no spillover group at $\alpha = .05$.

Utilization of Information:

The test statistic is $z = \dfrac{\bar{x}_1 - \bar{x}_2}{\sqrt{\left(\dfrac{s_1^2}{n_1} + \dfrac{s_2^2}{n_2}\right)}} = \dfrac{5.213 - 4.657}{\sqrt{\dfrac{1.719^2}{47} + \dfrac{2.226^2}{67}}} = 1.50$

Since the observed value of the test statistic does not fall in the rejection region ($z = 1.50 \not> 1.96$), H_0 is not rejected. There is insufficient evidence to indicate a difference in the mean utilization of information between the positive spillover group and the no spillover group at $\alpha = .05$.

Decisions Regarding Personnel Matters:

The test statistic is $z = \dfrac{\bar{x}_1 - \bar{x}_2}{\sqrt{\left(\dfrac{s_1^2}{n_1} + \dfrac{s_2^2}{n_2}\right)}} = \dfrac{3.319 - 2.746}{\sqrt{\dfrac{2.023^2}{47} + \dfrac{1.964^2}{67}}} = 1.51$

Since the observed value of the test statistic does not fall in the rejection region ($z = 1.51 \not> 1.96$), H_0 is not rejected. There is insufficient evidence to indicate a difference in the mean participation in decisions regarding personnel matters between the positive spillover group and the no spillover group at $\alpha = .05$.

Good Use of Job Skills:

The test statistic is $z = \dfrac{\bar{x}_1 - \bar{x}_2}{\sqrt{\left(\dfrac{s_1^2}{n_1} + \dfrac{s_2^2}{n_2}\right)}} = \dfrac{5.851 - 4.821}{\sqrt{\dfrac{1.161^2}{47} + \dfrac{1.100^2}{67}}} = 4.76$

Since the observed value of the test statistic falls in the rejection region
($z = 4.76 \not< 1.96$), H_0 is rejected. There is sufficient evidence to indicate a difference in the mean good use of skills between the positive spillover group and the no spillover group at $\alpha = .05$.

Task Identity:

The test statistic is $z = \dfrac{\bar{x}_1 - \bar{x}_2}{\sqrt{\left(\dfrac{s_1^2}{n_1} + \dfrac{s_2^2}{n_2}\right)}} = \dfrac{4.809 - 4.239}{\sqrt{\dfrac{2.028^2}{47} + \dfrac{1.156^2}{67}}} = 1.74$

Since the observed value of the test statistic does not fall in the rejection region
($z = 1.74 \not< 1.96$), H_0 is not rejected. There is insufficient evidence to indicate a difference in the mean task identity between the positive spillover group and the no spillover group at $\alpha = .05$.

Of the 5 job-related characteristics, there are only 2 that have significantly different means for the 2 groups. They are "use of creative ideas" and "good use of skills". For both of these characteristics, the mean for the positive spillover group was significantly higher than the mean for the no spillover group.

Analysis of Variance: Comparing More than Two Means

10.1   Since only one factor is utilized, the treatments are the four levels (A, B, C, D) of the qualitative factor.

10.3   One has no control over the levels of the factors in an observational experiment. One does have control of the levels of the factors in a designed experiment.

10.5   a.   This is an observational experiment. The economist has no control over the factor levels or unemployment rates.

   b.   This is a designed experiment. The psychologist selects the feedback programs of interest and randomly assigns five rats to each program.

   c.   This is a designed experiment. The marketer has control of the selection of the national publications.

   d.   This is an observational experiment. The load on the generators is not controlled by the utility.

   e.   This is an observational experiment. One has no control over the distance of the haul, the goods hauled, or the price of diesel fuel.

   f.   This is an observational experiment. The student does not control which state is assigned to a portion of the country.

10.7   a.   The response variable is the age when cancer is first detected.

   b.   The experimental units are the people.

   c.   The factor in the study is type of screening method. There are 2 levels of screening method.

   d.   The treatments are the same as the factor levels in this problem. There are 2 factor levels or treatments: CT scan or chest X-ray.

10.9   a.   The experimental units are the cockatiels.

   b.   This experiment is a designed experiment. The birds were randomly divided into 3 groups and each group received a different treatment.

   c.   There is one factor in this study. The factor is the group.

   d.   There are three levels of the group variable – Group 1 received purified water, Group 2 received purified water and liquid sucrose, and Group 3 received purified water and liquid sodium chloride.

e.   There are 3 treatments in the study. Because there is only one factor, the treatments are the same as the factor levels.

f.   The response variable is the water consumption.

10.11 a.   There are 2 factors in this experiment – Temperature and Type of yeast. Temperature has 4 levels – 45, 48, 51, and 54°C. Type of yeast has 2 levels – Baker's and Brewer's.

b.   The response variable is the autolysis yield (recorded as a percentage).

c.   There are a total of 4 x 2 = 8 treatments for this experiment.

d.   This is a designed experiment. The levels of temperature and type of yeast are controlled by the researcher.

10.13 For a completely randomized design, independent random samples are selected for each treatment.

10.15 The conditions required for a valid ANOVA F-test in a completely randomized design are:

1.   The samples are randomly selected in an independent manner from k treatment populations.

2.   All k sampled populations have distributions that are approximately normal.

3.   The k population variances are equal (i.e., $\sigma_1^2 = \sigma_2^2 = ... = \sigma_k^2$)

10.17 a.   Using Table IX, $F_{.05}$ = 6.59 with v_1 = 3, v_2 = 4.

b.   Using Table XI, $F_{.01}$ = 16.69 with v_1 = 3, v_2 = 4.

c.   Using Table VIII, $F_{.10}$ = 1.61 with v_1 = 20, v_2 = 40.

d.   Using Table X, $F_{.025}$ = 3.87 with v_1 = 12, v_2 = 9.

10.19 a.   In the second dot diagram **B**, the difference between the sample means is small relative to the variability within the sample observations. In the first dot diagram **A**, the values in each of the samples are grouped together with a range of 4, while in the second diagram **B**, the range of values is 8.

b. For diagram **A**,

$$\bar{x}_1 = \frac{\sum x_1}{n} = \frac{7+8+9+9+10+11}{6} = \frac{54}{6} = 9$$

$$\bar{x}_2 = \frac{\sum x_2}{n} = \frac{12+13+14+14+15+16}{6} = \frac{84}{6} = 14$$

For diagram **B**,

$$\bar{x}_1 = \frac{\sum x_1}{n} = \frac{5+5+7+11+13+13}{6} = \frac{54}{6} = 9$$

$$\bar{x}_2 = \frac{\sum x_2}{n} = \frac{10+10+12+16+18+18}{6} = \frac{84}{6} = 14$$

c. For diagram **A**,

$$SST = \sum_{i=1}^{2} n_i(\bar{x}_i - \bar{x})^2 = 6(9 - 11.5)^2 + 6(14 - 11.5)^2 = 75$$

$$\left(\bar{x} = \frac{\sum x}{n} = \frac{54 + 84}{12} = 11.5 \right)$$

For diagram **B**,

$$SST = \sum_{i=1}^{2} n_i(\bar{x}_i - \bar{x})^2 = 6(9 - 11.5)^2 + 6(14 - 11.5)^2 = 75$$

d. For diagram **A**,

$$s_1^2 = \frac{\sum x_1^2 - \dfrac{\left(\sum x_1\right)^2}{n_1}}{n_1 - 1} = \frac{496 - \dfrac{54^2}{6}}{6 - 1} = 2$$

$$s_2^2 = \frac{\sum x_2^2 - \dfrac{\left(\sum x_2\right)^2}{n_2}}{n_2 - 1} = \frac{1186 - \dfrac{84^2}{6}}{6 - 1} = 2$$

$$SSE = (n_1 - 1)s_1^2 + (n_2 - 1)s_2^2 = (6 - 1)2 + (6 - 1)2 = 20$$

For diagram **B**,

$$s_1^2 = \frac{\sum x_1^2 - \dfrac{\left(\sum x_1\right)^2}{n_1}}{n_1 - 1} = \frac{558 - \dfrac{54^2}{6}}{6 - 1} = 14.4$$

$$s_2^2 = \frac{\sum x_2^2 - \dfrac{\left(\sum x_2\right)^2}{n_2}}{n_2 - 1} = \frac{1248 - \dfrac{84^2}{6}}{6 - 1} = 14.4$$

$$SSE = (n_1 - 1)s_1^2 + (n_2 - 1)s_2^2 = (6 - 1)14.4 + (6 - 1)14.4 = 144$$

e. For diagram **A**, SS(Total) = SST + SSE = 75 + 20 = 95

$$\text{SST is } \frac{\text{SST}}{\text{SS(Total)}} \times 100\% = \frac{75}{95} \times 100\% = 78.95\% \text{ of SS(Total)}$$

For diagram **B**, SS(Total) = SST + SSE = 75 + 144 = 219

$$\text{SST is } \frac{\text{SST}}{\text{SS(Total)}} \times 100\% = \frac{75}{219} \times 100\% = 34.25\% \text{ of SS(Total)}$$

f. For diagram **A**, $\text{MST} = \frac{\text{SST}}{k-1} = \frac{75}{2-1} = 75$

$$\text{MSE} = \frac{\text{SSE}}{n-k} = \frac{20}{12-2} = 2 \qquad F = \frac{\text{MST}}{\text{MSE}} = \frac{75}{2} = 37.5$$

For diagram **B**, $\text{MST} = \frac{\text{SST}}{k-1} = \frac{75}{2-1} = 75$

$$\text{MSE} = \frac{\text{SSE}}{n-k} = \frac{144}{12-2} = 14.4 \qquad F = \frac{\text{MST}}{\text{MSE}} = \frac{75}{14.4} = 5.21$$

g. The rejection region for both diagrams requires $\alpha = .05$ in the upper tail of the F distribution with numerator df $= k - 1 = 2 - 1 = 1$ and denominator df $= n - k = 12 - 2 = 10$. From Table IX, Appendix A, $F_{.05} = 4.96$. The rejection region is $F > 4.96$.

For diagram **A**, the observed value of the test statistic falls in the rejection region ($F = 37.5 > 4.96$). Thus, H_0 is rejected. There is sufficient evidence to indicate the samples were drawn from populations with different means at $\alpha = .05$.

For diagram **B**, the observed value of the test statistic falls in the rejection region ($F = 5.21 > 4.96$). Thus, H_0 is rejected. There is sufficient evidence to indicate the samples were drawn from populations with different means at $\alpha = .05$.

h. We must assume both populations are normally distributed with common variances.

10.21 Refer to Exercise 10.19, the ANOVA table is:

For diagram **A**:

Source	df	SS	MS	F
Treatment	1	75	75	37.5
Error	10	20	2	
Total	11	95		

For diagram **B**:

Source	df	SS	MS	F
Treatment	1	75	75	5.21
Error	10	144	14.4	
Total	11	219		

10.23   For all parts, the hypotheses are:

H_0: $\mu_1 = \mu_2 = \mu_3 = \mu_4 = \mu_5$
H_a: At least two treatment means differ

The rejection region for all parts is the same.

The rejection region requires $\alpha = .10$ in the upper tail of the F distribution with $\nu_1 = k - 1 = 5 - 1 = 4$ and $\nu_2 = n - k = 30 - 5 = 25$. From Table VIII, Appendix A, $F_{.10} = 2.18$. The rejection region is $F > 2.18$.

a.   SST $= .2(500) = 100$ SSE $=$ SS(Total) $-$ SST $= 500 - 100 = 400$

$$MST = \frac{SST}{k-1} = \frac{100}{5-1} = 25 \qquad MSE = \frac{SSE}{n-k} = \frac{400}{30-5} = 16$$

$$F = \frac{MST}{MSE} = \frac{25}{16} = 1.5625$$

Since the observed value of the test statistic does not fall in the rejection region ($F = 1.5625 \not> 2.18$), H_0 is not rejected. There is insufficient evidence to indicate differences among the treatment means at $\alpha = .10$.

b.   SST $= .5(500) = 250$ SSE $=$ SS(Total) $-$ SST $= 500 - 250 = 250$

$$MST = \frac{SST}{k-1} = \frac{250}{5-1} = 62.5 \qquad MSE = \frac{SSE}{n-k} = \frac{250}{30-5} = 10$$

$$F = \frac{MST}{MSE} = \frac{62.5}{10} = 6.25$$

Since the observed value of the test statistic falls in the rejection region ($F = 6.25 > 2.18$), H_0 is rejected. There is sufficient evidence to indicate differences among the treatment means at $\alpha = .10$.

c.   SST $= .8(500) = 400$ SSE $=$ SS(Total) $-$ SST $= 500 - 400 = 100$

$$MST = \frac{SST}{k-1} = \frac{400}{5-1} = 100 \qquad MSE = \frac{SSE}{n-k} = \frac{100}{30-5} = 4$$

$$F = \frac{MST}{MSE} = \frac{100}{4} = 25$$

Since the observed value of the test statistic falls in the rejection region, ($F = 25 > 2.18$), H_0 is rejected. There is sufficient evidence to indicate differences among the treatment means at $\alpha = .10$.

d.   The F ratio increases as the treatment sum of squares increases.

10.25   a.   The experimental units are the NCAA tennis coaches. The dependent variable is the rating on a 7-point scale of how important the coaches think the web site was. There is one factor and it is the division of the college/university that employs the tennis coach. Since there is only one factor, the treatments are the different levels of the factor. The factor has 3 levels and thus, 3 treatments which are Division I, Division II and Division III.

        b.   To determine if the mean ratings of the web sites are different for the different levels of coaches, we test:

             H_0: $\mu_I = \mu_{II} = \mu_{III}$
             H_a: At least one mean differs

        c.   Since the p-value is less than $\alpha = .05$ ($p < .003$), H_0 is rejected. There is sufficient evidence to indicate a difference in mean responses among the different divisions of the colleges and universities at $\alpha = .05$.

10.27   a.   The experimental design used is a completely randomized design.

        b.   There are 4 treatments in this experiment. The four treatments are the 4 "colonies" to which the robots were assigned – 3, 6, 9, or 12 robots per colony.

        c.   To determine if the mean energy expended (per robot) of the four different colony sizes differed, we test:

             H_0: $\mu_1 = \mu_2 = \mu_3 = \mu_4$
             H_a: At least two treatment means differ

        d.   The test statistic is $F = 7.70$. The p-value is $p < .001$. Since the p-value is less than $\alpha = .05$, H_0 is rejected. There is sufficient evidence to indicate that mean energy expended (per robot) of the four different colony sizes differed at $\alpha = .05$

10.29   a.   To determine if the average heights of male Australian school children differ among the age groups, we test:

             H_0: $\mu_1 = \mu_2 = \mu_3$
             H_a: At least two treatment means differ

        b.   The test statistic is $F = 4.57$. The p-value is $p = .01$. Since the p-value is less than α ($p = .01 < .05$), H_0 is rejected. There is sufficient evidence to indicate the average heights of male Australian school children differ among the age groups at $\alpha = .05$.

        c.   To determine if the average heights of female Australian school children differ among the age groups, we test:

             H_0: $\mu_1 = \mu_2 = \mu_3$
             H_a: At least two treatment means differ

             The test statistic is $F = 0.85$. The p-value is $p = .43$. Since the p-value is not less than α ($p = .43 > .05$), H_0 is not rejected. There is insufficient evidence to indicate the average heights of female Australian school children differ among the age groups at $\alpha = .05$.

d. For the boys, differences were found in the mean standardized heights among the three tertiles. For the girls, no differences were found in the mean standardized heights among the three tertiles.

10.31 To determine if a driver's propensity to engage in road rage is related to his/her income, we test:

H_o: $\mu_1 = \mu_2 = \mu_3$
H_a: At least two means differ

The test statistic is $F = 3.90$.

The p-value is $p < .01$. Since the p-value is so small, H_0 would be rejected for any reasonable value of α. There is sufficient evidence to indicate that a driver's propensity to engage in road rage is related to his/her income at $\alpha \geq .01$.

We have just concluded that the mean road rage score is related to income group. For the Under $30,000 group, the mean road rage score is 4.60. For the $30,000 to $60,000 group, the mean road rage score is 5.08. For the Over $60,000 group, the mean road rage score is 5.15. Thus, as the income increases, the mean road rage score also increases.

10.33 a. The subjects were randomly divided into 3 treatment groups with one group receiving an injection of scopolamine, another group receiving an injection of glycopyrrolate, and the third group receiving nothing. Thus, this is a completely randomized design.

b. There are 3 treatments: injection of scopolamine, injection of glycopyrrolate, and nothing. The response variable is the number of pairs recalled.

c. Using MINITAB, the descriptive statistics are:

Descriptive Statistics: Scopolamine, Placebo, No Drug

Variable	N	N*	Mean	SE Mean	StDev	Minimum	Q1	Median	Q3
Scopolamine	12	0	6.167	0.366	1.267	4.000	5.250	6.000	7.500
Placebo	8	0	9.375	0.532	1.506	7.000	8.250	9.500	10.000
No Drug	8	0	10.625	0.532	1.506	8.000	9.250	11.000	12.000

Variable	Maximum
Scopolamine	8.000
Placebo	12.000
No Drug	12.000

The sample means for the 3 groups are: 6.167, 9.375, and 10.625. There is not sufficient information to support the researcher's theory. We need to take into account the variability within each group.

d.   Using MINITAB, the ANOVA table is:

One-way ANOVA: Scopolamine, Placebo, No Drug

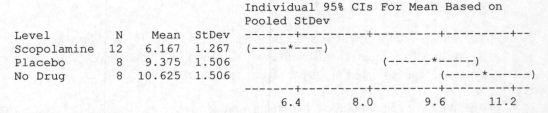

```
Source   DF      SS     MS      F      P
Factor    2   107.01  53.51  27.07  0.000
Error    25    49.42   1.98
Total    27   156.43

S = 1.406   R-Sq = 68.41%   R-Sq(adj) = 65.88%


                                      Individual 95% CIs For Mean Based on
                                      Pooled StDev
Level          N    Mean   StDev   -------+---------+---------+---------+--
Scopolamine   12   6.167   1.267   (-----*----)
Placebo        8   9.375   1.506                     (------*-----)
No Drug        8  10.625   1.506                               (-----*------)
                                   -------+---------+---------+---------+--
                                        6.4       8.0       9.6      11.2
```

Pooled StDev = 1.406

To determine if the mean number of words recalled differed among the 3 groups, we test:

H_0: $\mu_1 = \mu_2 = \mu_3$
H_a: At least two means differ

The test statistic is $F = 27.07$.

The p-value is $p = .000$. Since the p-value is so small, H_0 would be rejected for any reasonable value of α. There is sufficient evidence to indicate the mean number of words recalled differed among the 3 groups at $\alpha = .05$.

Prior to the experiment, the researchers theorized that the mean number of word pairs recalled for the scopolamine group would be less that the corresponding means for the other 2 groups. From the printout, the sample mean for the scopolamine group is 6.167. The means for the placebo group and the nothing group are 9.375 and 10.625.

Since the sample mean for the scopolamine group is much smaller than the other two, it appears that the researchers' theory was correct.

10.35   Using MINITAB, the results of the analysis are:

One-way ANOVA: UMRB-1, UMRB-2, UMBR-3, SWRA, SD

```
Source   DF       SS      MS      F       P
Factor    4    5.836   1.459   7.25   0.001
Error    21    4.225   0.201
Total    25   10.061

S = 0.4486    R-Sq = 58.00%    R-Sq(adj) = 50.00%
```

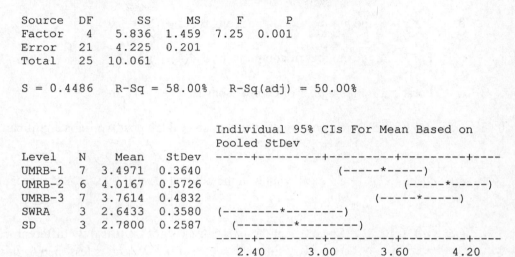

```
                                   Individual 95% CIs For Mean Based on
                                   Pooled StDev
Level    N    Mean    StDev   -----+---------+---------+---------+----
UMRB-1   7  3.4971   0.3640                      (-----*-----)
UMRB-2   6  4.0167   0.5726                              (-----*-----)
UMRB-3   7  3.7614   0.4832                         (-----*-----)
SWRA     3  2.6433   0.3580   (--------*--------)
SD       3  2.7800   0.2587     (--------*--------)
                               -----+---------+---------+---------+----
                                  2.40      3.00      3.60      4.20
```

Pooled StDev = 0.4486

To determine if there are differences among the mean Al/Be ratios for the 5 boreholes, we test:

H_0: $\mu_1 = \mu_2 = \mu_3 = \mu_4 = \mu_5$
H_a: At least 2 means differ

From the printout, the test statistic is $F = 7.25$ and the p-value is $p = 0.001$. Since the p-value is less than $\alpha = .10$ ($p = .001 < .10$), H_0 is rejected. There is sufficient evidence to indicate a difference in the mean Al/Be ratios among the 5 boreholes at $\alpha = .10$.

10.37   The experiment-wise error rate is the probability of declaring at least one pair of means different when they are not among all comparisons in the experiment.

10.39   a.   The confidence interval $(-10, 5)$ for $(\mu_1 - \mu_2)$ means that we are confident that the difference between $\mu_1 - \mu_2$ is between -10, and 5. Since this interval spans both positive and negative numbers, we cannot conclude that either mean is significantly larger.

   b.   The confidence interval $(-10, -5)$ for $(\mu_1 - \mu_2)$ means that we are confident that the difference between $\mu_1 - \mu_2$ is between -10, and -5. Since this interval spans only negative numbers, we can conclude that μ_2 is significantly larger than μ_1.

   c.   The confidence interval $(5, 10)$ for $(\mu_1 - \mu_2)$ means that we are confident that the difference between $\mu_1 - \mu_2$ is between 5, and 10. Since this interval spans only positive numbers, we can conclude that μ_1 is significantly larger than μ_2.

10.41 The number of pairwise comparisons is equal to $k(k-1)/2$.

    a. For $k = 3$, the number of comparisons is $3(3-1)/2 = 3$.

    b. For $k = 5$, the number of comparisons is $5(5-1)/2 = 10$.

    c. For $k = 4$, the number of comparisons is $4(4-1)/2 = 6$.

    d. For $k = 10$, the number of comparisons is $10(10-1)/2 = 45$.

10.43 $(\mu_1 - \mu_2)$: (2, 15)   Since all values in the interval are positive, μ_1 is significantly greater than μ_2.

     $(\mu_1 - \mu_3)$: (4, 7)   Since all values in the interval are positive, μ_1 is significantly greater than μ_3.

     $(\mu_1 - \mu_4)$: (-10, 3)   Since 0 is in the interval, μ_1 is not significantly different from μ_4. However, since the center of the interval is less than 0, μ_4 is larger than μ_1.

     $(\mu_2 - \mu_3)$: (-5, 11)   Since 0 is in the interval, μ_2 is not significantly different from μ_3. However, since the center of the interval is greater than 0, μ_2 is larger than μ_3.

     $(\mu_2 - \mu_4)$: (-12, -6)   Since all values in the interval are negative, μ_4 is significantly greater than μ_2.

     $(\mu_3 - \mu_4)$: (-8, -5)   Since all values in the interval are negative, μ_4 is significantly greater than μ_3.

Thus, the largest mean is μ_4 followed by μ_1, μ_2, and μ_3.

10.45   a. The total number of pairwise comparisons made in the Bonferroni analysis is $k(k-1)/2 = 4(4-1)/2 = 6$.

     b. The Sourdough treatment yielded the significantly highest mean soluble magnesium level. There is no significant difference in the mean soluble magnesium level between the Control and the Yeast groups. These two treatments yielded the lowest mean soluble magnesium level.

     c. The experiment-wise error rate is .05. This means that the probability of declaring at least one pair significantly different when it is not is .05.

10.47   a. There are $c = \dfrac{k(k-1)}{2} = \dfrac{4(4-1)}{2} = 6$ pairwise comparisons.

     b. There are no significant differences in the mean energy expended among the colony sizes of 3, 6, and 9. However, the mean energy expended for the colony containing 12 robots was significantly less than that for all other colony sizes.

10.49   a.   To determine if there are differences in the mean dental fear scores among the three groups, we test:

H_0: $\mu_1 = \mu_2 = \mu_3$
H_a: At least two treatment means differ

The test statistic is $F = 4.43$ and the p-value is $p < .05$. If we use $\alpha = .05$, then the rejection region will be p-value $< .05$. Since the p-value is less than .05, we reject H_0. There is sufficient evidence to indicate there are differences in the mean dental fear scores among the three groups at $\alpha = .05$.

b.   First, we arrange the means in order from the largest to the smallest. Then we draw a line between the means that are not significantly different. The results are:

Questionnaire   53.8

Slide   43.1 |
                     |
Control   41.8 |

10.51   a.   For $\alpha = .01$, the level of confidence is 99%. We are 99% confident that the three confidence intervals collectively include the true differences in the means.

b.   There are a total of $k(k - 1)/2 = 3(3 - 1)/2 = 3$ pair-wise comparisons. The pairs are:

$(\mu_1 - \mu_2)$, $(\mu_1 - \mu_3)$, and $(\mu_2 - \mu_3)$.

c.   A table comparing the means is:

Income Group	Mean
Under $30,000	4.60
$30,000 - $60,000	5.08 |
Over $60,000	5.15 |

The means for the $30,000 – $60,000 group and the Over $60,000 group are connected by a bar. This indicates that these means are not significantly different.

d.   The comparison involving μ_2 and μ_3 will result in a confidence interval that contain 0.

10.53 Using MINITAB, the Tukey's multiple comparison procedure yielded:

```
Tukey 90% Simultaneous Confidence Intervals
All Pairwise Comparisons

Individual confidence level = 98.44%


UMRB-1 subtracted from:

         Lower    Center   Upper   --------+---------+---------+---------+-
UMRB-2  -0.1369   0.5195   1.1760                    (----*-----)
UMRB-3  -0.3664   0.2643   0.8950                   (----*----)
SWRA    -1.6680  -0.8538  -0.0396         (------*------)
SD      -1.5314  -0.7171   0.0971          (------*------)
                                  --------+---------+---------+---------+-
                                       -1.2      0.0       1.2       2.4


UMRB-2 subtracted from:

         Lower    Center   Upper   --------+---------+---------+---------+-
UMRB-3  -0.9117  -0.2552   0.4012                 (------*----)
SWRA    -2.2077  -1.3733  -0.5390   (-------*------)
SD      -2.0710  -1.2367  -0.4023    (------*------)
                                  --------+---------+---------+---------+-
                                       -1.2      0.0       1.2       2.4


UMRB-3 subtracted from:

       Lower    Center   Upper   --------+---------+---------+---------+-
SWRA  -1.9323  -1.1181  -0.3039   (------*-----)
SD    -1.7957  -0.9814  -0.1672    (------*------)
                                --------+---------+---------+---------+-
                                     -1.2      0.0       1.2       2.4


SWRA subtracted from:

     Lower   Center  Upper   --------+---------+---------+---------+-
SD  -0.8267  0.1367  1.1001               (-------*-------)
                             --------+---------+---------+---------+-
                                  -1.2      0.0       1.2       2.4
```

UMRB-2	UMRB-3	UMRB-1	SD	SWRA
4.0167	3.7614	3.4971	2.7800	2.6433

The mean AI/Be ratios for UMRB-2 and UMRB-3 are significantly greater than the mean AI/Be ratios for SD and SWRA. The mean AI/Be ratio for UMRB-1 is significantly greater than the mean AI/Be ratio for SWRA. No other differences exist.

10.55 $\alpha^* = \dfrac{2\alpha}{k(k-1)} = \dfrac{2(.03)}{3(3-1)} = .01$

$t_{\alpha^*/2} = t_{.01/2} = t_{.005} \approx 2.66$ (From Table VI, Appendix A, with df $= 26 + 25 + 25 - 3 = 73$)

$s = \sqrt{MSE} = \sqrt{56.7671} = 7.5344$ (From Exercise 10.36)

Group T and Group V:

$$\left(\bar{x}_i - \bar{x}_j\right) \pm t_{\alpha^*/2}(s)\sqrt{\dfrac{1}{n_i} + \dfrac{1}{n_j}} \Rightarrow (10.5 - 3.9) \pm 2.66(7.5344)\sqrt{\dfrac{1}{26} + \dfrac{1}{25}}$$

$$\Rightarrow 6.6 \pm 5.614 \Rightarrow (.986,\ 12.214)$$

Group T and Group C:

$$(10.5 - 1.4) \pm 2.66(7.5344)\sqrt{\dfrac{1}{26} + \dfrac{1}{25}} \Rightarrow 9.1 \pm 5.614 \Rightarrow (3.486,\ 14.714)$$

Group V and Group C:

$$(3.9 - 1.4) \pm 2.66(7.5344)\sqrt{\dfrac{1}{25} + \dfrac{1}{25}} \Rightarrow 2.5 \pm 5.668 \Rightarrow (-3.168,\ 8.168)$$

We are 97% confident that the intervals collectively contain all the differences between the true means. Intervals that contain 0 do not support a conclusion that the true means differ. The interval comparing the means for the V and C groups contains 0. Thus, there is no significant difference in the drops between the group having a volunteer only and the control group.

If both of the endpoints are positive, then the first mean is significantly greater than the second mean. From above, the mean drop for group T (Volunteer + trained dog) is significantly greater than the mean drops of Groups V (Volunteer only) and C (Control).

A convenient summary of the results is below, where the treatment means are listed from highest to lowest with a solid line connecting those means that are not significantly different.

```
        Group T  Group V       Group C
         10.5      3.9            1.4
```

10.57 A paired difference experiment is a randomized block design with only 2 experimental units per block. In the paired difference design, the pairs are the blocks. However, a randomized block design is not limited to 2 experimental units per block.

10.59 a. There are 3 blocks used since Block df $= b - 1 = 2$ and 5 treatments since the treatment df $= k - 1 = 4$.

b. There were 15 observations since the Total df $= n - 1 = 14$.

c. H_0: $\mu_1 = \mu_2 = \mu_3 = \mu_4 = \mu_5$
H_a: At least two treatment means differ

d. The test statistic is $F = \dfrac{MST}{MSE} = 9.109$

e. The rejection region requires $\alpha = .01$ in the upper tail of the F distribution with $\nu_1 = k - 1 = 5 - 1 = 4$ and $\nu_2 = n - k - b + 1 = 15 - 5 - 3 + 1 = 8$. From Table XI, Appendix A, $F_{.01} = 7.01$. The rejection region is $F > 7.01$.

f. Since the observed value of the test statistic falls in the rejection region ($F = 9.109 > 7.01$), H_0 is rejected. There is sufficient evidence to indicate that at least two treatment means differ at $\alpha = .01$.

g. The assumptions necessary to assure the validity of the test are as follows:

1. The probability distributions of observations corresponding to all the block-treatment combinations are normal.
2. The variances of all the probability distributions are equal.
3. The b blocks are randomly selected and all k treatments are applied (in random order) to each block.

10.61 a. The ANOVA Table is as follows:

Source	df	SS	MS	F	p
Treatment	2	12.0317	6.0158	50.96	0.000
Block	3	71.7492	23.9164	202.59	0.000
Error	6	.7083	.1181		
Total	11	84.4892			

b. To determine if the treatment means differ, we test:

H_0: $\mu_A = \mu_B = \mu_C$
H_a: At least two treatment means differ

The test statistic is $F = \dfrac{MST}{MSE} = 50.986$

The rejection region requires $\alpha = .05$ in the upper tail of the F distribution with $\nu_1 = k - 1 = 3 - 1 = 2$ and $\nu_2 = n - k - b + 1 = 12 - 3 - 4 + 1 = 6$. From Table IX, Appendix A, $F_{.05} = 5.14$. The rejection region is $F > 5.14$.

Since the observed value of the test statistic falls in the rejection region ($F = 50.958 > 5.14$), H_0 is rejected. There is sufficient evidence to indicate that the treatment means differ at $\alpha = .05$.

c. To see if the blocking was effective, we test:

H_0: $\mu_1 = \mu_2 = \mu_3 = \mu_4$
H_a: At least two block means differ

The test statistic is $F = \dfrac{\text{MSB}}{\text{MSE}} = 202.586$

The rejection region requires $\alpha = .05$ in the upper tail of the F distribution with $v_1 = b - 1 = 4 - 1 = 3$ and $v_2 = n - k - b + 1 = 12 - 3 - 4 + 1 = 6$. From Table IX, Appendix A, $F_{.05} = 4.76$. The rejection region is $F > 4.76$.

Since the observed value of the test statistic falls in the rejection region ($F = 202.586 > 4.76$), H_0 is rejected. There is sufficient evidence to indicate that blocking was effective in reducing the experimental error at $\alpha = .05$.

d. From the printout, we are given the differences in the sample means. The difference between treatment B and treatment A is positive (1.1125). Thus, treatment B has a larger mean than treatment A. The difference between treatment C and treatment A is negative (-1.325). Thus, treatment A has a larger mean than treatment C. The difference between treatment C and treatment B is negative (-2.450). Thus, treatment B has a larger mean than treatment C. Finally, treatment B has the largest mean, treatment A has the next largest, and treatment C has the smallest. Since 0 is not contained in any of the confidence intervals, all the means are statistically different.

e. The assumptions necessary to assure the validity of the inferences above are:

1. The probability distributions of observations corresponding to all the block-treatment combinations are normal.
2. The variances of all the probability distributions are equal.
3. The b blocks are randomly selected and all k treatments are applied (in random order) to each block.

10.63 a. A randomized block design is preferred over a completely randomized design because there is large variation in the number of oil rigs running among the months. By blocking for months, we can reduce this variation.

b. The treatments in the experiment are the states. There are 3 states: California, Utah, and Alaska.

c. The blocks in this experiment are the selected months. There are 3 blocks: Month 1, Month 2, and Month 3.

d. To determine if there is a difference in the mean number of oil rigs running in each state per month, we test:

H_0: $\mu_C = \mu_U = \mu_A$
H_a: At least two treatment means differ

e. From the printout, the test statistics is $F = 38.07$ and the p-value is 0.002.

f. From the SPSS output, there is no difference in the mean number of oil rigs running between Alaska and Utah. All other means are different. Thus, the mean number of oil rigs running for California is significantly greater than the mean number of oil rigs running for Alaska and Utah.

10.65 a. This is a randomized block design. The blocks are the 12 plots of land. The treatments are the three methods used on the shrubs: fire, clipping, and control. The response variable is the mean number of flowers produced. The experimental units are the 36 shrubs.

b.

Treatment

Plot	Fire	Clipping	Control
1	Shrub 2	Shrub 1	Shrub 3
2	Shrub 3	Shrub 2	Shrub 1
. . .	. . .	. . .	. . .
12	Shrub 1	Shrub 3	Shrub 2

c. To determine if there is a difference in the mean number of flowers produced among the three treatments, we test:

H_0: $\mu_1 = \mu_2 = \mu_3$
H_a: The mean number of flowers produced differ for at least two of the methods.

The test statistic is $F = 5.42$ and $p = .009$. We can reject the null hypothesis at the $\alpha > .009$ level of significance. At least two of the methods differ with respect to mean number of flowers produced by pawpaws.

d. The mean number of flowers produced by the Control is significantly less than the mean number of flowers produced by the Clipping method and the Burning method. The mean number of flowers produced by the Clipping method does not differ from the mean number of flowers produced by the Burning method.

10.67 a. The experimenters expected there to be much variation in the number of participants from week to week (more participants at the beginning and fewer as time goes on). Thus, by blocking on weeks, this extraneous source of variation can be controlled.

b.   Using MINITAB, the ANOVA results are:

Two-way ANOVA: Walkers versus Week, Group

```
Source   DF      SS      MS      F       P
Group    4    1185.0   296.25  39.87   0.000
Week     5     386.4    77.28  10.40   0.000
Error   20     148.6     7.43
Total   29    1720.0
```

S = 2.726   R-Sq = 91.36%   R-Sq(adj) = 87.47%

```
                   Individual 95% CIs For Mean Based on
                   Pooled StDev
Group    Mean    ---------+---------+---------+---------+
1       2.6667   (--*---)
2      17.0000                             (---*---)
3      20.6667                                   (--*---)
4      10.5000                  (---*--)
5       9.1667               (---*---)
                 ---------+---------+---------+---------+
                     6.0       12.0      18.0      24.0
```

The ANOVA table is:

Source	df	SS	MS	F	p
Prompt	4	1185.0	296.25	39.87	0.000
Week	5	386.4	77.28	10.40	0.000
Error	20	148.6	7.43		
Total	29	1720.0			

c.   To determine if a difference exists in the mean number of walkers per week among the five walker groups, we test:

H_0: $\mu_1 = \mu_2 = \mu_3 = \mu_4 = \mu_5$
H_a: At least two treatment means differ

where μ_i represents the mean number of walkers in group i.

The test statistic is $F = 39.87$.

The rejection region requires $\alpha = .05$ in the upper tail of the F distribution with $\nu_1 = k - 1 = 5 - 1 = 4$ and $\nu_2 = n - k - b + 1 = 30 - 4 - 6 + 1 = 20$. From Table IX, Appendix A, $F_{.05} = 2.69$. The rejection region is $F > 2.69$.

Since the observed value of the test statistic falls in the rejection region ($F = 39.87 > 2.69$), H_0 is rejected. There is sufficient evidence to indicate differences exist among the mean number of walkers per week among the 5 walker groups at $\alpha = .05$.

d.   Using SAS, the Tukey multiple comparison results are:

```
            Tukey's Studentized Range (HSD) Test for walkers

NOTE: This test controls the Type I experimentwise error rate, but it generally has a
                    higher Type II error rate than REGWQ.


                Alpha                                     0.05
                Error Degrees of Freedom                    20
                Error Mean Square                         7.43
                Critical Value of Studentized Range    4.23186
                Minimum Significant Difference          4.7092


            Means with the same letter are not significantly different.


            Tukey Grouping          Mean      N      group

                         A        20.667      6      FREQ/HIGH
                         A
                         A        17.000      6      FREQ/LOW

                         B        10.500      6      INFREQ/LOW
                         B
                         B         9.167      6      INFREQ/HIGH

                         C         2.667      6      CONTROL
```

There is no difference in the mean number of walkers between the FREQ/HIGH group and the FREQ/LOW group. There is no difference in the mean number of walkers between the INFREQ/LOW group and the INFREQ/HIGH group.

The mean number of walkers in the Control group is significantly less than the means for the other 4 groups. The mean number of walkers in the INFREQ/LOW and INFRE/HIGH groups is significantly less than the mean number of walkers in the FREQ/LOW and FREQ/HIGH groups.

e.   In order for the above inferences to be valid, the following assumptions must hold:

1)   The probability distributions of observations corresponding to all block-treatment conditions are normal.
2)   The variances of all the probability distributions are equal.
3)   The b blocks are randomly selected and all k treatments are applied (in random order) to each block.

10.69   Using MINITAB, the ANOVA results are:

Two-way ANOVA: Temp versus Plant, Student

```
Source    DF      SS       MS      F      P
Plant      2   0.122  0.06100   0.02  0.981
Student    9  18.415  2.04611   0.63  0.754
Error     18  58.038  3.22433
Total     29  76.575
```

```
S = 1.796    R-Sq = 24.21%    R-Sq(adj) = 0.00%
```

```
                    Individual 95% CIs For Mean Based on
                    Pooled StDev
Plant    Mean    -------+---------+---------+---------+--
Live    95.21    (----------------*----------------)
None    95.20    (----------------*----------------)
Photo   95.34      (----------------*----------------)
                 -------+---------+---------+---------+--
                    94.50     95.20     95.90     96.60
```

To determine if there are differences among the mean temperatures among the three treatments, we test:

H_0: $\mu_1 = \mu_2 = \mu_3$
H_a: At least two treatment means differ

The test statistic is $F = 0.02$. The associated p-value is $p = .981$. Since the p-value is very large, there is no evidence of a difference in mean temperature among the three treatments. Since there is no difference, we do not need to compare the means. It appears that the presence of plants or pictures of plants does not reduce stress.

10.71   a.   The data were collected using a randomized complete block design and must be analyzed as a randomized complete block. In this case, each treatment was measured on each gene. Within each gene, the observations are not independent.

        b.   To determine if the mean growth differs among the 3 light/dark conditions, we test:

             H_0: $\mu_1 = \mu_2 = \mu_3$
             H_a: At least two treatment means differ

c. Using MINITAB, the results are:

Two-way ANOVA: Response versus Treatment, Block

```
Source       DF       SS         MS       F      P
Treatment     2    9.093    4.54660    5.33  0.006
Block       102   50.715    0.49720    0.58  0.999
Error       204  174.138    0.85362
Total       308  233.946
```

```
S = 0.9239   R-Sq = 25.56%   R-Sq(adj) = 0.00%
```

```
                            Individual 95% CIs For Mean Based on
                            Pooled StDev
Treatment       Mean    -----+---------+---------+---------+----
1          -0.318164    (---------*--------)
2           0.101547                            (---------*--------)
3          -0.090780                 (--------*--------)
                        -----+---------+---------+---------+----
                          -0.40     -0.20     -0.00      0.20
```

The test statistic is $F = 5.33$ and the p-value is $p = 0.006$. Since the p-value is less than $\alpha = .05$ ($p = .006 < .05$), H_0 is rejected. There is sufficient evidence to indicate a difference in mean growth among the 3 light/dark conditions at $\alpha = .05$.

d. We will use the Bonferroni method to compare the treatment means.

$$\alpha^* = \frac{2\alpha}{k(k-1)} = \frac{2(.05)}{3(3-1)} = .0167$$

$$t_{\alpha^*/2} = t_{.0167/2} = t_{.0083} \approx 2.39$$

FULL-DARK and TR-LIGHT:

$$\left(\bar{x}_i - \bar{x}_j\right) \pm t_{\alpha^*/2}(s)\sqrt{\frac{1}{n_i} + \frac{1}{n_j}} \Rightarrow (-.318164 - .101547) \pm 2.39(.9239)\sqrt{\frac{1}{103} + \frac{1}{103}}$$

$$\Rightarrow -.419711 \pm .307694 \Rightarrow (-.727405, \ -.112017)$$

FULL-DARK and TR-DARK:

$$(-.318164 - (-.090780)) \pm 2.39(.9239)\sqrt{\frac{1}{103} + \frac{1}{103}} \Rightarrow -.2273841 \pm .307694$$

$$\Rightarrow (-.535078, \ .080310)$$

TR-LIGHT and TR-DARK:

$$(.101547 - (-.090780)) \pm 2.39(.9239)\sqrt{\frac{1}{103} + \frac{1}{103}} \Rightarrow .192327 \pm .307694$$

$$\Rightarrow (-.115367, \ .500021)$$

A convenient summary of the results is below, where the treatment means are listed from highest to lowest with a solid line connecting those means that are not significantly different.

TR-LIGHT	TR-DARK	FULL-DARK
.101547	-.090780	-.318164

The mean growth for TR-LIGHT is significantly greater than the mean growth for FULL-DARK.

10.73   In a complete factorial experiment, the treatments are formed by taking every combination of all levels of both factors. The total number of treatments is equal to the total number of factor-level combinations.

10.75   The conditions that are required for valid inferences for a factorial ANOVA are:

1.   The response distribution for each factor-level combination (treatment) is normal.

2.   The response variance is constant for all treatments.

3.   Random and independent samples of experimental units are associated with each treatment.

10.77   a.   There are two factors.

b.   No, we cannot tell whether the factors are qualitative or quantitative.

c.   Yes. There are 3 levels of factor A and 5 levels of factor B.

d.   A treatment would consist of a combination of one level of factor A and one level of factor B. There are a total of $3 \times 5 = 15$ treatments.

e.   One problem with only one replicate is there are no degrees of freedom for error. This is overcome by having at least two replicates.

10.79   The ANOVA table is:

Source	df	SS	MS	F
A	2	.8	24.4000	2.00
B	3	5.3	1.7667	8.83
AB	6	9.6	1.6000	8.00
Error	12	2.4	.2000	
Total	23	18.1		

df for A is $a - 1 = 3 - 1 = 2$
df for B is $b - 1 = 4 - 1 = 3$
df for AB is $(a - 1)(b - 1) = 2(3) = 6$
df for Error is $n - ab = 24 - 3(4) = 12$
df for Total is $n - 1 = 24 - 1 = 23$

$$SSE = SS(Total) - SSA - SSB - SSAB = 18.1 - .8 - 5.3 - 9.6 = 2.4$$

$$MSA = \frac{SSA}{a-1} = \frac{.8}{3-1} = .40 \qquad MSB = \frac{SSB}{b-1} = \frac{5.3}{4-1} = 1.7667$$

$$MSAB = \frac{SSAB}{(a-1)(b-1)} = \frac{9.6}{(3-1)(4-1)} = 1.60$$

$$MSE = \frac{MSE}{n-ab} = \frac{2.4}{24-3(4)} = .2000$$

$$F_A = \frac{MSA}{MSE} = \frac{.4000}{.2000} = 2.00 \qquad F_B = \frac{MSB}{MSE} = \frac{1.7667}{.2000} = 8.83$$

$$F_{AB} = \frac{MSAB}{MSE} = \frac{1.6000}{.2000} = 8.00$$

b. Sum of Squares for Treatment $= SSA + SSB + SSAB = .8 = 5.3 + 2.6 = 15.7$.

$$MST = \frac{SST}{ab-1} = \frac{15.7}{3(4)-1} = 1.4273$$

$$F_T = \frac{MST}{MSE} = \frac{1.4273}{.2000} = 7.14$$

To determine if the treatment means differ, we test:

H_0: $\mu_1 = \mu_2 = \cdots = \mu_{12}$
H_a: At least two treatment means differ

The test statistic is $F = 7.14$.

The rejection region requires $\alpha = .05$ in the upper tail of the F distribution with $\nu_1 = ab - 1 = 3(4) - 1 = 11$ and $\nu_2 = n - ab = 24 - 3(4) = 12$. From Table IX, Appendix A, $F_{.05} \approx 2.75$. The rejection region is $F > 2.75$.

Since the observed value of the test statistic falls in the rejection region ($F = 7.14 > 2.75$), H_0 is rejected. There is sufficient evidence to indicate the treatment means differ at $\alpha = .05$.

c. Yes. We need to partition the Treatment Sum of Squares into the Main Effects and Interaction Sum of Squares. Then we test whether factors A and B interact. Depending on the conclusion of the test for interaction, we either test for main effects or compare the treatment means.

d. Two factors are said to interact if the effects of one factor on the dependent variable are not the same at different levels of the second factor. If the factors interact, then tests for main effects are not necessary. We need to compare the treatment means for one factor at each level of the second.

e. To determine if the factors interact, we test:

H_0: Factors A and B do not interact to affect the response mean
H_a: Factors A and B do interact to affect the response mean

The test statistic is $F = \dfrac{\text{MS}AB}{\text{MSE}} = 8.00$

The rejection region requires $\alpha = .05$ in the upper tail of the F distribution with $v_1 = (a-1)(b-1) = (3-1)(4-1) = 6$ and $v_2 = n - ab = 24 - 3(4) = 12$. From Table IX, Appendix A, $F_{.05} = 3.00$. The rejection region is $F > 3.00$.

Since the observed value of the test statistic falls in the rejection region ($F = 8.00 > 3.00$), H_0 is rejected. There is sufficient evidence to indicate the two factors interact to affect the response mean at $\alpha = .05$.

f. No. Testing for main effects is not warranted. Instead, we compare the treatment means of one factor at each level of the second factor.

10.81 a. $\text{SS}A = .2(1000) = 200$, $\text{SS}B = .1(1000) = 100$, $\text{SS}AB = .1(1000) = 100$
$\text{SSE} = \text{SS(Total)} - \text{SS}A - \text{SS}B - \text{SS}AB = 1000 - 200 - 100 - 100 = 600$
$\text{SST} = \text{SS}A + \text{SS}B + \text{SS}AB = 200 + 100 + 100 = 400$

$\text{MS}A = \dfrac{\text{SS}A}{a-1} = \dfrac{200}{3-1} = 100 \qquad \text{MS}B = \dfrac{\text{SS}B}{b-1} = \dfrac{100}{3-1} = 50$

$\text{MS}AB = \dfrac{\text{SS}AB}{(a-1)(b-1)} = \dfrac{100}{(3-1)(3-1)} = 25$

$\text{MSE} = \dfrac{\text{SSE}}{n-ab} = \dfrac{600}{27-3(3)} = 33.333 \qquad \text{MST} = \dfrac{\text{SST}}{ab-1} = \dfrac{400}{3(3)-1} = 50$

$F_A = \dfrac{\text{MS}A}{\text{MSE}} = \dfrac{100}{33.333} = 3.00 \qquad F_B = \dfrac{\text{MS}B}{\text{MSE}} = \dfrac{50}{33.333} = 1.50$

$F_{AB} = \dfrac{\text{MS}AB}{\text{MSE}} = \dfrac{25}{33.333} = .75 \qquad F_T = \dfrac{\text{MST}}{\text{MSE}} = \dfrac{50}{33.333} = 1.50$

Source	df	SS	MS	F
A	2	200	100	3.00
B	2	100	50	1.50
AB	4	100	25	.75
Error	18	600	33.333	
Total	26	1000		

To determine whether the treatment means differ, we test:

H_0: $\mu_1 = \mu_2 = \cdots = \mu_9$
H_a: At least two treatment means differ

The test statistic is $F = \dfrac{\text{MST}}{\text{MSE}} = 1.50$

Suppose $\alpha = .05$. The rejection region requires $\alpha = .05$ in the upper tail of the F distribution with $\nu_1 = ab - 1 = 3(3) - 1 = 8$ and $\nu_2 = n - ab = 27 - 3(3) = 18$. From Table IX, Appendix A, $F_{.05} = 2.51$. The rejection region is $F > 2.51$.

Since the observed value of the test statistic does not fall in the rejection region ($F = 1.50 \ngtr 2.51$), H_0 is not rejected. There is insufficient evidence to indicate the treatment means differ at $\alpha = .05$. Since there are no treatment mean differences, we have nothing more to do.

b. $\text{SS}A = .1(1000) = 100$, $\text{SS}B = .1(1000) = 100$, $\text{SS}AB = .5(1000) = 500$

$\text{SSE} = \text{SS(Total)} - \text{SS}A - \text{SS}B - \text{SS}AB = 1000 - 100 - 100 - 500 = 300$

$\text{SST} = \text{SS}A + \text{SS}B + \text{SS}AB = 100 + 100 + 500 = 700$

$$\text{MS}A = \frac{\text{SS}A}{a-1} = \frac{100}{3-1} = 50 \qquad \text{MS}B = \frac{\text{SS}B}{b-1} = \frac{100}{3-1} = 50$$

$$\text{MS}AB = \frac{\text{SS}AB}{(a-1)(b-1)} = \frac{500}{(3-1)(3-1)} = 125$$

$$\text{MSE} = \frac{\text{SSE}}{n-ab} = \frac{300}{27-3(3)} = 16.667 \qquad \text{MST} = \frac{\text{SST}}{ab-1} = \frac{700}{9-1} = 87.5$$

$$F_A = \frac{\text{MS}A}{\text{MSE}} = \frac{50}{16.667} = 3.00 \qquad F_B = \frac{\text{MS}B}{\text{MSE}} = \frac{50}{16.667} = 3.00$$

$$F_{AB} = \frac{\text{MS}AB}{\text{MSE}} = \frac{125}{16.667} = 7.50 \qquad F_T = \frac{\text{MST}}{\text{MSE}} = \frac{87.5}{16.667} = 5.25$$

Source	df	SS	MS	F
A	2	100	50	3.00
B	2	100	50	3.00
AB	4	500	125	7.50
Error	18	300	16.667	
Total	26	1000		

To determine if the treatment means differ, we test:

H_0: $\mu_1 = \mu_2 = \cdots = \mu_9$
H_a: At least two treatment means differ

The test statistic is $F = \dfrac{\text{MST}}{\text{MSE}} = 5.25$

The rejection region requires $\alpha = .05$ in the upper tail of the F distribution with $\nu_1 = ab - 1 = 3(3) - 1 = 8$ and $\nu_2 = n - ab = 27 - 3(3) = 18$. From Table IX, Appendix A, $F_{.05} = 2.51$. The rejection region is $F > 2.51$.

Since the observed value of the test statistic falls in the rejection region ($F = 5.25 >$ 2.51), H_0 is rejected. There is sufficient evidence to indicate the treatment means differ at $\alpha = .05$.

Since the treatment means differ, we next test for interaction between factors A and B. To determine if factors A and B interact, we test:

H_0: Factors A and B do not interact to affect the mean response
H_a: Factors A and B do interact to affect the mean response

The test statistic is $F = \dfrac{MSAB}{MSE} = 7.50$

The rejection region requires $\alpha = .05$ in the upper tail of the F distribution with $v_1 = (a-1)(b-1) = (3-1)(3-1) = 4$ and $v_2 = n - ab = 27 - 3(3) = 18$. From Table IX, Appendix A, $F_{.05} = 2.93$. The rejection region is $F > 2.93$.

Since the observed value of the test statistic falls in the rejection region ($F = 7.50 >$ 2.93), H_0 is rejected. There is sufficient evidence to indicate the factors A and B interact at $\alpha = .05$. Since interaction is present, no tests for main effects are necessary.

c. $SSA = .4(1000) = 400$, $SSB = .1(1000) = 100$, $SSAB = .2(1000) = 200$

$SSE = SS(\text{Total}) - SSA - SSB - SSAB = 1000 - 400 - 100 - 200 = 300$

$SST = SSA + SSB + SSAB = 400 + 100 + 200 = 700$

$$MSA = \frac{SSA}{a-1} = \frac{400}{3-1} = 50 \qquad\qquad MSB = \frac{SSB}{b-1} = \frac{100}{3-1} = 50$$

$$MSAB = \frac{SSAB}{(a-1)(b-1)} = \frac{200}{(3-1)(3-1)} = 50$$

$$MSE = \frac{SSE}{n-ab} = \frac{300}{27-3(3)} = 16.667 \qquad MST = \frac{SST}{ab-1} = \frac{700}{3(3)-1} = 87.5$$

$$F_A = \frac{MSA}{MSE} = \frac{200}{16.667} = 12.00 \qquad\qquad F_B = \frac{MSB}{MSE} = \frac{50}{16.667} \; 3.00$$

$$F_{AB} = \frac{MSAB}{MSE} = \frac{50}{16.667} = 3.00 \qquad\qquad F_T = \frac{MST}{MSE} = \frac{87.5}{16.667} = 5.25$$

Source	df	SS	MS	F
A	2	400	200	12.00
B	2	100	50	3.00
AB	4	200	50	3.00
Error	18	300	16.667	
Total	26	1000		

To determine if the treatment means differ, we test:

H_0: $\mu_1 = \mu_2 = \cdots = \mu_9$
H_a: At least two treatment means differ

The test statistic is $F = \dfrac{\text{MST}}{\text{MSE}} = 5.25$

The rejection region requires $\alpha = .05$ in the upper tail of the F distribution with $v_1 = ab - 1 = 3(3) - 1 = 8$ and $v_2 = n - ab = 27 - 3(3) = 18$. From Table IX, Appendix A, $F_{.05} = 2.51$. The rejection region is $F > 2.51$.

Since the observed value of the test statistic falls in the rejection region ($F = 5.25 > 2.51$), H_0 is rejected. There is sufficient evidence to indicate the treatment means differ at $\alpha = .05$.

Since the treatment means differ, we next test for interaction between factors A and B. To determine if factors A and B interact, we test:

H_0: Factors A and B do not interact to affect the mean response
H_a: Factors A and B do interact to affect the mean response

The test statistic is $F = \dfrac{\text{MS}AB}{\text{MSE}} = 3.00$

The rejection region requires $\alpha = .05$ in the upper tail of the F distribution with $v_1 = (a - 1)(b - 1) = (3 - 1)(3 - 1) = 4$ and $v_2 = n - ab = 27 - 3(3) = 18$. From Table IX, Appendix A, $F_{.05} = 2.93$. The rejection region is $F > 2.93$.

Since the observed value of the test statistic falls in the rejection region ($F = 3.00 > 2.93$), H_0 is rejected. There is sufficient evidence to indicate the factors A and B interact at $\alpha = .05$. Since interaction is present, no tests for main effects are necessary.

d. $\text{SS}A = .4(1000) = 400$, $\text{SS}B = .4(1000) = 400$, $\text{SS}AB = .1(1000) = 100$

$\text{SSE} = \text{SS(Total)} - \text{SS}A - \text{SS}B - \text{SS}AB = 1000 - 400 - 400 - 100 = 100$

$\text{SST} = \text{SS}A + \text{SS}B + \text{SS}AB = 400 + 400 + 100 = 900$

$$\text{MS}A = \frac{\text{SS}A}{a - 1} = \frac{400}{3 - 1} = 200 \qquad\qquad \text{MS}B = \frac{\text{SS}B}{b - 1} = \frac{400}{3 - 1} = 200$$

$$\text{MS}AB = \frac{\text{SS}AB}{(a - 1)(b - 1)} = \frac{100}{(3 - 1)(3 - 1)} = 25$$

$$\text{MSE} = \frac{\text{SSE}}{n - ab} = \frac{100}{27 - 3(3)} = 5.556 \qquad\qquad \text{MST} = \frac{\text{SST}}{ab - 1} = \frac{900}{3(3) - 1} = 112.5$$

$$F_A = \frac{\text{MS}A}{\text{MSE}} = \frac{200}{5.556} = 36.00 \qquad\qquad F_B = \frac{\text{MS}B}{\text{MSE}} = \frac{200}{5.556} = 36.00$$

$$F_{AB} = \frac{\text{MS}AB}{\text{MSE}} = \frac{25}{5.556} = 4.50 \qquad\qquad F_T = \frac{\text{MST}}{\text{MSE}} = \frac{112.5}{5.556} = 20.25$$

Source	df	SS	MS	F
A	2	400	200	36.00
B	2	400	200	36.00
AB	4	100	25	4.50
Error	18	100	5.556	
Total	26	1000		

To determine if the treatment means differ, we test:

H_0: $\mu_1 = \mu_2 = \cdots = \mu_9$

H_a: At least two treatment means differ

The test statistic is $F = \dfrac{\text{MST}}{\text{MSE}} = 20.25$

The rejection region requires $\alpha = .05$ in the upper tail of the F distribution with $v_1 = ab - 1 = 3(3) - 1 = 8$ and $v_2 = n - ab = 27 - 3(3) = 18$. From Table IX, Appendix A, $F_{.05} = 2.51$. The rejection region is $F > 2.51$.

Since the observed value of the test statistic falls in the rejection region ($F = 20.25 > 2.51$), H_0 is rejected. There is sufficient evidence to indicate the treatment means differ at $\alpha = .05$.

Since the treatment means differ, we next test for interaction between factors A and B. To determine if factors A and B interact, we test:

H_0: Factors A and B do not interact to affect the mean response

H_a: Factors A and B do interact to affect the mean response

The test statistic is $F = \dfrac{\text{MS}AB}{\text{MSE}} = 4.50$

The rejection region requires $\alpha = .05$ in the upper tail of the F distribution with $v_1 = (a - 1)(b - 1) = (3 - 1)(3 - 1) = 4$ and $v_2 = n - ab = 27 - 3(3) = 18$. From Table IX, Appendix A, $F_{.05} = 2.93$. The rejection region is $F > 2.93$.

Since the observed value of the test statistic falls in the rejection region ($F = 4.50 > 2.93$), H_0 is rejected. There is sufficient evidence to indicate the factors A and B interact at $\alpha = .05$. Since interaction is present, no tests for main effects are necessary.

10.83　a.　This is a 6×6 factorial design.

　　　b.　There are 2 factors – level of coagulant and acidity level. The factor "level of coagulation" has 6 levels – 5, 10, 20, 50, 100, and 200 milligrams per liter. The factor "acidity level" has 6 levels – 4.0, 5.0, 6.0, 7.0, 8.0, and 9.0. There are $6 \times 6 = 36$ treatments for this study. Each treatment is a combination of one level of coagulation level and acidity level. An example of a treatment is coagulation level 5 and acidity level 4.0.

10.85   a.   There are 2 factors in this experiment. The first factor is wash-up event. There are 3 levels of this factor. The second factor is strata. There are 4 levels of strata: coarse-branching, medium-branching, fine-branching, and hydroid algae.

       b.   There are a total of $3 \times 4 = 12$ treatments in the experiment.

       c.   There are 2 replicates for each treatment in the experiment.

       d.   The total sample size is $12 \times 2 = 24$.

       e.   The response variable is the mussel density (percent per square centimeter).

       f.   The first ANOVA F-test is the test for differences among the treatments. If this test is significant, then the test for interaction follows.

Since not enough information is provided to test for differences among the treatment means, we will test for interaction.

To determine if Event and Strata interact to affect the mean mussel density, we test:

H_0: Event and Strata do not interact
H_a: Event and Strata interact

The test statistic is $F = 1.91$.

The p-value is $p > .05$. Since the p-value ($> .05$) is not less than $\alpha = .05$, H_0 is not rejected. There is insufficient evidence to indicate Event and Strata interact at $\alpha = .05$.

       g.   Since the interaction is not significant, we will test for the main effects.

Event:

To determine if differences exist in the mean mussel densities among the three events, we test:

H_0: $\mu_1 = \mu_2 = \mu_3$
H_a: At least two event means differ

The test statistic is $F = 0.35$.

The p-value is $p > .05$. Since the p-value ($> .05$) is not less than $\alpha = .05$, H_0 is not rejected. There is insufficient evidence to indicate differences exist in the mean mussel densities among the three events at $\alpha = .05$.

Strata:

To determine if differences exist in the mean mussel densities among the four strata, we test:

H_0: $\mu_1 = \mu_2 = \mu_3 = \mu_4$
H_a: At least two strata means differ

The test statistic is $F = 217.33$.

The p-value is $p < .05$. Since the p-value ($< .05$) is less than $\alpha = .05$, H_0 is rejected. There is sufficient evidence to indicate differences exist in the mean mussel densities among the four strata at $\alpha = .05$.

h.   The mean density for Strata Hydroid is significantly larger than the mean densities of the other 3 strata. The mean density for Strata Fine is significantly larger than the mean densities of Strata Medium and Strata Coarse. There is no difference in the mean densities of Strata Medium and Strata Coarse.

10.87   a.   A 2×2 factorial design was used for this study. The two factors are color and type of questions, each at 2 levels. There are $2 \times 2 = 4$ treatments. The 4 treatments are "blue, difficult", "blue, simple", "red, difficult", and "red, simple".

b.   Since the p-value is small (p-value $< .03$), H_0 is rejected. There is sufficient evidence to indicate that color and type of questions interact to affect mean exam scores. This means that the effect of color on the mean exam scores depends on the difficulty of the questions.

c.   Using MINITAB, the graph is:

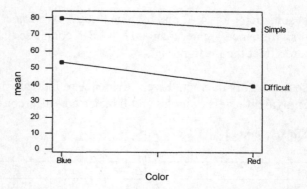

Looking at the graph, the lines are not parallel. The difference between the Blue and Red "simple" questions is not as great as the difference between the Blue and Red "difficult" questions. The effect of color on the exam scores depends on the level of difficulty.

10.89   a.

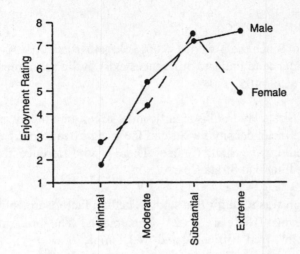

The pattern in the graph suggests interaction between suspense and gender. The difference between the mean ratings of the two genders varies depending on the suspense category. Notice that the lines cross twice.

b.   To determine if interaction exists between suspense and gender, we test:

H_0: Gender and Suspense do not interact
H_a: Gender and Suspense interact

The test statistic is $F = 4.42$ and p-value = .007. We reject the null hypothesis. For any level of significance greater than .007, we can conclude that Gender and Suspense interact to affect the mean rating.

c.   Because the factors interact, there is evidence to indicate the difference between the mean enjoyment levels of males and females are different for different suspense levels.

10.91   a.   The ANOVA table is:

Source	df	F	p-value
AGE	3 – 1 = 2	11.93	< .001
BOOK	3 – 1 = 2	23.64	< .001
AGE x BOOK	2 x 2 = 4	2.99	< .05
Error	108 – 9 = 99		
Total	107		

b.   The total number of treatments is 3 x 3 = 9. The treatments are:

18, Photos   24, Photos   30, Photos
18, Drawings   24, Drawings   30, Drawings
18, Control   24, Control   30, Control

c. To determine if the interaction between AGE and BOOK exists, we test:

H_0: AGE and BOOK do not interact to affect reenactment scores
H_a: AGE and BOOK do interact to affect reenactment scores

The test statistic is $F = 2.99$ and the p-value is $p < .05$. Since the p-value is less than $\alpha = .05$, H_0 is rejected. There is sufficient evidence to indicate that AGE and BOOK interact to affect target action scores at $\alpha = .05$. This means that the effect of AGE on the reenactment scores depends on the level of BOOK.

d. We do not need to conduct the main effect tests because the interaction between BOOK and AGE is significant. If the interaction is significant, it is possible that it could cover up any main effects.

e. Using MINITAB, a graph of the results is:

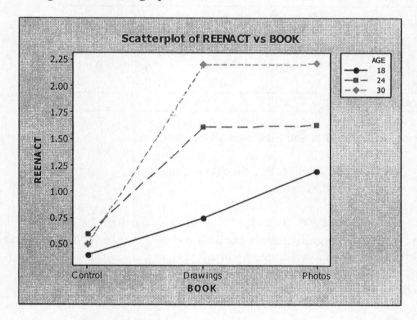

From the multiple comparisons, we can conclude:

Children at age 18 months: The mean reenactment score for children using books with Photos is significantly greater than the mean reenactment score for children using books with no photos or drawings. There are no other differences.

Children at age 24 months: The mean reenactment scores for children using books with Photos and Drawings are significantly greater than the mean reenactment score for children using books with no photos or drawings. There is no difference between the mean reenactment scores for children using books with Photos and children using books with Drawings.

Children at age 30 months: The mean reenactment scores for children using books with Photos and Drawings are significantly greater than the mean reenactment score for children using books with no photos or drawings. There is no difference between   the mean reenactment scores for children using books with Photos and children using books with Drawings.

10.93 a. Using MINITAB, the ANOVA results are:

ANOVA: Weight versus Diet, Size

```
Factor   Type   Levels   Values
Diet     fixed      2    1, 2
Size     fixed      2    1, 2


Analysis of Variance for Weight

Source      DF      SS       MS       F        P
Diet         1   0.0124   0.0124     0.22   0.645
Size         1   8.0679   8.0679   141.18   0.000
Diet*Size    1   0.0364   0.0364     0.64   0.432
Error       24   1.3715   0.0571
Total       27   9.4883
```

Source	df	SS	MS	F	P
Diet	1	0.0124	0.0124	0.22	0.645
Size	1	8.0679	8.0679	141.18	0.000
Interaction	1	0.0364	0.0364	0.64	0.432
Error	24	1.3715	0.0571		
Total	27	9.4883			

b. To determine if Diet and Size interact, we test:

H_0: The factors Diet and Size do not interact
H_a: Diet and Size interact

The test statistic is $F = .64$ and p-value $= .432$. We do not reject the null hypothesis. For any level of significance lower than $\alpha = .432$, we could not conclude that the factors interact. We therefore test the main effects.

To determine if the mean weight differs for the two diets, we test:

H_0: $\mu_{regular} = \mu_{supplement}$
H_a: $\mu_{regular} \neq \mu_{supplement}$

The test statistic is $F = .22$ and p-value $= .645$. We do not reject the null hypothesis. For any significance level lower than $\alpha = .645$, we cannot conclude that the mean weight of the kidney differed for the two diets.

To determine if the mean weight differs between the two sizes, we test:

H_0: $\mu_{lean} = \mu_{obese}$
H_a: $\mu_{lean} \neq \mu_{obese}$

The test statistic is $F = 141.18$ and p-value $= .000$. We reject the null hypothesis. For any significance level greater than or equal to $\alpha = .000$, we conclude that the mean weight of the kidney differs for the two sizes.

10.95   a.   The treatment totals for the 4 treatments are:

Low Load, Ambiguous: $25(18.0) = 450$

Low Load, Common: $25(7.8) = 195$

High Load, Ambiguous: $25(6.1) = 152.5$

High Load, Common: $25(6.3) = 157.5$

b. $CM = \dfrac{\left(\sum y_i\right)^2}{n} = \dfrac{(450 + 152.5 + 195 + 157.5)^2}{25 + 25 + 25 + 25} = \dfrac{955^2}{100} = 9,120.25$

c. $SS(LOAD) = \dfrac{\sum A_i^2}{br} - CM = \dfrac{(450 + 195)^2 + (152.5 + 157.5)^2}{2(25)} - 9,120.25$

$\qquad = \dfrac{645^2 + 310^2}{50} - 9,120.25 = \dfrac{512,125}{50} - 9,120.25$

$\qquad = 10,242.5 - 9,120.25 = 1,122.25$

$SS(NAME) = \dfrac{\sum B_i^2}{ar} - CM = \dfrac{(450 + 152.5)^2 + (195 + 157.5)^2}{2(25)} - 9,120.25$

$\qquad = \dfrac{602.5^2 + 352.5^2}{50} - 9,120.25 = \dfrac{487,262.5}{50} - 9,120.25$

$\qquad = 9,745.25 - 9,120.25 = 625$

$SS(LOADxNAME) = \dfrac{\sum\sum AB_{ij}^2}{r} - SS(NAME) - SS(LOAD) - CM$

$\qquad = \dfrac{450^2 + 195^2 + 152.5^2 + 157.5^2}{25} - 625 - 1,122.25 - 9,120.25$

$\qquad = 11,543.5 - 625 - 1,122.25 - 9,120.25 = 676$

d.   The treatment variances and sums of squares are:

Low Load, Ambiguous: $s_{LA}^2 = 15^2 = 225$,

$\sum (x_i - \bar{x}_{LA})^2 = (n_{LA} - 1)s_{LA}^2 = (25-1)225 = 5,400$

Low Load, Common: $s_{LC}^2 = 9.5^2 = 90.25$,

$\sum (x_i - \bar{x}_{LC})^2 = (n_{LC} - 1)s_{LC}^2 = (25-1)90.25 = 2,166$

High Load, Ambiguous: $s_{HA}^2 = 9.5^2 = 90.25$,

$\sum (x_i - \bar{x}_{HA})^2 = (n_{HA} - 1)s_{HA}^2 = (25-1)90.25 = 2,166$

High Load, Common: $s_{HC}^2 = 10^2 = 100$,

$$\sum(x_i - \bar{x}_{HC})^2 = (n_{HC} - 1)s_{HC}^2 = (25 - 1)100 = 2,400$$

e.   SSE = 5,400 + 2,166 + 2,166 + 2,400 = 12,132

f.   SS(Total) = SS(LOAD) + SS(NAME) + SS(LOAD x NAME) + SSE

$$= 1,122.25 + 625 + 676 + 12,132 = 14,555.25$$

g.   The ANOVA table is:

Source	df	SS	MSE	F
LOAD	$2 - 1 = 1$	1,122.25	1,122.25	8.88
NAME	$2 - 1 = 1$	625	625	4.95
LOAD x NAME	$1 \times 1 = 1$	676	676	5.35
Error	$100 - 4 = 96$	12,132	126.375	
Total	99	14,555.25		

h.   The researchers calculated the F value to be $F = 5.34$. We calculated the F to be $F = 5.35$. The difference could be due to some round-off error.

i.   To determine if LOAD and NAME interact to affect the number of jelly beans taken, we test:

H_0: LOAD and NAME do not interact to affect the number of jelly beans taken
H_a: LOAD and NAME do interact to affect the number of jelly beans taken

The test statistic is $F = 5.35$.

The rejection region requires $\alpha = .05$ in the upper tail of the F distribution with numerator df $= \nu_1 = (a - 1)(b - 1) = (2 - 1)(2 - 1) = 1$ and denominator df $= \nu_2 = n - ab$ $= 100 - 2(2) = 96$. From Table IX, Appendix A, $F \approx 3.96$. The rejection region is $F > 3.96$.

Since the observed value of the test statistic falls in the rejection region ($F = 5.35 > 3.96$), H_0 is rejected. There is sufficient evidence to indicate that LOAD and NAME interact to affect the number of jelly beans taken at $\alpha = .05$.

Since the interaction is significant, there is no need to run the main effects test. Using MINITAB, a graphical display of the results is:

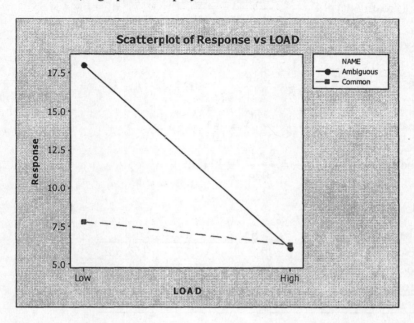

From the graph, it appears that when the NAME of the jelly beans is Ambiguous, there is a significant difference in the mean number of jelly beans taken for the two loads. When LOAD is Low, there is a significantly greater mean number of jelly beans taken than when LOAD is high. When the NAME of the jelly beans is Common, there appears to be no significant difference in the mean number of jelly beans taken between the 2 levels of load.

j.   In order for the techniques to be valid, the following assumptions are necessary:

   1. We must assume that the response distribution for each factor-level combination (treatment) is normal

   2. We must assume that the response variance is constant for all treatments.

   3. We must assume that we have random and independent samples of experimental units associated with each treatment.

10.97 In a completely randomized design, independent random selection of treatments to be assigned to experimental units is required; in a randomized block design, sets (blocks) of matched experimental units are employed, with an experimental unit in each block randomly assigned a treatment.

10.99 When the overall level of significance of a multiple comparisons procedure is α, the level of significance for each comparison is less than α.

10.101 a. SS(Treatment) = SS(Total) − SS(Block) − SSE = 22.308 − 10.688 − .288 = 11.332

$$MS(Treatment) = \frac{SS(Treatment)}{k-1} = \frac{11.332}{4-1} = 3.777, \ df = k - 1 = 3$$

$$MS(Block) = \frac{SS(Block)}{b-1} = \frac{10.688}{5-1} = 2.672, \ df = b - 1 = 4$$

$$MSE = \frac{SSE}{n-k-b+1} = \frac{.288}{20-4-5+1} = .024, \ df = n - k - b + 1 = 12$$

$$Treatment \ F = \frac{MS(Treatment)}{MSE} = \frac{3.777}{.024} = 157.375$$

$$Block \ F = \frac{MS(Block)}{MSE} = \frac{2.672}{.024} = 111.33$$

The ANOVA Table is:

Source	df	SS	MS	F
Treatment	3	11.332	3.777	157.38
Block	4	10.688	2.672	111.33
Error	12	.288	.024	
Total	19	22.308		

b. To determine if there is a difference among the treatment means, we test:

H_0: $\mu_A = \mu_B = \mu_C = \mu_D$
H_a: At least two treatment means differ

The test statistic is $F = \dfrac{MS(Treatment)}{MSE} = 157.38$

The rejection region requires $\alpha = .05$ in the upper tail of the F distribution with $v_1 = k - 1 = 4 - 1 = 3$ and $v_2 = n - k - b + 1 = 20 - 4 - 5 + 1 = 12$. From Table IX, Appendix A, $F_{.05} = 3.49$. The rejection region is $F > 3.49$.
Since the observed value of the test statistic falls in the rejection region ($F = 157.38 > 3.49$), H_0 is rejected. There is sufficient evidence to indicate a difference among the treatment means at $\alpha = .05$.

c. Since there is evidence of differences among the treatment means, we need to compare the treatment means. The number of pairwise comparisons is $\dfrac{k(k-1)}{2} = \dfrac{4(4-1)}{2} = 6$.

d. To determine if there are difference among the block means, we test:

H_0: All the block means are the same
H_a: At least two block means differ

The test statistic is $F = \dfrac{MS(Block)}{MSE} = 111.33$

The rejection region requires $\alpha = .05$ in the upper tail of the F distribution with $v_1 = b - 1 = 5 - 1 = 4$ and $v_2 = n - k - b + 1 = 20 - 4 - 5 + 1 = 12$. From Table IX, Appendix A, $F_{.05} = 3.26$. The rejection region is $F > 3.26$.

Since the observed value of the test statistic falls in the rejection region ($F = 111.33 > 3.26$), H_0 is rejected. There is sufficient evidence that the block means differ at $\alpha = .05$.

10.103   a.   The data are collected as a completely randomized design because five boxes of each size were randomly selected and tested.

b.   Yes. The confidence intervals surrounding each of the means do not overlap. This would indicate that there is a difference in the means for the two sizes.

c.   No. Several of the confidence intervals overlap. This would indicate that the mean compression strengths of the sizes that have intervals that overlap are not significantly different.

10.105   a.   The ethical behavior of the salesperson.

b.   There are two factors (with two levels at each factor):
Type of sales job (high tech versus low tech) and
Sales task (new account development).

c.   The treatments are the $2 \times 2 = 4$ factor-level combinations of type of sales job and sales task.

d.   The college students are the experimental units.

10.107   To determine if differences in the mean percent of pronoun errors exist among the three groups, we test:

H_0: $\mu_1 = \mu_2 = \mu_3$
H_a: At least two treatment means differ

where μ_i represents the mean percent of pronoun errors for the ith group.

The test statistic is $F = 8.295$.

Since the p-value is so small ($p = .0016$), H_0 is rejected for any $\alpha > .0016$. There is sufficient evidence to indicate a difference in the mean percent of pronoun errors among the three groups.

A multiple comparison was run on the three group means. A line connects the means of the SLI and the YND groups. The mean percent of pronoun errors is not different for these two groups. The mean percent of pronoun errors for the OND group is significantly smaller than the mean percent of pronoun errors for both the SLI and the YND groups.

10.109 a. To determine if the mean scores of the four groups differ, we test:

H_0: $\mu_1 = \mu_2 = \mu_3 = \mu_4$
H_a: At least two treatment means differ

The test statistic is $F = 39.1$.

The rejection region requires $\alpha = .05$ in the upper tail of the F distribution with $\nu_1 = k - 1 = 4 - 1 = 3$ and $\nu_2 = n - k = 278 - 4 = 274$. Using Table IX, Appendix A, $F_{.05} \approx 2.60$. The rejection region is $F > 2.60$.

Since the observed value of the test statistic falls in the rejection region ($F = 39.1 > 2.60$), H_0 is rejected. There is sufficient evidence to indicate a difference in the mean scores among the four groups at $\alpha = .05$.

b. No. No information was provided on the sample means. The test of hypotheses performed in part a just indicates that differences exist, but not where. Further information is needed.

c. There is no significant difference in the mean scores between the forensic psychiatric patients and the nonforensic psychiatric patients. There is no significant difference in the mean scores between the honest college students and the "fake bad" college students. However, the mean scores for the two college student groups are significantly greater than the mean scores for the psychiatric patients. Again the higher the score, the more deliberately distorted responses.

10.111 Using MINITAB, the ANOVA results are:

One-way ANOVA: Spillage versus Cause

```
Source   DF       SS     MS      F      P
Cause     3     6576   2192   0.74  0.535
Error    44   130565   2967
Total    47   137140

S = 54.47    R-Sq = 4.80%    R-Sq(adj) = 0.00%
```

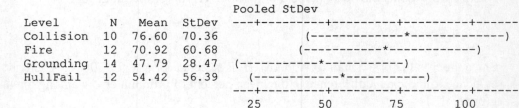

```
                             Individual 95% CIs For Mean Based on
                             Pooled StDev
Level        N    Mean   StDev  ---+---------+---------+---------+------
Collision   10   76.60   70.36               (--------------*--------------)
Fire        12   70.92   60.68             (------------*------------)
Grounding   14   47.79   28.47  (------------*------------)
HullFail    12   54.42   56.39    (-------------*------------)
                                ---+---------+---------+---------+------
                                  25        50        75       100

Pooled StDev = 54.47
```

The ANOVA table is:

Source	df	SS	MS	F	p
Cause	3	6,576	2,192	0.74	0.535
Error	44	130,565	2,967	2,967	
Total	47	137,140			

To determine if differences exist among the mean spillage amounts of the four accident types, we test:

H_0: $\mu_1 = \mu_2 = \mu_3 = \mu_4$
H_a: At least two treatment means differ

The test statistic is $F = 0.74$.

The p-value is $p = .535$. Since the p-value is greater than α ($p = .535 > .01$), H_0 is not rejected. There is insufficient evidence to indicate differences exist among the mean spillage amounts of the four accident types at $\alpha = .01$.

10.113 a.

Source	df	F-value	p-value
Time Period	3	11.25	.0001
Station	9		
Error	27		
Total	39		

b. To determine if the mean biomass densities differ among the three time periods, we test:

H_0: $\mu_1 = \mu_2 = \mu_3 = \mu_4$
H_a: The mean biomass density differ for at least two of the time periods

The test statistic is $F = 11.25$ and $p = .0001$. We can reject the null hypothesis at the $\alpha > .0001$ level of significance. At least two of the time periods differ with respect to the mean biomass of water boatmen nymphs.

c. The time period with the largest mean biomass is either 7/9-7/23 or 7/24-8/8. There is no significant difference between the two means. The time period with the smallest mean biomass is 8/24-8/31.

10.115 a. Using MINITAB, the ANOVA table is:

Source	df	SS	MS	F	P
Expressions	5	23.09	4.62	3.96	0.007
Error	30	34.99	1.17		
Total	35	58.07			

To determine if differences exist among the mean dominance ratings of the six facial expressions, we test:

H_0: $\mu_1 = \mu_2 = \mu_3 = \mu_4 = \mu_5 = \mu_6$
H_a: At least two treatment means differ

where μ_i represents the mean dominance rating for facial expression i.
The test statistic is $F = 3.96$.

The rejection region requires $\alpha = .10$ in the upper tail of the F distribution with $\nu_1 = k -1 = 6 - 1 = 5$ and $\nu_2 = n - k = 36 - 6 = 30$. Using Table VIII, Appendix A, $F_{.10} = 2.05$. The rejection region is $F > 2.05$.

Since the observed value of the test statistic falls in the rejection region ($F = 3.96 > 2.05$), H_0 is rejected. There is sufficient evidence to indicate that differences exists among the mean dominance rating for facial expressions at $\alpha = .10$.

b.   Using MINITAB, the results of Tukey's multiple comparison procedure using $\alpha = .05$ is:

```
Tukey 95% Simultaneous Confidence Intervals
All Pairwise Comparisons among Levels of FACES

Individual confidence level = 99.51%


FACES = 1 subtracted from:

FACES   Lower   Center   Upper     +---------+---------+---------+---------
2      -2.144   -0.248   1.647              (---------*--------)
3      -3.487   -1.592   0.304         (--------*----------)
4      -1.731    0.165   2.061              (---------*--------)
5      -3.794   -1.898  -0.003       (---------*--------)
6      -3.069   -1.173   0.722          (--------*---------)
                                     +---------+---------+---------+---------
                                    -4.0      -2.0       0.0       2.0


FACES = 2 subtracted from:

FACES   Lower   Center   Upper     +---------+---------+---------+---------
3      -3.239   -1.343   0.552          (--------*----------)
4      -1.482    0.413   2.309              (--------*----------)
5      -3.546   -1.650   0.246        (----------*--------)
6      -2.821   -0.925   0.971           (--------*---------)
                                     +---------+---------+---------+---------
                                    -4.0      -2.0       0.0       2.0


FACES = 3 subtracted from:

FACES   Lower   Center   Upper     +---------+---------+---------+---------
4      -0.139    1.757   3.652                  (---------*--------)
5      -2.202   -0.307   1.589            (--------*---------)
6      -1.477    0.418   2.314               (--------*----------)
                                     +---------+---------+---------+---------
                                    -4.0      -2.0       0.0       2.0
```

```
FACES = 4 subtracted from:

FACES   Lower   Center   Upper     +---------+---------+---------+---------
5      -3.959   -2.063  -C.168     (---------*--------)
6      -3.234   -1.338   C.557              (--------*---------)
                                   +---------+---------+---------+---------
                                 -4.0      -2.0      0.0       2.0


FACES = 5 subtracted from:

FACES   Lower   Center   Upper     +---------+---------+---------+---------
6      -1.171    0.725   2.621                      (---------*--------)
                                   +---------+---------+---------+---------
                                 -4.0      -2.0      0.0       2.0
```

A convenient summary of the results is below, where the treatment means are listed from highest to lowest with a solid line connecting those means that are not significantly different.

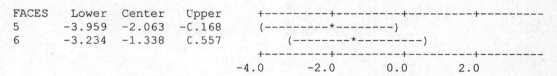

Happy	Angry	Disgusted	Neutral	Fearful	Sad
1.018	0.853	0.605	-0.320	-0.738	-1.045

The mean dominance rating for the Sad face is significantly lower than the mean dominance ratings for the Happy and Angry faces. There are no other significant differences.

10.117 a. Using MINITAB, the ANOVA results are:

ANOVA: Sales versus Price, Display

```
Factor    Type    Levels  Values
Price     fixed        3  1, 2, 3
Display   fixed        3  1, 2, 3

Analysis of Variance for Sales

Source         DF       SS       MS       F       P
Price           2  3089054  1544527  3121.89  0.000
Display         2  1691393   845696  1709.37  0.000
Price*Display   4   5_0705   127676   258.07  0.000
Error          18     8905      495
Total          26  5300057

S = 22.2428   R-Sq = 99.83%   R-Sq(adj) = 99.76%
```

b. Some preliminary calculations are:

$$SST = 3{,}089{,}054 + 1{,}691{,}393 + 510{,}705 = 5{,}291{,}152$$

$$MST = \frac{5{,}291{,}152}{8} = 661{,}394$$

To determine whether the treatment means differ, we test:

H_0: $\mu_1 = \mu_2 = \cdots = \mu_9$
H_a: At least two treatment means differ

The test statistic is $F = \dfrac{\text{MST}}{\text{MSE}} = \dfrac{661,394}{495} = 1336.85$

The rejection region requires $\alpha = .10$ in the upper tail of the F distribution with $v_1 = ab - 1 = 3(3) - 1 = 8$ and $v_2 = n - ab = 27 - 3(3) = 18$. From Table VIII, Appendix A, $F_{.10} = 2.04$. The rejection region is $F > 2.04$.

Since the observed value of the test statistic falls in the rejection region ($F = 1336.85 > 2.04$), H_0 is rejected. There is sufficient evidence to indicate the treatment means differ at $\alpha = .10$.

c. Since there are differences among the treatment means, we next test for the presence of interaction.

H_0: Factors A and B do not interact to affect the response means
H_a: Factors A and B do interact to affect the response means

The test statistic is $F = \dfrac{\text{MS}AB}{\text{MSE}} = 258.07$

The rejection region requires $\alpha = .10$ in the upper tail of the F distribution with $v_1 = (a - 1)(b - 1) = (3 - 1)(3 - 1) = 4$ and $v_2 = n - ab = 17 - 3(3) = 18$. From Table VIII, Appendix A, $F_{.10} = 2.29$. The rejection region is $F > 2.29$.

Since the observed value of the test statistic falls in the rejection region ($F = 258.07 > 2.29$), H_0 is rejected. There is sufficient evidence to indicate the two factors interact at $\alpha = .10$.

d. The main effect tests are not warranted since interaction is present in part c.

e. Since the interaction is significant, the nine treatment means need to be compared. To reduce the number of comparisons, we will compare the means of the prices at each level of the displays.

Using SAS, the Tukey's multiple comparisons are:

```
------------------------------ DISPLAY=NORMAL -------------------------------

                          The ANOVA Procedure

                Tukey's Studentized Range (HSD) Test for SALES

    NOTE: This test controls the Type I experimentwise error rate, but it
          generally has a higher Type II error rate than REGWQ.


                    Alpha                                  0.05
                    Error Degrees of Freedom                  6
                    Error Mean Square                   401.8889
                    Critical Value of Studentized Range  4.33902
                    Minimum Significant Difference        50.221


    Means with the same letter are not significantly different.


         Tukey Grouping            Mean      N     PRICE

                     A          1578.00      3     COST

                     B          1202.67      3     REDUCED

                     C          1014.67      3     REGULAR

------------------------------ DISPLAY=PLUS ---------------------------------

                          The ANOVA Procedure

                Tukey's Studentized Range (HSD) Test for SALES

    NOTE: This test controls the Type I experimentwise error rate, but it
          generally has a higher Type II error rate than REGWQ.


                    Alpha                                  0.05
                    Error Degrees of Freedom                  6
                    Error Mean Square                   653.4444
                    Critical Value of Studentized Range  4.33902
                    Minimum Significant Difference        64.038


    Means with the same letter are not significantly different.


         Tukey Grouping            Mean      N     PRICE

                     A          2510.00      3     COST

                     B          1898.67      3     REDUCED

                     C          1215.00      3     REGULAR
```

```
------------------------------ DISPLAY=TWICE ------------------------------

                          The ANOVA Procedure

                  Tukey's Studentized Range (HSD) Test for SALES

        NOTE: This test controls the Type I experimentwise error rate, but it
              generally has a higher Type II error rate than REGWQ.


                    Alpha                                  0.05
                    Error Degrees of Freedom                  6
                    Error Mean Square                   428.8889
                    Critical Value of Studentized Range  4.33902
                    Minimum Significant Difference         51.88


              Means with the same letter are not significantly different.


             Tukey Grouping         Mean       N    PRICE

                         A        1828.67       3    COST

                         B        1505.00       3    REDUCED

                         C        1202.67       3    REGULAR
```

For Display = Normal, there are significant differences among mean sales for all prices. The mean sales for those priced at "cost to supermarket" is significantly greater than the mean sales for those priced at "reduced" price and "regular" price. The mean sales for those priced at "reduced" price is significantly greater than those priced at "regular" price.

For Display = Normal Plus, there are significant differences among mean sales for all prices. The mean sales for those priced at "cost to supermarket" is significantly greater than the mean sales for those priced at "reduced" price and "regular" price. The mean sales for those priced at "reduced" price is significantly greater than those priced at "regular" price.

For Display = Twice Normal, there are significant differences among mean sales for all prices. The mean sales for those priced at "cost to supermarket" is significantly greater than the mean sales for those priced at "reduced" price and "regular" price. The mean sales for those priced at "reduced" price is significantly greater than those priced at "regular" price.

10.119 a. df Photoperiod = $a - 1 = 2 - 1 = 1$; df Gender = $b - 1 = 2 - 1 = 1$
df $P \times G = (a - 1)(b - 1) = (2 - 1)(2 - 1) = 1$
df Error = $n - ab = 124 - 2(2) = 120$; df Total = $n - 1 = 124 - 1 = 123$

The ANOVA table is:

Source	df
Photoperiod	1
Gender	1
$P \times G$	1
Error	120
Total	123

b. The test for interaction was not significant. This implies that the weight gain for each gender did not depend on the photoperiod.

c. The p-value for the main effect photoperiod was less than .001. Since this p-value is so small, H_0 is rejected. There is sufficient evidence to indicate that the weight gain differs for the two photoperiods.

The p-value for the main effect gender was less than .001. Since this p-value is so small, H_0 is rejected. There is sufficient evidence to indicate that the weight gain differs for the two genders.

10.121 a. This is a 2×2 factorial experiment.

b. The two factors are the tent type (treated or untreated) and location (inside or outside). There are $2 \times 2 = 4$ treatments. The four treatments are (treated, inside), (treated, outside), (untreated, inside), and (untreated, outside).

c. The response variable is the number of mosquito bites received in a 20 minute interval.

d. There is sufficient evidence to indicate interaction is present. This indicates that the effect of the tent type on the number of mosquito bites depends on whether the person is inside or outside.

10.123 Using SAS, the ANOVA results are:

```
                              The SAS System

                            The ANOVA Procedure

Dependent Variable: CORROSION

                                        Sum of
        Source             DF           Squares    Mean Square   F Value   Pr > F

        Model               5       72.68833333   14.53766667    155.30   <.0001

        Error               6        0.56166667    0.09361111

        Corrected Total    11       73.25000000


                  R-Square    Coeff Var    Root MSE    CORROSION Mean

                  0.992332     3.170563    0.305959          9.650000


        Source             DF        Anova SS    Mean Square   F Value   Pr > F

        EXPOSURE            2      63.10500000   31.55250000    337.06   <.0001
        SYSTEM              3       9.58333333    3.19444444     34.12   0.0004
```

```
                          The ANOVA Procedure

                 Tukey's Studentized Range (HSD) Test for CORROSION

     NOTE: This test controls the Type I experimentwise error rate, but it generally has a higher
     Type II
                                 error rate than REGWQ.


                     Alpha                                 0.05
                     Error Degrees of Freedom                 6
                     Error Mean Square                    0.093611
                     Critical Value of Studentized Range  4.89559
                     Minimum Significant Difference        0.8648


                Means with the same letter are not significantly different.


                Tukey Grouping          Mean      N    SYSTEM

                         A           11.0667      3    3

                         B            9.7333      3    2
                         B
                   C     B            9.0667      3    1
                   C
                   C                  8.7333      3    4
```

To determine if there are differences in the epoxy treatment means, we test:

H_0: $\mu_1 = \mu_2 = \mu_3 = \mu_4$
H_a: At least two treatment means differ

From the printout, the test statistic is $F = 34.12$ and the p-value is $p = 0.0004$.

Since the p-value is so small, H_0 is rejected. There is sufficient evidence to indicate a difference in the epoxy treatment means among the 4 groups for any reasonable value of α.

From the Tukey multiple comparison procedure, the mean corrosion rate for System 3 is significantly greater than the mean corrosion rate for any of the other 3 Systems. The mean corrosion rate for System 2 is significantly greater than the mean corrosion rate for System 4. No other significant differences exist. The System with the lowest corrosion rate is either System 1 or System 4. There is no significant difference in the mean corrosion rate between these two systems and they are both in the lowest group.

Simple Linear Regression

11.1   A deterministic model does not allow for random error or variation, whereas a probabilistic model does. An example where a deterministic model would be appropriate is:

> Let y = cost of a 2 × 4 piece of lumber and
> x = length (in feet)

> The model would be $y = \beta_1 x$. There should be no variation in price for the same length of wood.

An example where a probabilistic model would be appropriate is:

> Let y = sales per month of a commodity and
> x = amount of money spent advertising

> The model would be $y = \beta_0 + \beta_1 x + \varepsilon$. The sales per month will probably vary even if the amount of money spent on advertising remains the same.

11.3   The "line of means" is the deterministic component in a probabilistic model.

11.5

a.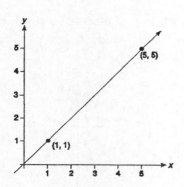

b.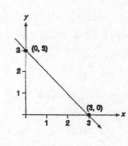

c.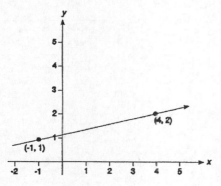

d.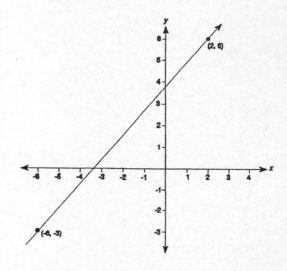

11.7   The two equations are:

$$4 = \beta_0 + \beta_1(-2) \text{ and } 6 = \beta_0 + \beta_1(4)$$

Subtracting the first equation from the second, we get

$$\begin{aligned} 6 &= \beta_0 + 4\beta_1 \\ -(4 &= \beta_0 - 2\beta_1) \\ \hline 2 &= \qquad 6\beta_1 \Rightarrow \beta_1 = \frac{2}{6} = \frac{1}{3} \end{aligned}$$

Substituting $\beta_1 = \dfrac{1}{3}$ into the first equation, we get:

$$4 = \beta_0 + \frac{1}{3}(-2) \Rightarrow \beta_0 = 4 + \frac{2}{3} = \frac{14}{3}$$

The equation for the line is $y = \dfrac{14}{3} + \dfrac{1}{3}x$

11.9   To graph a line, we need two points. Pick two values for x, and find the corresponding y values by substituting the values of x into the equation.

a.   Let $x = 0 \Rightarrow y = 4 + (0) = 4$
and $x = 2 \Rightarrow y = 4 + (2) = 6$

b.   Let $x = 0 \Rightarrow y = 5 \Rightarrow 2(0) = 5$
and $x = 2 \Rightarrow y = 5 \Rightarrow 2(2) = 1$

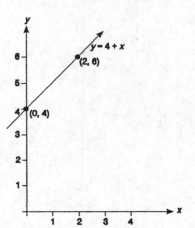

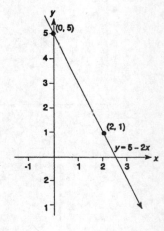

c.   Let $x = 0 \Rightarrow y = -4 + 3(0) = -4$
    and $x = 2 \Rightarrow y = -4 + 3(2) = 2$

d.   Let $x = 0 \Rightarrow y = -2(0) = 0$
    and $x = 2 \Rightarrow y = -2(2) = -4$

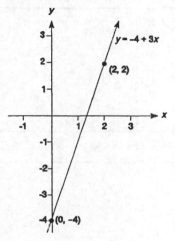

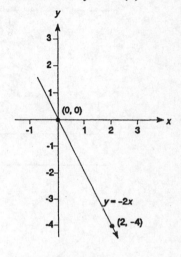

e.   Let $x = 0 \Rightarrow y = 0$
    and $x = 2 \Rightarrow y = 2$

f.   Let $x = 0 \Rightarrow y = .5 + 1.5(0) = .5$
    and $x = 2 \Rightarrow y = .5 + 1.5(2) = 3.5$

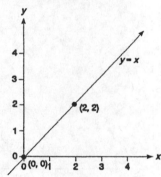

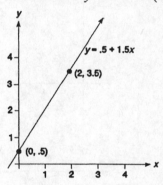

11.11   In regression, the error of prediction is the difference between the observed and predicted values of the dependent variable.

11.13   The statement "The estimates of β_0 and β_1 should be interpreted only within the sampled range of the independent variable, x." is true. We only have information about the relationship between y and x within the observed range of x.

11.15   From Exercise 11.14, $\hat{\beta}_0 = 7.10$ and $\hat{\beta}_1 = -.78$.

The fitted line is $\hat{y} = 7.10 - .78x$. To obtain values for $\hat{y}$, we substitute values of x into the equation and solve for $\hat{y}$.

a.

x	y	$\hat{y} = 7.10 - .78x$	$(y - \hat{y})$	$(y - \hat{y})^2$
7	2	1.64	.36	.1296
4	4	3.98	.02	.0004
6	2	2.42	−.42	.1764
2	5	5.54	−.54	.2916
1	7	6.32	.68	.4624
1	6	6.32	−.32	.1024
3	5	4.76	.24	.0576

$$\sum(y - \hat{y}) = 0.02 \qquad SSE = \sum(y - \hat{y})^2 = 1.2204$$

b.

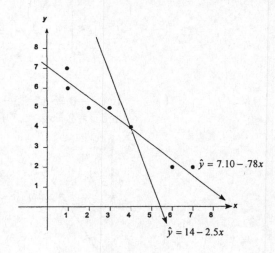

c.

x	y	$\hat{y} = 14 - 2.5x$	$(y - \hat{y})$	$(y - \hat{y})^2$
7	2	−3.5	5.5	30.25
4	4	4	0	0
6	2	−1	3	9
2	5	9	−4	16
1	7	11.5	−4.5	20.25
1	6	11.5	−5.5	30.25
3	5	6.5	−1.5	2.25
			= −7	SSE = 108.00

11.17   a.

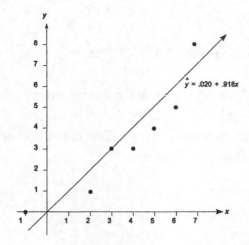

b.   As x increases, y tends to increase. Thus, there appears to be a positive, linear relationship between y and x.

c. $\hat{\beta}_1 = \dfrac{SS_{xy}}{SS_{xx}} = \dfrac{39.8571}{43.4286} = .9177616 \approx .918$

$\hat{\beta}_0 = \bar{y} - \hat{\beta}_1 \bar{x} = 3.4286 - .9177616(3.7143) = .0197581 \approx .020$

d.   The line appears to fit the data quite well.

e. $\hat{\beta}_0 = .020$ The estimated mean value of y when $x = 0$ is .020.

       $\hat{\beta}_1 = .918$ The estimated change in the mean value of y for each unit change in x is .918.

These interpretations are valid only for values of x in the range from -1 to 7.

11.19   a.   **Radiata Pine**: For each unit increase in the natural logarithm of the number of blade cycles, the mean stress is estimated to decrease by 2.50.

       Hoop Pine: For each unit increase in the natural logarithm of the number of blade cycles, the mean stress is estimated to decrease by 2.36.

b.   **Radiata Pine**: The mean stress is estimated to be 97.37 when the natural logarithm of the number of blade cycles is zero.

       Hoop Pine: The mean stress is estimated to be 122.03 when the natural logarithm of the number of blade cycles is zero.

c.   Based on these results, it appears that the Hoop Pine is stronger and more fatigue resistant. When the natural logarithm of the number of blade cycles is zero, the mean stress for the Hoop Pine is greater than the mean stress for the Radiata Pine. As the natural logarithm of the number of blade cycles increases, the mean stress for the Radiata Pine decreases at a faster rate than does the Hoop Pine.

11.21   a.   The straight-line model is $y = \beta_0 + \beta_1 x + \varepsilon$

b.   The least squares prediction equation is $\hat{y} = 2.522 + 7.261x$

c. $\hat{\beta}_1 = 7.261$. For each additional one day duration, the mean number of arrests is estimated to increase by 7.261.

$\hat{\beta}_0 = 2.522$. Since x = 0 is not in the observed range of the duration in days, $\hat{\beta}_0$ has no interpretation other than the y-intercept.

11.23   a.   Using MINITAB, the least squares line is:

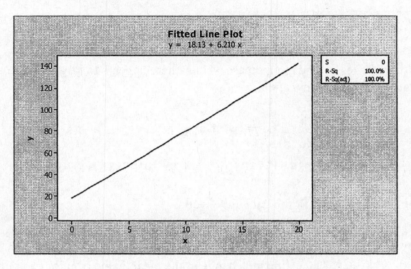

The slope of the line is positive – as x increases, y also increases.

b. $\hat{\beta}_0 = 18.13$. Since we have no information as to whether x = 0 is within the observed range, we will assume that it is. Then, the estimate of the mean magnitude when redshift level is 0 is 18.13.

c. $\hat{\beta}_1 = 6.21$. For each unit increase in redshift level, the mean value of magnitude is estimated to increase by 6.21.

11.25 a.   The straight-line model is $y = \beta_o + \beta_1 x + \varepsilon$.

b.   Using MINITAB, the results are:

Regression Analysis: Accuracy versus Distance

```
The regression equation is
Accuracy = 250 - 0.629 Distance


Predictor      Coef   SE Coef      T       P
Constant     250.14     14.23   17.58   0.000
Distance    -0.62944   0.04759  -13.23   0.000


S = 2.23639    R-Sq = 82.2%    R-Sq(adj) = 81.7%


Analysis of Variance

Source          DF      SS       MS       F       P
Regression       1   874.99   874.99   174.95   0.000
Residual Error  38   190.06     5.00
Total           39  1065.04
```

The least squares prediction equation is: $\hat{y} = 250.14 - .6294x$

c.   The estimated y-intercept is $\hat{\beta}_o = 250.14$. Since 0 is not in the observed range for values of driving distance, the y-intercept has no meaning.

d.   The estimated slope of the line is $\hat{\beta}_1 = -.6294$. For each additional yard of driving distance, the mean driving accuracy value will decrease by an estimated .6294.

e.   The slope will help determine if the golfer's concern is valid. If there is no significant linear relationship between driving accuracy and driving distance, then the golfer would have nothing to worry about. However, if there is a significant linear relationship between driving accuracy and driving distance and the relationship is negative, then as the driving distance increases, the mean driving accuracy will decrease.

11.27 a.   The straight-line model is $y = \beta_0 + \beta_1 x + \varepsilon$

b.   Some preliminary calculations are:

$$\sum x_i = 51.4 \qquad \sum y_i = 45.5 \qquad \sum x_i y_i = 210.49$$
$$\sum x_i^2 = 227.5 \qquad \sum y_i^2 = 214.41$$

$$SS_{xy} = \sum x_i y_i - \frac{\sum x_i \sum y_i}{n} = 210.49 - \frac{51.4(45.5)}{15} = 54.5766667$$

$$SS_{xx} = \sum x_i^2 - \frac{\left(\sum x_i\right)^2}{n} = 227.5 - \frac{51.4^2}{15} = 51.3693333$$

$$\hat{\beta}_1 = \frac{SS_{xy}}{SS_{xx}} = \frac{54.5766667}{51.3693333} = 1.062436734$$

$$\hat{\beta}_0 = \bar{y} - \hat{\beta}_1 \bar{x} = \frac{45.5}{15} - (1.062436734)\left(\frac{51.4}{15}\right) = -.607283208$$

The least squares prediction equation is $\hat{y} = -.607 + 1.062x$

c.   Using MINITAB, the graph is:

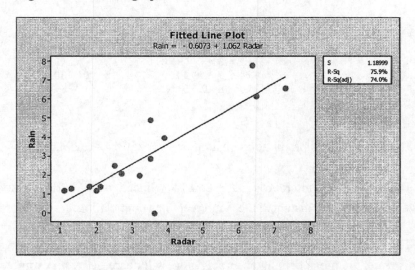

There appears to be a positive linear relationship between the two variables. As the radar rainfall increases, the rain gauge values also increase.

d. $\hat{\beta}_0 = -.607$. Since $x = 0$ is not in the observed range, $\hat{\beta}_0$ has no meaning other than the y-intercept.

   $\hat{\beta}_1 = 1.062$. For each unit increase in radar rainfall, the mean rain gauge rainfall increases by an estimated 1.062.

e.   The straight-line model is $y = \beta_0 + \beta_1 x + \varepsilon$

Some preliminary calculations are:

$$\sum x_i = 46.7 \qquad\qquad \sum y_i = 45.5 \qquad\qquad \sum x_i y_i = 207.89$$
$$\sum x_i^2 = 210.21 \qquad\qquad \sum y_i^2 = 214.41$$

$$SS_{xy} = \sum x_i y_i - \frac{\sum x_i \sum y_i}{n} = 207.89 - \frac{46.7(45.5)}{15} = 66.2333333$$

$$SS_{xx} = \sum x_i^2 - \frac{\left(\sum x_i\right)^2}{n} = 210.21 - \frac{46.7^2}{15} = 64.8173333$$

$$\hat{\beta}_1 = \frac{SS_{xy}}{SS_{xx}} = \frac{66.2333333}{64.8173333} = 1.021846008$$

$$\hat{\beta}_0 = \bar{y} - \hat{\beta}_1 \bar{x} = \frac{45.5}{15} - (1.021846008)\left(\frac{46.7}{15}\right) = -.148013904$$

The least squares prediction equation is $\hat{y} = -.148 + 1.022x$

Using MINITAB, the graph is:

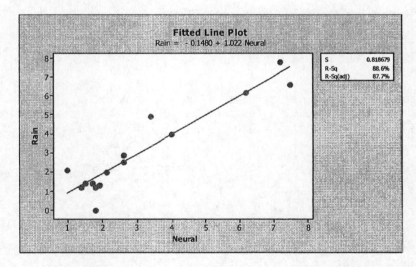

There appears to be a positive linear relationship between the two variables. As the neural rainfall increases, the rain gauge values also increase.

11.29 a. Using MINITAB, the results are:

Regression Analysis: Cost versus JIF

```
The regression equation is
Cost = 560 + 63 JIF


Predictor   Coef  SE Coef     T     P
Constant   560.1    168.4  3.33  0.003
JIF         63.3    113.2  0.56  0.581


S = 454.248   R-Sq = 1.2%   R-Sq(adj) = 0.0%


Analysis of Variance

Source          DF       SS       MS     F     P
Regression       1    64381    64381  0.31  0.581
Residual Error  26  5364863   206341
Total           27  5429244
```

The fitted regression equation is $\hat{y} = 560.1 + 63.3x$.

For each additional unit increase in the Journal Impact Factor (JIF), the mean cost of the journal is estimated to increase by 63.3 dollars.

b. Using MINITAB, the results are:

Regression Analysis: Cost versus Cites

```
The regression equation is
Cost = 326 + 1.48 Cites


Predictor    Coef  SE Coef     T     P
Constant    326.5    109.1  2.99  0.006
Cites      1.4801   0.3953  3.74  0.001


S = 368.331   R-Sq = 35.0%   R-Sq(adj) = 32.5%


Analysis of Variance

Source          DF        SS       MS      F     P
Regression       1   1901880  1901880  14.02  0.001
Residual Error  26   3527364   135668
Total           27   5429244
```

The fitted regression equation is $\hat{y} = 326.5 + 1.48x$.

For each additional unit increase in the number of citations for the journal over the past 5 years, the mean cost of the journal is estimated to increase by 1.48 dollars.

c. Using MINITAB, the results are:

Regression Analysis: Cost versus RPI

```
The regression equation is
Cost = 339 + 197 RPI


27 cases used, 1 cases contain missing values


Predictor     Coef  SE Coef      T      P
Constant     338.9    156.6   2.16  0.040
RPI         197.21    86.24   2.29  0.031


S = 423.363    R-Sq = 17.3%   R-Sq(adj) = 14.0%


Analysis of Variance

Source           DF        SS       MS     F      P
Regression        1    937326   937326  5.23  0.031
Residual Error   25   4480916   179237
Total            26   5418242
```

The fitted regression equation is $\hat{y} = 338.9 + 197.21x$.

For each additional unit increase in the Relative Price Index (RPI), the mean cost of the journal is estimated to increase by 197.21 dollars.

11.31 Using MINITAB, the results are:

Regression Analysis: Mass versus Time

```
The regression equation is
Mass = 5.22 - 0.114 Time


Predictor      Coef  SE Coef       T      P
Constant     5.2207   0.2960   17.64  0.000
Time        -0.11402  0.01032  -11.05  0.000


S = 0.857257    R-Sq = 85.3%   R-Sq(adj) = 84.6%


Analysis of Variance

Source           DF        SS       MS      F      P
Regression        1    89.794   89.794  122.19  0.000
Residual Error   21    15.433    0.735
Total            22   105.227
```

The fitted regression line is $\hat{y} = 5.2207 - .11402x$. Since the estimate of the slope is negative, the data indicate that the mass of the spill tends to diminish as time increases. For each additional minute, the mean mass is estimated to decrease by .11402 pounds.

11.33   See Figure 11.7 in the text.

11.35   a. $\text{SSE} = \text{SS}_{yy} - \hat{\beta}_1 \text{SS}_{xy} = 95 - .75(50) = 57.5$

$$s^2 = \frac{\text{SSE}}{n-2} = \frac{57.5}{20-2} = 3.19444$$

b. $\text{SS}_{yy} = \sum y^2 - \frac{\left(\sum y\right)^2}{n} = 860 - \frac{50^2}{40} = 797.5$

$\text{SSE} = \text{SS}_{yy} - \hat{\beta}_1 \text{SS}_{xy} = 797.5 - .2(2700) = 257.5$

$$s^2 = \frac{\text{SSE}}{n-2} = \frac{257.5}{40-2} = 6.776315789 \approx 6.7763$$

c. $\text{SS}_{yy} = \sum (y_i - \bar{y})^2 = 58 \qquad \hat{\beta}_1 = \frac{\text{SS}_{xy}}{\text{SS}_{xx}} = \frac{91}{170} = .535294117$

$\text{SSE} = \text{SS}_{yy} - \hat{\beta}_1 \text{SS}_{xy} = 58 - .535294117(91) = 9.2882353 \approx 9.288$

$$s^2 = \frac{\text{SSE}}{n-2} = \frac{9.2882353}{10-2} = 1.161029413 \approx 1.1610$$

11.37 $\text{SSE} = \text{SS}_{yy} - \hat{\beta}_1 \text{SS}_{xy}$

where $\text{SS}_{yy} = \sum y_i^2 - \frac{\left(\sum y_i\right)^2}{n}$

For Exercise 11.14,

$\sum y_i^2 = 159 \qquad\qquad \sum y_i = 31$

$\text{SS}_{yy} = 159 - \frac{31^2}{7} = 159 - 137.2857143 = 21.7142857$

$\text{SS}_{xy} = -26.2857143 \qquad \hat{\beta}_1 = -.779661017$

Therefore, $\text{SSE} = 21.7142857 - (-.779661017)(-26.2857143) = 1.22033896 \approx 1.2203$

$$s^2 = \frac{\text{SSE}}{n-2} = \frac{1.22033896}{7-2} = .244067792, \quad s = \sqrt{.244067792} = .4960$$

We would expect most of the observations to fall within 2s or 2(.4940) or .988 units of the least squares prediction line.

For Exercise 11.17,

$$\sum x_i = 26 \quad \sum y_i = 24 \quad \sum x_i y_i = 129 \quad \sum x_i^2 = 140 \quad \sum y_i^2 = 124$$

$$SS_{xy} = \sum x_i y_i - \frac{\left(\sum x_i \sum y_i\right)}{n} = 129 - \frac{(26)(24)}{7} = 129 - 89.14285714 = 39.85714286$$

$$SS_{xx} = \sum x_i^2 - \frac{\left(\sum x_i\right)^2}{n} = 140 - \frac{(26)^2}{7} = 140 - 96.57142857 = 43.42857143$$

$$S_{yy} = \sum y_i^2 - \frac{\left(\sum y_i\right)^2}{n} = 124 - \frac{(24)^2}{7} = 124 - 82.28571429 = 41.71428571$$

$$\hat{\beta}_1 = \frac{SS_{xy}}{SS_{xx}} = \frac{39.85714286}{43.42857143} = .917763157$$

$$SSE = SS_{yy} - \hat{\beta}_1 SS_{xy} = 41.71428571 - (.917763157)(39.85714286)$$

$$= 41.71428571 - 36.57941726 = 5.13486841 \approx 5.1349$$

$$s^2 = \frac{SSE}{n-2} = \frac{5.13486841}{7-2} = 1.026973682 \quad s = \sqrt{1.026973682} = 1.0134$$

We would expect most of the observations to fall within $2s$ or $2(1.0134)$ or 2.0268 units of the least squares prediction line.

11.39 Based on the plots, the women's linear regression model will have the smallest estimate of σ. The observations around the women's line are closer to the line than the men's observations are to the men's line.

11.41 a. From the printout, $SSE = 2760$, $s^2 = 387.0$, and $s = 17.5111$.

b. We would expect most of the observations to be within $2s = 2(17.5111) = 35.0222$ of the least squares line.

11.43 a. From Exercise 11.26,

$$\sum y_i = 3,781.1 \quad \sum y_i^2 = 651,612 \quad SS_{xy} = -3,881.9986 \quad \hat{\beta}_1 = -.305416114$$

$$SS_{yy} = \sum y_i^2 - \frac{\left(\sum y_i\right)^2}{n} = 651,612 - \frac{3,781.1^2}{22} = 1,761.2177$$

$$SSE = SS_{yy} - \hat{\beta}_1 SS_{xy} = 1,761.2177 - (-.305416114)(-3,881.9986) = 575.592773$$

$$s^2 = \frac{SSE}{n-2} = \frac{575.592773}{22-2} = 28.77963865$$

$$s = \sqrt{s^2} = \sqrt{28.77963865} = 5.364665754$$

We would expect most of the observations to be within $2s = 2(5.3647) = 10.7294$ of the least squares line.

b. From Exercise 11.26,

$$\sum y_i = 3,764.2 \qquad \sum y_i^2 = 645,221 \qquad SS_{xy} = -3,441.97 \qquad \hat{\beta}_1 = -.270796878$$

$$SS_{yy} = \sum y_i^2 - \frac{\left(\sum y_i\right)^2}{n} = 645,221 - \frac{3,764.2^2}{22} = 1,166.38$$

$$SSE = SS_{yy} - \hat{\beta}_1 SS_{xy} = 1,166.38 - (-.270796878)(-3,441.97) = 234.3052698$$

$$s^2 = \frac{SSE}{n-2} = \frac{234.3052698}{22-2} = 11.71526349$$
$$s = \sqrt{s^2} = \sqrt{11.71526349} = 3.422756709$$

We would expect most of the observations to be within $2s = 2(3.4228) = 6.8456$ of the least squares line.

c. Since the standard deviation ($s = 3.423$) for the reading scores is smaller than the standard deviation for the math scores ($s = 5.365$), the reading score can be more accurately predicted than the math score.

11.45 a. From Exercise 11.29a, s = 454.248. We would expect to predict cost to within 2s = 2(454.248) = 908.596 dollars.

b. From Exercise 11.29b, s = 368.331. We would expect to predict cost to within 2s = 2(368.331) = 736.662 dollars.

c. From Exercise 11.29c, s = 423.363. We would expect to predict cost to within 2s = 2(423.363) = 846.726 dollars.

11.47 In the equation, $E(y) = \beta_0 + \beta_1 x$, the value of β_1 is 0 if x has no linear relationship with y.

11.49 If you are running a one-tailed test of β_1 in simple linear regression, you divide the p-value from the computer printout by 2 to obtain the correct p-value of the one-tailed test.

11.51 a. For confidence coefficient .95, $\alpha = 1 - .95 = .05$ and $\alpha/2 = .05/2 = .025$. From Table VI, Appendix A, with df $= n - 2 = 12 - 2 = 10$, $t_{.025} = 2.228$.

The 95% confidence interval for β_1 is:

$$\hat{\beta}_1 \pm t_{.025} s_{\hat{\beta}_1} \text{ where } s_{\hat{\beta}_1} = \frac{s}{\sqrt{SS_{xx}}} = \frac{3}{\sqrt{35}} = .5071$$

$$\Rightarrow 31 \pm 2.228(.5071) \Rightarrow 31 \pm 1.13 \Rightarrow (29.87, 32.13)$$

For confidence coefficient .90, $\alpha = 1 - .90 = .10$ and $\alpha/2 = .10/2 = .05$. From Table VI, Appendix A, with df = 10, $t_{.05} = 1.812$.

The 90% confidence interval for β_1 is:

$$\hat{\beta}_1 \pm t_{.05} s_{\hat{\beta}_1} \Rightarrow 31 \pm 1.812(.5071) \Rightarrow 31 \pm .92 \Rightarrow (30.08, 31.92)$$

b. $s^2 = \dfrac{\text{SSE}}{n-2} = \dfrac{1960}{18-2} = 122.5, s = \sqrt{s^2} = 11.0680$

For confidence coefficient, .95, $\alpha = 1 - .95 = .05$ and $\alpha/2 = .05/2 = .025$. From Table VI, Appendix A, with df $= n - 2 = 18 - 2 = 16$, $t_{.025} = 2.120$. The 95% confidence interval for β_1 is:

$$\hat{\beta}_1 \pm t_{.025}\, s_{\hat{\beta}_1} \quad \text{where} \quad s_{\hat{\beta}_1} = \dfrac{s}{\sqrt{\text{SS}_{xx}}} = \dfrac{11.0680}{\sqrt{30}} = 2.0207$$

$$\Rightarrow 64 \pm 2.120(2.0207) \Rightarrow 64 \pm 4.28 \Rightarrow (59.72, 68.28)$$

For confidence coefficient .90, $\alpha = 1 - .90 = .10$ and $\alpha/2 = .10/2 = .05$. From Table VI, Appendix A, with df $= 16$, $t_{.05} = 1.746$.

The 90% confidence interval for β_1 is:

$$\hat{\beta}_1 \pm t_{.05}\, s_{\hat{\beta}_1} \Rightarrow 64 \pm 1.746(2.0207) \Rightarrow 64 \pm 3.53 \Rightarrow (60.47, 67.53)$$

c. $s^2 = \dfrac{\text{SSE}}{n-2} = \dfrac{146}{24-2} = 6.6364, s = \sqrt{s^2} = 2.5761$

For confidence coefficient .95, $\alpha = 1 - .95 = .05$ and $\alpha/2 = .05/2 = .025$. From Table VI, Appendix A, with df $= n - 2 = 24 - 2 = 22$, $t_{.025} = 2.074$. The 95% confidence interval for β_1 is:

$$\hat{\beta}_1 \pm t_{.025}\, s_{\hat{\beta}_1} \quad \text{where} \quad s_{\hat{\beta}_1} = \dfrac{s}{\sqrt{\text{SS}_{xx}}} = \dfrac{2.5761}{\sqrt{64}} = .3220$$

$$\Rightarrow -8.4 \pm 2.074(.322) \Rightarrow -8.4 \pm .67 \Rightarrow (-9.07, -7.73)$$

For confidence coefficient .90, $\alpha = 1 - .90 = .10$ and $\alpha/2 = .10/2 = .05$. From Table VI, Appendix A, with df $= 22$, $t_{.05} = 1.717$.

The 90% confidence interval for β_1 is:

$$\hat{\beta}_1 \pm t_{.05}\, s_{\hat{\beta}_1} \Rightarrow -8.4 \pm 1.717(.322) \Rightarrow -8.4 \pm .55 \Rightarrow (-8.95, -7.85)$$

11.53   a & c.   Using MINITAB, a scatterplot of the data is:

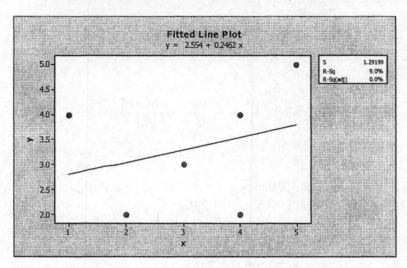

b.   Some preliminary calculations are:

$$\sum x_i = 19 \qquad \sum y_i = 20 \qquad \sum x_i y_i = 66$$
$$\sum x_i^2 = 71 \qquad \sum y_i^2 = 74$$

$$SS_{xy} = \sum x_i y_i - \frac{\sum x_i \sum y_i}{n} = 66 - \frac{19(20)}{6} = 2.666666667$$

$$SS_{xx} = \sum x_i^2 - \frac{\left(\sum x_i\right)^2}{n} = 71 - \frac{19^2}{6} = 10.833333333$$

$$\hat{\beta}_1 = \frac{SS_{xy}}{SS_{xx}} = \frac{2.66666667}{10.833333333} = .246153846$$

$$\hat{\beta}_0 = \overline{y} - \hat{\beta}_1 \overline{x} = \frac{20}{6} - (.246153846)\left(\frac{19}{6}\right) = 2.553846155$$

The least squares prediction equation is $\hat{y} = 2.554 + .246x$

d.   Some preliminary calculations are:

$$SS_{yy} = \sum y_i^2 - \frac{\left(\sum y_i\right)^2}{n} = 74 - \frac{20^2}{6} = 7.3333333333$$

$$SSE = SS_{yy} - \hat{\beta}_1 SS_{xy} = 7.3333333333 - (.246153846)(2.666666667) = 6.6769$$

$$s^2 = \frac{SSE}{n-2} = \frac{6.6769}{6-2} = 1.6692 \qquad s = \sqrt{s^2} = \sqrt{1.6692} = 1.292$$

The test statistic is $t = \dfrac{\hat{\beta}_1 - 0}{s_{\hat{\beta}_1}} = \dfrac{.246}{\dfrac{1.292}{\sqrt{10.83333333}}} = .627$

e.   To determine if x and y are linearly related, we test:

H_0: $\beta_1 = 0$
H_a: $\beta_1 \neq 0$

The test statistic is $t = .627$.

The rejection region requires $\alpha/2 = .01/2 = .005$ in each tail of the t distribution with $df = n - 2 = 6 - 2 = 4$. From Table VI, Appendix A, $t_{.005} = 4.604$. The rejection region is $t < -4.604$ or $t > 4.604$.

Since the observed value of the test statistic does not fall in the rejection region ($t = .627 \not> 4.604$), H_0 is not rejected. There is insufficient evidence to indicate that x and y are linearly related at $\alpha = .01$.

f.   For confidence coefficient .99, $\alpha = 1 - .99 = .01$ and $\alpha/2 = .01/2 = .005$. From Table VI, Appendix A, with $df = n - 2 = 6 - 2 = 4$, $t_{.005} = 4.604$. The 99% confidence interval is:

$$\hat{\beta}_1 \pm t_{.005} s_{\hat{\beta}_1} \Rightarrow \hat{\beta}_1 \pm t_{.005} \frac{s}{\sqrt{SS_{xx}}} \Rightarrow .246 \pm 4.604 \frac{1.292}{\sqrt{10.8333333}}$$

$$\Rightarrow .246 \pm 1.807 \Rightarrow (-1.561, 2.053)$$

We are 99% confident that the true value of β_1 is between -1.561 and 2.053.

11.55 a.   To determine whether driving accuracy decreases linearly as driving distance increases, we test:

H_0: $\beta_1 = 0$
H_a: $\beta_1 < 0$

b.   From the results in Exercise 11.25, the test statistic is $t = -13.23$ and the p-value is $p = .000$.

c.   Since the p-value is less than $\alpha = .01$ ($p = .000 < .01$), H_0 is rejected. There is sufficient evidence to indicate driving accuracy decreases linearly as driving distance increases at $\alpha = .01$.

11.57 a. From Exercises 11.27 and 11.44, $SS_{xx} = 64.8173333$, $\hat{\beta}_1 = 1.022$, $s = .8187$,

To determine if the rain gauge amount and artificial neural network rain amount are positively linearly related, we test:

H_0: $\beta_1 = 0$
H_a: $\beta_1 > 0$

The test statistic is $t = \dfrac{\hat{\beta}_1 - 0}{s_{\hat{\beta}_1}} = \dfrac{1.022}{\dfrac{.8187}{\sqrt{64.8173333}}} = 10.050$

The rejection region requires $\alpha = .10$ in the upper tail of the t distribution with df = $n - 2 = 15 - 2 = 13$. From Table VI, Appendix A, $t_{.10} = 1.350$. The rejection region is $t > 1.350$.

Since the observed value of the test statistic falls in the rejection region ($t = 10.050 > 1.350$), H_0 is rejected. There is sufficient evidence to indicate the rain gauge amount and artificial neural network rain amount are positively linearly related at $\alpha = .10$.

b. For confidence coefficient .90, $\alpha = 1 - .90 = .10$ and $\alpha/2 = .10/2 = .05$. From Table VI, Appendix A, with df = $n - 2 = 15 - 2 = 13$, $t_{.05} = 1.771$. The 90% confidence interval is:

$$\hat{\beta}_1 \pm t_{.05} s_{\hat{\beta}_1} \Rightarrow \hat{\beta}_1 \pm t_{.05} \frac{s}{\sqrt{SS_{xx}}} \Rightarrow 1.022 \pm 1.771 \frac{.8187}{\sqrt{64.817333}}$$

$$\Rightarrow 1.022 \pm .180 \Rightarrow (.842, \quad 1.202)$$

We are 90% confident that the change in the mean value of rain gauge amount for each unit change in neural network rain is between .842 and 1.202.

11.59 Some preliminary calculations are:

$$SS_{yy} = \sum y_i^2 - \frac{\left(\sum y_i\right)^2}{n} = 769.72 - \frac{(135.8)^2}{24} = 1.3183333$$

$$SSE = SS_{yy} - \hat{\beta}_1 SS_{xy} = 1.3183333 - (-.002301769)(-129.94167) = 1.019237592$$

$$s^2 = \frac{SSE}{n-2} = \frac{1.019237592}{22} = .046329$$

$$s_{\hat{\beta}_1} = \sqrt{\frac{s^2}{SS_{xx}}} = \sqrt{\frac{.046329}{56452.958}} = .000906$$

For confidence level .90, $\alpha = .10$ and $\alpha/2 = .10/2 = .05$. From Table VI, Appendix A with df = $n - 2 = 24 - 2 = 22$, $t_{.05} = 1.717$.

The confidence interval is:

$$\hat{\beta}_1 \pm t_{.05} s_{\hat{\beta}_1} \Rightarrow -.0023 \pm 1.717(.000906)$$

$$\Rightarrow (-.0039, -.0007)$$

We are 90% confident that the change in the mean sweetness index for each one unit change in the pectin is between $-.0039$ and $-.0007$.

11.61   a.   The model relating Anthropogenic Index, y, to Natural Origin Index, x, is:

$$y = \beta_0 + \beta_1 x + \varepsilon$$

     b.   Some preliminary calculations are:

$$\sum x_i = 1,047.22 \qquad \sum y_i = 1,283.46 \qquad \sum x_i y_i = 32,048.2$$
$$\sum x_i^2 = 24,607.5 \qquad \sum y_i^2 = 54,184.8$$

$$SS_{xy} = \sum x_i y_i - \frac{\sum x_i \sum y_i}{n} = 32,048.2 - \frac{1,047.22(1,283.46)}{54} = 7,158.10776$$

$$SS_{xx} = \sum x_i^2 - \frac{\left(\sum x_i\right)^2}{n} = 24,607.5 - \frac{1,047.22^2}{54} = 4,298.80133$$

$$\hat{\beta}_1 = \frac{SS_{xy}}{SS_{xx}} = \frac{7,158.10776}{4,298.80133} = 1.665140399$$

$$\hat{\beta}_0 = \bar{y} - \hat{\beta}_1 \bar{x} = \frac{1,283.46}{54} - (1.665140399)\left(\frac{1,047.22}{54}\right) = -8.524228315$$

The least squares prediction equation is $\hat{y} = -8.524 + 1.665x$

     c. $\hat{\beta}_1 = 1.665$. For each unit increase in the Natural Origin Index, the mean Anthropogenic Index is estimated to increase by 1.665.

             $\hat{\beta}_0 = 2.522$. Since $x = 0$ is not in the observed range of the Natural Origin Index, $\hat{\beta}_0$ has no interpretation other than the y-intercept.

     d.   Some preliminary calculations are:

$$SS_{yy} = \sum y_i^2 - \frac{\left(\sum y_i\right)^2}{n} = 54,184.8 - \frac{1,283.46^2}{54} = 23,679.80793$$

$$SSE = SS_{yy} - \hat{\beta}_1 SS_{xy} = 23,679.80793 - (1.665140399)(7,158.10776) = 11,760.55352$$

$$s^2 = \frac{\text{SSE}}{n-2} = \frac{11,760.55352}{54-2} = 226.1644908$$

$$s = \sqrt{s^2} = \sqrt{226.1644908} = 15.03876627$$

To determine if the natural origin index and anthropogenic index are positively linearly related, we test:

H_0: $\beta_1 = 0$
H_a: $\beta_1 > 0$

The test statistic is $t = \dfrac{\hat{\beta}_1 - 0}{s_{\hat{\beta}_1}} = \dfrac{1.665}{\dfrac{15.0688}{\sqrt{4,298.80133}}} = 7.245$

The rejection region requires $\alpha = .05$ in the upper tail of the t distribution with df $= n - 2$ $= 54 - 2 = 52$. From Table VI, Appendix A, $t_{.05} \approx 1.678$. The rejection region is $t > 1.678$.
Since the observed value of the test statistic falls in the rejection region ($t = 7.245 > 1.678$), H_0 is rejected. There is sufficient evidence to indicate the natural origin index and anthropogenic index are positively linearly related at $\alpha = .05$.

e. For confidence coefficient .95, $\alpha = 1 - .95 = .05$ and $\alpha/2 = .05/2 = .025$. From Table VI, Appendix A, with df $= n - 2 = 54 - 2 = 52$, $t_{.025} \approx 2.011$. The 95% confidence interval is:

$$\hat{\beta}_1 \pm t_{.025}s_{\hat{\beta}_1} \Rightarrow \hat{\beta}_1 \pm t_{.025}\frac{s}{\sqrt{\text{SS}_{xx}}} \Rightarrow 1.665 \pm 2.011\frac{15.0688}{\sqrt{4,298.80133}}$$

$$\Rightarrow 1.665 \pm .462 \Rightarrow (1.203, \quad 2.127)$$

We are 95% confident that the change in the anthropogenic index for each unit increase in the natural origin index is between 1.203 and 2.127.

11.63 a. To determine if body plus head rotation and active head movement are positively linearly related, we test:

H_0: $\beta_1 = 0$
H_a: $\beta_1 > 0$

The test statistic is $t = \dfrac{\hat{\beta}_1 - 0}{s_{\hat{\beta}_1}} = \dfrac{.88 - 0}{.14} = 6.286$

The rejection region requires $\alpha = .05$ in each tail of the t distribution with df $= n - 2$ $= 39 - 2 = 37$. From Table VI, Appendix A, $t_{.05} \approx 1.687$. The rejection region is $t > 1.687$.

Since the observed value of the test statistic falls in the rejection region ($t = 6.286 > 1.687$), H_0 is rejected. There is sufficient evidence to indicate that the two variables are positively linearly related at $\alpha = .05$.

b.   For confidence level .90, $\alpha = .10$ and $\alpha/2 = .10/2 = .05$. From Table VI, Appendix A, with df $= n - 2 = 39 - 2 = 37$, $t_{.05} \approx 1.687$. The confidence interval is:

$$\hat{\beta}_1 \pm t_{.05} s_{\hat{\beta}_1} \Rightarrow .88 \pm 1.687(.14)$$
$$\Rightarrow .88 \pm .23618$$
$$\Rightarrow (.6438, 1.1162)$$

We are 90% confident that the true value of β_1 is between .6438 and 1.1162.

c.   Because the interval in part b contains the value 1, there is no evidence that the true slope of the line differs from 1.

11.65   a.   From the printout, $\hat{\beta}_o = .515$, $\hat{\beta}_1 = .000021$, and s $= .0370$.

b.   To determine if there is a positive linear relationship between elevation and slugging percentage, we test:

H_0: $\beta_1 = 0$
H_a: $\beta_1 > 0$

From the printout, the test statistic is $t = 2.89$ and the p-value is $p = .008$.

Since the p-value is less than $\alpha = .01$ ($p = .008 < .01$), H_o is rejected. There is sufficient evidence to indicate a positive linear relationship between elevation and slugging percentage at $\alpha = .01$.

c.   Using MINITAB, the scatterplot is:

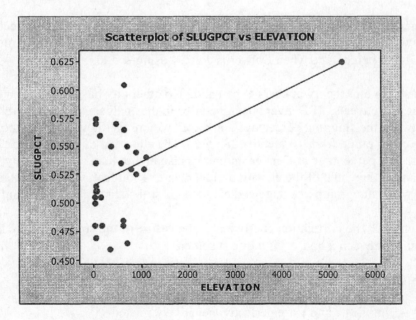

The data point corresponding to Denver is the data point furthest to the right. This point is very influential. If this point were deleted, there probably would not be a relationship between slugging percentage and elevation.

d.   Using MINITAB with Denver removed, the results are:

Regression Analysis: SLUGPCT2 versus ELEVATION2

```
The regression equation is
SLUGPCT2 = 0.515 + 0.000020 ELEVATION2


Predictor          Coef      SE Coef      T       P
Constant        0.51537      0.01066   48.33   0.000
ELEVATION2   0.00002012   0.00002034    0.99   0.332


S = 0.0376839    R-Sq = 3.6%    R-Sq(adj) = 0.0%


Analysis of Variance

Source          DF        SS         MS       F       P
Regression       1   0.001389   0.001389    0.98   0.332
Residual Error  26   0.036922   0.001420
Total           27   0.038311
```

From the printout, $\hat{\beta}_o = .515$, $\hat{\beta}_1 = .000020$, and $s = .0377$.

To determine if there is a positive linear relationship between elevation and slugging percentage, we test:

H_0: $\beta_1 = 0$
H_a: $\beta_1 > 0$

From the printout, the test statistic is $t = 0.99$ and the p-value is $p = .332$.

Since the p-value is not less than $\alpha = .01$ ($p = .332 > .01$), H_0 is not rejected. There is insufficient evidence to indicate a positive linear relationship between elevation and slugging percentage when Denver is deleted using $\alpha = .01$.

The "thin air" theory appears to be valid. No other city has an elevation that is close to that of Denver's. If Denver is not included in the analysis, there is no relationship between the slugging percentage and the elevation. Without Denver, the elevations of the cities range from 10 feet to 1082 feet. The elevation of Denver is 5277. The slugging percentage at Denver stadium is also higher than any other city. Even if we changed the value of the elevation of Denver to 1500 feet, the linear relationship between slugging percentage and elevation is still statistically significant.

11.67   The statement "The correlation coefficient is a measure of the strength of the linear relationship between x and y." is a true statement.

11.69   a. $r = 1$ implies x and y are perfectly, positively related.

b. $r = -1$ implies x and y are perfectly, negatively related.

c. $r = 0$ implies x and y are not related.

d. $r = .90$ implies x and y are positively related. Since r is close to 1, the strength of the relationship is very high.

e. $r = .10$ implies x and y are positively related. Since r is close to 0, the relationship is fairly weak.

f. $r = -.88$ implies x and y are negatively related. Since r is close to -1, the relationship is fairly strong.

11.71   From Exercises 11.17 and 11.33,

$$r^2 = 1 - \frac{SSE}{SS_{yy}} = 1 - \frac{5.13486841}{41.7142857} = .8769$$

87.69% of the total sample variability around $\bar{y}$ is explained by the linear relationship between y and x.

11.73   a.   Since the p-value is somewhat small ($p = .07$) there is some evidence to indicate a positive linear relationship between baseline and follow-up physical activity among the obese adults. The correlation coefficient of $r = .50$ indicates that the strength of the positive linear relationship is moderate.

b.   Using MINITAB, a possible scatterplot with 13 data points that would yield a value of $r = .50$ is:

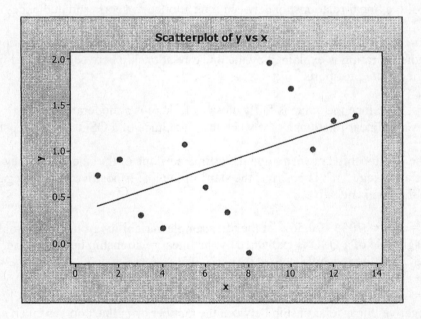

c.   Since the p-value is not small ($p = .66$) there is no evidence to indicate a linear relationship between baseline and follow-up physical activity among the normal adults. The correlation coefficient of $r = -.12$ indicates that the strength of the negative linear relationship is extremely weak.

d. Using MINITAB, a possible scatterplot with 15 data points that would yield a value of $r = -.12$ is:

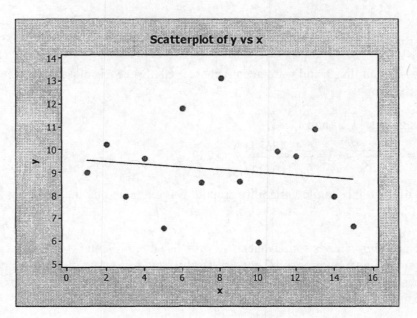

Scatterplot of y vs x

11.75 a. $r = .43$. Since the value is not particularly close to 1, there is a moderately weak positive linear relationship between time allotted to sports and audience ratings.

b. $r^2 = .43^2 = .1849$. 18.49% of the total sample variability around the sample mean audience rating is explained by the linear relationship between audience rating and total time allotted to sports.

11.77 a. $r = .84$. Since the value is fairly close to 1, there is a moderately strong positive linear relationship between the magnitude of a QSO and the redshift level.

b. The relationship between r and the estimated slope of the line is that they will both have the same sign. If r is positive, the slope of the line is positive. If r is negative, the slope of the line is negative.

c. $r^2 = .84^2 = .7056$. 70.56% of the total sample variability around the sample mean magnitude of a QSO is explained by the linear relationship between magnitude of a QSO and the redshift level.

11.79 a. $r = -.726$. Since this value is somewhat close to -1, there is a moderately strong negative linear relationship between the number of online courses taken and the weekly quiz grade.

b. In order to state that there is a significant negative correlation, we must run a test of hypothesis to see if the relationship is statistically significant.

11.81   a.   Piano: $r = .447$
Because this value is near .5, there is s slight positive linear relationship between recognition exposure time and goodness of view for piano.

Bench: $r = -.057$
Because this value is extremely close to 0, there is an extremely weak negative linear relationship between recognition exposure time and goodness of view for bench.

Motorbike: $r = .619$
Because this value is near .5, there is a moderate positive linear relationship between recognition exposure time and goodness of view for motorbike.

Armchair: $r = .294$
Because this value is fairly close to 0, there is a weak positive linear relationship between recognition exposure time and goodness of view for armchair.

Teapot: $r = .949$
Because this value is very close to 1, there is a strong positive linear relationship between recognition exposure time and goodness of view for teapot.

b.   Piano: $r^2 = (.447)^2 = .1998$
19.98% of the total sample variability around the sample mean recognition exposure time is explained by the linear relationship between the recognition exposure time and the goodness of view for piano.

Bench: $r^2 = (-.057)^2 = .0032$
.32% of the total sample variability around the sample mean recognition exposure time is explained by the linear relationship between the recognition exposure time and the goodness of view for bench.

Motorbike: $r^2 = (.619)^2 = .3832$
38.32% of the total sample variability around the sample mean recognition exposure time is explained by the linear relationship between the recognition exposure time and the goodness of view for motorbike.

Armchair: $r^2 = (.294)^2 = .0864$
8.64% of the total sample variability around the sample mean recognition exposure time is explained by the linear relationship between the recognition exposure time and the goodness of view for armchair.

Teapot: $r^2 = (.949)^2 = .9006$
90.06% of the total sample variability around the sample mean recognition exposure time is explained by the linear relationship between the recognition exposure time and the goodness of view for teapot.

c.   The test is:

H_0: $\beta_1 = 0$
H_a: $\beta_1 \neq 0$

Following are the values of α and $t_{\alpha/2}$ that correspond to df $= n - 2 = 25 - 2 = 23$.

α	.20	.10	.05	.02	.01	.002	.001
$t_{\alpha/2}$	1.319	1.714	2.069	2.500	2.807	3.485	3.767

Piano: $t = 2.40$
$2.069 < 2.40 < 2,500, p \approx .025$
For levels of significance greater than $\alpha = .025$, H_0 can be rejected. There is sufficient evidence to indicate that there is a linear relationship between goodness of view and recognition exposure time for piano for $\alpha > .025$.

Bench: $t = .27$
$.27 < 1.319, p > .2$
H_0 is not rejected. There is insufficient evidence to indicate that there is a linear relationship between goodness of view and recognition exposure time for bench for $\alpha \leq .2$.

Motorbike: $t = 3.78$
$3.78 > 3.767, p < .001$
H_0 can be rejected for $\alpha \geq .001$. There is sufficient evidence to indicate that there is a linear relationship between goodness of view and recognition exposure time for motorbike for $\alpha \geq .001$.

Armchair: $t = 1.47$
$1.319 < 1.47 < 1.714, p \approx .15$
H_0 cannot be rejected for levels of significance $\alpha < .15$. There is insufficient evidence to indicate that there is a linear relationship between goodness of view and recognition exposure time for armchair for $\alpha < .15$.

Teapot: $t = 14.50$
$14.50 > 3.767, p < .001$
H_0 can be rejected for $\alpha \geq .001$. There is sufficient evidence to indicate that there is a linear relationship between goodness of view and recognition exposure time for teapot for $\alpha \geq .001$.

11.83 a.   A straight line model is $y = \beta_0 + \beta_1 x + \varepsilon$.

b.   The researcher hypothesized that therapists with more years of formal dance training will report a higher perceived success rate in cotherapy relationships. This indicates that $\beta_1 > 0$.

c. $r = -.26$. Because this value is fairly close to 0, there is a weak negative linear relationship between years of formal training and reported success rate.

d. To determine if there is a positive linear relationship between years of formal training and reported success rate, we test:

H_0: $\beta_1 = 0$
H_0: $\beta_1 > 0$

The test statistic is $t = \dfrac{r}{\sqrt{(1-r^2)/(n-2)}} = \dfrac{-.26}{\sqrt{\left(1-(-.26^2)\right)/(136-2)}} = -3.12$

The rejection region requires $\alpha = .05$ in the upper tail of the t distribution with df $= n - 2 = 136 - 2 = 134$. From Table VI, Appendix A, $t_{.05} \approx 1.645$. The rejection region is $t > 1.645$.

Since the observed value of the test statistic does not fall in the rejection region ($t = -8.66 \not> 1.645$), H_0 is not rejected. There is insufficient evidence to indicate that there is a positive linear relationship between years of formal training and perceived success rates at $\alpha = .05$.

11.85 Using the values computed in Exercise 11.60:

$$r = \frac{SS_{xy}}{\sqrt{SS_{xx}SS_{yy}}} = \frac{48.125}{\sqrt{379.9375(18.75)}} = .5702$$

Because r is moderately small, there is a rather weak positive linear relationship between blood lactate concentration and perceived recovery.

$r^2 = .5702^2 = .3251$.

32.51% of the sample variance of blood lactate concentration around the sample mean is explained by the linear relationship between blood lactate concentration and perceived recovery.

11.87 For a given x, y is the actual value of the dependent variable and $E(y)$ is the mean value of y at the given value of x.

11.89 The statement "The greater the deviation between x and $\bar{x}$, the wider the prediction interval for y will be." is true. The further x gets from $\bar{x}$, the less precise the interval will be.

11.91 a. $\hat{\beta}_1 = \dfrac{SS_{xy}}{SS_{xx}} = \dfrac{28}{32} = .875$

$\hat{\beta}_0 = \bar{y} - \hat{\beta}_1\bar{x} = 4 - .875(3) = 1.375$

The least squares line is $\hat{y} = 1.375 + .875x$.

b. The least squares line is:

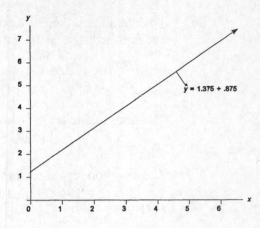

$\hat{y} = 1.375 + .875$

c. $\text{SSE} = \text{SS}_{yy} - \hat{\beta}_1 \text{SS}_{xy} = 26 - .875(28) = 1.5$

d. $s^2 = \dfrac{\text{SSE}}{n-2} = \dfrac{1.5}{10-2} = .1875$

e. $s = \sqrt{.1875} = .4330$

The form of the confidence interval is $\hat{y} \pm t_{\alpha/2} s \sqrt{\dfrac{1}{n} + \dfrac{(x_p - \bar{x})^2}{\text{SS}_{xx}}}$

For $x_p = 2.5$, $\hat{y} = 1.375 + .875(2.5) = 3.5625$

For confidence coefficient .95, $\alpha = .05$ and $\alpha/2 = .025$. From Table VI, Appendix A, with df $= n - 2 = 10 - 2 = 8$, $t_{.025} = 2.306$. The confidence interval is:

$$3.5625 \pm 2.306(.4330) \sqrt{\dfrac{1}{10} + \dfrac{(2.5-3)^2}{32}} \Rightarrow 3.5625 \pm .3279 \Rightarrow (3.2346, 3.8904)$$

f. The form of the prediction interval is $\hat{y} \pm t_{\alpha/2} s \sqrt{1 + \dfrac{1}{n} + \dfrac{(x_p - \bar{x})^2}{\text{SS}_{xx}}}$

For $x_p = 4$, $\hat{y} = 1.375 + .875(4) = 4.875$

For confidence coefficient .95, $\alpha = .05$ and $\alpha/2 = .025$. From Table VI, Appendix A, with df $= n - 2 = 10 - 2 = 8$, $t_{.025} = 2.306$. The prediction interval is:

$$4.875 \pm 2.306(.4330) \sqrt{1 + \dfrac{1}{10} + \dfrac{(4-3)^2}{32}} \Rightarrow 4.875 \pm 1.062 \Rightarrow (3.813, 5.937)$$

11.93   a, b. The scattergram is:

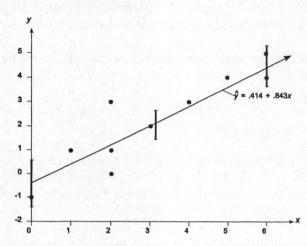

c. $SSE = SS_{yy} - \hat{\beta}_1 SS_{xy} = 33.6 - .84318766(32.8) = 5.94344473$

$$s^2 = \frac{SSE}{n-2} = \frac{5.94344473}{10-2} = .742930591 \qquad\qquad s = \sqrt{.742930591} = .8619$$

$$\bar{x} = \frac{31}{10} = 3.1$$

The form of the confidence interval is $\hat{y} \pm t_{\alpha/2}\sqrt{\dfrac{1}{n} + \dfrac{(x_p - \bar{x})^2}{SS_{xx}}}$

For $x_p = 6$, $\hat{y} = -.414 + .843(6) = 4.644$

For confidence coefficient .95, $\alpha = .05$ and $\alpha/2 = .025$. From Table VI, Appendix A, with df $= n - 2 = 10 - 2 = 8$, $t_{.025} = 2.306$. The confidence interval is:

$$4.644 \pm 2.306(.8619)\sqrt{\frac{1}{10} + \frac{(6-3.1)^2}{38.9}} \Rightarrow 4.644 \pm 1.118 \Rightarrow (3.526,\ 5.762)$$

d.   For $x_p = 3.2$, $\hat{y} = -.414 + .843(3.2) = 2.284$

The confidence interval is:

$$2.284 \pm 2.306(.8619)\sqrt{\frac{1}{10} + \frac{(3.2-3.1)^2}{38.9}} \Rightarrow 2.284 \pm .629 \Rightarrow (1.655,\ 2.913)$$

For $x_p = 0$, $\hat{y} = -.414 + .843(0) = -.414$

The confidence interval is:

$$-.414 \pm 2.306(.8619)\sqrt{\frac{1}{10} + \frac{(0-3.1)^2}{38.9}} \Rightarrow -.414 \pm 1.717 \Rightarrow (-1.585,\ .757)$$

e. The width of the confidence interval for the mean value of y depends on the distance x_p is from . The width of the interval for $x_p = 3.2$ is the smallest because 3.2 is the closest to $= 3.1$. The width of the interval for $x_p = 0$ is the widest because 0 is the farthest from $= 3.1$.

11.95 a. The researchers should use a prediction interval to estimate the actual ELSR score based on a value of the independent variable of $x = 50\%$.

b. The researchers should use a confidence interval for the mean ELSR score based on a value of the independent variable of $x = 70\%$.

11.97 For observation 7, the 95% prediction interval for the actual annual rainfall when the maximum daily temperature is 11.4 is (92.298, 125.104). We are 95% confident that the actual rainfall in a site with a maximum daily temperature of 11.4 is between 92.298 and 125.104.

11.99 a. The equation for the straight-line model relating duration to frequency is $y = \beta_0 + \beta_1 x + \varepsilon$.

b. Some preliminary calculations are:

$$\bar{y} = \frac{\sum y}{n} = \frac{1,287}{11} = 117$$

$$\bar{x} = \frac{\sum x}{n} = \frac{394}{11} = 35.818$$

$$SS_{xy} = \sum xy - \frac{\sum x \sum y}{n} = 30,535 - \frac{394(1,287)}{11} = -15,563$$

$$SS_{xx} = \sum x^2 - \frac{\left(\sum x\right)^2}{n} = 28,438 - \frac{394^2}{11} = 14,325.63636$$

$$\hat{\beta}_1 = \frac{SS_{xy}}{SS_{xx}} = \frac{-15,563}{14,325.63636} = -1.086374079$$

$$\hat{\beta}_0 = \bar{y} - \hat{\beta}_1 \bar{x} = \frac{1,287}{11} - (-1.086374079)\left(\frac{394}{11}\right) = 155.9119443$$

The least squares prediction equation is $\hat{y} = 155.912 - 1.086x$.

c. Some preliminary calculations are:

$$SS_{yy} = \sum y^2 - \frac{\left(\sum y\right)^2}{n} = 203{,}651 - \frac{1{,}287^2}{11} = 53{,}072$$

$$SSE = SS_{yy} - \hat{\beta}_1\left(SS_{xy}\right) = 53{,}072 - \left(-1.086374079\right)\left(-15{,}563\right) = 36{,}164.76021$$

$$s^2 = \frac{SSE}{n-2} = \frac{36{,}164.76021}{11-2} = 4{,}018.30669$$

$$s = \sqrt{s^2} = \sqrt{4{,}018.30669} = 63.39011508$$

To determine if there is a linear relationship between duration and frequency, we test:

H_0: $\beta_1 = 0$
H_a: $\beta_1 \neq 0$

The test statistic is $t = \dfrac{\hat{\beta}_1 - 0}{s_{\hat{\beta}}} = \dfrac{\hat{\beta}_1 - 0}{s/\sqrt{SS_{xx}}} = \dfrac{-1.086 - 0}{63.3901/\sqrt{14{,}325.63636}} = -2.051$

The rejection region requires $\alpha/2 = .05/2 = .025$ in each tail of the t distribution. From Table VI, Appendix A, with df $= n - 2 = 11 - 2 = 9$, $t_{.025} = 2.262$. The rejection region is $t < -2.262$ or $t > 2.262$.

Since the observed test statistic does not fall in the rejection region ($t = -2.051 \not< -2.262$), H_0 is not rejected. There is insufficient evidence to indicate that duration and frequency are linearly related at $\alpha = .05$.

d. For $x = 25$, the predicted duration is $\hat{y} = 155.912 - 1.086\,(25) = 128.762$.

For confidence coefficient .95, $\alpha = .05$ and $\alpha/2 = .05/2 = .025$. From Table VI, Appendix A, with df $= n - 2 = 11 - 2 = 9$, $t_{.025} = 2.262$. The 95% prediction interval is:

$$\hat{y} \pm t_{\alpha/2}s\sqrt{1 + \frac{1}{n} + \frac{\left(x_p - \bar{x}\right)^2}{SS_{xx}}} \Rightarrow 128.762 \pm 2.262\left(63.3901\right)\sqrt{1 + \frac{1}{11} + \frac{\left(25 - 35.818\right)^2}{14{,}325.63636}}$$

$$\Rightarrow 128.762 \pm 150.324 \Rightarrow (-21.562,\ 279.086)$$

We are 95% confident that the actual duration of a person who participates 25 times a year is between –21.562 and 279.086 days. Since the duration cannot be negative, the actual duration will be between 0 and 279.086.

11.101 a.   Using MINITAB, the output is:

```
Predicted Values for New Observations

New
Obs    Fit   SE Fit      99% CI          99% PI
  1  3.510   0.196  (2.955, 4.066)  (1.020, 6.000)


Values of Predictors for New Observations

New
Obs  Time
  1  15.0
```

From the printout, the 99% confidence interval for the mean mass of all spills with an elapsed time of 15 minutes is (2.955, 4.066). We are 99% confident that the mean mass of all spills with an elapsed time of 15 minutes is between 2.955 and 4.066.

b.   From the printout, the 99% prediction interval for the actual mass of a spill with an elapsed time of 15 minutes is (1.020, 6.000). We are 99% confident that the actual mass of a spill with an elapsed time of 15 minutes is between 1.020 and 6.000.

c.   The prediction interval in part b is larger than the confidence interval in part a. This will always be the case. The confidence interval gives a range of values for the mean of a distribution. The prediction interval gives a range of values for the actual value of a distribution. Once a mean is located the actual values can still vary around it. This is why the prediction interval is always wider than the confidence interval for the mean.

11.103 a.   From Exercise 11.46, $SS_{xx} = 3000$ and $= 50$.

Also, for Brand A, $s = 1.211$; for Brand B, $s = .610$.

For Brand A, $\hat{y} = 6.62 - .0727(45) = 3.349$, while for Brand B, $\hat{y} = 9.31 - .1077(45) = 4.464$.

The degrees of freedom for both brands is $n - 2 = 15 - 2 = 13$. For confidence coefficient .90, (i.e., for all parts of this question), $\alpha = .10$ and $\alpha/2 = .05$. From Table VI, Appendix A, with df = 13, $t_{.05} = 1.771$.

The form of both confidence intervals is $\hat{y} \pm t_{\alpha/2}s\sqrt{\dfrac{1}{n} + \dfrac{(x_p - \bar{x})^2}{SS_{xx}}}$

For Brand A, we obtain:

$$3.349 \pm 1.771(1.211)\sqrt{\frac{1}{15} + \frac{(45-50)^2}{3000}} \Rightarrow 3.349 \pm .587 \Rightarrow (2.762, 3.936)$$

For Brand B, we obtain:

$$4.464 \pm 1.771(.610)\sqrt{\frac{1}{15} + \frac{(45-50)^2}{3000}} \Rightarrow 4.464 \pm .296 \Rightarrow (4.168, 4.760)$$

The first interval is wider, caused by the larger value of s.

b. The form of both prediction intervals is $t_{\alpha/2}s\sqrt{1+\dfrac{1}{n}+\dfrac{(x_p-\bar{x})^2}{SS_{xx}}}$

For Brand A, we obtain:

$$3.349 \pm 1.771(1.211)\sqrt{1+\frac{1}{15}+\frac{(45-50)^2}{3000}} \Rightarrow 3.349 \pm 2.224 \Rightarrow (1.125, 5.573)$$

For Brand B, we obtain:

$$4.464 \pm 1.771(.610)\sqrt{1+\frac{1}{15}+\frac{(40+50)^2}{3000}} \Rightarrow 4.464 \pm 1.120 \Rightarrow (3.344, 5.584)$$

Again, the first interval is wider, caused by the larger value of s. Each of these intervals is wider than its counterpart from part **a**, since, for the same x, a prediction interval for an individual y is always wider than a confidence interval for the mean of y. This is due to an individual observation having a greater variance than the variance of the mean of a set of observations.

c. To obtain a confidence interval for the life of a brand A cutting tool that is operated at 100 meters per minute, we use:

$$\hat{y} \pm t_{\alpha/1}s\sqrt{1+\frac{1}{n}+\frac{(x_p-\bar{x})}{SS_{xx}}}$$

For $x = 100$, $\hat{y} = 6.62 - .0727(100) = -.65$.

The degrees of freedom are $n - 2 = 15 - 2 = 13$. For confidence coefficient .95, $\alpha = .05$ and $\alpha/2 = .025$. From Table VI, Appendix A, with df $= 13$, $t_{.025} = 2.160$.

Here, we obtain:

$$-.65 \pm 2.160(1.211)\sqrt{1+\frac{1}{15}+\frac{(100-50)^2}{3000}} \Rightarrow -.65 \pm 3.606 \Rightarrow (-4.256, 2.956)$$

The additional assumption would be that the straight line model fits the data well for the x's actually observed all the way up to the value under consideration, 100. Clearly from the estimated value of $-.65$, this is not true (usually, negative "useful lives" are not found).

11.105 The general form of the straight-line model for $E(y)$ is $E(y) = \beta_0 + \beta_1 x$.

11.107 The statement "In simple linear regression, about 95% of the y-values in the sample will fall within 2s of their respective predicted values." is a true statement.

11.109 a.

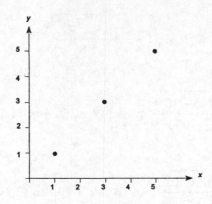

b.   One possible line is $\hat{y} = x$.

x	y	$\hat{y}$	$y - \hat{y}$
1	1	1	0
3	3	3	0
5	5	5	0
			0

For this example $\sum(y - \hat{y}) = 0$

A second possible line is $\hat{y} = 3$.

x	y	$\hat{y}$	$y - \hat{y}$
1	1	3	−2
3	3	3	0
5	5	3	2
			0

For this example $\sum(y - \hat{y}) = 0$

c.   Some preliminary calculations are:

$$\sum x_i = 9 \qquad \sum x_i^2 = 35 \qquad \sum x_i y_i = 35$$
$$\sum y_i = 9 \qquad \sum y_i^2 = 35$$

$$SS_{xy} = \sum x_i \sum y_i - \frac{\sum x_i \sum y_i}{n} = 35 - \frac{9(9)}{3} = 8$$

$$SS_{xx} = \sum x_i^2 - \frac{\left(\sum x_i\right)^2}{n} = 35 - \frac{9^2}{3} = 8$$

$$SS_{yy} = \sum y_i^2 - \frac{\left(\sum y_i\right)^2}{n} = 35 - \frac{9^2}{3} = 8$$

$$\hat{\beta}_1 = \frac{SS_{xy}}{SS_{xx}} = \frac{8}{8} = 1 \qquad\qquad \hat{\beta}_0 = \bar{y} - \hat{\beta}_1 \bar{x} = \frac{9}{3} - 1\left(\frac{9}{3}\right) = 0$$

The least squares line is $\hat{y} = 0 + 1x = x$.

d.   For $\hat{y} = x$, $SSE = SS_{yy} - \hat{\beta}_1 SS_{xy} = 8 - 1(8) = 0$

For $\hat{y} = 3$, $SSE = \sum(y_i - \hat{y}_i)^2 = (1 - 3)^2 + (3 - 3)^2 + (5 - 3)^2 = 8$

The least squares line has the smallest SSE of all possible lines.

11.111 a.   The straight-line model is $y = \beta_0 + \beta_1 x + \varepsilon$. Based on the theory, we would expect the metal level to decrease as the distance increases. Thus, the slope should be negative.

  b.   Yes. As the distance from the plant increases, the concentration of calcium tends to decrease.

  c.   No. As the distance from the plant increases, the concentration of arsenic tends to increase.

11.113 a.   We would expect the slope of the line from modeling the relationship between the mean number of games won and the team's batting average to be positive. As the team's batting average increases, we would expect that the number of games won should also increase.

  b.   Using MINITAB, a scattergram of the data is:

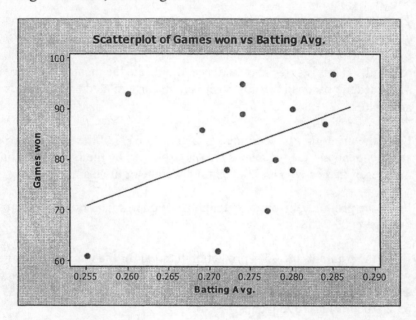

The graph indicates that batting average and games won have a positive linear relationship. As batting average increases, the number of games won tends to increase.

  c.   From the printout, $\hat{\beta}_0 = -85.68421$ and $\hat{\beta}_1 = 614.03509$. The least squares line is:
$\hat{y} = -85.684 + 614.035x$

  d.   The least squares line is plotted on the graph in part **a**. The data appear to fit the data fairly well.

e. $\hat{\beta}_0 = -85.684$. Since there is no team with a batting average of 0, this has no meaning other than the y-intercept.

$\hat{\beta}_1 = 614.035$. For each unit increase in batting average, the mean number of games won is estimated to increase by 614.035. This means very little since the batting average cannot increase by one unit. To make this interpretation more practical, we will look at only an increase of .01 in the batting average. For every .01 increase in batting average, the mean number of games won is estimated to increase by 6.140.

f. To determine if the mean number of games won by a major league baseball team is positively linearly related to the team's batting average, we test:
H_0: $\beta_1 = 0$
H_a: $\beta_1 > 0$

The test statistic is $t = 1.82$ and the p-value is $p = 0.0936$. Since we are running only a one-tailed test, we must divide the p-value by 2. Thus, the p-value of this test is $p = 0.0936/2 = 0.0468$. Since the p-value is less than $\alpha = .05$ ($p = 0.0468 < .05$), H_0 is rejected. There is sufficient evidence to indicate the mean number of games won by a major league baseball team is positively linearly related to the team's batting average at $\alpha = .05$

g. From the printout, $r^2 = $ R-Square $= .2165$. 21.65% of the sample variability around the sample mean number of games won is explained by the linear relationship between the number of games won and the batting average of the team.

h. From the printout, the predicted number of games won by a team with a .285 batting average is 89.3158.

i. From the printout, the 95% prediction interval for the number of games won by a team with a .285 batting average is (63.2620 and 115.3696). We are 95% confident that the actual number of games won by a team with a batting average of .285 is between 63.2620 and 115.3696 games.

11.115 a. The equation for the straight-line model is $y = \beta_0 + \beta_1 x + \varepsilon$.

b. Some preliminary calculations are:

$$\bar{y} = \frac{\sum y}{n} = \frac{398}{9} = 44.2222$$

$$\bar{x} = \frac{\sum x}{n} = \frac{1,444}{9} = 160.4444$$

$$SS_{xy} = \sum xy - \frac{\sum x \sum y}{n} = 60,428 - \frac{1,444(398)}{9} = -3,428.88889$$

$$SS_{xx} = \sum x^2 - \frac{(\sum x)^2}{n} = 235,866 - \frac{1,444^2}{9} = 4,184.2222$$

$$\hat{\beta}_1 = \frac{SS_{xy}}{SS_{xx}} = \frac{-3,428.88889}{4,184.2222} = -.819480593$$

$$\hat{\beta}_0 = \bar{y} - \hat{\beta}_1 \bar{x} = \frac{398}{9} - (-.819480593)\left(\frac{1,444}{9}\right) = 175.7033307$$

The fitted model is $\hat{y} = 175.7033 - .8195x$.

c. $\hat{\beta}_0 = 175.7033$. Since $x = 0$ (age of fish) is not in the observed range, $\hat{\beta}_0$ has no practical meaning.

d. $\hat{\beta}_1 = -.8195$. For each additional day of age, the mean number of strikes is estimated to decrease by .8195 strikes.

e.   Some additional calculations are:

$$\sum y_i = 398 \qquad \sum y_i^2 = 22,078$$

$$SS_{yy} = \sum y_i^2 - \frac{\left(\sum y_i\right)^2}{n} = 22,078 - \frac{398^2}{9} = 4,477.55556$$

$$SSE = SS_{yy} - \hat{\beta}_1 SS_{xy} = 4,477.55556 - (-.819480593)(-3,428.888889) = 1,667.647661$$

$$s^2 = \frac{SSE}{n-2} = \frac{1,667.647661}{9-2} = 238.2353801$$

$$s = \sqrt{s^2} = \sqrt{238.2353801} = 15.43487545$$

H_o: $\beta_1 = 0$
H_a: $\beta_1 < 0$

The test statistic is $t = \dfrac{\hat{\beta}_1 - 0}{s_{\hat{\beta}_1}} = \dfrac{-.8195 - 0}{15.4349 \Big/ \sqrt{4,184.2222}} = -3.434$

The rejection region requires $\alpha = .10$ in the lower tail of the t distribution with df $= n - 2 = 9 - 2 = 7$. From, Table VI, Appendix A, $t_{.10} = 1.415$. The rejection region is $t < -1.415$.

Since the observed value of the test statistic falls in the rejection region ($t = -3.434 < -1.415$), H_0 is rejected. There is sufficient evidence to indicate a negative linear relationship exists between the number of strikes and the age of the fish.

11.117 a. $r = .70$. Since this value is moderately close to 1, there is a moderately strong positive linear relationship between self-knowledge skill level and goal-setting ability.

b. $p = .001$. Since the p-value is so small, we would reject H_0. There is sufficient evidence to indicate a linear relationship between self-knowledge skill level and goal-setting ability for any reasonable level of α.

c. $r^2 = .70^2 = .49$. 49% of the total sample variability around the sample mean goal-setting ability is explained by the linear relationship between goal-setting ability and self-knowledge skill level.

11.119 Dependent Variable: VALUE

Analysis of Variance

Source	DF	Sum of Squares	Mean Square	F Value	Pr > F
Model	1	865746	865746	10.55	0.0021
Error	48	3939797	82079		
Corrected Total	49	4805543			

Root MSE	286.49451	R-Square	0.1802
Dependent Mean	128.90000	Adj R-Sq	0.1631
Coeff Var	222.26106		

Parameter Estimates

Variable	DF	Parameter Estimate	Standard Error	t Value	Pr > |t|
Intercept	1	-92.45768	79.29106	-1.17	0.2494
AGE	1	8.34682	2.57005	3.25	0.0021

Answers may vary. One possible answer may include:

The least squares line is $\hat{y} = -92.45768 + 8.34682x$. To determine if age can be used to predict market value, we test:

H_0: $\beta_1 = 0$
H_a: $\beta_1 \neq 0$

The test statistic is $t = 3.25$ with p-value = .0021. Reject the null hypothesis for levels of significance $\alpha > .0021$. There is sufficient evidence to indicate that age contributes information for the prediction of market value (y) at $\alpha > .0021$.

$r = .42$; Since this value is near .5, there is a moderate positive linear relationship between the value and age of the Beanie Baby.

11.121 a.   Using MINITAB, the scattergram of the data is:

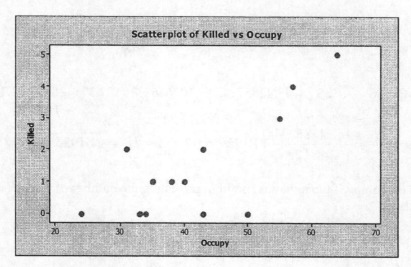

From the graph, it appears that as nest box tit occupancy increases, the number of flycatchers killed also increases.

b.   Some preliminary calculations are:

$$\sum x_i = 582 \qquad \sum y_i = 20 \qquad \sum x_i y_i = 1,009$$
$$\sum x_i^2 = 25,844 \qquad \sum y_i^2 = 62$$

$$SS_{xy} = \sum x_i y_i - \frac{\sum x_i \sum y_i}{n} = 1,009 - \frac{582(20)}{14} = 177.5714286$$

$$SS_{xx} = \sum x_i^2 - \frac{\left(\sum x_i\right)^2}{n} = 25,844 - \frac{582^2}{14} = 1,649.42857$$

$$\hat{\beta}_1 = \frac{SS_{xy}}{SS_{xx}} = \frac{177.5714286}{1,649.42857} = .107656331$$

For each unit increase in the nest box tit occupancy percent, the mean number of flycatchers killed is estimated to increase by .1077.

$$\hat{\beta}_0 = \bar{y} - \hat{\beta}_1 \bar{x} = \frac{20}{14} - (.107656331)\left(\frac{582}{14}\right) = -3.046856046$$

Since $x = 0$ is not in the observed range, $\hat{\beta}_0$ has no meaning other than the y-intercept.

c. Some additional calculations are:

$$SS_{yy} = \sum y_i^2 - \frac{\left(\sum y_i\right)^2}{n} = 62 - \frac{20^2}{14} = 33.42857143$$

$$SSE = SS_{yy} - \hat{\beta}_1 SS_{xy} = 33.42857143 - (.107656331)(177.5714286) = 14.31188294$$

$$s^2 = \frac{SSE}{n-2} = \frac{14.31188294}{14-2} = 1.192656912 \qquad s = \sqrt{s^2} = \sqrt{1.192656912} = 1.0921$$

To determine if the model is useful in predicting the number of flycatchers killed, we test:

H_0: $\beta_1 = 0$
H_a: $\beta_1 \neq 0$

The test statistic is $t = \dfrac{\hat{\beta}_1 - 0}{s_{\hat{\beta}_1}} = \dfrac{.1077 - 0}{1.0921 \Big/ \sqrt{1,649.42857}} = 4.005$

The rejection region requires $\alpha/2 = .05/2 = .025$ in each tail of the t distribution. From Table VI, Appendix A, with df $= n - 2 = 14 - 2 = 12$, $t_{.025} = 2.179$. The rejection region is $t < -2.179$ or $t > 2.179$.

Since the observed value of the test statistic falls in the rejection region ($t = 4.005 > 2.179$), H_0 is rejected. There is sufficient evidence to indicate that the model is useful in predicting the number of flycatchers killed at $\alpha = .05$.

d. $r = \dfrac{SS_{xy}}{\sqrt{SS_{xx}SS_{yy}}} = \dfrac{177.5714286}{\sqrt{1,649.42857(33.42857143)}} = .7562$

The value of r is moderately close to 1. Thus, there is a moderately strong positive linear relationship between the number of flycatchers killed and the nest box tit occupancy.

$r^2 = .7562^2 = .572$. 57.2% of the sample variability around the sample mean number of flycatchers killed is explained by the linear relationship between the number of flycatchers killed and the nest box tit occupancy.

e. From part c, $s = 1.0921$. We would expect most of the observed number of flycatchers killed to fall within $2s$ or $2(1.0921) = 2.1842$ units of their predicted values.

f. Since there is a significant linear relationship between the number of flycatchers killed and the nest box tit occupancy, we would recommend using this model. In addition, the r^2 value is .572. Although this is not great, over half of the sample variation is explained by the model.

11.123 Using MINITAB, the output for fitting the least squares line is:

Regression Analysis: Impur versus Temp

```
The regression equation is
Impur = - 13.5 - 0.0528 Temp


Predictor       Coef    SE Coef       T       P
Constant      -13.490     2.074    -6.51   0.000
Temp         -0.052829  0.007728   -6.84   0.000


S = 0.133347   R-Sq = 85.4%   R-Sq(adj) = 83.6%


Analysis of Variance

Source         DF       SS        MS      F       P
Regression      1    0.83089   0.83089  46.73   0.000
Residual Error  8    0.14225   0.01778
Total           9    0.97315


Unusual Observations

Obs   Temp   Impur     Fit   SE Fit  Residual  St Resid
  2   -265  0.2020  0.5093   0.0492   -0.3073    -2.48R
  3   -256  0.2040  0.0339   0.1038    0.1701     2.03RX

R denotes an observation with a large standardized residual.
X denotes an observation whose X value gives it large influence.


Predicted Values for New Observations

New
Obs    Fit   SE Fit      95% CI             95% PI
  1  0.9320  0.0558  (0.8034, 1.0605)  (0.5987, 1.2652)


Values of Predictors for New Observations

New
Obs   Temp
  1   -273
```

a. $\hat{\beta}_0 = -13.490$ –   has no meaning since $x = 0$ is not in the observed range. It is the y-intercept.

$\hat{\beta}_1 = -.052829$ –   is the estimated change in the mean proportion of impurity passing through helium for each additional degree.

b. For confidence coefficient .95, $\alpha = .05$ and $\alpha/2 = .025$. From Table VI, Appendix A, with df $= n - 2 = 10 - 2 = 8$, $t_{.025} = 2.306$. The confidence interval is:

$$\hat{\beta}_1 \pm t_{\alpha/2} s_{\hat{\beta}_1} \Rightarrow -.0528 \pm 2.306(.007728) \Rightarrow -.0528 \pm .0178 \Rightarrow (-.0706, -.0350)$$

We are 95% confident that the change in mean proportion of impurity passing through helium for each additional degree is between $-.0706$ and $-.0350$. Since 0 is not in the interval, there is evidence to indicate that temperature contributes information about the proportion of impurity passing through helium.

c. $r^2 = R\text{-sq} = .854$. 85.4% of the total sample variation around the mean proportion of impurity is explained by the linear relationship between proportion of impurity and temperature.

d. From the printout, the 95% prediction interval is (.5987, 1.2652).

e. We cannot be sure that the relationship between the proportion of impurity passing through helium and temperature is the same outside the observed range.

11.125 a. A scattergram of the data is:

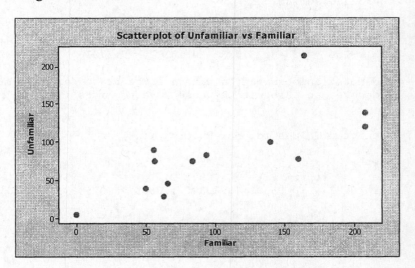

It appears that there is a positive linear relationship between total number of words with unfamiliar partner and with familiar partner.

b. A straight line model would be $y = \beta_0 + \beta_1 x + \varepsilon$.

c. Some preliminary calculations are:

$$\sum x_i = 1,341 \qquad \sum y_i = 1,099 \qquad \sum x_i y_i = 145,607$$
$$\sum x_i^2 = 189,963 \qquad \sum y_i^2 = 127,397$$

$$SS_{xy} = \sum x_i y_i - \frac{\sum x_i \sum y_i}{n} = 145,607 - \frac{1,341(1,099)}{13} = 32,240.9231$$

$$SS_{xx} = \sum x_i^2 - \frac{\left(\sum x_i\right)^2}{n} = 189,963 - \frac{1,341^2}{13} = 51,633.6923$$

$$\hat{\beta}_1 = \frac{SS_{xy}}{SS_{xx}} = \frac{32,240.9231}{51,633.6923} = .624416377$$

$$\hat{\beta}_0 = \bar{y} - \hat{\beta}_1\bar{x} = \frac{1,099}{13} - (.624416377)\left(\frac{1,341}{13}\right) = 20.12751061$$

The least squares prediction equation is $\hat{y} = 20.1275 + .6244x$

d. $\hat{\beta}_0 = 20.1275$. Since $x = 0$ is in the observed range, the mean number of words in conversation with an unfamiliar partner is estimated to be 20.1275 when the number of words in conversation with a familiar partner is 0.

$\hat{\beta}_1 = .62442$. For each additional word in conversation with a familiar partner, the mean number of words in conversation with an unfamiliar partner is estimated to increase by .62442.

11.127 Some preliminary calculations are:

$$\sum x = 4305 \qquad\qquad \sum x^2 = 1,652,025 \qquad\qquad \sum xy = 76,652,695$$
$$\sum y = 201,558 \qquad\qquad \sum y^2 = 3,571,211,200$$

a. $\hat{\beta}_1 = \dfrac{\sum xy}{\sum x^2} = \dfrac{76,652,695}{1,652,025} = 46.39923427 \approx 46.3992$

The least squares line is $\hat{y} = 46.3992x$.

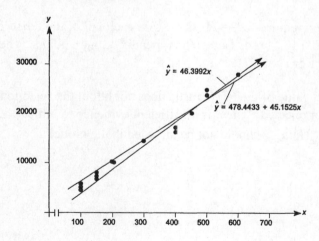

b. $SS_{xy} = \sum xy - \dfrac{\sum x \sum y}{n} = 76,652,695 = \dfrac{4305(201,558)}{15} = 18,805,549$

$SS_{xx} = \sum x^2 - \dfrac{(\sum x)^2}{n} = 1,652,025 - \dfrac{4305^2}{15} = 416,490$

$\hat{\beta}_1 = \dfrac{SS_{xy}}{SS_{xx}} = \dfrac{18,805,549}{416,490} = 45.15246224 \approx 45.1525$

$\hat{\beta}_0 = \bar{y} - \hat{\beta}_1 \bar{x} = \dfrac{201,558}{15} - 45.15246224\left(\dfrac{4305}{15}\right) = 478.4433$

The least squares line is $\hat{y} = 478.4433 + 45.1525x$.

c.   Because $x = 0$ is not in the observed range, we are trying to represent the data on the observed interval with the best fitting line. We are not concerned with whether the line goes through (0, 0) or not.

d.   Some preliminary calculations are:

$$SS_{yy} = \sum y^2 - \dfrac{(\sum y)^2}{n} = 3,571,211,200 - \dfrac{201,558^2}{15} = 862,836,042$$

$$SSE = SS_{yy} - \hat{\beta}_1 SS_{xy} = 862,836,042 - 45.15246224(18,805,549) = 13,719,200.88$$

$$s^2 = \dfrac{SSE}{n-2} = \dfrac{13,719,200.88}{15-2} = 1,055,323.145 \quad s = 1027.2892$$

H_0: $\beta_0 = 0$
H_a: $\beta_0 \neq 0$

The test statistic is $t = \dfrac{\hat{\beta}_0 - 0}{s\sqrt{\dfrac{1}{n} + \dfrac{\bar{x}^2}{SS_{xx}}}} = \dfrac{478.443}{1027.2892\sqrt{\dfrac{1}{15} + \dfrac{287^2}{416,490}}} = .906$

The rejection region requires $\alpha/2 = .10/2 = .05$ in each tail of the t distribution with df $= n - 2 = 15 - 2 = 13$. From Table VI, Appendix A, $t_{.05} = 1.771$. The rejection region is $t < -1.771$ or $t > 1.771$.

Since the observed value of the test statistic does not fall in the rejection region ($t = .906 \not> 1.771$), H_0 is not rejected. There is insufficient evidence to indicate β_0 is different from 0 at $\alpha = .10$. Thus, β_0 should not be included in the model.

11.129 a.   Some preliminary calculations are:

$$\sum x_i = 1{,}098 \qquad\qquad \sum y_i = 121.8 \qquad\qquad \sum x_i y_i = 4{,}294.6$$
$$\sum x_i^2 = 56{,}712 \qquad\qquad \sum y_i^2 = 597.52$$

$$SS_{xy} = \sum x_i y_i - \frac{\sum x_i \sum y_i}{n} = 4{,}294.6 - \frac{1{,}098(121.8)}{27} = -658.6$$

$$SS_{xx} = \sum x_i^2 - \frac{\left(\sum x_i\right)^2}{n} = 56{,}712 - \frac{1{,}098^2}{27} = 12{,}060$$

$$\hat{\beta}_1 = \frac{SS_{xy}}{SS_{xx}} = \frac{-658.6}{12{,}060} = -.054610281$$

$$\hat{\beta}_o = \bar{y} - \hat{\beta}_1 \bar{x} = \frac{121.8}{27} - (-.054610281)\left(\frac{1{,}098}{27}\right) = 6.731929244$$

$$SS_{yy} = \sum y_i^2 - \frac{\left(\sum y_i\right)^2}{n} = 597.52 - \frac{121.8^2}{27} = 48.06666667$$

$$SSE = SS_{yy} - \hat{\beta}_1 SS_{xy} = 48.06666667 - (-.054610281)(-658.6) = 12.10033563$$

$$r^2 = 1 - \frac{SSE}{SS_{yy}} = 1 - \frac{12.10033563}{48.066666667} = .7483 \quad 74.83\% \text{ of the total sample variability around}$$

the sample mean fertility rate is explained by the linear relationship between the fertility rate and the contraceptive prevalence. This is not 90%.

b.   From above, the fitted regression line is $\hat{y} = 6.731 - .0546x$

If the contraceptive use increases by 18 percent, then the mean fertility rate will decrease by an estimated $(.0546)(18) = .9828$. This is approximately 1. The researchers' statement is supported by the data.

Multiple Regression

12.1   a. $E(y) = \beta_0 + \beta_1 x_1 + \beta_2 x_2$

b. $E(y) = \beta_0 + \beta_1 x_1 + \beta_2 x_2 + \beta_3 x_3 + \beta_4 x_4$

c. $E(y) = \beta_0 + \beta_1 x_1 + \beta_2 x_2 + \beta_3 x_3 + \beta_4 x_4 + \beta_5 x_5$

12.3   The 6 steps in a multiple regression analysis are:

1.   Hypothesize the deterministic component of the model. This model relates the mean, $E(y)$, to the independent variables $x_1, x_2, ..., x_k$. This involves the choice of the independent variables to be included in the model.

2.   Use the sample data to estimate the unknown model parameters $\beta_0, \beta_1, \beta_2, ..., \beta_k$ in the model.

3.   Specify the probability distribution of the random error term, ε, and estimate the standard deviation of the distribution, σ.

4.   Check the assumptions on ε are satisfied, and make model modifications if necessary.

5.   Statistically evaluate the usefulness of the model.

6.   When satisfied that the model is useful, use it for prediction, estimation, and other purposes.

12.5   To test the overall adequacy of a multiple regression model, you would use a global test. The null hypothesis would be that all of the beta parameters are 0 and the alternative hypothesis would be that at least one beta parameter is not 0.

12.7   a.   We are given $\hat{\beta}_2 = 2.7$, $s_{\hat{\beta}_2} = 1.86$, and $n = 30$.

H_0: $\beta_2 = 0$
H_a: $\beta_2 \neq 0$

The test statistic is $t = \dfrac{\hat{\beta}_2 - 0}{s_{\hat{\beta}_2}} = \dfrac{2.7}{1.86} = 1.45$

The rejection region requires $\alpha/2 = .05/2 = .025$ in each tail of the t distribution with df $= n - (k + 1) = 30 - (3 + 1) = 26$. From Table VI, Appendix A, $t_{.025} = 2.056$. The rejection region is $t < -2.056$ or $t > 2.056$.

Since the observed value of the test statistic does not fall in the rejection region ($t = 1.45$ $\not> 2.056$), H_0 is not rejected. There is insufficient evidence to indicate $\beta_2 \neq 0$ at $\alpha = .05$.

b. We are given $\hat{\beta}_3 = .93$, $s_{\hat{\beta}_3} = .29$, and $n = 30$.

Test H_0: $\beta_3 = 0$
        H_a: $\beta_3 \neq 0$

The test statistic is $t = \dfrac{\hat{\beta}_3 - 0}{s_{\hat{\beta}_3}} = \dfrac{.93}{.29} = 3.21$

The rejection region is the same as part **a**, $t < -2.056$ or $t > 2.056$.

Since the observed value of the test statistic falls in the rejection region ($t = 3.21 > 2.056$), H_0 is rejected. There is sufficient evidence to indicate $\beta_3 \neq 0$ at $\alpha = .05$.

c. $\hat{\beta}_3$ has a smaller estimated standard error than $\hat{\beta}_2$. Therefore, the test statistic is larger for $\hat{\beta}_3$ even though $\hat{\beta}_3$ is smaller than $\hat{\beta}_2$.

12.9   The number of degrees of freedom available for estimating σ^2 is $n - (k + 1)$ where k is the number of variables in the regression model. Each additional independent variable placed in the model causes a corresponding decrease in the degrees of freedom.

12.11   a. Yes. Since $R^2 = .92$ is close to 1, this indicates the model provides a good fit. Also, SSE = .33 is fairly small which indicates the model provides a good fit.

b. H_0: $\beta_1 = \beta_2 = \cdots = \beta_5 = 0$
H_a: At least one of the parameters is not 0

The test statistic is $F = \dfrac{R^2/k}{(1-R^2)/[n-(k+1)]} = \dfrac{.92/5}{(1-.92)/[30-(5+1)]} = 55.2$

The rejection region requires $\alpha = .05$ in the upper tail of the F distribution with $v_1 = k = 5$ and $v_2 = n - (k + 1) = 30 - (5 + 1) = 24$. From Table IX, Appendix A, $F_{.05} = 2.62$. The rejection region is $F > 2.62$.

Since the observed value of the test statistic falls in the rejection region ($F = 55.2 > 2.62$), H_0 is rejected. There is sufficient evidence to indicate the model is useful in predicting y at $\alpha = .05$.

12.13   a. The least squares prediction equation for the interstate highway model is:

$\hat{y} = 1.81231 + .10875x_1 + .00017x_2$

b. $\hat{\beta}_o = 1.81231$. This value has no meaningful interpretation because $x_1 = 0$ and $x_2 = 0$ are not in the observed range.

$\hat{\beta}_1 = .10875$. We estimate the mean number of crashes per 3 years will increase by .10875 for each additional mile of roadway, holding AADT constant.

$\hat{\beta}_2 = .00017$. We estimate the mean number of crashes per 3 years will increase by .00017 for each additional vehicle per day, holding roadway mileage constant.

c. For confidence coefficient .99, $\alpha = .01$ and $\alpha/2 = .01/2 = .005$. From Table VI, Appendix A, with df = n − (k + 1) = 100 − (2 + 1) = 97, $t_{.005} \approx 2.62$. The 99% confidence interval is:

$$\hat{\beta}_1 \pm t_{.005} s_{\hat{\beta}_1} \Rightarrow .10875 \pm 2.62(.03166) \Rightarrow .10875 \pm .08295 \Rightarrow (.02580, \ .19170)$$

We are 99% confident that the increase in mean number of crashes for each additional mile of roadway is between .02580 and .19170, holding AADT constant.

d. The 99% confident interval is:

$$\hat{\beta}_2 \pm t_{.005} s_{\hat{\beta}_2} \Rightarrow .00017 \pm 2.62(.00003) \Rightarrow .00017 \pm .00008 \Rightarrow (.00009, \ .00025)$$

We are 99% confident that the increase in mean number of crashes for each additional vehicle per day is between .00009 and .00025, holding roadway mileage constant.

e. The least squares prediction equation for the non-interstate highway model is:

$$\hat{y} = 1.20785 + .06343x_1 + .00056x_2$$

$\hat{\beta}_o = 1.20785$. This value has no meaningful interpretation because $x_1 = 0$ and $x_2 = 0$ are not in the observed range.

$\hat{\beta}_1 = .06343$. We estimate the mean number of crashes per 3 years will increase by .06343 for each additional mile of roadway, holding AADT constant.

$\hat{\beta}_2 = .00056$. We estimate the mean number of crashes per 3 years will increase by .00056 for each additional vehicle per day, holding roadway mileage constant.

For confidence coefficient .99, $\alpha = .01$ and $\alpha/2 = .01/2 = .005$. From Table VI, Appendix A, with df = $n - (k + 1) = 100 - (2 + 1) = 97$, $t_{.005} \approx 2.62$. The 99% confidence interval is:

$$\hat{\beta}_1 \pm t_{.005} s_{\hat{\beta}_1} \Rightarrow .06343 \pm 2.62(.01809) \Rightarrow .06343 \pm .04740 \Rightarrow (.01603, \ .11083)$$

We are 99% confident that the increase in mean number of crashes for each additional mile of roadway is between .01603 and .11083, holding AADT constant.

The 99% confident interval is:

$$\hat{\beta}_2 \pm t_{.005} s_{\hat{\beta}_2} \Rightarrow .00056 \pm 2.62(.00012) \Rightarrow .00056 \pm .00031 \Rightarrow (.00025, \ .00087)$$

We are 99% confident that the increase in mean number of crashes for each additional vehicle per day is between .00025 and .00087, holding roadway mileage constant.

12.15 a. From the output, the least squares prediction equation is
$$\hat{y} = 3.70 + .34x_1 + .49x_2 + .72x_3 + 1.14x_4 + 1.51x_5 + .26x_6 - .14x_7 - .10x_8 - .10x_9.$$

b. $\hat{\beta}_0 = 3.70.$ This value has no meaningful interpretation.

$\hat{\beta}_1 = .34.$ We estimate the mean number of runs scored by a team in a season to increase by .34 for each unit increase in the number of walks, when all the other independent variables are held fixed.

$\hat{\beta}_2 = .49.$ We estimate the mean number of runs scored by a team in a season to increase by .49 for each additional single, when all the other independent variables are held fixed.

$\hat{\beta}_3 = .72.$ We estimate the mean number of runs scored by a team in a season to increase by .72 for each additional double, when all the other independent variables are held fixed.

$\hat{\beta}_4 = 1.14.$ We estimate the mean number of runs scored by a team in a season to increase by 1.14 for each additional triple, when all the other independent variables are held fixed.

$\hat{\beta}_5 = 1.51.$ We estimate the mean number of runs scored by a team in a season to increase by 1.51 for each additional home run, when all the other independent variables are held fixed.

$\hat{\beta}_6 = .26.$ We estimate the mean number of runs scored by a team in a season to increase by .26 for each additional stolen base, when all the other independent variables are held fixed.

$\hat{\beta}_7 = -.14.$ We estimate the mean number of runs scored by a team in a season to decrease by .14 for each additional player caught stealing, when all the other independent variables are held fixed.

$\hat{\beta}_8 = -.10.$ We estimate the mean number of runs scored by a team in a season to decrease by .10 for each additional strikeout, when all the other independent variables are held fixed.

$\hat{\beta}_9 = -.10.$ We estimate the mean number of runs scored by a team in a season to decrease by .10 for each additional out, when all the other independent variables are held fixed.

c. H_0: $\beta_7 = 0$
H_a: $\beta_7 < 0$

The test statistic is $t = \dfrac{\hat{\beta}_7 - 0}{s_{\hat{\beta}_7}} = \dfrac{-.14 - 0}{.14} = -1.00$

The rejection region requires $\alpha = .05$ in the lower tail of the t distribution. From Table VI, Appendix A, with $df = n - (k + 1) = 234 - (9 + 1) = 224$, $t_{.05} \approx 1.645$. The rejection region is $t < -1.645$.

Since the observed value of the test statistic does not fall in the rejection region $(t = -1.00 \not< -1.645)$, H_0 is not rejected. There is insufficient evidence to indicate that there is a negative linear relationship between the number of runs scored and the number of times caught stealing at $\alpha = .05$.

d.   For confidence coefficient .95, $\alpha = .05$ and $\alpha/2 = .05/2 = .025$. From Table VI, Appendix A, with $df = n - (k + 1) = 234 - (9 + 1) = 224$, $t_{.025} \approx 1.96$. The 95% confidence interval is:

$$\hat{\beta}_5 \pm t_{.025} s_{\hat{\beta}_5} \Rightarrow 1.51 \pm 1.96(.05) \Rightarrow 1.51 \pm .098 \Rightarrow (1.412,\ 1.608)$$

We are 95% confident that for each additional home run, the mean number of runs scored will increase by anywhere from 1.412 to 1.608, holding all the other variables constant.

12.17   a.   The first order model is $y = \beta_0 + \beta_1 x_1 + \beta_2 x_2 + \beta_3 x_3 + \varepsilon$

b. $R^2 = .08$. 8% of the sample variation in the frequency of marijuana use in the past 6 months is explained by the model containing severity of inattention, severity of impulsivity-hyperactivity, and level of oppositional-defiant and conduct disorder.

c.   The global F-test has the hypotheses:

H_0: $\beta_1 = \beta_2 = \beta_3 = 0$
H_a: At least one $\beta_i \neq 0$

The global F-test yielded a p-value less than .01. Since the p-value is so small, there is evidence to reject H_0 for $\alpha > .01$. There is sufficient evidence to indicate that at least one of the independent variables contributes to the prediction of the frequency of marijuana use in the past 6 months at $\alpha > .01$.

d.	The test for β_1:

H_0: $\beta_1 = 0$
H_a: $\beta_1 \neq 0$

The p-value for the test is less than .01. Since the p-value is so small, there is evidence to reject H_o for $\alpha > .01$. There is sufficient evidence to indicate that severity of inattention contributes to the prediction of the frequency of marijuana use in the past 6 months, when holding severity of impulsivity-hyperactivity and level of oppositional-defiant and conduct disorder fixed at $\alpha > .01$.

e.	The test for β_2:

H_0: $\beta_2 = 0$
H_a: $\beta_2 \neq 0$

The p-value for the test is greater than .05. Since the p-value is not small, there is no evidence to reject H_0 for $\alpha \leq .05$. There is insufficient evidence to indicate that severity of impulsivity-hyperactivity contributes to the prediction of the frequency of marijuana use in the past 6 months, when holding severity of inattention and level of oppositional-defiant and conduct disorder fixed at $\alpha \leq .05$.

f.	The test for β_3:

H_0: $\beta_3 = 0$
H_a: $\beta_3 \neq 0$

The p-value for the test is greater than .05. Since the p-value is not small, there is no evidence to reject H_0 for $\alpha \leq .05$. There is insufficient evidence to indicate that level of oppositional-defiant and conduct disorder contributes to the prediction of the frequency of marijuana use in the past 6 months, when holding severity of inattention and severity of impulsivity-hyperactivity fixed at $\alpha \leq .05$.

12.19	a.	Let y = arsenic level, x_1 = latitude, x_2 = longitude, and x_3 = depth. The first-order model would be:

$$y = \beta_o + \beta_1 x_1 + \beta_2 x_2 + \beta_3 x_3 + \varepsilon$$

b. Using MINITAB, the results are:

Regression Analysis: ARSENIC versus LATITUDE, LONGITUDE, DEPTH-FT

```
The regression equation is
ARSENIC = - 86868 - 2219 LATITUDE + 1542 LONGITUDE - 0.350 DEPTH-FT


327 cases used, 1 cases contain missing values


Predictor      Coef   SE Coef       T      P
Constant     -86868     31224   -2.78  0.006
LATITUDE    -2218.8     526.8   -4.21  0.000
LONGITUDE    1542.2     373.1    4.13  0.000
DEPTH-FT    -0.3496    0.1566   -2.23  0.026


S = 103.301   R-Sq = 12.8%   R-Sq(adj) = 12.0%


Analysis of Variance

Source          DF        SS      MS      F      P
Regression       3    505770  168590  15.80  0.000
Residual Error 323   3446791   10671
Total          326   3952562


Source      DF   Seq SS
LATITUDE     1   132448
LONGITUDE    1   320144
DEPTH-FT     1    53179
```

The fitted regression model is:

$$\hat{y} = -86{,}868 - 2{,}218.8x_1 + 1{,}542.2x_2 - .3496x_3$$

c. $\hat{\beta}_o = -86{,}868$. This value has no meaningful interpretation because $x_1 = 0$, $x_2 = 0$, and $x_3 = 0$ are not in the observed range.

$\hat{\beta}_1 = -2{,}218.8$. We estimate the mean arsenic level will decrease by 2,218.8 for each unit increase in latitude, holding longitude and depth constant.

$\hat{\beta}_2 = 1{,}542.2$. We estimate the mean arsenic level will increase by 1,542.2 for each unit increase in longitude, holding latitude and depth constant.

$\hat{\beta}_3 = -.3496$. We estimate the mean arsenic level will decrease by .3496 for each unit increase in depth, holding latitude and longitude constant.

d. $S = 103.301$. We would expect about 95% of the observed values of arsenic levels to fall within $2s$ or $2(103.301) = 206.602$ units of their least squares predicted values.

e.   R-Sq = 12.8. 12.8% of the sample variation in the arsenic level is explained by the model containing latitude, longitude and depth.

R-Sq(adj) = 12.0. 12.0% of the sample variation in the arsenic level is explained by the model containing latitude, longitude and depth, adjusted for the sample size and the number of independent variables in the model.

f.   To determine if the overall model is useful for predicting y, we test:

H_0: $\beta_1 = \beta_2 = \beta_3 = 0$
H_a: At lease one $\beta_i \neq 0$

The test statistic is $F = 15.80$ and the p-value is p = .000.

Since the p-value is less than $\alpha = .05$ (p =.000 < .05), H_0 is rejected. There is sufficient evidence to indicate that at least one of the independent variables is useful in predicting arsenic levels at $\alpha = .05$.

g.   Even though the independent variables are significant in predicting arsenic levels, the R^2 value and adjusted R^2 values are very low. Not very mush of the sample variation in the arsenic levels can be explained by the model. In addition, the estimate of the standard deviation (s = 103.301) is very large compared with the actual values of the data. I would not recommend using this model.

12.21 a.   **Regression Analysis: WeightChg versus Digest, Fiber**

```
The regression equation is
WeightChg = 12.2 - 0.0265 Digest - 0.458 Fiber

Predictor         Coef      SE Coef          T          P
Constant        12.180        4.402       2.77      0.009
Digest         -0.02654      0.05349      -0.50      0.623
Fiber          -0.4578       0.1283       -3.57      0.001

S = 3.519       R-Sq = 52.9%      R-Sq(adj) = 50.5%

Analysis of Variance

Source            DF           SS          MS          F          P
Regression         2       542.03      271.02      21.88      0.000
Residual Error    39       483.08       12.39
Total             41      1025.12
```

$\hat{y} = 12.180 - .02654x_1 - .4578x_2$.

b. $\hat{\beta}_0$ = the estimate of the y-intercept

$\hat{\beta}_1 = -.0265$. We estimate that the mean weight change will decrease by .0265% for each additional increase of 1% in digestion efficiency, with acid-detergent fibre held constant.

$\hat{\beta}_2 = -.458$. We estimate that the mean weight change will decrease by .458% for each additional increase of 1% in acid-detergent fibre, with digestion efficiency held constant.

c. To determine if digestion efficiency is a useful predictor of weight change, we test:

$H_0: \beta_1 = 0$
$H_a: \beta_1 \neq 0$

The test statistic is $t = -.50$. The p-value is $p = .623$. Since the p-value is greater than α ($p = .623 > .01$), H_0 is not rejected. There is insufficient evidence to indicate that digestion efficiency is a useful linear predictor of weight change at $\alpha = .01$.

d. For confidence coefficient .99, $\alpha = 1 - .99 = .01$ and $\alpha/2 = .01/2 = .005$. From Table VI, Appendix A, with df $= n - (k + 1) = 42 - (2 + 1) = 39$, $t_{.005} \approx 2.704$. The 99% confidence interval is:

$$\hat{\beta}_2 \pm t_{.005} s_{\hat{\beta}} \Rightarrow -.4578 \pm 2.704 \,(.1283) \Rightarrow -.4578 \pm .3469 \Rightarrow (-.8047, -.1109)$$

We are 99% confident that the change in mean weight change for each unit change in acid-detergent fiber, holding digestion efficiency constant is between $-.8047$ and $-.1109$.

e. $R^2 = \text{R-Sq} = 52.9\%$. 52.9% of the total sample variation in the weight change scores is explained by the model containing the independent variables digestion efficiency and acid-detergent fibre. $R_a^2 = \text{R-Sq(adj)} = 50.5\%$. 50.5% of the total sample variation in the weight change scores is explained by the model containing the independent variables digestion efficiency and acid-detergent fibre, adjusting for the sample size and the number of independent variables in the model.

f. To determine if the overall model is useful in the prediction of weight change, we test:

$H_0: \beta_1 = \beta_2 = 0$
H_a: At least one $\beta_i \neq 0$, $i = 1, 2$

The test statistic is $F = 21.88$ and the p-value is $p = .000$. Since the p-value is less than α ($p = .000 < \alpha = .05$), H_0 is rejected. There is sufficient evidence to indicate at least one of the variables digestion efficiency and acid-detergent fibre is useful in the prediction weight change at $\alpha = .05$.

12.23 a. Let y = heat rate, x_1 = speed, x_2 = inlet temperature, x_3 = exhaust temperature, x_4 = cycle pressure ratio, and x_5 = air flow rate. The first-order model would be:

$$y = \beta_o + \beta_1 x_1 + \beta_2 x_2 + \beta_3 x_3 + \beta_4 x_4 + \beta_5 x_5 + \varepsilon$$

b. Using MINITAB, the results are:

Regression Analysis: HEATRATE versus RPM, INLET-TEMP, ...

```
The regression equation is
HEATRATE = 13614 + 0.0888 RPM - 9.20 INLET-TEMP + 14.4 EXH-TEMP + 0.4
CPRATIO - 0.848 AIRFLOW


Predictor      Coef  SE Coef        T       P
Constant    13614.5    870.0    15.65   0.000
RPM         0.08879  0.01391     6.38   0.000
INLET-TEMP   -9.201    1.499    -6.14   0.000
EXH-TEMP     14.394    3.461     4.16   0.000
CPRATIO        0.35    29.56     0.01   0.991
AIRFLOW     -0.8480   0.4421    -1.92   0.060


S = 458.828   R-Sq = 92.4%   R-Sq(adj) = 91.7%


Analysis of Variance

Source         DF         SS        MS       F       P
Regression      5  155055273  31011055  147.30   0.000
Residual Error 61   12841935    210524
Total          66  167897208


Source     DF    Seq SS
RPM         1  119598530
INLET-TEMP  1   26893467
EXH-TEMP    1    7784225
CPRATIO     1       4623
AIRFLOW     1     774427
```

The fitted regression model is:

$$\hat{y} = 13,614.5 + .08879x_1 - 9.201x_2 + 14.394x_3 + .35x_4 - .848x_5$$

c. $\hat{\beta}_o = 13,614.5$. This value has no meaningful interpretation because $x_1 = 0$, $x_2 = 0$, $x_3 = 0$, $x_4 = 0$, and $x_5 = 0$, are not in the observed range.

$\hat{\beta}_1 = .08879$. We estimate the mean heat rate will increase by .08879 for each unit increase in speed, holding inlet temperature, exhaust temperature, cycle pressure ratio, and air flow rate constant.

$\hat{\beta}_2 = -9.201$. We estimate the mean heat rate will decrease by 9.201 for each unit increase in inlet temperature, holding speed, exhaust temperature, cycle pressure ratio, and air flow rate constant.

$\hat{\beta}_3 = 14.394$. We estimate the mean heat rate will increase by 14.394 for each unit increase in exhaust temperature, holding speed, inlet temperature, cycle pressure ratio, and air flow rate constant.

$\hat{\beta}_4 = .35$. We estimate the mean heat rate will increase by .35 for each unit increase in cycle pressure ratio, holding speed, exhaust temperature, inlet temperature, and air flow rate constant.

$\hat{\beta}_5 = -.8480$. We estimate the mean heat rate will decrease by .848 for each unit increase in air flow rate constant, holding speed, inlet temperature, exhaust temperature, and cycle pressure ratio.

d. $s = 458.828$. We would expect about 95% of the observed values of heat rate to fall within $2s$ or $2(458.828) = 917.656$ units of their least squares predicted values.

e. R-Sq(adj) = 91.7% 91.7% of the sample variation in the heat rate is explained by the model containing speed, inlet temperature, exhaust temperature, cycle pressure ratio, and air flow rate, adjusted for the sample size and the number of independent variables in the model.

f. To determine if the overall model is useful for predicting y, we test:

H_0: $\beta_1 = \beta_2 = \beta_3 = \beta_4 = \beta_5 = 0$
H_a: At lease one $\beta_i \neq 0$

The test statistic is $F = 147.30$ and the p-value is $p = 0.000$.

Since the p-value is less than $\alpha = .01$ ($p = 0.000 < .01$), H_0 is rejected. There is sufficient evidence to indicate that at least one of the independent variables is useful in predicting heat rate at $\alpha = .01$.

12.25 a. $R^2 = .362$. 36.2% of the variability in the AC scores can be explained by the model containing the variables self-esteem score, optimism score, and group cohesion score.

b. To test the utility of the model, we test:

H_0: $\beta_1 = \beta_2 = \beta_3 = 0$
H_a: At least one $\beta_i \neq 0$, $i = 1, 2, 3$

The test statistic is:
$$F = \frac{R^2 / k}{(1 - R^2)[n - (k+1)]} = \frac{.362/3}{(1 - .362)/[31 - (3+1)]} = 5.11$$

The rejection region requires $\alpha = .05$ in the upper tail of the F distribution with $v_1 = k = 3$ and $v_2 = n - (k+1) = 31 - (3+1) = 27$. From Table IX, Appendix A, $F_{.05} = 2.96$. The rejection region is $F > 2.96$.

Since the observed value of the test statistic falls in the rejection region ($F = 5.11 > 2.96$), H_0 is rejected. There is sufficient evidence that the model is useful in predicting AC score at $\alpha = .05$.

12.27   a.   **Model 1:**

H_0: $\beta_1 = 0$
H_a: $\beta_1 \neq 0$

The test statistic is $t = \dfrac{\hat{\beta}_1 - 0}{s_{\hat{\beta}_1}} = \dfrac{.0354}{.0137} = 2.58$.

Since no α was given, we will use $\alpha = .05$. The rejection region requires $\alpha/2 = .05/2 = .025$ in each tail of the t distribution. From Table VI, Appendix A, with df $= n - (k + 1) = 12 - (1 + 1) = 10$, $t_{.025} = 2.228$. The rejection region is $t < -2.228$ or $t > 2.228$.

Since the observed value of the test statistic falls in the rejection region ($t = 2.58 > 2.228$), H_0 is rejected. There is sufficient evidence to indicate that there is a linear relationship between vintage and the logarithm of price.

Model 2:

H_0: $\beta_1 = 0$
H_a: $\beta_1 \neq 0$

The test statistic is $t = \dfrac{\hat{\beta}_1 - 0}{s_{\hat{\beta}_1}} = \dfrac{.0238}{.00717} = 3.32$

Since no α was given, we will use $\alpha = .05$. The rejection region requires $\alpha/2 = .05/2 = .025$ in each tail of the t distribution. From Table VI, Appendix A, with df $= n - (k + 1) = 12 - (4 + 1) = 7$, $t_{.025} = 2.365$. The rejection region is $t < -2.365$ or $t > 2.365$.

Since the observed value of the test statistic falls in the rejection region ($t = 3.32 > 2.365$), H_0 is rejected. There is sufficient evidence to indicate that there is a linear relationship between vintage and the logarithm of price, adjusting for all other variables.

H_0: $\beta_2 = 0$
H_a: $\beta_2 \neq 0$

The test statistic is $t = \dfrac{\hat{\beta}_2 - 0}{s_{\hat{\beta}_2}} = \dfrac{.616}{.0952} = 6.47$

The rejection region is $t < -2.365$ or $t > 2.365$.

Since the observed value of the test statistic falls in the rejection region ($t = 6.47 > 2.365$), H_0 is rejected. There is sufficient evidence to indicate that there is a linear relationship between average growing season temperature and the logarithm of price, adjusting for all other variables.

H_0: $\beta_3 = 0$
H_a: $\beta_3 \neq 0$

The test statistic is $t = \dfrac{\hat{\beta}_3 - 0}{s_{\hat{\beta}_3}} = \dfrac{-.00386}{.00081} = -4.77$

The rejection region is $t < -2.365$ or $t > 2.365$.

Since the observed value of the test statistic falls in the rejection region ($t = -4.77 < -2.365$), H_0 is rejected. There is sufficient evidence to indicate that there is a linear relationship between Sept./ Aug. rainfall and the logarithm of price, adjusting for all other variables.

H_0: $\beta_4 = 0$
H_a: $\beta_4 \neq 0$

The test statistic is $t = \dfrac{\hat{\beta}_4 - 0}{s_{\hat{\beta}_4}} = \dfrac{.0001173}{.000482} = 0.24$.

The rejection region is $t < -2.365$ or $t > 2.365$.

Since the observed value of the test statistic does not fall in the rejection region ($t = 0.24 \ngtr 2.365$), H_0 is not rejected. There is insufficient evidence to indicate that there is a linear relationship between rainfall in months preceding vintage and the logarithm of price, adjusting for all other variables.

Model 3:

H_0: $\beta_1 = 0$
H_a: $\beta_1 \neq 0$

The test statistic is $t = \dfrac{\hat{\beta}_1 - 0}{s_{\hat{\beta}_1}} = \dfrac{.0240}{.00747} = 3.21$

Since no α was given, we will use $\alpha = .05$. The rejection region requires $\alpha/2 = .05/2 = .025$ in each tail of the t distribution. From Table VI, Appendix A, with df $= n - (k + 1) = 12 - (5 + 1) + 7$, $t_{.025} = 2.447$. The rejection region is $t < -2.447$ or $t > 2.447$.

Since the observed value of the test statistic falls in the rejection region ($t = 3.21 > 2.447$), H_0 is rejected. There is sufficient evidence to indicate that there is a linear relationship between vintage and the logarithm of price, adjusting for all other variables.

H_0: $\beta_2 = 0$
H_a: $\beta_2 \neq 0$

The test statistic is $t = \dfrac{\hat{\beta}_2 - 0}{s_{\hat{\beta}_2}} = \dfrac{.608}{.116} = 5.24$.

The rejection region is $t < -2.447$ or $t > 2.447$.

Since the observed value of the test statistic falls in the rejection region ($t = 5.24 >$ 2.447), H_0 is rejected. There is sufficient evidence to indicate that there is a linear relationship between average growing season temperature and the logarithm of price, adjusting for all other variables.

H_0: $\beta_3 = 0$
H_a: $\beta_3 \neq 0$

The test statistic is $t = \dfrac{\hat{\beta}_3 - 0}{s_{\hat{\beta}_3}} = \dfrac{-.00380}{.00095} = -4.00$

The rejection region is $t < -2.447$ or $t > -2.447$.

Since the observed value of the test statistic falls in the rejection region ($t = -4.00 <$ -2.447), H_0 is rejected. There is sufficient evidence to indicate that there is a linear relationship between Sept./Aug. rainfall and the logarithm of price, adjusting for all other variables.

H_0: $\beta_4 = 0$
H_a: $\beta_4 \neq 0$

The test statistic is $t = \dfrac{\hat{\beta}_4 - 0}{s_{\hat{\beta}_4}} = \dfrac{.00115}{.000505} = 2.28$

The rejection region is $t < -2.447$ or $t > 2.447$.

Since the observed value of the test statistic does not fall in the rejection region ($t = 2.28$ $\not> 2.365$), H_0 is not rejected. There is insufficient evidence to indicate that there is a linear relationship between rainfall in months preceding vintage and the logarithm of price, adjusting for all other variables.

H_0: $\beta_5 = 0$
H_a: $\beta_5 \neq 0$

The test statistic is $t = \dfrac{\hat{\beta}_5 - 0}{s_{\hat{\beta}_5}} = \dfrac{.00765}{.0565} = 0.14.$

The rejection region is $t < -2.447$ or $t > 2.447$.

Since the observed value of the test statistic does not fall in the rejection region ($t = 0.14$ $\not> 2.365$), H_0 is not rejected. There is insufficient evidence to indicate that there is a linear relationship between average September temperature and the logarithm of price, adjusting for all other variables.

b. **Model 1:**

$$\hat{\beta}_1 = .0354, \; e^{.0354} - 1 = .036$$

We estimate that the mean price will increase by 3.6% for each additional increase of unit of x_1, vintage year.

Model 2:

$$\hat{\beta}_1 = .0238, \; e^{.0238} - 1 = .024$$

We estimate that the mean price will increase by 2.4% for each additional increase of 1 unit of x_1, vintage year (with all other variables held constant).

$$\hat{\beta}_2 = .616, \; e^{.616} - 1 = .852$$

We estimate that the mean price will increase by 85.2% for each additional increase of 1 unit of x_2, average growing season temperature °C (with all other variables held constant).

$$\hat{\beta}_3 = -.00386, \; e^{.00386} - 1 = -.004$$
We estimate that the mean price will decrease by .4% for each additional increase of 1 unit of x_3, Sept./Aug. rainfall in cm (with all other variables held constant).

$$\hat{\beta}_4 = .0001173, \; e^{.0001173} - 1 = .0001$$

We estimate that the mean price will increase by .01% for each additional increase of 1 unit of x_4, rainfall in months prededing vintage in cm (with all other variables held constant).

Model 3:

$$\hat{\beta}_1 = .0240, \; e^{.0240} - 1 = .024$$

We estimate that the mean price will increase by 2.4% for each additional increase of 1 unit of x_1, vintage year (with all other variables held constant).

$$\hat{\beta}_2 = .608, \; e^{.608} - 1 = .837$$

We estimate that the mean price will increase by 83.7% for each additional increase of 1 unit of x_2, average growing season temperatures in °C (with all other variables held constant).

$$\hat{\beta}_3 = -.00380, \; e^{.00380} - 1 = -.004$$
We estimate that the mean price will decrease by .4% for each additional increase of 1 unit of x_3, Sept./Aug. rainfall in cm, (with all other variables held constant).

$\hat{\beta}_4 = .00115$, $e^{.00115} - 1 = .001$

We estimate that the average mean price will increase by .1% for each additional increase of 1 unit of x_4, rainfall in months preceding vintage in cm (with all other variables held constant).

$\hat{\beta}_5 = .00765$, $e^{.00765} - 1 = .008$

We estimate that the average mean price will increase by .8% for each additional increase of 1 unit of x_5, average Sept. temperature in °C (with all other variables held constant).

c.   For model 1, $R^2 = .212$ and $s = .575$. For model 2, $R^2 = .828$ and $s = .287$. For Model 3, $R^2 = .828$ and $s = .293$.

Based on the R^2 and s values, the best model is model 2. Model 2 has the highest R^2 value and the lowest value of s.

12.29   The 95% confidence interval for E(y) will be narrower than the 95% prediction interval for y. When estimating the value of E(y) or mean value of y, there is some variance. However, when predicting the actual value of y, we must first find the mean (with some variance above). Once the mean is located, the actual values of y can vary around this mean.

12.31   The 95% prediction interval is (90.69, 158.57). With 95% confidence, we conclude that the equivalent width for an individual quasar with a redshift of 3.07, line flux of -13.56, line luminosity of 45.30, and AB_{1450} of 19.59 will be between 90.69 and 158.57.

12.33   Using MINITAB, the results of the prediction are:

```
Predicted Values for New Observations

New
Obs    Fit   SE Fit       95% CI              95% PI
  1  11.75   16.66   (-21.19, 44.68)   (-183.76, 207.25)


Values of Predictors for New Observations

New
Obs  Miles  Length  Weight
  1    100    40.0     800
```

The 95% prediction interval is $(-183.76, 207.25)$. We are 95% confident that the actual DDT level of a fish caught 100 miles upstream with a length of 40 centimeters and a weight of 800 grams will be between -186.76 and 207.25 parts per million. Since DDT levels cannot be negative, the actual interval would be 0 to 207.25.

12.35   Yes, we agree. The estimated coefficients for latitude and depth are both negative. Thus, to maximize arsenic levels, we would want latitude and depth to be as low as possible. The estimated coefficient for longitude is positive. To maximize arsenic levels, we would want longitude to be as large as possible.

Using MINTAB to find the prediction interval:

```
Predicted Values for New Observations

New
Obs     Fit   SE Fit       95% CI            95% PI
  1   232.33   23.23   (186.63, 278.04)   (24.03, 440.64)X

X denotes a point that is an outlier in the predictors.


Values of Predictors for New Observations

New
Obs   LATITUDE   LONGITUDE   DEPTH-FT
  1    23.755     90.662      25.0
```

The 95% prediction interval is (24.03, 440.64). We are 95% confident that the actual value of arsenic level is between 24.03 and 440.61 when latitude is minimized at 23.803, longitude is maximized at 90.662, and depth is minimized at 25.

12.37   a. $E(y) = \beta_0 + \beta_1 x_1 + \beta_2 x_2 + \beta_3 x_1 x_2$

b. $E(y) = \beta_0 + \beta_1 x_1 + \beta_2 x_2 + \beta_3 x_3 + \beta_4 x_1 x_2 + \beta_5 x_1 x_3 + \beta_6 x_2 x_3$

12.39   a.   The response surface is a twisted surface in three-dimensional space.

b.   For $x_1 = 0$, $E(y) = 3 + 0 + 2x_2 - 0x_2 = 3 + 2x_2$
For $x_1 = 1$, $E(y) = 3 + 1 + 2x_2 - 1x_2 = 4 + x_2$
For $x_1 = 2$, $E(y) = 3 + 2 + 2x_2 - 2x_2 = 5$

The plot of the lines is:

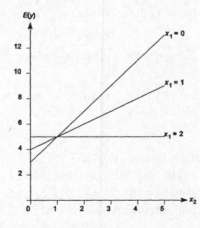

c.   The lines are not parallel because interaction between x_1 and x_2 is present. Interaction between x_1 and x_2 means that the effect of x_2 on y depends on what level x_1 takes on.

d. For $x_1 = 0$, as x_2 increases from 0 to 5, $E(y)$ increases from 3 to 13.
 For $x_1 = 1$, as x_2 increases from 0 to 5, $E(y)$ increases from 4 to 9.
 For $x_1 = 2$, as x_2 increases from 0 to 5, $E(y) = 5$.

e. For $x_1 = 2$ and $x_2 = 4$, $E(y) = 5$
 For $x_1 = 0$ and $x_2 = 5$, $E(y) = 13$

 Thus, $E(y)$ changes from 5 to 13.

12.41 a. The prediction equation is:

$$\hat{y} = -2.55 + 3.82x_1 + 2.63x_2 - 1.29x_1x_2$$

 b. The response surface is **a** twisted plane, since the equation contains an interaction term.

 c. For $x_2 = 1$, $= -2.55 + 3.82x_1 + 2.63(1) - 1.29x_1(1)$
 $= .08 + 2.53x_1$
 For $x_2 = 3$, $= -2.55 + 3.82x_1 + 2.63(3) - 1.29x_1(3)$
 $= 5.34 - .05x_1$
 For $x_2 = 5$, $= -2.55 + 3.82x_1 + 2.63(5) - 1.29x_1(5)$
 $= 10.6 - 2.63x_1$

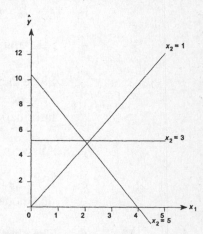

 d. If x_1 and x_2 interact, the effect of x_1 on y is different at different levels of x_2. When $x_2 = 1$, as x_1 increases, $\hat{y}$ also increases. When $x_2 = 5$, as x_1 increases, $\hat{y}$ decreases.

 e. The hypotheses are:

 H_0: $\beta_3 = 0$
 H_a: $\beta_3 \neq 0$

 f. The test statistic is $t = \dfrac{\hat{\beta_3}}{s_{\hat{\beta_3}}} = \dfrac{-1.285}{.159} = -8.06$

 The rejection region requires $\alpha/2 = .01/2 = .005$ in each tail of the t distribution with df $= n - (k + 1) = 15 - (3 + 1) = 11$. From Table VI, Appendix A, $t_{.005} = 3.106$. The rejection region is $t < -3.106$ or $t > 3.106$.

 Since the observed value of the test statistic falls in the rejection region ($t = -8.06 < -3.106$), H_0 is rejected. There is sufficient evidence to indicate that x_1 and x_2 interact at $\alpha = .01$.

12.43   a.   A regression model containing the interaction between blade position and turntable speed would be:

$$y = \beta_o + \beta_1 x_1 + \beta_2 x_2 + \beta_3 x_1 x_2 + \varepsilon$$

        b.   "Blade position and turntable speed interact" means that the effect of turntable speed on the number of defects (y) depends on the blade position.

        c.   This means that the interaction term in the model in part **a** would be significant. The slope of the line relating the number of defects and the turntable speed is different for different levels of blade position.

12.45   a. $E(y) = \beta_0 + \beta_1 x_1 + \beta_2 x_2 + \beta_3 x_3 + \beta_4 x_4 + \beta_5 x_5$

        b. H_0: $\beta_4 = 0$

        c. $t = 4.408$, p-value $= .001$

             Since the p-value is so small, there is strong evidence to reject H_0. There is sufficient evidence to indicate that the strength of client-therapist relationship contributes information for the prediction of a client's reaction for any $\alpha > .001$.

        d.   Answers may vary.

        e. $R^2 = .2946$. 29.46% of the variability in the client's reaction scores can be explained by this model.

12.47   a.   The phrase "client credibility and linguistic delivery style interact" means that the affect of client credibility on likelihood depends on the level of linguistic delivery style.

        b.   To determine if the overall model is adequate for predicting likelihood, we test:

             H_0: $\beta_1 = \beta_2 = \beta_3 = 0$
             H_a: At least one $\beta_i \neq 0$, $i = 1, 2, 3$

        c.   From the table, the test statistic is $F = 55.35$ and the p-value is $p < 0.0005$. Since the p value is so small, H_0 is rejected. There is sufficient evidence to indicate the model is adequate in predicting likelihood for any reasonable level of α.

        d.   To determine if client credibility and linguistic delivery style interact, we test:

             H_0: $\beta_3 = 0$
             H_a: $\beta_3 \neq 0$

        e.   From the table, $t = 4.008$ and the p-value is $p < .005$. Since the p-value is so small, H_0 is rejected. There is sufficient evidence to indicate the that client credibility and linguistic delivery style interact for any reasonable level of α.

f. The least squares prediction equation is $\hat{y} = 15.865 + .037x_1 - .678x_2 + .036x_1x_2$.

When client credibility is 22, the least squares line is:

$$\hat{y} = 15.865 + .037(22) - .678x_2 + .036(22)x_2 = 16.679 + .114x_2$$

When client credibility is 22, for each unit increase in linguistic delivery score, the mean likelihood score is estimated to increase by .114.

g. When client credibility is 46, the least squares line is:

$$\hat{y} = 15.865 + .037(46) - .678x_2 + .036(46)x_2 = 17.567 + .978x_2$$

When client credibility is 46, for each unit increase in linguistic delivery score, the mean likelihood score is estimated to increase by .978.

12.49 a. The model would be:

$$y = \beta_o + \beta_1 x_1 + \beta_2 x_2 + \beta_3 x_3 + \beta_4 x_4 + \beta_5 x_5 + \beta_6 x_2 x_5 + \beta_7 x_3 x_5 + \varepsilon$$

b. Using MINITAB, the results are:

Regression Analysis: HEATRATE versus RPM, INLET-TEMP

```
The regression equation is
HEATRATE = 13646 + 0.0460 RPM - 12.7 INLET-TEMP + 23.0 EXH-TEMP - 3.0
CPRATIO + 1.29 AIRFLOW + 0.0161 INxAIRF - 0.0414 EXxAIRF


Predictor        Coef     SE Coef       T       P
Constant        13646        1068    12.77   0.000
RPM           0.04599     0.01602     2.87   0.006
INLET-TEMP    -12.675       1.542    -8.22   0.000
EXH-TEMP       23.003       3.768     6.11   0.000
CPRATIO         -3.02       26.42    -0.11   0.909
AIRFLOW         1.288       3.563     0.36   0.719
INxAIRF       0.016149    0.003673    4.40   0.000
EXxAIRF       -0.04143     0.01098   -3.77   0.000


S = 404.693   R-Sq = 94.2%   R-Sq(adj) = 93.6%
```

```
Analysis of Variance

Source          DF          SS          MS         F       P
Regression       7    158234406    22604915    138.02    0.000
Residual Error  59      9662802      163776
Total           66    167897208


Source        DF      Seq SS
RPM            1    119598530
INLET-TEMP     1     26893467
EXH-TEMP       1      7784225
CPRATIO        1         4623
AIRFLOW        1       774427
INxAIRF        1       849347
EXxAIRF        1      2329786
```

The least squares prediction equation is:

$$\hat{y} = 13,646 + .04599x_1 - 12.675x_2 + 23.003x_3 - 3.02x_4 + 1.288x_5 + .016149x_2x_5 - .04143x_3x_5$$

c. To determine if inlet temperature and air flow rate interact to affect heat rate, we test:

H_0: $\beta_6 = 0$
H_a: $\beta_6 \neq 0$

From the printout, the test statistic is $t = 4.40$ and the p-value is $p = .000$.

Since the p-value is less than α ($p = .000 < .05$), H_0 is rejected. There is sufficient evidence to indicate that inlet temperature and air flow rate interact to affect heat rate at $\alpha = .05$.

d. To determine if exhaust temperature and air flow rate interact to affect heat rate, we test:

H_0: $\beta_7 = 0$
H_a: $\beta_7 \neq 0$

From the printout, the test statistic is $t = -3.37$ and the p-value is $p = .000$.

Since the p-value is less than α ($p = .000 < .05$), H_0 is rejected. There is sufficient evidence to indicate that exhaust temperature and air flow rate interact to affect heat rate at $\alpha = .05$.

e. Both of the interaction tests were significant. This means that the effect of inlet temperature on the heat rate is different at different levels of air flow and the effect of exhaust temperature on heat rate is different at different levels of air flow.

12.51 a. $E(y) = \beta_0 + \beta_1x_1 + \beta_2x_1^2$

b. $E(y) = \beta_0 + \beta_1x_1 + \beta_2x_2 + \beta_3x_1^2 + \beta_4x_2^2 + \beta_5x_1x_2$

c. $E(y) = \beta_0 + \beta_1x_1 + \beta_2x_2 + \beta_3x_3 + \beta_4x_1^2 + \beta_5x_2^2 + \beta_6x_3^2 + \beta_7x_1x_2 + \beta_8x_1x_3 + \beta_9x_2x_3$

12.53 a. H_0: $\beta_2 = 0$
H_a: $\beta_2 \neq 0$

The test statistic is $t = \dfrac{\hat{\beta}_2 - 0}{s_{\hat{\beta}_2}} = \dfrac{.47 - 0}{.15} = 3.133$

The rejection region requires $\alpha/2 = .05/2 = .025$ in each tail of the t distribution with df $= n - (k + 1) = 25 - (2 + 1) = 22$. From Table VI, Appendix A, $t_{.025} = 2.074$. The rejection region is $t < -2.074$ or $t > 2.074$.

Since the observed value of the test statistic falls in the rejection region ($t = 3.133 > 2.074$), H_0 is rejected. There is sufficient evidence to indicate the true relationship is given by the quadratic model at $\alpha = .05$.

b. H_0: $\beta_2 = 0$
H_a: $\beta_2 > 0$

The test statistic is the same as in part **a**, $t = 3.133$.

The rejection region requires $\alpha = .05$ in the upper tail of the t distribution with df $= 22$. From Table VI, Appendix A, $t_{.05} = 1.717$. The rejection region is $t > 1.717$.

Since the observed value of the test statistic falls in the rejection region ($t = 3.133 > 1.717$), H_0 is rejected. There is sufficient evidence to indicate the quadratic curve opens upward at $\alpha = .05$.

12.55 a.

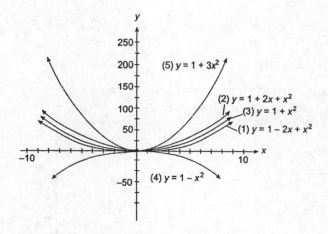

b. It moves the graph to the right ($-2x$) or to the left ($+2x$) compared to the graph of $y = 1 + x^2$.

c. It controls whether the graph opens up ($+x^2$) or down ($-x^2$). It also controls how steep the curvature is, i.e., the larger the absolute value of the coefficient of x^2, the narrower the curve is.

12.57 a.

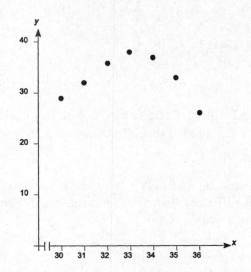

b.   If information was available only for $x = 30$, 31, 32, and 33, we would suggest a first-order model where $\beta_1 > 0$. If information was available only for $x = 33$, 34, 35, and 36, we would again suggest a first-order model where $\beta_1 < 0$. If all the information was available, we would suggest a second-order model.

12.59 a.   The complete second-order model is:

$$y = \beta_o + \beta_1 x_1 + \beta_2 x_2 + \beta_3 x_1^2 + \beta_4 x_4^2 + \beta_5 x_1 x_2 + \varepsilon$$

b.   The terms that allow for a curvilinear relationship are $\beta_3 x_1^2$ and $\beta_4 x_4^2$.

12.61 a.   Using MINITAB, the graph is:

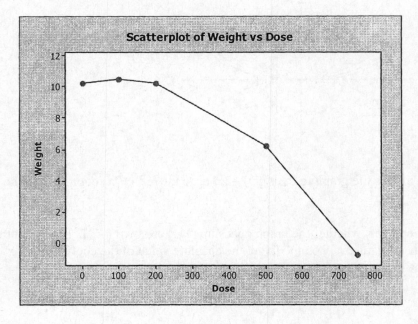

b. For a dosage of 500 mg/kg, the estimated weight change would be:

$$\hat{y} = 10.25 + .0053(500) - .0000266(500^2) = 6.25$$

c. For a dosage of 0 mg/kg, the estimated weight change would be:

$$\hat{y} = 10.25 + .0053(0) - .0000266(0^2) = 10.25$$

d. For a dosage of 100 mg/kg, the estimated weight change would be:

$$\hat{y} = 10.25 + .0053(100) - .0000266(100^2) = 10.514$$

For a dosage of 200 mg/kg, the estimated weight change would be:

$$\hat{y} = 10.25 + .0053(200) - .0000266(200^2) = 10.246$$

For any dosage greater than 200 mg/kg, the weight change will be less than 10.246. Thus, a dosage of 200 mg/kg will be the largest dosage that yields an estimated weight change that is closest to, but below the estimated weight change for the control group (10.25).

12.63   a. The quadratic model would be:

$$E(y) = \beta_0 + \beta_1 x + \beta_2 x^2$$

b. From the plot, the β_2 would be positive because the points appear to form an upward curve.

c. This value of r^2 applies to the linear model, not the quadratic model. The linear model is:

$$E(y) = \beta_0 + \beta_1 x$$

12.65   a.   Using MINITAB, the results are:

Regression Analysis: RATE versus EST, EST-SQ

```
The regression equation is
RATE = - 288 + 1.39 EST + 0.000035 EST-SQ


Predictor          Coef      SE Coef       T       P
Constant           -288         8049   -0.04   0.972
EST               1.395        3.651    0.38   0.706
EST-SQ      0.00003509   0.00009724    0.36   0.722


S = 31901.1   R-Sq = 45.9%   R-Sq(adj) = 40.8%


Analysis of Variance

Source          DF            SS           MS      F      P
Regression       2   18138955261   9069477631   8.91   0.002
Residual Error  21   21371254395   1017678781
Total           23   39510209656


Source   DF       Seq SS
EST       1   18006405335
EST-SQ    1     132549926
```

To determine if the incidence rate is curvilinearly related to estimated rate, we test:

H_0: $\beta_2 = 0$
H_a: $\beta_2 \neq 0$

From the printout, the test statistic is $t = .36$ and the p-value is $p = .722$.

Since the p-value is not less than α ($p = .722 \not< .05$), H_0 is not rejected. There is insufficient evidence that the incidence rate is curvilinearly related to estimated rate at $\alpha = .05$.

b. Using MINITAB, the scatterplot is:

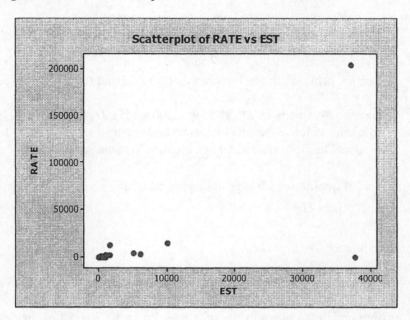

Scatterplot of RATE vs EST

The point associated with Botulism is in the lower right corner of the graph. This point looks to be out of line with the other data points.

c. Using MINITAB the new results are:

Regression Analysis: RATE versus EST, EST-SQ

```
The regression equation is
RATE = 735 - 0.081 EST + 0.000151 EST-SQ


Predictor          Coef      SE Coef       T       P
Constant          735.0        695.9    1.06   0.303
EST             -0.0810       0.3167   -0.26   0.801
EST-SQ       0.00015052   0.00000868   17.34   0.000


S = 2756.80   R-Sq = 99.6%   R-Sq(adj) = 99.6%


Analysis of Variance

Source          DF            SS            MS         F       P
Regression       2   39251490541   19625745270   2582.35   0.000
Residual Error  20     151998825       7599941
Total           22   39403489366


Source   DF       Seq SS
EST       1   36967483627
EST-SQ    1    2284006914
```

Now, if we test to see if the incidence rate is curvilinearly related to estimated rate:

H_0: $\beta_2 = 0$
H_a: $\beta_2 \neq 0$

From the printout, the test statistic is $t = 17.34$ and the p-value is $p = 0.000$.

Since the p-value is less than α ($p = .000 < .05$), H_0 is rejected. There is sufficient evidence that the incidence rate is curvilinearly related to estimated rate at $\alpha = .05$. The fit of the model has improved dramatically.

12.67   Let $x = \begin{cases} 1 & \text{if qualitative variable assumes 2nd level} \\ 0 & \text{otherwise} \end{cases}$

The model is $E(y) = \beta_0 + \beta_1 x$.

β_0 = mean value of y when the qualitative variable assumes the first level.

β_1 = difference in the mean values of y between levels 2 and 1 of the qualitative variable.

12.69   a.   Level 1 implies $x_1 = x_2 = x_3 = 0$. $\hat{y} = 10.2$
Level 2 implies $x_1 = 1$ and $x_2 = x_3 = 0$. $\hat{y} = 10.2 - 4(1) = 6.2$
Level 3 implies $x_2 = 1$ and $x_1 = x_3 = 0$. $\hat{y} = 10.2 + 12(1) = 22.2$
Level 4 implies $x_3 = 1$ and $x_1 = x_2 = 0$. $\hat{y} = 10.2 + 2(1) = 12.2$

   b.   The hypotheses are:

H_0: $\beta_1 = \beta_2 = \beta_3 = 0$
H_a: At least one $\beta_i \neq 0$, $i = 1, 2, 3$

12.71   a.   Let $x_1 = \begin{cases} 1 & \text{if player is white} \\ 0 & \text{otherwise} \end{cases}$ $x_2 = \begin{cases} 1 & \text{if card availability is low} \\ 0 & \text{otherwise} \end{cases}$

$x_3 = \begin{cases} 1 & \text{if player position is running back} \\ 0 & \text{otherwise} \end{cases}$ $x_4 = \begin{cases} 1 & \text{if player position is wide receiver} \\ 0 & \text{otherwise} \end{cases}$

$x_5 = \begin{cases} 1 & \text{if player position is tight end} \\ 0 & \text{otherwise} \end{cases}$ $x_6 = \begin{cases} 1 & \text{if player position is defensive lineman} \\ 0 & \text{otherwise} \end{cases}$

$x_7 = \begin{cases} 1 & \text{if player position is linebacker} \\ 0 & \text{otherwise} \end{cases}$ $x_8 = \begin{cases} 1 & \text{if player position is defensive back} \\ 0 & \text{otherwise} \end{cases}$

$x_9 = \begin{cases} 1 & \text{if player position is offensive lineman} \\ 0 & \text{otherwise} \end{cases}$

b. The model for price (y) as a function of race is:

$$E(y) = \beta_0 + \beta_1 x_1 .$$

β_0 = mean price of card of black player
β_1 = difference in mean price of card between a white player and a black player

c. The model for price (y) as a function of card availability is:

$$E(y) = \beta_0 + \beta_1 x_2 .$$

β_0 = mean price of card of high availability
β_1 = difference in mean price of card between a card of low availability and a card of high availability

d. The model for price (y) as a function of position is:

$$E(y) = \beta_0 + \beta_1 x_3 + \beta_2 x_4 + \beta_3 x_5 + \beta_4 x_6 + \beta_5 x_7 + \beta_6 x_8 + \beta_7 x_9$$

β_0 = mean price of card of quarterback
β_1 = difference in mean price of card between a running back and a quarterback
β_2 = difference in mean price of card between a wide receiver and a quarterback
β_3 = difference in mean price of card between a tight end and a quarterback
β_4 = difference in mean price of card between a defensive lineman and a quarterback
β_5 = difference in mean price of card between a linebacker and a quarterback
β_6 = difference in mean price of card between a defensive back and a quarterback
β_7 = difference in mean price of card between a offensive lineman and a quarterback

12.73 a. There would be a total of $2 \times 2 = 4$ different gene marker combinations at the two locations. They are:

$$AA \quad AB \quad BA \quad BB$$

b. Let $x_1 = \begin{cases} 1 & \text{if gene marker is AB} \\ 0 & \text{otherwise} \end{cases}$ $\qquad x_2 = \begin{cases} 1 & \text{gene marker is BA} \\ 0 & \text{otherwise} \end{cases}$

$$x_3 = \begin{cases} 1 & \text{gene marker is BB} \\ 0 & \text{otherwise} \end{cases}$$

A model for $E(y)$ as a function of the gene marker combination is:

$$E(y) = \beta_0 + \beta_1 x_1 + \beta_2 x_2 + \beta_3 x_3$$

c. β_0 = mean extent of disease for gene marker combination AA
β_1 = difference in mean extent of disease between gene marker combinations AB and AA
β_2 = difference in mean extent of disease between gene marker combinations BA and AA
β_3 = difference in mean extent of disease between gene marker combinations BB and AA

d.   To determine if the overall model is useful for predicting extent of the disease, we test:

H_0: $\beta_1 = \beta_2 = \beta_3 = 0$
H_a: At least one $\beta_i \neq 0$

12.75 a.   The model is $E(y) = \beta_0 + \beta_1 x_1 + \beta_2 x_2$

where $x_1 = \begin{cases} 1 & \text{if political status is democratic} \\ 0 & \text{otherwise} \end{cases}$

$x_2 = \begin{cases} 1 & \text{is political status is dictatorship} \\ 0 & \text{otherwise} \end{cases}$

b. β_0 = mean value of level of assassination risk (y) for political status communist.

β_1 = difference in mean value of level of assassination risk (y) between political status democratic and political status communist.

β_2 = difference in mean value of level of assassination risk (y) between political status dictatorship and political status communist.

12.77 Let $x_1 = \begin{cases} 1 & \text{if Juror is female} \\ 0 & \text{otherwise} \end{cases}$ $x_2 = \begin{cases} 1 & \text{if no expert testimony is present} \\ 0 & \text{otherwise} \end{cases}$

To allow for the likelihood that female jurors would be more likely than male jurors to change a verdict from not guilty to guilty after deliberations if expert testimony is present, the model would have to include an interaction term between gender and expert testimony. The model would be:

$E(y) = \beta_0 + \beta_1 x_1 + \beta_2 x_2 + \beta_3 x_1 x_2$

A possible sketch of this relationship would be:

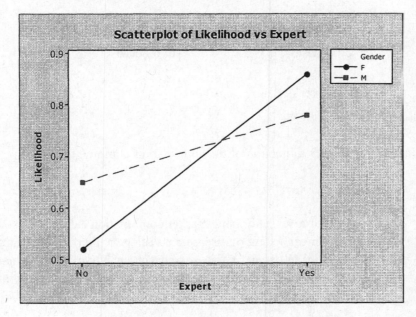

12.79   a.   Let y = mean body mass and $x_1 = \begin{cases} 1 & \text{if flightless} \\ 0 & \text{otherwise} \end{cases}$

The model is $y = \beta_0 + \beta_1 x_1 + \varepsilon$

b.   Let y = mean body mass and

$x_1 = \begin{cases} 1 & \text{if vertebrates} \\ 0 & \text{otherwise} \end{cases}$ $x_2 = \begin{cases} 1 & \text{if vegetables} \\ 0 & \text{otherwise} \end{cases}$ $x_3 = \begin{cases} 1 & \text{if invertebrates} \\ 0 & \text{otherwise} \end{cases}$

The model is $y = \beta_o + \beta_1 x_1 + \beta_2 x_2 + \beta_3 x_3 + \varepsilon$

c.   Let y = mean egg length and

$x_1 = \begin{cases} 1 & \text{if cavity within ground} \\ 0 & \text{otherwise} \end{cases}$ $x_2 = \begin{cases} 1 & \text{if tree} \\ 0 & \text{otherwise} \end{cases}$ $x_3 = \begin{cases} 1 & \text{if cavity above ground} \\ 0 & \text{otherwise} \end{cases}$

The model is $y = \beta_0 + \beta_1 x_1 + \beta_2 x_2 + \beta_3 x_3 + \varepsilon$

d.   Using MINTAB, the results are:

Regression Analysis: Body Mass versus Flt

```
The regression equation is
Body Mass = 641 + 30647 Flt


Predictor    Coef   SE Coef      T       P
Constant      641      2665   0.24   0.810
Flt         30647      5331   5.75   0.000


S = 26521.3    R-Sq = 20.3%    R-Sq(adj) = 19.7%


Analysis of Variance

Source           DF            SS           MS       F       P
Regression        1   23246186220   23246186220   33.05   0.000
Residual Error  130   91439576421     703381357
Total           131   1.14686E+11
```

The fitted model is $\hat{y} = 641 + 30,647 x_1$

$\hat{\beta}_0 = 641$. The mean body mass for Volant birds is estimated to be 641.

$\hat{\beta}_1 = 30,647$. The difference in the mean body mass between Flightless birds and Volant birds is estimated to be 30,647.

e.   To determine if the model is useful in estimating body mass, we test:

H_0: $\beta_1 = 0$
H_a: $\beta_1 \neq 0$

The test statistic is $F = 33.05$ and the p-value is $p = .000$. Since the p-value is less than α ($p = .000 < .01$), H_0 is rejected. There is sufficient evidence to indicate a difference in mean body mass between the two types of flight capacity birds at $\alpha = .01$.

f.   Using MINITAB, the results are:

Regression Analysis: Body Mass versus DX1, DX2, DX3

```
The regression equation is
Body Mass = 903 + 2997 DX1 + 26206 DX2 - 660 DX3


Predictor    Coef   SE Coef       T      P
Constant      903      4171    0.22  0.829
DX1          2997     14298    0.21  0.834
DX2         26206      6090    4.30  0.000
DX3          -660      5772   -0.11  0.909


S = 27352.1   R-Sq = 16.5%   R-Sq(adj) = 14.5%


Analysis of Variance

Source           DF            SS          MS     F      P
Regression        3   18924196564  6308065521  8.43  0.000
Residual Error  128   95761566077   748137235
Total           131   1.14686E+11


Source   DF       Seq SS
X1        1     79962608
X2        1  18834456334
X3        1      9777622
```

The fitted model is $\hat{y} = 903 + 2,997x_1 + 26,206x_2 - 660x_3$

$\hat{\beta}_0 = 903$. The mean body mass for birds on a fish diet is estimated to be 903.

$\hat{\beta}_1 = 2,997$. The difference in the mean body mass between birds on a vertebrate diet and birds on a fish diet is estimated to be 2,997.

$\hat{\beta}_2 = 26,206$. The difference in the mean body mass between birds on a vegetable diet and birds on a fish diet is estimated to be 26,206.

$\hat{\beta}_3 = -660$. The difference in the mean body mass between birds on an invertebrate diet and birds on a fish diet is estimated to be -660.

g.   To determine if the model is useful in estimating body mass, we test:

H_0: $\beta_1 = \beta_2 = \beta_3 = 0$
H_a: At least one $\beta_i \neq 0$, $i = 1, 2, 3$

The test statistic is $F = 8.43$ and the p-value is $p = .000$. Since the p-value is less than α ($p = .000 < .01$), H_0 is rejected. There is sufficient evidence to indicate a difference in mean body mass among the four types of diets at $\alpha = .01$.

h.   Using MINITAB, the results are:

Regression Analysis: Egg Length versus NX1, NX2, NX3

```
The regression equation is
Egg Length = 73.7 - 9.13 NX1 - 45.0 NX2 - 39.5 NX3


130 cases used, 2 cases contain missing values


Predictor      Coef  SE Coef       T      P
Constant     73.732    4.907   15.03  0.000
NX1          -9.132    9.515   -0.96  0.339
NX2          -45.01    14.45   -3.12  0.002
NX3         -39.510    9.253   -4.27  0.000


S = 40.7622   R-Sq = 16.1%   R-Sq(adj) = 14.1%


Analysis of Variance

Source          DF       SS      MS      F      P
Regression       3    40230   13410   8.07  0.000
Residual Error 126   209356    1662
Total          129   249586


Source  DF  Seq SS
NX1      1     482
NX2      1    9455
NX3      1   30293
```

The fitted model is $\hat{y} = 73.732 - 9.132x_1 - 45.01x_2 - 39.51x_3$

$\hat{\beta}_o = 73.732$. The mean egg length for nesting on the ground is estimated to be 73.732 mm.

$\hat{\beta}_1 = -9.132$. The difference in the mean egg length between birds nesting in a cavity within the ground and birds nesting on the ground is estimated to be -9.132.

$\hat{\beta}_2 = -45.01$. The difference in the mean egg length between birds nesting in a tree and birds nesting on the ground is estimated to be -45.01.

$\hat{\beta}_3 = -39.51$. The difference in the mean egg length between birds nesting in a cavity above ground and birds nesting on the ground is estimated to be -39.51.

i. To determine if the model is useful in estimating body mass, we test:

H_0: $\beta_1 = \beta_2 = \beta_3 = 0$
H_a: At least one $\beta_i \neq 0$, $i = 1, 2, 3$

The test statistic is $F = 8.07$ and the p-value is $p = .000$. Since the p-value is less than α ($p = .000 < .01$), H_0 is rejected. There is sufficient evidence to indicate a difference in mean egg length among the four types of nesting sites at $\alpha = .01$.

12.81 a. Let $x_1 = \begin{cases} 1 & \text{if adult female with offspring} \\ 0 & \text{if otherwise} \end{cases}$

$x_2 = \begin{cases} 1 & \text{if independent sub - adult female} \\ 0 & \text{if otherwise} \end{cases}$

$x_3 = \begin{cases} 1 & \text{if adult male} \\ 0 & \text{if otherwise} \end{cases}$

$x_4 = \begin{cases} 1 & \text{if independent sub - adult male} \\ 0 & \text{if otherwise} \end{cases}$

The model would be: $E(y) = \beta_0 + \beta_1 x_1 + \beta_2 x_2 + \beta_3 x_3 + \beta_4 x_4$

b. $\hat{\beta}_0 = \hat{\mu}_1 = 38$

$\hat{\beta}_1 = \hat{\mu}_2 - \hat{\mu}_1 = 16 - 38 = -22$

$\hat{\beta}_2 = \hat{\mu}_3 - \hat{\mu}_1 = 89 - 38 = 51$

$\hat{\beta}_3 = \hat{\mu}_4 - \hat{\mu}_1 = 43 - 38 = 5$

$\hat{\beta}_4 = \hat{\mu}_5 - \hat{\mu}_1 = 58 - 38 = 20$

c. To determine if the mean percent use of HTZ differs among the grizzly bear classes, we test:

H_0: $\beta_1 = \beta_2 = \beta_3 = \beta_4 = 0$
H_a: At least one $\beta_i \neq 0$, $i = 1, 2, 3, 4$

d. The p-value is .15. Since the p-value is not small, there is no evidence to indicate the mean percent use of HTZ differs among the grizzly bear classes for $\alpha < .15$.

12.83 a.   The first-order model is $E(y) = \beta_0 + \beta_1 x_1$

b.   The new model is $E(y) = \beta_0 + \beta_1 x_1 + \beta_2 x_2 + \beta_3 x_3$

where $x_2 = \begin{cases} 1 & \text{if level 2} \\ 0 & \text{otherwise} \end{cases}$ $x_3 = \begin{cases} 1 & \text{if level 3} \\ 0 & \text{otherwise} \end{cases}$

c.   To allow for interactions, the model is:

$$E(y) = \beta_0 + \beta_1 x_1 + \beta_2 x_2 + \beta_3 x_3 + \beta_4 x_1 x_2 + \beta_5 x_1 x_3$$

d.   The response lines will be parallel if $\beta_4 = \beta_5 = 0$

e.   There will be one response line if $\beta_2 = \beta_3 = \beta_4 = \beta_5 = 0$

12.85 The model is $E(y) = \beta_0 + \beta_1 x_1 + \beta_2 x_1^2 + \beta_3 x_2 + \beta_4 x_3 + \beta_5 x_4$
where x_1 is the quantitative variable and

$x_2 = \begin{cases} 1 & \text{if level 2 of qualitative variable} \\ 0 & \text{otherwise} \end{cases}$

$x_3 = \begin{cases} 1 & \text{if level 3 of qualitative variable} \\ 0 & \text{otherwise} \end{cases}$

$x_4 = \begin{cases} 1 & \text{if level 4 of qualitative variable} \\ 0 & \text{otherwise} \end{cases}$

12.87 a.   For $x_2 = 0$ and $x_3 = 0$,

$$\hat{y} = 48.8 - 3.4 x_1 + .07 x_1^2 - 2.4(0) - 7.5(0) + 3.7 x_1(0) + 2.7 x_1(0) - .02 x_1^2(0) - .04 x_1^2(0)$$

$$= 48.8 - 3.4 x_1 + .07 x_1^2$$

For $x_2 = 1$ and $x_3 = 0$,

$$\hat{y} = 48.8 - 3.4 x_1 + .07 x_1^2 - 2.4(1) - 7.5(0) + 3.7 x_1(1) + 2.7 x_1(0) - .02 x_1^2(1) - .04 x_1^2(0)$$
$$= 46.4 + .3 x_1 + .05 x_1^2$$

For $x_2 = 0$ and $x_3 = 1$,

$$\hat{y} = 48.8 - 3.4 x_1 + .07 x_1^2 - 2.4(0) - 7.5(1) + 3.7 x_1(0) + 2.7\, x_1(1) - .02 x_1^2(0) - .04 x_1^2(1)$$

$$= 41.3 - .7 x_1 + .03 x_1^2$$

Multiple Regression

b.   The plots of the lines are:

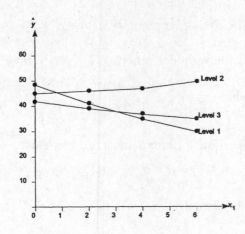

12.89 a.   Let $x_1 = \begin{cases} 1 & \text{if water is from sub-surface flow} \\ 0 & \text{otherwise} \end{cases}$ $x_2 = \begin{cases} 1 & \text{if water is from overground flow} \\ 0 & \text{otherwise} \end{cases}$

and x_3 = silica concentration.

A model for $E(y)$ as a function of the water source and silica concentration is:

$E(y) = \beta_0 + \beta_1 x_1 + \beta_2 x_2 + \beta_3 x_3$

A possible graph would be:

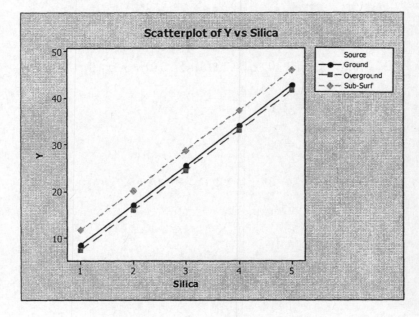

b. A model allowing for the relationship between nitrate concentrations and silica concentrations to differ for the three water sources is:

$$E(y) = \beta_0 + \beta_1 x_1 + \beta_2 x_2 + \beta_3 x_3 + \beta_4 x_1 x_3 + \beta_5 x_2 x_3$$

A possible sketch of the relationships is:

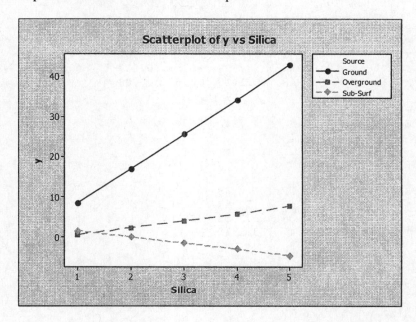

12.91 a. For obese smokers, $x_2 = 0$. The equation of the hypothesized line relating mean REE to time after smoking for obese smokers is:

$$E(y) = \beta_0 + \beta_1 x_1 + \beta_2(0) + \beta_3 x_1(0) = \beta_0 + \beta_1 x_1$$

The slope of the line is β_1.

b. For normal smokers, $x_2 = 1$. The equation of the hypothesized line relating mean REE to time after smoking for normal smokers is:

$$E(y) = \beta_0 + \beta_1 x_1 + \beta_2(1) + \beta_3 x_1(1) = (\beta_0 + \beta_2) + (\beta_1 + \beta_3)x_1$$

The slope of the line is $\beta_1 + \beta_3$.

c. The reported p-value is .044. Since the p-value is so small, there is evidence to indicate that interaction between time and weight is present for $\alpha > .044$.

12.93 a. The first-order model is $E(y) = \beta_0 + \beta_1 x_1 + \beta_2 x_2 + \beta_3 x_1 x_2$

$$\text{where } x_2 = \begin{cases} 1 & \text{if plants} \\ 0 & \text{otherwise} \end{cases}$$

b. Using MINITAB, the results are:

Regression Analysis: WeightChg versus X1, X2, X1X2

```
The regression equation is
WeightChg = 8.14 - 0.016 X1 - 10.4 X2 + 0.095 X1X2


Predictor      Coef   SE Coef      T      P
Constant      8.144     8.446   0.96  0.341
X1          -0.0165    0.1329  -0.12  0.902
X2          -10.354     8.538  -1.21  0.233
X1X2         0.0948    0.1418   0.67  0.508


S = 3.88175   R-Sq = 44.1%   R-Sq(adj) = 39.7%


Analysis of Variance

Source            DF       SS      MS      F      P
Regression         3   452.54  150.85  10.01  0.000
Residual Error    38   572.58   15.07
Total             41  1025.12


Source  DF  Seq SS
X1       1  384.24
X2       1   61.57
X1X2     1    6.73
```

The least squares prediction equation is $\hat{y} = 8.14 - .016x_1 - 10.4x_2 + .095x_1x_2$.

c.   For diet = plants, $x_2 = 1$. The prediction equation is

$\hat{y} = 8.14 - .016x_1 - 10.4(1) + .095x_1(1) = -2.26 + .079x_1$. The slope of the line is .079.

For each unit increase is diet efficiency, the mean gosling weight change is estimated to increase by .079 for goslings fed plants.

d.   For diet = duck chow, $x_2 = 0$. The prediction equation is

$\hat{y} = 8.14 - .016x_1 - 10.4(0) + .095x_1(0) = 8.14 - .016x_1$. The slope of the line is −.016.

For each unit increase is diet efficiency, the mean gosling weight change is estimated to decrease by .016 for goslings fed duck chow.

e.   To determine if the slopes associated with the two diets are different, we test:

H_0: $\beta_3 = 0$
H_a: $\beta_3 \neq 0$

From the printout, the test statistic is $t = .67$ and the p-value is $p = .508$. Since the p-value is greater than α ($p = .508 > .05$), H_0 is not rejected. There is insufficient evidence to indicate the slopes associated with the two diets differ at $\alpha = .05$.

12.95   a.   Let $x_2 = \begin{cases} 1 & \text{if intervention group} \\ 0 & \text{if otherwise} \end{cases}$

The first-order model would be:

$$E(y) = \beta_0 + \beta_1 x_1 + \beta_2 x_2$$

b.   For the control group, $x_2 = 0$. The first-order model is:

$$E(y) = \beta_0 + \beta_1 x_1 + \beta_2(0) = \beta_0 + \beta_1 x_1$$

For the intervention group, $x_2 = 1$. The first-order model is:

$$E(y) = \beta_0 + \beta_1 x_1 + \beta_2(1) = \beta_0 + \beta_1 x_1 + \beta_2 = (\beta_0 + \beta_2) + \beta_1 x_1$$

In both models, the slope of the line is β_1.

c.   If pretest score and group interact, the first-order model would be:

$$E(y) = \beta_0 + \beta_1 x_1 + \beta_2 x_2 + \beta_3 x_1 x_2$$

d.   For the control group, $x_2 = 0$. The first-order model including the interaction is:

$$E(y) = \beta_0 + \beta_1 x_1 + \beta_2(0) + \beta_3 x_1(0) = \beta_0 + \beta_1 x_1$$

For the intervention group, $x_2 = 1$. The first-order model including the interaction is:

$$E(y) = \beta_0 + \beta_1 x_1 + \beta_2(1) + \beta_3 x_1(1) = \beta_0 + \beta_1 x_1 + \beta_2 + \beta_3 x_1$$
$$= (\beta_0 + \beta_2) + (\beta_1 + \beta_3)x_1$$

The slope of the model for the control group is β_1. The slope of the model for the intervention group is $\beta_1 + \beta_3$.

12.97   The models in parts **a** and **b** are nested:

The complete model is $E(y) = \beta_0 + \beta_1 x_1 + \beta_2 x_2$
The reduced model is $E(y) = \beta_0 + \beta_1 x_1$

The models in parts **a** and **d** are nested.

The complete model is $E(y) = \beta_0 + \beta_1 x_1 + \beta_2 x_2 + \beta_3 x_1 x_2$
The reduced model is $E(y) = \beta_0 + \beta_1 x_1 + \beta_2 x_2$
The models in parts **a** and **e** are nested.

The complete model is $E(y) = \beta_0 + \beta_1 x_1 + \beta_2 x_2 + \beta_3 x_1 x_2 + \beta_4 x_1^2 + \beta_5 x_2^2$
The reduced model is $E(y) = \beta_0 + \beta_1 x_1 + \beta_2 x_2$

The models in parts **b** and **c** are nested.

The complete model is $E(y) = \beta_0 + \beta_1 x_1 + \beta_2 x_1^2$
The reduced model is $E(y) = \beta_0 + \beta_1 x_1$

The models in parts **b** and **d** are nested.

The complete model is $E(y) = \beta_0 + \beta_1 x_1 + \beta_2 x_2 + \beta_3 x_1 x_2$
The reduced model is $E(y) = \beta_0 + \beta_1 x_1$

The models in parts **b** and **e** are nested.

The complete model is $E(y) = \beta_0 + \beta_1 x_1 + \beta_2 x_2 + \beta_3 x_1 x_2 + \beta_4 x_1^2 + \beta_5 x_2^2$
The reduced model is $E(y) = \beta_0 + \beta_1 x_1$

The models in parts **c** and **e** are nested.

The complete model is $E(y) = \beta_0 + \beta_1 x_1 + \beta_2 x_2 + \beta_3 x_1 x_2 + \beta_4 x_1^2 + \beta_5 x_2^2$
The reduced model is $E(y) = \beta_0 + \beta_1 x_1 + \beta_2 x_1^2$

The models in parts **d** and **e** are nested.

The complete model is $E(y) = \beta_0 + \beta_1 x_1 + \beta_2 x_2 + \beta_3 x_1 x_2 + \beta_4 x_1^2 + \beta_5 x_2^2$
The reduced model is $E(y) = \beta_0 + \beta_1 x_1 + \beta_2 x_2 + \beta_3 x_1 x_2$

12.99   A parsimonious model is a general linear model with a small number of β parameters. In situations where two competing models have essentially the same predictive power (as determined by an F-test), choose the model with the fewer number of βs.

12.101 a.   There are five β parameters in the complete model and three in the reduced model.

b.   The hypotheses are:

H_0: $\beta_3 = \beta_4 = 0$
H_a: At least one $\beta_i \neq 0$, $i = 3, 4$

c.   The test statistic is $F = \dfrac{(SSE_R - SSE_C)/(k-g)}{SSE_C/[n-(k+1)]}$

$$= \frac{(160.44 - 152.66)/(4-2)}{152.66/[20-(4+1)]} = \frac{3.89}{10.1773} = .38$$

The rejection region requires $\alpha = .05$ in the upper tail of the F distribution with numerator df $= k - g = 4 - 2 = 2$ and denominator df $= n - (k+1) = 20 - (4+1) = 15$. From Table IX, Appendix A, $F_{.05} = 3.68$. The rejection region is $F > 3.68$.

Since the observed value of the test statistic does not fall in the rejection region ($F = .38$ $\not> 3.68$), H_0 is not rejected. There is insufficient evidence to indicate the complete model is better than the reduced model at $\alpha = .05$.

12.103 a. $E(y) = \beta_0 + \beta_1x_1 + \beta_2x_2 + \beta_3x_3 + \beta_4x_4 + \beta_5x_5 + \beta_6x_6 + \beta_7x_7 + \beta_8x_8 + \beta_9x_9$
              $+ \beta_{10}x_{10} + \beta_{11}x_{11}$

       b. $E(y) = \beta_0 + \beta_1x_1 + \beta_2x_2 + \beta_3x_3 + \beta_4x_4 + \beta_5x_5 + \beta_6x_6 + \beta_7x_7$
              $+ \beta_8x_8 + \beta_9x_9 + \beta_{10}x_{10} + \beta_{11}x_{11}$
              $+ \beta_{12}x_1x_9 + \beta_{13}x_1x_{10} + \beta_{14}x_1x_{11}$
              $+ \beta_{15}x_2x_9 + \beta_{16}x_2x_{10} + \beta_{17}x_2x_{11}$
              $+ \beta_{18}x_3x_9 + \beta_{19}x_3x_{10} + \beta_{20}x_3x_{11}$
              $+ \beta_{21}x_4x_9 + \beta_{22}x_4x_{10} + \beta_{23}x_4x_{11}$
              $+ \beta_{24}x_5x_9 + \beta_{25}x_5x_{10} + \beta_{26}x_5x_{11}$
              $+ \beta_{27}x_6x_9 + \beta_{28}x_6x_{10} + \beta_{29}x_6x_{11}$
              $+ \beta_{30}x_7x_9 + \beta_{31}x_7x_{10} + \beta_{32}x_7x_{11}$
              $+ \beta_{33}x_8x_9 + \beta_{34}x_8x_{10} + \beta_{35}x_8x_{11}$

       c.   To test for interaction, we would use the hypothesis:

              H_0: $\beta_{12} = \beta_{13} = \beta_{14} = ... = \beta_{35} = 0$
              H_a: At least one $\beta_i \neq 0$, $i = 12, 13, 14, ..., 35$

              We would compare the complete model in part **b** to the reduce model in part **a**.

12.105 a.   Model 1: $R^2 = .101$. 10.1% of the total sample variation in supervisor-directed aggression score is explained by the model including the 4 independent variables.

              Model 2: $R^2 = .555$. 55.5% of the total sample variation in supervisor-directed aggression score is explained by the model including the 8 independent variables.

       b.   To determine if the additional 4 variables are useful in the prediction of supervisor-directed aggression score, we test:

              H_0: $\beta_5 = \beta_6 = \beta_7 = \beta_8 = 0$
              H_a: At least one $\beta_i \neq 0$

       c.   Yes, the models are nested. All of the variables included in Model 1 are also included in model 2.

       d.   Since the p-value is very small ($p < .001$), there is evidence to reject H_0. There is sufficient evidence to indicate at least 1 of the additional 4 variables is useful in the prediction of supervisor-directed aggression score at $\alpha > .001$.

e. Model 3 would be:

$$E(y) = \beta_o + \beta_1(\text{Age}) + \beta_2(\text{Gender})$$
$$+ \beta_3(\text{Interactional injustice at 2nd job}) + \beta_4(\text{Abusive supervisor at 2nd job})$$
$$+ \beta_5(\text{Self} - \text{esteem}) + \beta_6(\text{History of agression})$$
$$+ \beta_7(\text{Interactional injustice at primary job})$$
$$+ \beta_8(\text{Abusive supervisor at primary job})$$
$$+ \beta_9(\text{Self} - \text{esteem})(\text{History of agression})$$
$$+ \beta_{10}(\text{Self} - \text{esteem})(\text{Interactional injustice at primary job})$$
$$+ \beta_{11}(\text{Self} - \text{esteem})(\text{Abusive supervisor at primary job})$$
$$+ \beta_{12}(\text{History of agression})(\text{Interactional injustice at primary job})$$
$$+ \beta_{13}(\text{History of agression})(\text{Abusive supervisor at primary job})$$
$$+ \beta_{14}(\text{Interactional injustice at primary job})(\text{Abusive supervisor at primary job})$$

f. Since the p-value is $p > .10$, there is no evidence to indicate that any of the interaction terms are useful in the prediction of supervisor-directed aggression score for $\alpha \leq .10$.

12.107 a. The hypothesized alternative model is

$$E(y) = \beta_0 + \beta_1 x_1 + \beta_2 x_2 + \beta_3 x_3 + \beta_4 x_4 + \beta_5 x_5 + \beta_6 x_6 + \beta_7 x_7 + \beta_8 x_8 + \beta_9 x_9 + \beta_{10} x_{10}$$

b. The null hypothesis would be $H_0: \beta_3 = \beta_4 = \beta_5 = \beta_6 = \beta_7 = \beta_8 = \beta_9 = \beta_{10} = 0$

c. The test was statistically significant. Thus, H_o was rejected. There is sufficient evidence to indicate that at least one of the "control" variables contributes to the prediction of SAT-Math scores.

d. $R^2_{adj} = .79$. 79% of the sample variability of the SAT-Math scores around their means is explained by the proposed model relating SAT-Math scores to the 10 independent variables, adjusting for the sample size and the number of β parameters in the model.

e. For confidence coefficient .95, $\alpha = .05$ and $\alpha/2 = .05/2 = .025$. From Table VI, Appendix A, with df $= n - (k+1) = 3,492 - (10+1) = 3,481$, $t_{.025} = 1.96$. The 95% confidence interval is:

$$\hat{\beta}_2 \pm t_{.025} s_{\hat{\beta}_2} \Rightarrow 14 \pm 1.96(3) \Rightarrow 14 \pm 5.88 \Rightarrow (8.12, \ 19.88)$$

We are 95% confident that the difference in the mean SAT-Math scores between students who were coached and those who were not is between 8.12 and 19.88 points, holding all the other variables constant.

f. Yes. From Exercise 12.74, the confidence interval for β_2 was (13.12, 24.88). In part e, the confidence interval for β_2 was (8.12, 19.88). Even though coaching is significant in both models, the change in the mean SAT-Math scores is not as great if the control variables are added to the model.

g. The complete model would be:

$$E(y) = \beta_0 + \beta_1 x_1 + \beta_2 x_2 + \beta_3 x_3 + \beta_4 x_4 + \beta_5 x_5 + \beta_6 x_6 + \beta_7 x_7 + \beta_8 x_8 + \beta_9 x_9 + \beta_{10} x_{10} + \beta_{11} x_1 x_2$$
$$+ \beta_{12} x_3 x_2 + \beta_{13} x_4 x_2 + \beta_{14} x_5 x_2 + \beta_{15} x_6 x_2 + \beta_{16} x_7 x_2 + \beta_{17} x_8 x_2 + \beta_{18} x_9 x_2 + \beta_{19} x_{10} x_2$$

h. The null hypothesis would be H_0: $\beta_{11} = \beta_{12} = \beta_{13} = \beta_{14} = \beta_{15} = \beta_{16} = \beta_{17} = \beta_{18} = \beta_{19} = 0$. To perform this test, you would fit the complete model specified in part **g**. You would also fit the reduced model specified in part **a**. Then, you would perform the test comparing the complete and reduced models.

12.109 a. To determine whether the rate of increase of emotional distress with experience is different for the two groups, we test:

H_0: $\beta_4 = \beta_5 = 0$
H_a: At least one $\beta_i \neq 0$, $i = 4, 5$

b. To determine whether there are differences in mean emotional distress levels that are attributable to exposure group, we test:

H_0: $\beta_3 = \beta_4 = \beta_5 = 0$
H_a: At least one $\beta_i \neq 0$, $i = 3, 4, 5$

12.111 Two caveats with using the stepwise regression results as the final model are:

1. An extremely large number of t tests are performed, inflating the overall probability of a Type I error.

2. The stepwise procedure does not include any higher order or interaction terms.

12.113 a. In Step 1, all one-variable models are fit to the data. These models are of the form:

$$E(y) = \beta_0 + \beta_1 x_i$$

Since there are 7 independent variables, 7 models are fit. (Note: There are actually only 6 independent variables. One of the qualitative variables has three levels and thus two dummy variables. Some statistical packages will allow one to bunch these two variables together so that they are either both in or both out. In this answer, we are assuming that each x_i stands by itself.

b. In Step 2, all two-varirable models are fit to the data, where the variable selected in Step 1 is one of the variables. These models are of the form:

$$E(y) = \beta_0 + \beta_1 x_1 + \beta_2 x_i$$

Since there are 6 independent variables remaining, 6 models are fit.

c. In Step 3, all three-variable models are fit to the data, where the variables selected in Step 2 are two of the variables. These models are of the form:

$$E(y) = \beta_0 + \beta_1 x_1 + \beta_2 x_2 + \beta_3 x_i$$

Since there are 5 independent variables remaining, 5 models are fit.

d.   The procedure stops adding independent variables when none of the remaining variables, when added to the model, have a p-value less than some predetermined value. This predetermined value is usually $\alpha = .05$.

e.   Two major drawbacks to using the final stepwise model as the "best" model are:

   (1)   An extremely large number of single β parameter t-tests have been conducted. Thus, the probability is very high that one or more errors have been made in including or excluding variables.

   (2)   Often the variables selected to be included in a stepwise regression do not include the high-order terms. Consequently, we may have initially omitted several important terms from the model.

12.115 a.   In the first step of stepwise regression, 11 models are fit. These would be all the one-variable models.

   b.   In step two, all two variable models that include the variable selected in step one are fit. There will be a total of 10 models fit in this step.

   c.   In step 11, there is one model fit – the model containing all the independent variables.

   d.   The equation for $E(y)$ is:

   $$E(y) = \beta_0 + \beta_1 x_{11} + \beta_2 x_4 + \beta_3 x_2 + \beta_4 x_7 + \beta_5 x_{10} + \beta_6 x_1 + \beta_7 x_9 + \beta_8 x_3$$

   e. $R^2 = .677$. 67.7% of the sample variation in the overall satisfaction with BRT scores is explained by the model containing the 8 independent variables.

   f.   In any stepwise regression, many t-tests are performed. This inflates the probability of committing a Type I error. In addition, there are certain combinations of variables that may never be reached because of the first few variables put in the model. Finally, you should consider including interaction and higher order terms in the model.

12.117 A regression residual, $\hat{\varepsilon}$, is defined as the difference between an observed y value and its corresponding predicted value:

$$\hat{\varepsilon} = (y - \hat{y}) = y - (\hat{\beta}_0 + \hat{\beta}_1 x_1 + \hat{\beta}_2 x_2 + \dots + \hat{\beta}_k x_k)$$

12.119 A residual that is larger than 3s (in absolute value) is considered to be an outlier.

12.121 Three indicators of a multicollinearity problem are:

1.   Significant correlations between pairs of independent variables.

2.   Nonsignificant t-tests for all (or nearly all) of the individual β parameters when the F-test for overall model adequacy is significant.

3.   Signs opposite from what is expected in the estimated β parameters.

12.123 The statement "Regression models fit to time series data typically result in uncorrelated errors." is false. Time series data tend to be correlated.

12.125 Yes. x_2 and x_4 are highly correlated (.93), as well as x_4 and x_5 (.86). When highly correlated independent variables are present in a regression model, the results can be confusing. The researcher may want to include only one of the variables.

12.127 a.   No, there is no evidence of extreme multicollinearity. The independent variables would be irritability, trait anger, and narcissism. The largest pairwise correlation (in absolute value) between independent variables is .57. This is not extremely large and would probably not lead to extreme multicollinearity.

b.   Yes, there would be some evidence of multicollinearity. The independent variables would be aggressive behavior, irritability, and trait anger. The largest pairwise correlation (in absolute value) between independent variables is .77. This is fairly large and would probably lead to a problem with multicollinearity.

12.129 The residual for the last observation is $y - \hat{y} = 87 - 29.63 = 57.37$. The standard deviation is $s = 24.68$. The number of standard deviations $\hat{y}$ is from y is found by dividing the residual by the value of the standard deviation or $57.37 / 24.68 = 2.32$. This point is more than 2 standard deviations from 0 but less than 3. Thus, this point is a possible outlier.

12.131 a.   The main effects model for $E(y)$ would be:

$$E(y) = \beta_0 + \beta_1 x_1 + \beta_2 x_8$$

b. $\hat{\beta}_1 = -.28$ The difference in mean relative error between a developer and a project leader is estimated to be $-.28$, holding previous accuracy constant.

$\hat{\beta}_2 = .27$ The difference in mean relative error between a previous accuracy of $> 20\%$ and previous accuracy of $< 20\%$ is estimated to be .27, holding company role of estimator constant.

c.   One possible reason could be that company role of estimator and previous accuracy interact to affect the relative error and the interaction term was not included in the model.

12.133 Using MINITAB, the results of fitting the model are:

Regression Analysis: DDT versus Miles, Length, Weight

```
The regression equation is
DDT = - 108 + 0.0851 Miles + 3.77 Length - 0.0494 Weight


Predictor      Coef   SE Coef       T      P
Constant    -108.07     62.70   -1.72  0.087
Miles       0.08509   0.08221    1.03  0.302
Length        3.771     1.619    2.33  0.021
Weight     -0.04941   0.02926   -1.69  0.094


S = 97.4756    R-Sq = 3.9%   R-Sq(adj) = 1.8%


Analysis of Variance

Source           DF        SS      MS     F      P
Regression        3     53794   17931  1.89  0.135
Residual Error  140   1330210    9501
Total           143   1384003


Source  DF  Seq SS
Miles    1    1973
Length   1   24725
Weight   1   27095
Unusual Observations

Obs  Miles      DDT     Fit  SE Fit  Residual  St Resid
 33    280     3.00   -1.91   29.86      4.91     0.05 X
 80    305    22.00   92.76   32.78    -70.76    -0.77 X
105    320   360.00   43.80   12.74    316.20     3.27R
115    325  1100.00   50.40   14.29   1049.60    10.89R
143    345     0.35  -21.27   31.10     21.62     0.23 X

R denotes an observation with a large standardized residual.
X denotes an observation whose X value gives it large influence.
```

First, we will look at the plots of the residuals versus the 3 independent variables, miles captured upstream, length, and weight. The standardized residual are used in the plots. The standardized residuals have a mean of 0 and a standard deviation of 1. Using MINITAB, the plots are:

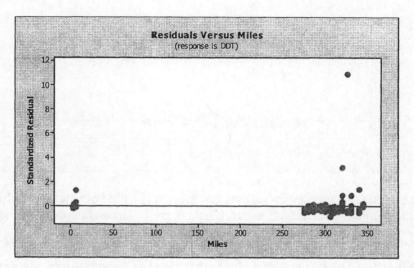

From this plot, there is no indication that the relationship between level of DDT and miles caught upstream is anything other than linear. There is no 'U' or 'upside-down U' shape to the plot.

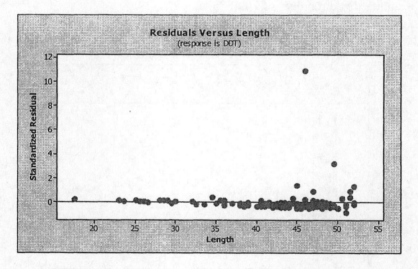

From this plot, there is no indication that the relationship between level of DDT and fish length is anything other than linear. There is no 'U' or 'upside-down U' shape to the plot.

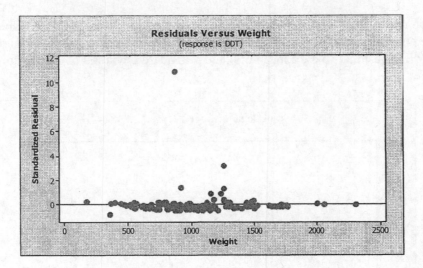

From this plot, there is no indication that the relationship between level of DDT and fish weight is anything other than linear. There is no 'U' or 'upside-down U' shape to the plot.

From the above plots, there is evidence that there are some outliers. Any standardized residual with values greater than 3 in absolute value are considered outliers. There are 2 observations with standardized residuals greater than 3 in absolute value. They are observation number 105 with a standardized residual of 3.2719 and observation # 115 with a standardized residual of 10.8855.

The histogram and normal probability plots of the standardized residuals are:

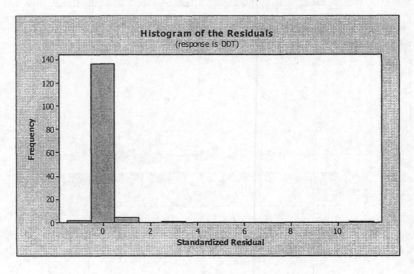

This histogram does not look very mound shaped. This indicates that the data may not be normal.

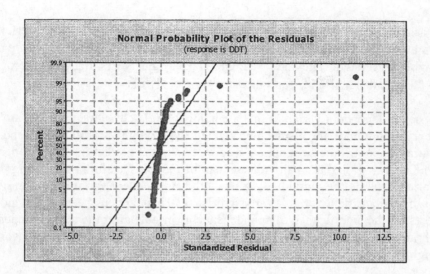

This normal probability plot does not look like a straight line. This indicates that the data are not normal.

To check for common variance, we look at the pot of the standardized residual versus the fitted values.

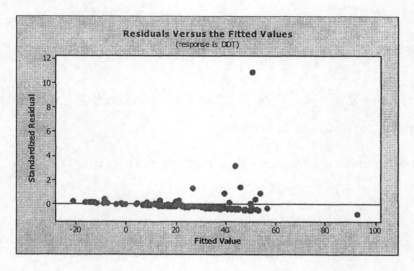

The data appear to be more spread out as the fitted values increase. Thus, our assumption of common variance does not appear to be valid.

12.135 a.	Variables that are moderately or highly correlated have correlation coefficients of .5 or higher in absolute value. There are only two pairs of variables which are moderately or highly correlated. They are "year GRE taken" and "years in graduate program" ($r = -.602$), and "race" and "foreign status" ($r = -.515$).

b.	When independent variables that are highly correlated with each other are included in a regression model, the results may be confusing. Highly correlated independent variables contribute overlapping information in the prediction of the dependent variable. The overall global test can indicate that the model is useful in predicting the dependent variable, while the individual t-tests on the independent variables can indicate that none of the independent variables are significant. This happens because the individual t-tests tests for the significance of an independent variable after the other independent variables are taken into account. Usually, only one of the independent variables that are highly correlated with each other are included in the regression model.

12.137	$E(y) = \beta_0 + \beta_1 x_1 + \beta_2 x_2 + \beta_3 x_3$

where $x_1 = \begin{cases} 1, & \text{if level 2} \\ 0, & \text{otherwise} \end{cases}$ $x_2 = \begin{cases} 1, & \text{if level 3} \\ 0, & \text{otherwise} \end{cases}$ $x_3 = \begin{cases} 1, & \text{if level 4} \\ 0, & \text{otherwise} \end{cases}$

12.139 a.	$E(y) = \beta_0 + \beta_1 x_1 + \beta_2 x_2 + \beta_3 x_3$

where $x_2 = \begin{cases} 1, & \text{if level 2} \\ 0, & \text{otherwise} \end{cases}$ $x_3 = \begin{cases} 1, & \text{if level 3} \\ 0, & \text{otherwise} \end{cases}$

b.	$E(y) = \beta_0 + \beta_1 x_1 + \beta_2 x_1^2 + \beta_3 x_2 + \beta_4 x_3 + \beta_5 x_1 x_2 + \beta_6 x_1 x_3 + \beta_7 x_1^2 x_2 + \beta_8 x_1^2 x_3$

where x_1, x_2, and x_3 are as in part **a**.

12.141 a.	To determine if at least one of the β parameters is not zero, we test:

H_0: $\beta_1 = \beta_2 = \beta_3 = \beta_4 = 0$
H_a: At least one $\beta_i \neq 0$

The test statistic is $F = \dfrac{R^2 / k}{(1 - R^2)/[n - (k+1)]} = \dfrac{.83/4}{(1 - .83)/[25 - (4+1)]} = 24.41$

The rejection region requires $\alpha = .05$ in the upper tail of the F distribution with numerator df $= k = 4$ and denominator df $= n - (k + 1) = 25 - (4 + 1) = 20$. From Table IX, Appendix A, $F_{.05} = 2.87$. The rejection region is $F > 2.87$.

Since the observed value of the test statistic falls in the rejection region ($F = 24.41 > 2.87$), H_0 is rejected. There is sufficient evidence to indicate at least one of the β parameters is nonzero at $\alpha = .05$.

b. H_0: $\beta_1 = 0$
    H_a: $\beta_1 < 0$

The test statistic is $t = \dfrac{\hat{\beta}_1 - 0}{s_{\hat{\beta}_1}} = \dfrac{-2.43 - 0}{1.21} = -2.01$

The rejection region requires $\alpha = .05$ in the lower tail of the t distribution with df = $n - (k + 1) = 25 - (4 + 1) = 20$. From Table VI, Appendix A, $t_{.05} = 1.725$. The rejection region is $t < -1.725$.

Since the observed value of the test statistic falls in the rejection region ($t = -2.01$ < -1.725), H_0 is rejected. There is sufficient evidence to indicate β_1 is less than 0 at $\alpha = .05$.

c. H_0: $\beta_2 = 0$
    H_a: $\beta_2 > 0$

The test statistic is $t = \dfrac{\hat{\beta}_2 - 0}{s_{\hat{\beta}_2}} = \dfrac{.05 - 0}{.16} = .31$

The rejection region requires $\alpha = .05$ in the upper tail of the t distribution. From part **b** above, the rejection region is $t > 1.725$.

Since the observed value of the test statistic does not fall in the rejection region ($t = .31 \not> 1.725$), H_0 is not rejected. There is insufficient evidence to indicate β_2 is greater than 0 at $\alpha = .05$.

d. H_0: $\beta_3 = 0$
    H_a: $\beta_3 \neq 0$

The test statistic is $t = \dfrac{\hat{\beta}_3 - 0}{s_{\hat{\beta}_3}} = \dfrac{.62 - 0}{.26} = 2.38$

The rejection region requires $\alpha/2 = .05/2 = .025$ in each tail of the t distribution with df = 20. From Table VI, Appendix A, $t_{.025} = 2.086$. The rejection region is $t < -2.086$ or $t > 2.086$.

Since the observed value of the test statistic falls in the rejection region ($t = 2.38 > 2.086$), H_0 is rejected. There is sufficient evidence to indicate β_3 is different from 0 at $\alpha = .05$.

12.143 The error of prediction is smallest when the values of x_1, x_2, and x_3 are equal to their sample means. The further x_1, x_2, and x_3 are from their means, the larger the error. When $x_1 = 60$, $x_2 = .4$, and $x_3 = 900$, the observed values are outside the observed ranges of the x values. When $x_1 = 30$, $x_2 = .6$, and $x_3 = 1300$, the observed values are within the observed ranges and consequently the x values are closer to their means. Thus, when $x_1 = 30$, $x_2 = .6$, and $x_3 = 1300$, the error of prediction is smaller.

12.145 Even though SSE = 0, we cannot estimate σ^2 because there are no degrees of freedom corresponding to error. With three data points, there are only two degrees of freedom available. The degrees of freedom corresponding to the model is $k = 2$ and the degrees of freedom corresponding to error is $n - (k + 1) = 3 - (2 + 1) = 0$. Without an estimate for σ^2, no inferences can be made.

12.147 a. $R^2 = .31$. 31% of the sample variation in the ln (level of CO_2 emissions in 1996) is explained by the model containing ln(foreign investments in 1980), gross domestic investment in 1980, trade exports in 1980, ln(GNP in 1980), agricultural production in 1980, African country, and ln(level of CO_2 emissions in 1980).

b. H_0: $\beta_1 = \beta_2 = \ldots = \beta_7 = 0$
H_a: At least one $\beta_i \neq 0$

The test statistic is $F = \dfrac{R^2/k}{(1-R^2)/[n-(k+1)]} = \dfrac{.31/7}{(1-.31)/[66-(7+1)]} = 3.723$

The rejection region requires $\alpha = .01$ in the upper tail of the F distribution. From Table XI, Appendix A, with $v_1 = k = 7$ and $v_2 = n - (k + 1) = 66 - (7 + 1) = 58$, $F_{.01} \approx 2.95$.

Since the observed value of the test statistic falls in the rejection region ($F = 3.723 > 2.95$), H_0 is rejected. There is sufficient evidence to indicate that at least one of the 7 independent variables contributes to the prediction of ln (level of CO_2 emissions in 1996) at $\alpha = .01$.

c.   We would not advise using individual t-tests on each of the independent variables to test for overall model adequacy. If we did, we would be very likely to make one or more errors in deciding which terms to retain in the model and which to exclude.

d.   To determine if foreign investments in 1980 is a statistically useful predictor of CO_2 emissions in 1996, we test:

H_0: $\beta_1 = 0$
H_a: $\beta_1 \neq 0$

e.   The test statistic is $t = 2.52$ and the p-value is $p < .05$. Since the p-value is less than $\alpha = .05$, H_0 is rejected. There is sufficient evidence to indicate that foreign investments in 1980 is a statistically useful predictor of CO_2 emissions in 1996.

f.   The independent variables that are highly correlated are x_4 (ln(GNP in 1980)) and x_5 (agricultural production in 1980). The correlation coefficient is $r = -.84$. Several other variables have moderate correlation: x_1 and x_3 with $r_{13} = .57$, x_4 and x_6 with $r_{46} = -.53$, and x_5 and x_7 with $r_{57} = -.50$. If these highly correlated variables are included in the model at the same time, signs of the estimates of the beta parameters may be different than what is expected. Also, the global test could indicate that at least one of the variables contributes to the prediction of y, but none of the individual variable tests are significant.

12.149 a. The first order model for $E(y)$ as a function of the first five independent variables is:

$$E(y) = \beta_0 + \beta_1 x_1 + \beta_2 x_2 + \beta_3 x_3 + \beta_4 x_4 + \beta_5 x_5$$

b. To test the utility of the model, we test:

H_0: $\beta_1 = \beta_2 = \beta_3 = \beta_4 = \beta_5 = 0$
H_a: At least one $\beta_i \neq 0$, $i = 1, 2, 3, 4, 5$

The test statistic is $F = 34.47$.

The p-value is $p < .001$. Since the p-value is so small, there is sufficient evidence to indicate the model is useful for predicting GSI at $\alpha > .001$.

$R^2 = .469$.   46.9% of the variability in the GSI scores is explained by the model including the first five independent variables.

c. The first order model for $E(y)$ as a function of the first seven independent variables is:

$$E(y) = \beta_0 + \beta_1 x_1 + \beta_2 x_2 + \beta_3 x_3 + \beta_4 x_4 + \beta_5 x_5 + \beta_6 x_6 + \beta_7 x_7$$

d. $R^2 = .603$ 60.3% of the variability in the GSI scores is explained by the model including the first seven independent variables.

e. Since the p-values associated with the variables DES and PDEQ-SR are both less than .001, there is evidence that both variables contribute to the prediction of GSI, adjusted for all the other variables already in the model for $\alpha > .001$.

12.151 a. Because the estimated coefficient of the quadratic term is negative, the prediction equation will open downward something like the following:

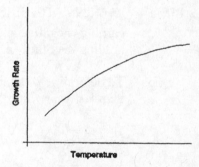

b. $R^2 = .67$. 67% of the variability in the daily growth rates is explained by the quadratic relationship between growth rate and temperature.

c. To test the utility of the model, we test:

H_0: $\beta_1 = \beta_2 = 0$
H_a: At least one $\beta_i \neq 0$, $i = 1, 2$

The test statistic is:

$$F = \frac{R^2/k}{(1-R^2)/[n-(k+1)]} = \frac{.67/2}{(1-.67)/[33-(2+1)]} = 30.45$$

The rejection region requires $\alpha = .05$ in the upper tail of the F distribution with $v_1 = k = 2$ and $v_2 = n - (k + 1) = 33 - (2 + 1) = 30$. From Table IX, Appendix A, $F_{.05} = 3.32$. The rejection region is $F > 3.32$.

Since the observed value of the test statistic falls in the rejection region ($F = 30.45 > 3.32$), H_0 is rejected. There is sufficient evidence that the model is useful for predicting algae growth rate at $\alpha = .05$.

12.153 a.   Let $\begin{cases} 1 & \text{if pond is enriched} \\ 0 & \text{if otherwise} \end{cases}$

The model is $E(y) = \beta_0 + \beta_1 x_1$

     b. β_0 = mean mosquito larvae in the natural pond
β_1 = difference in the mean mosquito larvae between the enriched pond and the natural pond

     c.   To determine if the mean larval density for the enriched pond exceeds the mean for the natural pond, we test:

H_0: $\beta_1 = 0$
H_a: $\beta_1 > 0$

     d.   The p-value for the global F test is $p = .004$. Thus, the p-value for this test is $p = .004/2 = .002$. Since the p-value is so small, there is evidence to reject H_0. There is sufficient evidence to indicate that the mean larval densities differ for the enriched pond and the natural pond at $\alpha > .002$.

12.155 a. $\hat{\beta}_4 = .296$. The difference in the mean DTVA between firms that had operating earnings negative and lower than last year and firms that did not is estimated to be .296, holding the other variables constant.

     b.   To determine if the mean DTVA for firms with negative earnings and earnings lower than last year will exceed the mean DTVA of other firms, we test:

H_0: $\beta_4 = 0$
H_a: $\beta_4 \neq 0$

The p-value for the test is $p = .001$. Since the p-value is less than $\alpha = .05$ ($p = .001 < .05$), H_o is rejected. There is sufficient evidence to indicate the mean DTVA for firms with negative earnings and earnings lower than last year will exceed the mean DTVA of other firms at $\alpha = .05$.

     c. $R_a^2 = .280$. 28% of the sample variation in DTVA scores is explained by the model containing the 5 independent variables leverage, bonus, market value, Bbath, and operating earnings, adjusted for the sample size and the number of independent variables in the model.

12.157 a. Let x_1 = ASO scale, x_2 = Burns scale, and x_3 = Rotter scale. The first order model relating depression (Zung scale) to self-acceptance (ASO scale), perfectionism (Burns scale), and reinforcement (Rotter scale) is:

$$E(y) = \beta_0 + \beta_1 x_1 + \beta_2 x_2 + \beta_3 x_3$$

b. $R^2 = .70$. 70% of the variability in the depression scores is explained by the model including self-acceptance, perfectionism, and reinforcement.

c. To test if the model is useful for predicting depression, we test:

H_0: $\beta_1 = \beta_2 = \beta_3 = 0$
H_a: At least one $\beta_i \neq 0$, $i = 1, 2, 3$

The test statistic is $F = \dfrac{R^2/k}{(1-R^2)/[n-(k+1)]} = \dfrac{.70/3}{(1-.70)/[76-(3+1)]} = 56$

The rejection region requires $\alpha = .05$ in the upper tail of the F distribution with $\nu_1 = k = 3$ and $\nu_2 = n - (k + 1) = 76 - (3 + 1) = 72$. From Table IX, Appendix A, $F_{.05} \approx 2.76$. The rejection region is $F > 2.76$.

Since the observed value of the test statistic falls in the rejection region ($F = 56 > 2.76$), H_0 is rejected. There is sufficient evidence that the model is useful in predicting depression at $\alpha = .05$.

d. The p-value for the perfectionism variable is .87. Since the p-value is so large, H_0 is not rejected. There is insufficient evidence to indicate perfectionism is useful for predicting depression, adjusting for self-acceptance and reinforcement already being in the model at $\alpha < .87$.

12.159 a. Since in the proposed model the funniness of the joke should increase and then decrease, the value of β_2 is expected to be negative.

b. To determine if the quadratic model relating pain to funniness rating is useful, we test:

H_0: $\beta_1 = \beta_2 = 0$
H_a: At least one $\beta_i \neq 0$, $i = 1, 2$

The test statistic is $F = 1.60$.

The rejection region requires $\alpha = .05$ in the upper tail of the F distribution with $\nu_1 = k = 2$ and $\nu_2 = n - (k + 1) = 32 - (2 + 1) = 29$. From Table IX, Appendix A, $F_{.05} = 3.33$. The rejection region is $F > 3.33$.

Since the observed value of the test statistic does not fall in the rejection region ($F = 1.60 \not> 3.33$), H_0 is not rejected. There is insufficient evidence to indicate that the quadratic model relating pain to funniness rating is useful at $\alpha = .05$.

c. To determine if the quadratic model relating aggression/hostility to funniness rating is useful, we test:

$$H_0:\ \beta_1 = \beta_2 = 0$$
$$H_a:\ \text{At least one } \beta_i \neq 0,\ i = 1, 2$$

The test statistic is $F = 1.61$.

The rejection region requires $\alpha = .05$ in the upper tail of the F distribution with $v_1 = k = 2$ and $v_2 = n - (k + 1) = 32 - (2 + 1) = 29$. From Table IX, Appendix A, $F_{.05} = 3.33$. The rejection region is $F > 3.33$.

Since the observed value of the test statistic does not fall in the rejection region ($F = 1.61 \not> 3.33$), H_0 is not rejected. There is insufficient evidence to indicate that the quadratic model relating aggression/hostility to funniness rating is useful at $\alpha = .05$.

12.161 a.

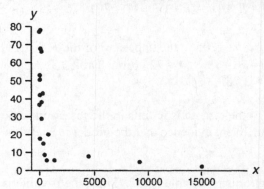

There is a curvilinear trend.

b. From MINITAB, the output is:

```
The regression equation is y = 42.2 - 0.0114x + 0.000001 xsq

Predictor          Coef          StDev             T             P
Constant         42.247          5.712          7.40         0.000
x              -0.011404       0.005053         -2.26         0.037
xsq           0.00000061      0.00000037         1.66         0.115

S = 21.81        R-Sq = 34.9%        R-Sq(adj) = 27.2%

Analysis of Variance

Source           DF            SS            MS          F           P
Regression        2        4325.4        2162.7       4.55       0.026
Residual Error   17        8085.5         475.6
Total            19       12410.9

Sourc            DF        Seq SS
e
x                 1        3013.3
xsq               1        1312.1

Unusual Observations

Obs       x1         y        Fit    StDev Fit    Residual    St Resid
16      9150      4.60     -11.21       16.24       15.81        1.09 x
17     15022      2.20       8.09       21.40       -5.89       -1.41 x

X denotes an observation whose X value gives it large influence.
```

The fitted model is $\hat{y} = 42.2 - .0114x + .00000061x^2$

c. To determine if a curvilinear relationship exists, we test:

H_0: $\beta_2 = 0$
H_a: $\beta_2 \neq 0$

From MINITAB, the test statistic is $t = 1.66$ with p-value $= .115$. Since the p-value is greater than $\alpha = .05$, do not reject H_0. There is insufficient evidence to indicate that a curvilinear relationship exists between dissolved phosphorus percentage and soil loss at $\alpha = .05$.

12.163 a. $E(y) = \beta_0 + \beta_1 x_1 + \beta_2 x_6 + \beta_3 x_7$

where $x_7 = \begin{cases} 1 & \text{if condition is good} \\ 0 & \text{otherwise} \end{cases}$

$x_6 = \begin{cases} 1 & \text{if condition is fair} \\ 0 & \text{otherwise} \end{cases}$

b. The model specified in part **a** seems appropriate. The points for E, F, and G cluster around three parallel lines.

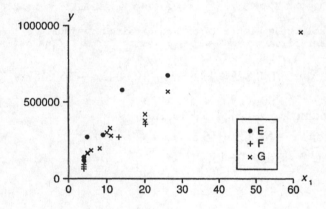

c. Using MINITAB, the output is

```
The regression equation is
y = 188875 + 15617 x1 - 103046 x6 - 152487 x7

Predictor       Coef       StDev          T         P
Constant      188875       28588       6.61     0.000
x1             15617        1066      14.66     0.000
x6           -103046       31784      -3.24     0.004
x7           -152487       39157      -3.89     0.001

S = 64624       R-Sq = 91.8%     R-Sq(adj) = 90.7%

Analysis of Variance

Source           DF          SS          MS         F         P
Regression        3   9.86170E+11   3.28723E+11     78.71     0.000
Residual Error   21   87700442851    4176211564
Total            24   1.07387E+12

Source       DF        SeqSS
x1            1   9.15776E+11
x6            1    7061463149
x7            1   63332198206

Unusual Observations
Obs     x1        y         Fit    StDev Fit     Residual    St Resid
 10   62.0   950000     1054078       53911      -104078     -2.92RX
 23   14.0   573200      407512       26670       165688      2.81R

R denotes an observation with a large standardized residual
X denotes an observation whose X value gives it large influence.
```

The fitted model is $\hat{y} = 188,875 + 15,617x_1 - 103,046x_6 - 152,487x_7$

For excellent condition, $\hat{y} = 188,875 + 15,617x_1$

For good condition, $\hat{y} = 85,829 + 15,617x_1$

For fair condition, $\hat{y} = 36,388 + 15,617x_1$

d.

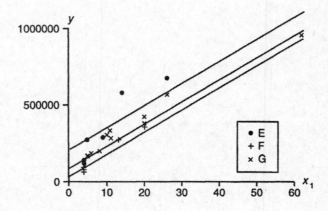

e. We must first fit a reduced model with just x_1, number of apartments. Using MINITAB, the output is:

```
The regression equation is
y = 101786 + 15525 x1

Predictor        Coef        StDev          T          P
Constant       101786        23291       4.37      0.000
x1              15525         1345      11.54      0.000

S = 82908       R-Sq = 85.3%    R-Sq(adj) = 84.6%

Analysis of Variance

Source           DF           SS           MS          F          P
Regression        1    9.15776E+11   9.15776E+11    133.23      0.000
Residual Error   23    1.58094E+11    6873656705
Total            24    1.07387E+12

Unusual Observations
Obs     x1        y         Fit    StDev Fit    Residual    St Resid
  4   26.0   676200      505433       24930      170757       2.16R
 10   62.0   950000     1064353       69058     -114353      -2.49RX
 23   14.0   573200      319140       16765      254060       3.13R

R denotes an observation with a large standardized residual
X denotes an observation whose X value gives it large influence.
```

To determine if the relationship between sale price and number of units differs depending on the physical condition of the apartments, we test:

H_0: $\beta_6 = \beta_7 = 0$
H_a: At least one $\beta_i \neq 0$, $i = 6, 7$

The test statistic is:

$$F = \frac{(\mathrm{SSE_R} - \mathrm{SSE_C})/(k-g)}{\mathrm{SSE_C}/[n-(k+1)]} = \frac{(1.58094 \times 10^{11} - 87700442851)/2}{4176211564} = 8.43$$

The rejection region requires $\alpha = .05$ in the upper tail of the F distribution with $v_1 = k - g = 3 - 1 = 2$ and $v_2 = n - (k+1) = 25 - (3+1) = 21$. From Table IX, Appendix A, $F_{.05} = 3.47$. The rejection region is $F > 3.47$.

Since the observed value of the test statistic falls in the rejection region ($F = 8.43 > 3.47$), H_0 is rejected. There is evidence to indicate that the relationship between sale price and number of units differs depending on the physical condition of the apartments at $\alpha = .05$.

f. We will look for high pairwise correlations

```
        x1       x2       x3     x4      x5      x6
x2    -0.014
x3     0.800  -0.188
x4     0.224  -0.363   0.166
x5     0.878   0.027   0.673  0.089
x6     0.175  -0.447   0.271  0.112   0.020
x7    -0.128   0.392  -0.118  0.050  -0.238  -0.564
```

When highly correlated independent variables are present in a regression model, the results are confusing. The researchers may only want to include one of the variables. This may be the case for the variables: x_1 and x_3, x_1 and x_5, x_3 and x_5

g. Use the following plots to check the assumptions on ε.

standardized residuals vs x_1
standardized residuals vs x_2
standardized residuals vs x_3
standardized residuals vs x_4
standardized residuals vs x_5
standardized resisduals vs predicted values
histogram of the standardized residuals
normal probability plot

From the plots of the residuals, there do not appear to be any outliers no standardized residuals are larger than 2.38 in magnitude. In all the plots of the residuals vs x_i, there is no trend that would indicate non-constant variance (no funnel shape). In addition, there is no U or upside-down U shape that would indicate that any of the variables should be squared. In the histogram of the residuals, the plot is fairly mound-shaped, that would indicate the residuals are approximately normally distributed. The normal probability plot is a fairly straight line. This indicates that the assumption of normality is valid. All of the assumptions appear to be met.

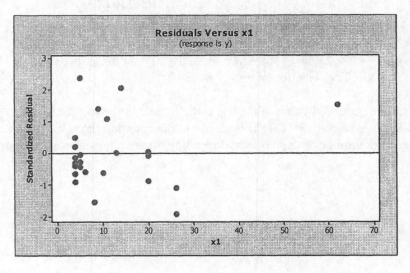

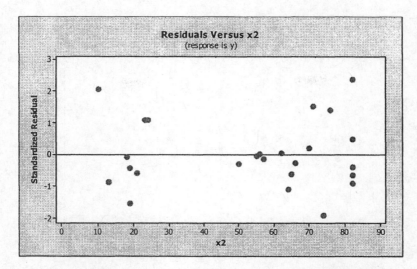

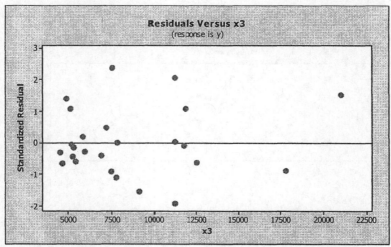

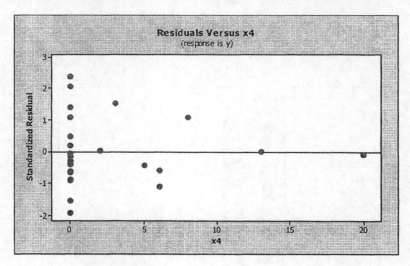

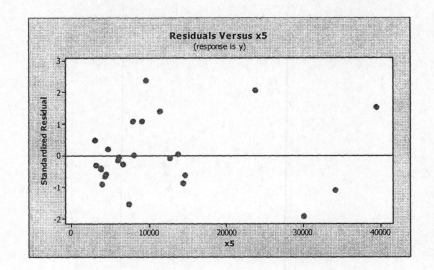

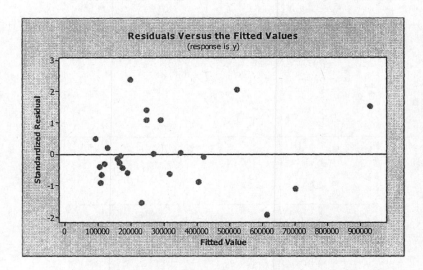

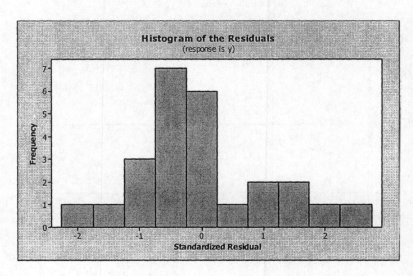

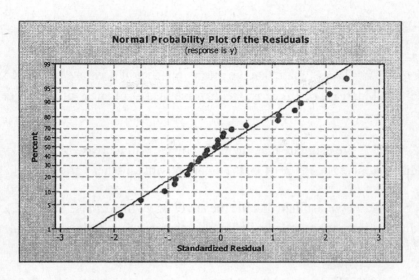

Normal Probability Plot of the Residuals
(response is y)

12.165 a. $E(y) = \beta_0 + \beta_1 x_1 + \beta_2 x_2 + \beta_3 x_3 + \beta_4 x_4 + \beta_5 x_5 + \beta_6 x_6 + \beta_7 x_7$

b. Using MINITAB, the output is:

The regression equation is y = 0.998 − 0.0224 x1 + 0.156x2 − 0.0172x3 − 0.00953x4
+ 0.421x5 + 0.417x6 − 0.155x7

Predictor	Coef	StDev	T	P
Constant	0.9981	0.2475	4.03	0.002
x1	−0.022429	0.005039	−4.45	0.001
x2	0.15571	0.07429	2.10	0.060
x3	−0.01719	0.01186	−1.45	0.175
x4	−0.009527	0.009619	−0.99	0.343
x5	0.4214	0.1008	4.18	0.002
x6	0.4171	0.4377	0.95	0.361
x7	−0.1552	0.1486	−1.04	0.319

S = 0.4365   R-Sq = 77.1%   R-Sq(adj) = 62.5%

Analysis of Variance

Source	DF	SS	MS	F	P
Regression	7	7.9578	1.1368	5.29	0.007
Residual Error	11	2.3632	0.2148		
Total	18	10.3210			

Source	DF	Seq SS
x1	1	1.4016
x2	1	1.9263
x3	1	0.1171
x4	1	0.0446
x5	1	4.0771
x6	1	0.1565
x7	1	0.2345

Unusual Observations

Obs	x1	y	Fit	StDev Fit	Residual	St Resid
14	80.0	0.120	0.628	0.328	0.748	2.28R

R denotes an observation with a large standardized residual.

The least squares model is $= .9981 − .0224x_1 + .1557x_2 − .0172x_3 − .0095x_4$
$+ .4214x_5 + .4171x_6 − .1552x_7$

c. $\hat{\beta}_0 = .9981 =$ the estimate of the y-intercept.

$\hat{\beta}_1 = -.0224$; We estimate that the mean voltage will decrease by .0224 kw/cm, for each additional increase of 1% of x_1, the disperse phase volume (with all other variables held constant).

$\hat{\beta}_2 = .1557$; We estimate that the mean voltage will increase by .1557 kw/cm for each additional increase of 1% of x_2, the salinity (with all other variables held constant).

$\hat{\beta}_3 = -.0172$; We estimate the the mean voltage will decrease by .0172 kw/cm for each additional increase of 1 degree of x_3, the temperature in Celcius (with all other variables held constant).

$\hat{\beta}_4 = -.0095$; We estimate that the mean voltage will decrease by .0095 kw/cm for each additional increase of 1 hour of x_4, the time delay (with all other variables held constant).

$\hat{\beta}_5 = .4214$; We estimate that the mean voltage will increase by .4124 kw/cm for each additional increase of 1% of x_5, sufficient concentration (with all other variables held constant).

$\hat{\beta}_6 = .4171$; We estimate that the mean voltage will increase by .4171 kw/cm for each additional increase of 1 unit of x_6, span: Triton (with all other variables held constant).

$\hat{\beta}_7 = -.1552$; We estimate that the mean voltage will decrease by .1552 kw/cm for each additional increase of 1% of x_7, the solid particles (with all other variables held constant).

d. The first order model is:

$E(y) = \beta_0 + \beta_1 x_1 + \beta_2 x_2 + \beta_3 x_5$

We want to find a 95% prediction interval for the actual voltage when the volume fraction of the disperse phase is at the high level ($x_1 = 80$), the salinity is at the low level ($x_2 = 1$), and the amount of surfactant is at the low level ($x_5 = 2$).

Using MINITAB, the output is:

```
The regression equation is
y = 0.993 - 0.243 x1 - 0.142 x2 + 0.385 x5


Predictor          Coef        StDev             T           P
Constant         0.9326       0.2482          3.76       0.002
x1             -0.024272     0.004900         -4.95       0.000
x2              0.14206       0.07573          1.88       0.080
x5              0.38457       0.09801          3.92       0.001


S = 0.4796        R-Sq = 66.6%      R-Sq(adj) = 59.9%

Analysis of Variance

Source             DF          SS           MS          F         P
Regression          3       6.8701       2.2900       9.95     0.001
Residual           15       3.4509       0.2301
Error
Total              18      10.3210

Source       DF      Seq SS
x1            1      1.4016
x2            1      1.9263
x5            1      3.5422

Unusual Observations
Obs         x1          y        Fit    StDev Fit    Residual    St Resid
  3        40.0      3.200      2.068      0.239       1.132       2.72R

R denotes an observation with a large standardized residual

Predicted Values for New Observations

New
Obs     Fit   SE Fit       95% CI            95% PI
  1   -0.098   0.232   (-0.592, 0.396)   (-1.233, 1.038)


Values of Predictors for New Observations

New
Obs    x1     x2      x5
  1   80.0   1.00    2.00
```

The 95% prediction interval is $(-1.233, 1.038)$. We are 95% confident that the actual voltage is between -1.233 and 1.038 when the volume fraction of the disperse phase is at the high level ($x_1 = 80$), the salinity is at the low level ($x_2 = 1$), and the amount of surfactant is at the low level ($x_5 = 2$).

e. By including the interaction terms, it implies that the relationship between voltage and volume fraction of the disperse phase depend on the levels of salinity and surfactant concentration.

A possible sketch of the relationship is:

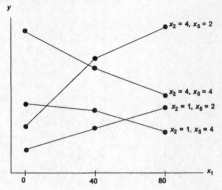

f. Using MINITAB, the results are:

Regression Analysis: y versus x1, x2, x5, x1x2, x1x5

```
The regression equation is
y = 0.906 - 0.0228 x1 + 0.305 x2 + 0.275 x5 - 0.00280 x1x2 + 0.00158 x1x5

Predictor        Coef    SE Coef       T      P
Constant       0.9057     0.2855    3.17  0.007
x1           -0.022753   0.008318   -2.74  0.017
x2            0.3047      0.2366    1.29  0.220
x5            0.2747      0.2270    1.21  0.248
x1x2         -0.002804   0.003790   -0.74  0.473
x1x5          0.001579   0.003947    0.40  0.696


S = 0.504650   R-Sq = 67.9%   R-Sq(adj) = 55.6%


Analysis of Variance

Source           DF        SS       MS      F      P
Regression        5    7.0103   1.4021   5.51  0.006
Residual Error   13    3.3107   0.2547
Total            18   10.3210


Source   DF   Seq SS
x1        1   1.4016
x2        1   1.9263
x5        1   3.5422
x1x2      1   0.0994
x1x5      1   0.0408
```

The fitted regression line is:

$$\hat{y} = .906 - .023x_1 + .305x_2 + .275x_5 - .003x_1x_2 + .002x_1x_5$$

To determine if the model is useful, we test:

H_0: $\beta_1 = \beta_2 = \beta_3 = \beta_4 = \beta_5 = 0$
H_a: At least one $\beta_i \neq 0$, for $i = 1, 2, ..., 5$

The test statistic is $F = 5.51$.

The rejection region requires $\alpha = .05$ in the upper tail of the F-distribution with $\nu_1 = k = 5$ and $\nu_2 = n - (k + 1) = 19 - (5 + 1) = 13$. From Table VIII, Appendix B, $F_{.05} = 3.03$. The rejection region is $F > 3.03$.

Since the observed value of the test statistic falls in the rejection region ($F = 5.51 > 3.03$), H_0 is rejected. There is sufficient evidence to indicate the model is useful for predicting voltage at $\alpha = .05$.

$R^2 = 67.9\%$. Thus, 67.9% of the sample variation of voltage is explained by the model containing the three independent variables and two interaction terms.

The estimate of the standard deviation is $s = .505$.

Comparing this model to that fit in part b the model in part b appears to fit the data better. The model in part b has a higher R^2 (77.1% vs 67.9%) and a smaller estimate of the standard deviation (.437 vs .505).

g. $\hat{\beta}_0 = .906$.   This is simply the estimate of the y-intercept.

$\hat{\beta}_1 = -.023$.   For each unit increase in disperse phase volume, we estimate that the mean voltage will decrease by .023 units, holding salinity and surfactant concentration at 0.

$\hat{\beta}_2 = .305$.   For each unit increase in salinity, we estimate that the mean voltage will increase by .305 units, holding disperse phase volume and surfactant concentration at 0.

$\hat{\beta}_3 = .275$.   For each unit increase in surfactant concentration, we estimate that the mean voltage will increase by .275 units, holding disperse phase volume and salinity at 0.

$\hat{\beta}_4 = -.003$.   This estimates the difference in the slope of the relationship between voltage and disperse phase volume for each unit increase in salinity, holding surfactant concentration constant.

$\hat{\beta}_5 = .002$.   This estimates the difference in the slope of the relationship between voltage and disperse phase volume for each unit increase in surfactant concentration, holding salinity constant.

h. Using MINITAB to fit the model without the interaction terms, the results are:

Regression Analysis: y versus x1, x2, x5
```
The regression equation is
y = 0.933 - 0.0243 x1 + 0.142 x2 + 0.385 x5


Predictor        Coef    SE Coef        T       P
Constant       0.9326     0.2482     3.76   0.002
x1          -0.024272   0.004900    -4.95   0.000
x2            0.14206    0.07573     1.88   0.080
x5            0.38457    0.09801     3.92   0.001


S = 0.479646    R-Sq = 66.6%    R-Sq(adj) = 59.9%


Analysis of Variance

Source            DF         SS        MS      F       P
Regression         3     6.8701    2.2900   9.95   0.001
Residual Error    15     3.4509    0.2301
Total             18    10.3210


Source  DF   Seq SS
x1       1   1.4016
x2       1   1.9263
x5       1   3.5422
```

To determine whether the interaction terms contribute significantly to the prediction of voltage, we test:

H_0: $\beta_4 = \beta_5 = 0$
H_a: At least one $\beta_i \neq 0$, $i = 4, 5$

The test statistic is $F = \dfrac{(\text{SSE}_R - \text{SSE}_C)/(k-g)}{\text{SSE}_C/[n-(k+1)]} = \dfrac{(3.4509 - 3.3017)/(5-3)}{3.3017/[19-(5+1)]} = .294$

The rejection region requires $\alpha = .05$ in the upper tail of the F distribution with $\nu_1 = k - g = 5 - 3 = 2$ and $\nu_2 = n - (k + 1) = 19 - (5 + 1) = 13$. From Table IX, Appendix A, $F_{.05} = 3.18$. The rejection region is $F > 3.18$.

Since the observed value of the test statistic does not fall in the rejection region ($F = .294 \not> 3.18$), H_0 is not rejected. There is insufficient evidence to indicate the interaction terms contribute to the prediction of voltage at $\alpha = .05$.

12.167 a. Using MINITAB, the scattergram is:

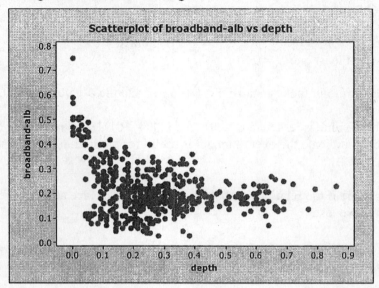

From the plot, it appears that the relationship between broadband surface albedo level and depth is curvilinear. The hypothesized model is:

$$E(y) = \beta_0 + \beta_1 x + \beta_2 x^2$$

b. Using MINITAB, the results are:

Regression Analysis: broadband-alb versus depth, depth-sq

```
The regression equaticn is
broadband-alb = 0.334 - 0.810 depth + 0.941 depth-sq


Predictor       Coef    SE Coef        T       P
Constant     0.33382    0.01182    28.24   0.000
depth       -0.80997    0.07935   -10.21   0.000
depth-sq      0.9406     0.1125     8.36   0.000


S = 0.0879081   R-Sq = 19.9%   R-Sq(adj) = 19.6%


Analysis of Variance

Source            DF        SS        MS       F       P
Regression         2   0.96092   0.48046   62.17   0.000
Residual Error   501   3.87164   0.00773
Total            503   4.83256


Source     DF   Seq SS
depth       1   0.42087
depth-sq    1   0.54005
```

The least squares prediction equation is:

$$\hat{y} = .33382 - .80997x + .9406x^2$$

c.   To determine if the overall model is useful for predicting broadband surface albedo level, we test:

H_o: $\beta_1 = \beta_2 = 0$
H_a: At least one $\beta_i \neq 0$

From the printout, the test statistic is F = 62.17 and the p-value is p = 0.000.

Since the p-value is less than $\alpha = .01$ ($p = 0.000 < 01$), H_0 is rejected. There is sufficient evidence to indicate the overall model is useful for predicting broadband surface albedo level at $\alpha = .01$.

d.   To determine if the relationship between broadband surface albedo level and depth is quadratic, we test:

H_0: $\beta_2 = 0$
H_a: $\beta_2 \neq 0$

From the printout, the test statistic is $t = 8.36$ and the p-value is $p = 0.000$.

Since the p-value is less than $\alpha = .01$ ($p = 0.000 < 01$), H_o is rejected. There is sufficient evidence to indicate the relationship between broadband surface albedo level and depth is quadratic at $\alpha = .01$.

Since the above test is significant, we do not need to run a test on the liner term.

e.   R-sq(adj) = 19.6%. 19.6% of the total sample variance is explained by the quadratic model containing depth, adjusted for the sample size and the number of variables in the model.

f.   Using MINITAB, plots of the residuals are:

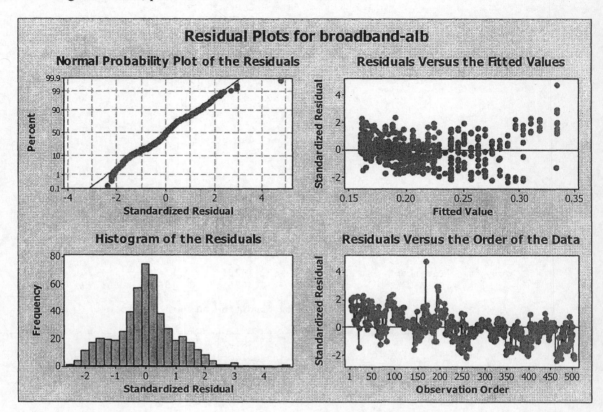

Residual Plots for broadband-alb

From the histogram of the residuals (lower left), it appears that the data are skewed to the right. This indicates that there may be a few outliers. Also, from the plot of the residuals against the fitted values (upper right), there are a couple residuals (standardized) greater than 3 in absolute value. These observations would be considered outliers.

12.169 All of the independent variables are continuous except District. Four dummy variables were created for District before using the stepwise regression. Also, note that several observations had missing values. Thus, only 217 observations were used in the analysis. Since Ratio was computed from the ratio of Price and DOT estimate, it was not used as a predictor variable. Using SAS and $\alpha = .10$ for keeping variables in the model, the Stepwise regression results are:

```
                        The REG Procedure
                          Model: MODEL1
                     Dependent Variable: PRICE

                     Stepwise Selection: Step 1

         Variable DOTEST Entered: R-Square = 0.9763 and C(p) = 14.2985


                         Analysis of Variance

                                  Sum of        Mean
         Source          DF      Squares       Square     F Value   Pr > F

         Model            1    687136244    687136244     8856.79   <.0001
         Error          215     16680334        77583
         Corrected Total 216    703816578


                     Parameter    Standard
         Variable     Estimate      Error    Type II SS  F Value  Pr > F

         Intercept     33.24768   22.19263      174129     2.24   0.1356
         DOTEST         0.90637    0.00963   687136244  8856.79   <.0001

                 Bounds on condition number: 1, 1

-------------------------------------------------------------------------------

                     Stepwise Selection: Step 2


         Variable STATUS Entered: R-Square = 0.9774 and C(p) = 5.9531


                         Analysis of Variance

                                  Sum of        Mean
         Source          DF      Squares       Square     F Value   Pr > F

         Model            2    687894714    343947357     4622.87   <.0001
         Error          214     15921865        74401
         Corrected Total 216    703816578

                        The REG Procedure
                          Model: MODEL1
                     Dependent Variable: PRICE

                     Stepwise Selection: Step 2

                     Parameter    Standard
         Variable     Estimate      Error    Type II SS  F Value  Pr > F

         Intercept    -12.92808   26.10499       18247     0.25   0.6209
         DOTEST         0.91233    0.00961   669924194  9004.21   <.0001
         STATUS       132.16624   41.39440      758469    10.19   0.0016

             Bounds on condition number: 1.0392, 4.1569
```

```
----------------------------------------------------------------------

                     Stepwise Selection: Step 3


          Variable DAYS Entered: R-Square = 0.9779 and C(p) = 3.4115


                         Analysis of Variance

                              Sum of          Mean
Source              DF        Squares         Square     F Value   Pr > F

Model               3        688228730      229409577    3134.76   <.0001
Error             213         15587848          73182
Corrected Total   216        703816578


                   Parameter      Standard
         Variable   Estimate        Error    Type II SS  F Value  Pr > F

         Intercept  -59.79987     33.93608      227239     3.11   0.0795
         DOTEST       0.88616      0.01552   238455655  3258.37   <.0001
         STATUS     139.40952     41.19370      838166    11.45   0.0008
         DAYS         0.35761      0.16739      334016     4.56   0.0338

              Bounds on condition number: 2.7653, 19.698
----------------------------------------------------------------------


     All variables left in the model are significant at the 0.1000 level.


                         The REG Procedure
                           Model: MODEL1
                      Dependent Variable: PRICE

                     Stepwise Selection: Step 3

No other variable met the 0.1500 significance level for entry into the model.


                   Summary of Stepwise Selection

       Variable  Variable  Number  Partial   Model
  Step Entered   Removed   Vars In R-Square R-Square  C(p)    F Value Pr > F

   1  DOTEST                  1     0.9763   0.9763  14.2985 8856.79 <.0001
   2  STATUS                  2     0.0011   0.9774   5.9631   10.19 0.0016
   3  DAYS                    3     0.0005   0.9779   3.4115    4.56 0.0338
```

From the results, only three independent variables are selected to predict price using the stepwise regression and $\alpha = .05$. These variables are DOT estimate, Status, and Days.

A new model was fit with just the variables DOT estimate, Status, and Days using SAS. The results are:

```
                         The REG Procedure
                          Model: MODEL1
                     Dependent Variable: PRICE

                        Analysis of Variance

                                  Sum of           Mean
Source               DF          Squares         Square    F Value    Pr > F

Model                 3        688228730      229409577    3134.76    <.0001
Error               213         15587848          73182
Corrected Total     216        703816578


            Root MSE              270.52243    R-Square    0.9779
            Dependent Mean       1126.66359    Adj R-Sq    0.9775
            Coeff Var              24.01093


                        Parameter Estimates

                        Parameter      Standard
Variable     DF          Estimate         Error    t Value    Pr > |t|

Intercept     1         -59.79987      33.93608      -1.76      0.0795
DOTEST        1           0.88616       0.01552      57.08      <.0001
STATUS        1         139.40952      41.19370       3.38      0.0008
DAYS          1           0.35761       0.16739       2.14      0.0338
```

We see that this is a fairly good model. The R-square value is .9779. 97.79% of the variation in the PRICE values is explained by the model containing DODEST, STATUS, and DAYS.

We note that the parameter estimate for the parameter associated Bid Status is 139.40952. Since Bid Status is either 1 (fixed) or 0 (competitive), the PRICE for the fixed bid is estimated to be 139.40952 units higher than the competitive bid, all other variables held constant.

To check the model assumptions, we will look at the plots. The plots of the residuals versus the independent variables are:

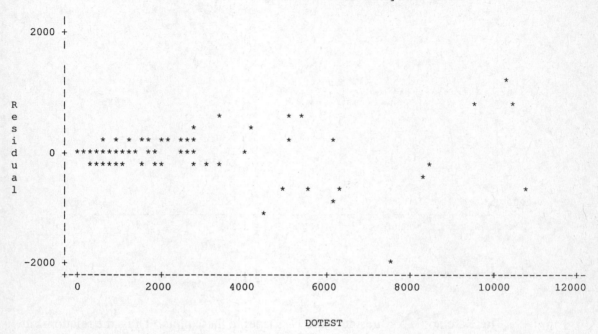

Plot of RES*DOTEST. Symbol used is '*'.

NOTE: 160 obs hidden.

There is no 'U' or 'upside-down U' shape to this plot

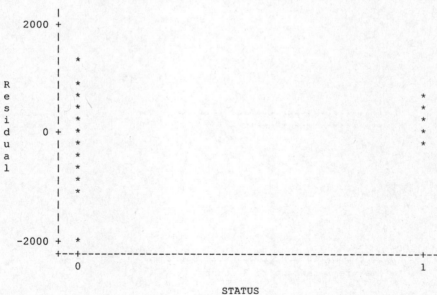

Plot of RES*STATUS. Symbol used is '*'.

NOTE: 200 obs hidden.

Plot of RES*DAYS. Symbol used is '*'.

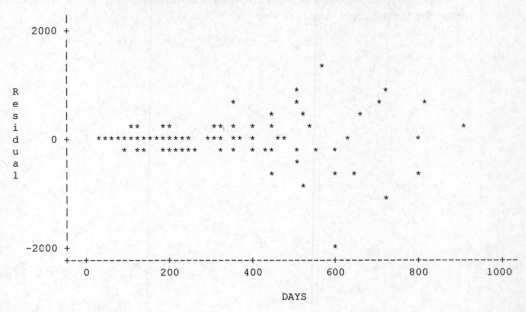

NOTE: 146 obs hidden.

There are no 'U' or 'upside-down U' shapes to these plots. Thus, the relationship between PRICE and the independent variables appear to be linear. There are a total of 6 observations that are more than 3 standard deviations from the mean. These 6 points could be outliers.

A histogram of the data is:

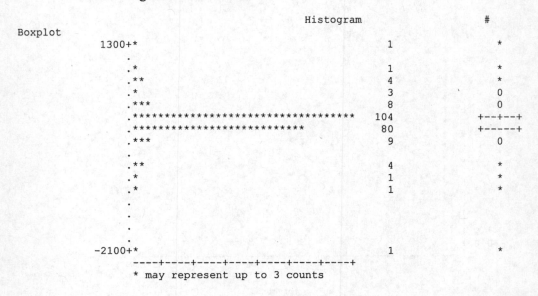

This appears to be fairly mound shaped, so the assumption that the data are normal appears to be OK.

The plot of the residuals against the fitted values is:

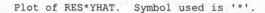

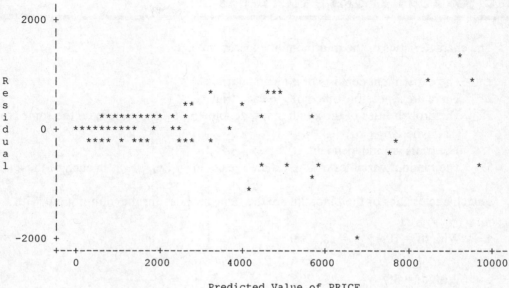

NOTE: 158 obs hidden.

The spread of the data points appears to increase as the predicted value of y increases. This indicates that there is non-constant variance.

Thus, even though the fit of this model appears to be fairly good ($R^2 = .9779$), some of the model assumptions are not met. The data do appear to be normally distributed. However, there appears that there are several (6) outliers and that the variance is not constant.

Additional models could be tried. One could look at interactions of the 3 variables included so far. When this is tried, the interaction term of DOTEST*STATUS is significant. However, the R-square value does not increase and the problems with the assumptions still exist.

Categorical Data Analysis

13.1   The characteristics of the multinomial experiment are:

1.   The experiment consists of n identical trials.
2.   There are k possible outcomes to each trial.
3.   The probabilities of the k outcomes, denoted $p_1, p_2, \ldots, p_k$, remain the same from trial to trial, where $p_1 + p_2 + \cdots + p_k = 1$.
4.   The trials are independent.
5.   The random variables of interest are the counts $n_1, n_2, \ldots, n_k$ in each of the k cells.

The characteristics of the binomial are the same as those for the multinomial with $k = 2$.

13.3   a.   With df $= 10$, $\chi^2_{.05} = 18.3070$

b.   With df $= 50$, $\chi^2_{.990} = 29.7067$

c.   With df $= 16$, $\chi^2_{.10} = 23.5418$

d.   With df $= 50$, $\chi^2_{.005} = 79.4900$

13.5   a.   The rejection region requires $\alpha = .05$ in the upper tail of the χ^2 distribution with df $= k - 1 = 3 - 1 = 2$. From Table VII, Appendix A, $\chi^2_{.05} = 5.99147$. The rejection region is $\chi^2 > 5.99147$.

b.   The rejection region requires $\alpha = .10$ in the upper tail of the χ^2 distribution with df $= k - 1 = 5 - 1 = 4$. From Table VII, Appendix A, $\chi^2_{.10} = 7.77944$. The rejection region is $\chi^2 > 7.77944$.

c.   The rejection region requires $\alpha = .01$ in the upper tail of the χ^2 distribution with df $= k - 1 = 4 - 1 = 3$. From Table VII, Appendix A, $\chi^2_{.10} = 11.3449$. The rejection region is $\chi^2 > 11.3449$.

13.7   Some preliminary calculations are:

If the probabilities are the same, $p_{1,0} = p_{2,0} = p_{3,0} = p_{4,0} = .25$

$$E_1 = np_{1,0} = 205(.25) = 51.25$$
$$E_2 = E_3 = E_4 = 205(.25) = 51.25$$

a.   To determine if the multinomial probabilities differ, we test:

H_0: $p_1 = p_2 = p_3 = p_4 = .25$
H_a: At least one of the probabilities differs from .25

The test statistic is $\chi^2 = \sum \dfrac{[n_i - E_i]^2}{E_i}$

$$= \frac{(43-51.25)^2}{51.25} + \frac{(56-51.25)^2}{51.25} + \frac{(59-51.25)^2}{51.25} + \frac{(47-51.25)^2}{51.25} = 3.293$$

The rejection region requires $\alpha = .05$ in the upper tail of the χ^2 distribution with df = k 1 = 4 − 1 = 3. From Table VII, Appendix A, $\chi^2_{.05} = 7.81473$. The rejection region is χ^2 > 7.81473.

Since the observed value of the test statistic does not fall in the rejection region ($\chi^2 = 3.293 \not> 7.81473$), H_0 is not rejected. There is insufficient evidence to indicate the multinomial probabilities differ at $\alpha = .05$.

b.   The Type I error is concluding the multinomial probabilities differ when, in fact, they do not.

The Type II error is concluding the multinomial probabilities are equal, when, in fact, they are not.

c.   For confidence coefficient .95, $\alpha = .05$ and $\alpha/2 = .05/2 = .025$. From Table IV, Appendix A, $z_{.025} = 1.96$.

$\hat{p}_3 = 59/205 = .288$

The confidence interval is:

$$\hat{p} \pm z_{.025}\sqrt{\frac{\hat{p}\hat{q}}{n}} \Rightarrow .288 \pm 1.96\sqrt{\frac{.288(.712)}{205}} \Rightarrow .288 \pm .062 \Rightarrow (.226, .350)$$

13.9   a.   The qualitative variable of interest is nonfunctional jaw habits. It has 4 levels: bruxism, teeth clenching, both bruxism and clenching, and neither.

b.   A one-way table of the data is:

	Cell			
	Bruxism	Clench	Bruxism & Clench	Neither
n_i	3	11	30	16

c.   To determine whether the percentages associated with the admitted habits differ, we test:

H_0: $p_1 = p_2 = p_3 = p_4 = .25$
H_a: At least one p_i differs from .25, $i = 1, 2, 3, 4$

d. $E_1 = E_2 = E_3 = E_4 = np_{i,0} = 60(.25) = 15$.

e. The test statistic is

$$\chi^2 = \sum \frac{[n_i - E_i]^2}{E_i} = \frac{[3-15]^2}{15} + \frac{[11-15]^2}{15} + \frac{[30-15]^2}{15} + \frac{[16-15]^2}{15} = 25.733$$

f. The rejection region requires $\alpha = .05$ in the upper tail of the χ^2 distribution with df = $k - 1 = 4 - 1 = 3$. From Table VII, Appendix A, $\chi^2_{.05} = 7.81473$. The rejection region is $\chi^2 > 7.81473$.

g. Since the observed value of the test statistic falls in the rejection region $(\chi^2 = 25.733 > 7.81473)$, H_0 is rejected. There is sufficient evidence to indicate the percentages associated with the admitted habits differ at $\alpha = .05$.

h. $\hat{p}_3 = \frac{30}{60} = .5$

For confidence coefficient .95, $\alpha = .05$ and $\alpha/2 = .05/2 = .025$. From Table IV, Appendix A, $z_{.025} = 1.96$. The 95% confidence interval is:

$$\hat{p}_i \pm z_{.025} \sqrt{\frac{\hat{p}_i \hat{q}_i}{n}} \Rightarrow .5 \pm 1.96 \sqrt{\frac{.5(.5)}{60}} \Rightarrow .5 \pm .127 \Rightarrow (.373, \ .627)$$

We are 95% confident that the true proportion of dental patients who admit to both habits is between .373 and .627.

13.11 a. The qualitative variable of interest in this problem is the type of pottery found. There are 4 levels of the variable: burnished, monochrome, painted, and other.

b. If all 4 types of pottery occur with equal probabilities, the values of p_1, p_2, p_3, and p_4 are all .25.

c. To determine if one type of pottery is more likely to occur at the site than any other, we test:

H_0: $p_1 = p_2 = p_3 = p_4 = .25$
H_a: At least one $p_i \neq .25$ for $i = 1, 2, 3, 4$

d. Some preliminary calculations are:

$E_1 = E_2 = E_3 = E_4 = np_{i,0} = 837(.25) = 209.25$.

$$\chi^2 = \sum \frac{[n_i - E_i]^2}{E_i} = \frac{[133 - 209.25]^2}{209.25} + \frac{[460 - 209.25]^2}{209.25} + \frac{[183 - 209.25]^2}{209.25} + \frac{[61 - 209.25]^2}{209.25}$$

$$= 436.591$$

e. The p-value is $p = P(\chi^2 \geq 436.591)$. Using Table VII, Appendix A, with df $= k - 1$ $= 4 - 1 = 3$, $p = P(\chi^2 \geq 436.591) < .005$. Since the p-value is less than $\alpha = .10$, H_0 is rejected. There is sufficient evidence to indicate at least one type of pottery is more likely to occur at the site than another at $\alpha = .10$.

13.13 a.   Some preliminary calculations are:

$E_1 = np_{1,0} = 1,000(.50) = 500$ $E_2 = np_{2,0} = 1,000(.22) = 220$
$E_3 = np_{3,0} = 1,000(.11) = 110$ $E_4 = np_{4,0} = 1,000(.17) = 170$

To determine if the data disagree with the percentages reported by Nielson/NetRatings, we test:

H_0: $p_1 = .50$, $p_2 = .22$, $p_3 = .11$, and $p_4 = .17$
H_a: At least one p_i differs from it hypothesized value

The test statistic is

$$\chi^2 = \sum \frac{[n_i - E_i]^2}{E_i}$$

$$= \frac{[487 - 500]^2}{500} + \frac{[245 - 220]^2}{220} + \frac{[121 - 110]^2}{110} + \frac{[147 - 170]^2}{170} = 7.39$$

The rejection region requires $\alpha = .05$ in the upper tail of the χ^2 distribution with df $= k - 1 = 4 - 1 = 3$. From Table VII, Appendix A, $\chi^2_{.05} = 7.81473$. The rejection region is $\chi^2 > 7.81473$.

Since the observed value of the test statistic falls in the rejection region ($\chi^2 = 7.39 > 7.81473$), H_0 is rejected. There is sufficient evidence to indicate the data disagree with the percentages reported by Nielson/NetRatings at $\alpha = .05$.

b. $\hat{p}_1 = \dfrac{487}{1000} = .487$

For confidence coefficient .95, $\alpha = .05$ and $\alpha/2 = .05/2 = .025$. From Table IV, Appendix A, $z_{.025} = 1.96$. The 95% confidence interval is:

$$\hat{p}_1 \pm z_{.025}\sqrt{\frac{\hat{p}_1\hat{q}_1}{n}} \Rightarrow .487 \pm 1.96\sqrt{\frac{.487(.513)}{1000}} \Rightarrow .487 \pm .031 \Rightarrow (.456,\ .518)$$

We are 95% confident that the true proportion of all Internet searches that use the Google Search engine is between .456 and .518.

13.15 a.   There were 1,470 responses that were missing. In addition, 14 responses were 8 = Don't know and 7 responses were 9 = Missing. Those responding with 8 or 9 were not included. The frequency table is:

Response	Frequency
1	450
2	627
3	219
4	23
Totals	1319

b. To determine if the true proportions in each category are equal, we test:

H_0: $p_1 = p_2 = p_3 = p_4 = .25$
H_a: At least one p_i differs

c. $E_1 = E_2 = E_3 = E_4 = np_{i,0} = 1,319(.25) = 329.75$

d. The test statistic is

$$\chi^2 = \sum \frac{[n_i - E_i]^2}{E_i}$$
$$= \frac{[450 - 329.75]^2}{329.75} + \frac{[627 - 329.75]^2}{329.75} + \frac{[219 - 329.75]^2}{329.75} + \frac{[23 - 329.75]^2}{329.75} = 634.36$$

e. The rejection region requires $\alpha = .10$ in the upper tail of the χ^2 distribution with df = $k - 1 = 4 - 1 = 3$. From Table VII, Appendix A, $\chi^2_{.10} = 6.25139$. The rejection region is $\chi^2 > 6.25139$.

    Since the observed value of the test statistic falls in the rejection region ($\chi^2 = 634.36 > 6.25139$), H_0 is rejected. There is sufficient evidence to indicate at least one proportion differs at $\alpha = .10$.

f. To determine if the true proportions follow the proportions given, we test:

H_0: $p_1 = .30, p_2 = .50, p_3 = .15, p_4 = .05$
H_a: At least one p_i differs from its hypothesized value

$E_1 = np_{1,0} = 1,319(.30) = 395.7$ $E_2 = np_{2,0} = 1,319(.50) = 659.5$
$E_3 = np_{3,0} = 1,319(.15) = 197.85$ $E_4 = np_{4,0} = 1,319(.05) = 65.95$

The test statistic is

$$\chi^2 = \sum \frac{[n_i - E_i]^2}{E_i}$$
$$= \frac{[450 - 395.7]^2}{395.7} + \frac{[627 - 659.5]^2}{659.5} + \frac{[219 - 197.85]^2}{197.85} + \frac{[23 - 65.95]^2}{65.95} = 39.29$$

The rejection region requires $\alpha = .10$ in the upper tail of the χ^2 distribution with df = $k - 1 = 4 - 1 = 3$. From Table VII, Appendix A, $\chi^2_{.10} = 6.25139$. The rejection region is $\chi^2 > 6.25139$.

Since the observed value of the test statistic falls in the rejection region ($\chi^2 = 39.29 > 6.25139$), H_0 is rejected. There is sufficient evidence to indicate at least one proportion differs from its hypothesized value at $\alpha = .10$.

13.17 a. Some preliminary calculations are:

$$E_A = np_{A,0} = 700(.09) = 63 \qquad E_B = np_{B,0} = 700(.02) = 14$$
$$E_C = np_{C,0} = 700(.02) = 14 \qquad E_D = np_{D,0} = 700(.04) = 28$$
$$E_E = np_{E,0} = 700(.04) = 28 \qquad E_F = np_{F,0} = 700(.02) = 14$$
$$E_G = np_{G,0} = 700(.02) = 14 \qquad E_H = np_{H,0} = 700(.02) = 14$$
$$E_I = np_{I,0} = 700(.02) = 14 \qquad E_J = np_{J,0} = 700(.01) = 7$$
$$E_K = np_{K,0} = 700(.01) = 7 \qquad E_L = np_{L,0} = 700(.04) = 28$$
$$E_M = np_{M,0} = 700(.02) = 14 \qquad E_N = np_{N,0} = 700(.06) = 42$$
$$E_O = np_{O,0} = 700(.08) = 56 \qquad E_P = np_{P,0} = 700(.02) = 14$$
$$E_Q = np_{Q,0} = 700(.01) = 7 \qquad E_R = np_{R,0} = 700(.06) = 42$$
$$E_S = np_{S,0} = 700(.04) = 28 \qquad E_T = np_{T,0} = 700(.06) = 42$$
$$E_U = np_{U,0} = 700(.04) = 28 \qquad E_V = np_{V,0} = 700(.02) = 14$$
$$E_W = np_{W,0} = 700(.02) = 14 \qquad E_X = np_{X,0} = 700(.01) = 7$$
$$E_Y = np_{Y,0} = 700(.02) = 14 \qquad E_Z = np_{Z,0} = 700(.01) = 7$$
$$E_{BLANK} = np_{BL,0} = 700(.02) = 14$$

To determine if the ScrabbleExpress™ presents the player with "unfair word selection opportunities" that are different from the Scrabble™ board game, we test:

H_0: $p_A = .09, p_B = .02, p_C = .02, p_D = .04, p_E = .12, p_F = .02, p_G = .03, p_H = .02, p_I = .09,$
$\quad p_J = .01, p_K = .01, p_L = .04, p_M = .02, p_N = .06, p_O = .08, p_P = .02, p_Q = .01, p_R = .06,$
$\quad p_S = .04, p_T = .06, p_U = .04, p_V = .02, p_W = .02, p_X = .01, p_Y = .02, p_Z = .01,$
$\quad p_{BLANK} = .02$
H_a: At least one p_i differs from its hypothesized value, $i = A, B, ..., BLANK$

The test statistic is

$$\chi^2 = \sum \frac{[n_i - E_i]^2}{E_i} = \frac{[36-63]^2}{63} + \frac{[18-14]^2}{14} + \frac{[30-14]^2}{14} + \cdots + \frac{[34-14]^2}{14} = 360.48$$

The rejection region requires $\alpha = .05$ in the upper tail of the χ^2 distribution with df $= k - 1 = 27 - 1 = 26$. From Table VII, Appendix A, $\chi^2_{.05} = 38.8852$. The rejection region is $\chi^2 > 38.8852$.

Since the observed value of the test statistic falls in the rejection region ($\chi^2 = 360.48 > 38.8852$), H_0 is rejected. There is sufficient evidence to indicate the ScrabbleExpress™ presents the player with "unfair word selection opportunities" that are different from the Scrabble™ board game at $\alpha = .05$.

b. The true proportion of letters drawn from the board game is

$$p_A + p_E + p_I + p_O + p_U = .09 + .12 + .09 + .08 + .04 = .42$$

The number of vowels drawn by the electronic game is $39 + 31 + 25 + 20 + 21 = 136$. The proportion of vowels drawn is $136 / 700 = .194$.

For confidence coefficient .95, $\alpha = .05$ and $\alpha/2 = .05/2 = .025$. From Table IV, Appendix A, $z_{.025} = 1.96$. The 95% confidence interval is:

$$\hat{p}_i \pm z_{.025}\sqrt{\frac{\hat{p}_i\hat{q}_i}{n}} \Rightarrow .194 \pm 1.96\sqrt{\frac{.194(.806)}{700}} \Rightarrow .194 \pm .029 \Rightarrow (.165, \ .223)$$

We are 95% confident that the true proportion of vowels drawn by the electronic game is between .165 and .223. The actual proportion of vowels drawn in the board game is .42. Since this value is not contained in the 95% confidence interval, one can conclude that the ScrabbleExpress™ game produces too few vowels in the 7-letter draws.

13.19   A two-way contingency table presents multinomial count data classified on two scales of classification.

13.21   The conditions that are required for a valid chi-square test of data from a contingency table are:

1.   The n observed counts are a random sample from the population of interest.
2.   The sample size, n, will be large enough so that, for every cell, the expected count, E_{ij}, will be equal to 5 or more.

13.23   a. H_0:   The row and column classifications are independent
             H_a: The row and column classifications are dependent

        b.   The test statistic is $\chi^2 = \sum\sum \dfrac{[n_{ij} - E_{ij}]^2}{E_{ij}}$

             The rejection region requires $\alpha = .01$ in the upper tail of the χ^2 distribution with df = $(r-1)(c-1) = (2-1)(3-1) = 2$. From Table VII, Appendix A, $\chi^2_{.01} = 9.21034$. The rejection region is $\chi^2 > 9.21034$.

        c.   The expected cell counts are:

             $E_{11} = \dfrac{R_1C_1}{n} = \dfrac{96(25)}{167} = 14.37$ $E_{21} = \dfrac{R_2C_1}{n} = \dfrac{71(25)}{167} = 10.63$

             $E_{12} = \dfrac{R_1C_2}{n} = \dfrac{96(64)}{167} = 36.79$ $E_{22} = \dfrac{R_2C_2}{n} = \dfrac{71(64)}{167} = 27.21$

             $E_{13} = \dfrac{R_1C_3}{n} = \dfrac{96(78)}{167} = 44.84$ $E_{23} = \dfrac{R_2C_3}{n} = \dfrac{71(78)}{167} = 33.16$

        d.   The test statistic is $\chi^2 = \sum\sum \dfrac{[n_{ij} - E_{ij}]^2}{E_{ij}}$

             $$= \frac{(9-14.37)^2}{14.37} + \frac{(34-36.79)^2}{36.79} + \frac{(53-44.84)^2}{44.84} + \frac{(16-10.63)^2}{10.63}$$

             $$+ \frac{(30-27.21)^2}{27.21} + \frac{(25-33.16)^2}{33.16} = 8.71$$

        Since the observed value of the test statistic does not fall in the rejection region ($\chi^2 = 8.71 \not> 9.21034$), H_0 is not rejected. There is insufficient evidence to indicate the row and column classifications are dependent at $\alpha = .01$.

13.25 Some preliminary calculations are:

$$E_{11} = \frac{R_1 C_1}{n} = \frac{154(134)}{439} = 47.007 \qquad E_{21} = \frac{186(134)}{439} = 56.774$$

$$E_{12} = \frac{154(163)}{439} = 57.180 \qquad E_{22} = \frac{186(163)}{439} = 69.062$$

$$E_{13} = \frac{154(142)}{439} = 49.813 \qquad E_{23} = \frac{186(142)}{439} = 60.164$$

$$E_{31} = \frac{99(134)}{439} = 30.219 \qquad E_{33} = \frac{99(142)}{439} = 32.023$$

$$E_{32} = \frac{99(163)}{439} = 36.759$$

To determine if the row and column classifications are dependent, we test:

H_0: The row and column classifications are independent
H_a: The row and column classifications are dependent

The test statistic is $\chi^2 = \sum\sum \dfrac{[n_{ij} - E_{ij}]^2}{E_{ij}}$

$$= \frac{(40-47.007)^2}{47.007} + \frac{(72-57.180)^2}{57.180} + \frac{(42-49.813)^2}{49.813} + \frac{(63-56.774)^2}{56.774}$$

$$+ \frac{(53-69.062)^2}{69.062} + \frac{(70-60.164)^2}{60.164} + \frac{(31-30.219)^2}{30.219}$$

$$+ \frac{(38-36.759)^2}{36.759} + \frac{(30-32.023)^2}{32.023} = 12.36$$

The rejection region requires $\alpha = .05$ in the upper tail of the χ^2 distribution with df = $(r-1)(c-1) = (3-1)(3-1) = 4$. From Table VII, Appendix A, $\chi^2_{.05} = 9.48773$. The rejection region is $\chi^2 > 9.48773$.

Since the observed value of the test statistic falls in the rejection region ($\chi^2 = 12.36 > 9.48773$), H_0 is rejected. There is sufficient evidence to indicate the row and column classification are dependent at $\alpha = .05$.

13.27. a. $\hat{p}_1 = \dfrac{6}{77} = .078$

b. $\hat{p}_2 = \dfrac{11}{70} = .157$

c. The proportion of girls drawing a dog is almost 2 times the proportion of boys drawing a dog.

d. To determine if the likelihood of drawing a dog depends on gender, we test:

H_0: Presence of a dog and gender are independent
H_a: Presence of a dog and gender are dependent

e.   From the printout, the test statistic is $\chi^2 = 2.250$ and the p-value is $p = 0.134$.

Since the p-value is not less than $\alpha = .05$ ($p = .134 \not< .05$), H_0 is not rejected. There is insufficient evidence to indicate the likelihood of drawing a dog depends on gender at $\alpha = .05$.

f.   Using MINITAB, the results are:

Tabulated statistics: TV, GENDER

```
Using frequencies in NUMBER


Rows: TV    Columns: GENDER

           Boy    Girl     All

No          66      61     127
          66.52   60.48  127.00

Yes         11       9      20
          10.48    9.52   20.00

All         77      70     147
          77.00   70.00  147.00

Cell Contents:          Count
                        Expected count


Pearson Chi-Square = 0.064, DF = 1, P-Value = 0.801
Likelihood Ratio Chi-Square = 0.064, DF = 1, P-Value = 0.801
```

To determine if the likelihood of drawing a TV in the bedroom depends on gender, we test:

H_0: Presence of TV and gender are independent
H_a: Presence of TV and gender are dependent

From the printout, the test statistic is $\chi^2 = .064$ and the p-value is $p = 0.801$.

Since the p-value is not less than $\alpha = .05$ ($p = .801 \not< .05$), H_0 is not rejected. There is insufficient evidence to indicate the likelihood of drawing TV depends on gender at $\alpha = .05$.

13.29 a.   The two categorical variables are Risk of masculinity and type of event. Risk of masculinity has two levels: High and Low. Type of event has two levels: Violent event and Avoided-Violent event.

b.   The experimental units are 1,507 newly incarcerated men.

c. $E_{11} = \dfrac{R_1 C_1}{n} = \dfrac{379(1,037)}{1,507} = 260.798$

d.　$E_{12} = \dfrac{R_1 C_2}{n} = \dfrac{379(470)}{1,507} = 118.202$　　　$E_{21} = \dfrac{R_2 C_1}{n} = \dfrac{1,128(1,037)}{1,507} = 776.202$

　　$E_{22} = \dfrac{R_2 C_2}{n} = \dfrac{1,128(470)}{1,507} = 351.798$

e.　$\chi^2 = \sum\sum \dfrac{[n_{ij} - E_{ij}]^2}{E_{ij}} = \dfrac{[236 - 260.798]^2}{260.798} + \dfrac{[143 - 118.202]^2}{118.202}$

　　$+ \dfrac{[801 - 776.202]^2}{776.202} + \dfrac{[327 - 351.798]^2}{351.798} = 10.101$

f.　To determine whether event type depends on high/low risk masculinity, we test:

H_0:　Event Type and High/Low Risk Masculinity are independent
H_a:　Event Type and High/Low Risk Masculinity are dependent

The test statistic is $\chi^2 = 10.101$.

The rejection region requires $\alpha = .05$ in the upper tail of the χ^2 distribution with df $= (r - 1)(c - 1) = (2 - 1)(2 - 1) = 1$. From Table VII, Appendix A, $\chi^2_{.05} = 3.84146$. The rejection region is $\chi^2 > 3.84146$.

Since the observed value of the test statistic falls in the rejection region $(\chi^2 = 10.101 > 3.84146)$, H_0 is rejected. There is sufficient evidence that event type depends on high/low risk masculinity at $\alpha = .05$.

13.31.　Some preliminary calculations are:

$E_{11} = \dfrac{R_1 C_1}{n} = \dfrac{192(115)}{526} = 41.98$　　　$E_{12} = \dfrac{R_1 C_2}{n} = \dfrac{192(411)}{526} = 150.02$

$E_{21} = \dfrac{R_2 C_1}{n} = \dfrac{157(115)}{526} = 34.33$　　　$E_{22} = \dfrac{R_2 C_2}{n} = \dfrac{157(411)}{526} = 122.67$

$E_{31} = \dfrac{R_3 C_1}{n} = \dfrac{177(115)}{526} = 38.70$　　　$E_{32} = \dfrac{R_3 C_2}{n} = \dfrac{177(411)}{526} = 138.30$

To determine if the proportion of students diagnosed with MR depends on the IQ test/retest method, we test:

H_0:　MR diagnosis and Test/Retest are independent
H_a:　MR diagnosis and Test/Retest are dependent

The test statistic is

$\chi^2 = \sum\sum \dfrac{[n_{ij} - E_{ij}]^2}{E_{ij}} = \dfrac{[25 - 41.98]^2}{41.98} + \dfrac{[54 - 34.33]^2}{34.33} + \dfrac{[36 - 38.70]^2}{38.70} + \dfrac{[167 - 150.02]^2}{150.02}$

$+ \dfrac{[103 - 122.67]^2}{122.67} + \dfrac{[141 - 138.30]^2}{138.30} = 23.46$

The rejection region requires $\alpha = .01$ in the upper tail of the χ^2 distribution with df $= (r-1)(c-1) = (3-1)(2-1) = 2$. From Table VII, Appendix A, $\chi^2_{.01} = 9.21034$. The rejection region is $\chi^2 > 9.21034$.

Since the observed value of the test statistic falls in the rejection region ($\chi^2 = 23.46 > 9.21034$), H_0 is rejected. There is sufficient evidence to indicate the proportion of students diagnosed with MR depends on the IQ test/retest method at $\alpha = .01$.

13.33   Some preliminary calculations are:

$$E_{11} = \frac{R_1 C_1}{n} = \frac{62(106)}{275} = 23.90 \qquad E_{12} = \frac{R_1 C_2}{n} = \frac{62(112)}{275} = 25.25$$

$$E_{13} = \frac{R_1 C_3}{n} = \frac{62(57)}{275} = 12.85 \qquad E_{21} = \frac{R_2 C_1}{n} = \frac{59(106)}{275} = 22.74$$

$$E_{22} = \frac{R_2 C_2}{n} = \frac{59(112)}{275} = 24.03 \qquad E_{23} = \frac{R_2 C_3}{n} = \frac{59(57)}{275} = 12.23$$

$$E_{31} = \frac{R_3 C_1}{n} = \frac{59(106)}{275} = 22.74 \qquad E_{32} = \frac{R_3 C_2}{n} = \frac{59(112)}{275} = 24.03$$

$$E_{33} = \frac{R_3 C_3}{n} = \frac{59(57)}{275} = 12.23 \qquad E_{41} = \frac{R_4 C_1}{n} = \frac{95(106)}{275} = 36.62$$

$$E_{42} = \frac{R_4 C_2}{n} = \frac{95(112)}{275} = 38.69 \qquad E_{43} = \frac{R_4 C_3}{n} = \frac{95(57)}{275} = 19.69$$

To determine if political strategy of ethnic groups depends on world region, we test:

H_0: Political strategy and world region are independent
H_a: Political strategy and world region are dependent

The test statistic is $\chi^2 = \sum\sum \dfrac{[n_{ij} - E_{ij}]^2}{E_{ij}} = \dfrac{(24-23.90)^2}{23.90} + \dfrac{(31-25.25)^2}{25.25}$

$+ \dfrac{(7-12.85)^2}{12.85} + \dfrac{(32-22.74)^2}{22.74} + \dfrac{(23-24.03)^2}{24.03} + \dfrac{(4-12.23)^2}{12.23} + \dfrac{(11-22.74)^2}{22.74}$

$+ \dfrac{(22-24.03)^2}{24.03} + \dfrac{(26-12.23)^2}{12.23} + \dfrac{(39-36.62)^2}{36.62} + \dfrac{(36-38.69)^2}{38.69} + \dfrac{(20-19.69)^2}{19.69}$

$= 35.409$

The rejection region requires $\alpha = .10$ in the upper tail of the χ^2 distribution with df $= (r-1)(c-1) = (4-1)(3-1) = 6$. From Table VII, Appendix A, $\chi^2_{.10} = 10.6446$. The rejection region is $\chi^2 > 10.6446$.

Since the observed value of the test statistic falls in the rejection region ($\chi^2 = 35.409 >$ 10.6446), H_0 is rejected. There is sufficient evidence to indicate that the political strategy of the ethnic groups depends on world region at $\alpha = .10$.

To graph the data, we will first compute the percent of observations in each category of Political Strategy for each World Region. To do this, we divide each cell frequency by the row total and then multiply by 100%. Using SAS, a graph of the data is:

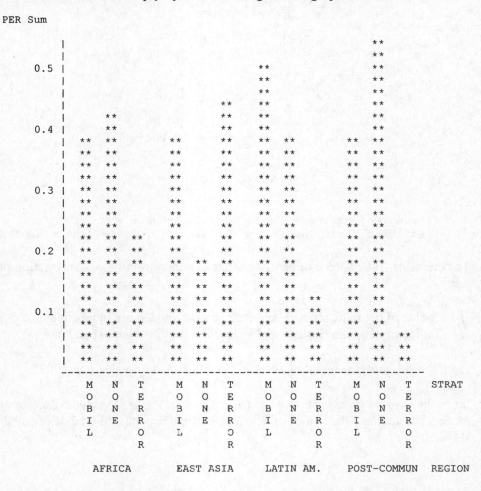

The graph supports the results of the test above. The percents in each level of Political Strategy differs for the different values of World Region.

13.35 Using MINITAB, the results of the analyses are:

Results for: AIRTHR~1.MTP

Tabulated statistics: Instruction, Strategy

```
Rows: Instruction    Columns: Strategy

              Guess     Other     TTBC      All

Cue               5         6        13       24
              7.000     8.500     8.500   24.000

Pattern           9        11         4       24
              7.000     8.500     8.500   24.000

All              14        17        17       48
             14.000    17.000    17.000   48.000

Cell Contents:          Count
                        Expected count


Pearson Chi-Square = 7.378, DF = 2, P-Value = 0.025
Likelihood Ratio Chi-Square = 7.668, DF = 2, P-Value = 0.022
```

To determine if the choice of heuristic strategy depends on type of instruction provided, we test:

H_0: Choice of heuristic strategy and instruction provided are independent
H_0: Choice of heuristic strategy and instruction provided are dependent

From the printout, the test statistic is $\chi^2 = 7.378$ and the p-value is $p = .025$.

Since the p-value is less than $\alpha = .05$ ($p = .025 < .05$), H_0 is rejected. There is sufficient evidence to indicate the choice of heuristic strategy depends on type of instruction provided at $\alpha = .05$.

13.37   a.   Using MINITAB, the results of the analyses are:

Tabulated statistics: SHSS, CAHS

```
Using frequencies in Fr

Rows: SHSS    Columns: CAHS

            H       L       M       V      All

H          19       6      14       3       42
        10.02   15.83   14.22    1.94    42.00

L           2      32      14       0       48
        11.45   18.09   16.25    2.22    48.00

M           6      11      14       0       31
         7.39   11.68   10.49    1.43    31.00

V           4       0       2       3        9
         2.15    3.39    3.05    0.42     9.00

All        31      49      44       6      130
        31.00   49.00   44.00    6.00   130.00

Cell Contents:       Count
                     Expected count


Pearson Chi-Square = 60.103, DF = 9
Likelihood Ratio Chi-Square = 59.638, DF = 9

* WARNING * 1 cells with expected counts less than 1
* WARNING * Chi-Square approximation probably invalid

* NOTE * 7 cells with expected counts less than 5
```

From the printout, 7 cells have expected counts less than 5. In order for the test to be valid, all of the cells should have expected counts greater than 5. Thus, we should not proceed with the analysis.

b.   Combining the High and Very High categories, the new table is:

		CAHS LEVEL		
		Low	**Medium**	**High/Very High**
SHSS: C LEVEL	**Low**	32	14	2
	Medium	11	14	6
	High/Very High	6	16	29

c. The expected cell counts are:

$$E_{11} = \frac{R_1C_1}{n} = \frac{48(49)}{130} = 18.09 \qquad E_{21} = \frac{R_2C_1}{n} = \frac{31(49)}{130} = 11.68$$

$$E_{12} = \frac{R_1C_2}{n} = \frac{48(44)}{130} = 16.25 \qquad E_{22} = \frac{R_2C_2}{n} = \frac{31(44)}{130} = 10.49$$

$$E_{13} = \frac{R_1C_3}{n} = \frac{48(37)}{130} = 13.66 \qquad E_{23} = \frac{R_2C_3}{n} = \frac{31(37)}{130} = 8.82$$

$$E_{31} = \frac{R_3C_1}{n} = \frac{51(49)}{130} = 19.22 \qquad E_{32} = \frac{R_3C_2}{n} = \frac{51(44)}{130} = 17.26$$

$$E_{33} = \frac{R_3C_3}{n} = \frac{51(37)}{130} = 14.52$$

Since all of the expected cell counts are now 5 or greater, the assumption is met.

d. To determine whether CAHS Levels and SHSS:C Levels are dependent, we test:

H_0: CAHS Levels and SHSS:C Levels are independent
H_a: CAHS Levels and SHSS:C Levels are dependent

The test statistic is $\chi^2 = \sum\sum \frac{[n_{ij} - E_{ij}]^2}{E_{ij}}$

$$= \frac{(32-18.09)^2}{18.09} + \frac{(14-16.25)^2}{16.25} + \frac{(2-13.66)^2}{13.66} + \frac{(11-11.68)^2}{11.68}$$

$$+ \frac{(14-10.49)^2}{10.49} + \frac{(6-8.82)^2}{8.82} + \frac{(6-19.22)^2}{19.22} + \frac{(16-17.26)^2}{17.26}$$

$$+ \frac{(29-14.52)^2}{14.52} = 46.70$$

The rejection region requires $\alpha = .05$ in the upper tail of the χ^2 distribution with df $=$ $(r-1)(c-1) = (3-1)(3-1) = 4$. From Table VII, Appendix A, $\chi^2_{.05} = 9.48773$. The rejection region is $\chi^2 > 9.48773$.

Since the observed value of the test statistic falls in the rejection region ($\chi^2 = 46.70 > 9.48773$), H_0 is rejected. There is sufficient evidence to indicate that CAHS Levels and SHSS:C Levels are dependent at $\alpha = .05$.

13.39 a. Some preliminary calculations are:

$$E_{11} = \frac{R_1C_1}{n} = \frac{31(24)}{38} = 19.58 \qquad E_{12} = \frac{R_1C_2}{n} = \frac{31(14)}{38} = 11.42$$

$$E_{21} = \frac{R_2C_1}{n} = \frac{7(24)}{38} = 4.42 \qquad E_{22} = \frac{R_2C_2}{n} = \frac{7(14)}{38} = 2.58$$

To determine whether the vaccine is effective in treating the MN strain of HIV, we test:

H_0: Vaccine and MN strain are independent
H_a: Vaccine and MN strain are dependent

The test statistic is $\chi^2 = \sum\sum \dfrac{[n_{ij} - E_{ij}]^2}{E_{ij}} = \dfrac{(22-19.58)^2}{19.58} + \dfrac{(9-11.42)^2}{11.42}$

$$+ \dfrac{(2-4.42)^2}{4.42} + \dfrac{(5-2.58)^2}{2.58} = 4.407$$

The rejection region requires $\alpha = .05$ in the upper tail of the χ^2 distribution with df = $(r-1)(c-1) = (2-1)(2-1) = 1$. From Table VII, Appendix A, $\chi^2_{.05} = 3.84146$. The rejection region is $\chi^2 > 3.84146$.

Since the observed value of the test statistic falls in the rejection region ($\chi^2 = 4.407 > 3.84146$), H_0 is rejected. There is sufficient evidence to indicate that the vaccine is effective in treating the MN strain of HIV at $\alpha = .05$.

b.   The necessary assumptions are:

1. The n observed counts are a random sample from the population of interest.

2. The sample size, n, will be large enough so that, for every cell, the expected count, $E(n_{ij})$, will be equal to 5 or more.

For this example, the second assumption is violated. Two of the expected cell counts are less than 5. What we have computed for the test statistic may not have a χ^2 distribution.

c.   For this contingency table:

$$p = \dfrac{\dbinom{7}{2}\dbinom{31}{22}}{\dbinom{38}{24}} = \dfrac{\dfrac{7!}{2!5!}\dfrac{31!}{22!9!}}{\dfrac{38!}{24!14!}} = .0438$$

d.   If vaccine and MN strain are independent, then the proportion of positive results should be relatively the same for both patient groups. In the two tables presented, the proportion of positive results for the vaccinated group is smaller than the proportion for the original table.

For the first table,

$$p = \dfrac{\dbinom{7}{1}\dbinom{31}{23}}{\dbinom{38}{24}} = \dfrac{\dfrac{7!}{1!6!}\dfrac{31!}{23!8!}}{\dfrac{38!}{24!14!}} = .0057$$

For the second table:

$$p = \frac{\binom{7}{0}\binom{31}{24}}{\binom{38}{24}} = \frac{\dfrac{7!}{0!7!}\dfrac{31!}{24!7!}}{\dfrac{38!}{24!14!}} = .0003$$

e. The p-value for Fisher's exact test is $p = .0438 + .0057 + .0003 = .0498$. Since the p-value is small, there is evidence to reject H_0. There is sufficient evidence to indicate that the vaccine is effective in treating the MN strain of HIV at $\alpha > .0498$.

13.41 The statement "Rejecting the null hypothesis in a chi-square test for independence implies that a causal relationship between the two categorical variables exists." is false. If the variables are dependent (reject H_0), it simply implies that they are related. One cannot infer causal relationships from contingency table analysis.

13.43 a. Some preliminary calculations are:

$$E_{11} = \frac{R_1 C_1}{n} = \frac{50(50)}{250} = 10 \qquad\qquad E_{21} = \frac{R_2 C_1}{n} = \frac{100(50)}{250} = 20$$

$$E_{12} = \frac{R_1 C_2}{n} = \frac{50(90)}{250} = 18 \qquad\qquad E_{22} = \frac{R_2 C_2}{n} = \frac{100(90)}{250} = 36$$

$$E_{13} = \frac{R_1 C_3}{n} = \frac{50(110)}{250} = 22 \qquad\qquad E_{23} = \frac{R_2 C_3}{n} = \frac{100(100)}{250} = 44$$

$$E_{31} = \frac{R_3 C_1}{n} = \frac{100(50)}{250} = 20$$

$$E_{32} = \frac{R_3 C_2}{n} = \frac{100(90)}{250} = 36$$

$$E_{33} = \frac{R_3 C_3}{n} = \frac{100(110)}{250} = 44$$

To determine if the rows and columns are dependent, we test:

H_0: Rows and columns are independent
H_a: Rows and columns are dependent

The test statistic is $\chi^2 = \sum\sum \dfrac{[n_{ij} - E_{ij}]^2}{E_{ij}} = \dfrac{(20-10)^2}{10} + \cdots + \dfrac{(30-44)^2}{44} = 54.14$

The rejection region requires $\alpha = .05$ in the upper tail of the χ^2 distribution with df = $(r-1)(c-1) = (3-1)(3-1) = 4$. From Table VII, Appendix A, $\chi^2_{.05} = 9.48773$. The rejection region is $\chi^2 > 9.48773$.

Since the observed value of the test statistic falls in the rejection region ($\chi^2 = 54.14 > 9.48773$), H_0 is rejected. There is sufficient evidence to indicate a dependence between rows and columns at $\alpha = .05$.

b. No, the analysis remains identical.

c. Yes, the assumptions on the sampling differ.

d. The percentages are in the table below.

	Column			
	1	**2**	**3**	**Totals**
1	$\frac{20}{50} \times 100\% = 40\%$	$\frac{20}{90} \times 100\% = 22.2\%$	$\frac{10}{110} \times 100\% = 9.1\%$	$\frac{50}{250} \times 100\% = 20\%$
Row 2	$\frac{10}{50} \times 100\% = 20\%$	$\frac{20}{90} \times 100\% = 22.2\%$	$\frac{70}{110} \times 100\% = 63.6\%$	$\frac{100}{250} \times 100\% = 40\%$
3	$\frac{20}{50} \times 100\% = 40\%$	$\frac{50}{90} \times 100\% = 55.6\%$	$\frac{30}{110} \times 100\% = 37.3\%$	$\frac{100}{250} \times 100\% = 40\%$

e. The graph is:

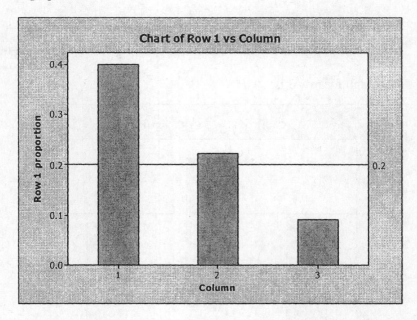

If the rows and columns are independent, then the proportion of observations for each column should be relatively the same. In this graph, the percentages are quite different, supporting the results of the test in part **a**.

f.   The graph for row 2 is:

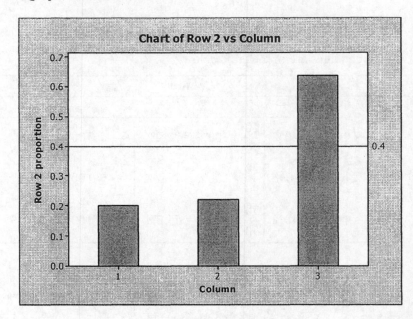

The percentages are quite different, supporting the results of the test in part **a**.

g.   The graph for row 3 is:

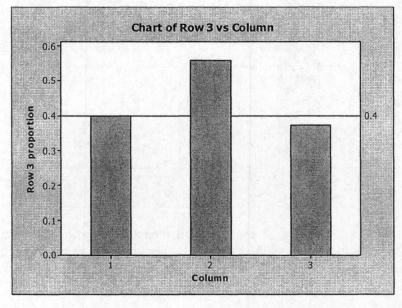

The percentages are quite different, supporting the results of the test in part **a**.

13.45   Some preliminary calculations are:

$$E_1 = E_2 = E_3 = E_4 = E_5 = np_{i,0} = 95(.20) = 19$$

To determine if the true percentages of ADEs in the 5 "cause" categories differ, we test:

H_0: $p_1 = p_2 = p_3 = p_4 = p_5 = .2$
H_a: At least one p_i differs from .2, $i = 1, 2, 3, 4, 5$

The test statistic is

$$\chi^2 = \sum \frac{[n_i - E_i]^2}{E_i} = \frac{[29-19]^2}{19} + \frac{[17-19]^2}{19} + \frac{[13-19]^2}{19} + \frac{[9-19]^2}{19} + \frac{[27-19]^2}{19} = 16$$

Since no α was given, we will use $\alpha = .05$. The rejection region requires $\alpha = .05$ in the upper tail of the χ^2 distribution with df $= k - 1 = 5 - 1 = 4$. From Table VII, Appendix A, $\chi^2_{.05} = 9.48773$. The rejection region is $\chi^2 > 9.48773$.

Since the observed value of the test statistic falls in the rejection region ($\chi^2 = 16 > 9.48773$), H_0 is rejected. There is sufficient evidence to indicate the true percentages of ADEs in the 5 "cause" categories differ at $\alpha = .05$.

13.47 a. The 2×2 contingency table is:

	LLD		
Genetic Trait	Yes	No	**Total**
Yes	21	15	36
No	150	306	456
Total	171	321	492

b. Some preliminary calculations are:

$$E_{11} = \frac{R_1 C_1}{n} = \frac{36(171)}{492} = 12.51 \qquad E_{12} = \frac{R_1 C_2}{n} = \frac{36(321)}{492} = 23.49$$

$$E_{21} = \frac{R_2 C_1}{n} = \frac{456(171)}{492} = 158.49 \qquad E_{22} = \frac{R_2 C_2}{n} = \frac{456(321)}{492} = 297.51$$

To determine if the genetic trait occurs at a higher rate in LDD patients than in the controls, we test:

H_0: Genetic trait and LDD group are independent
H_a: Genetic trait and LDD group are dependent

The test statistic is $\chi^2 = \sum\sum \frac{[n_{ij} - E_{ij}]^2}{E_{ij}} = \frac{(21-12.51)^2}{12.51} + \frac{(15-23.49)}{23.49}$

$$+ \frac{(150-158.49)^2}{158.49} + \frac{(306-297.51)^2}{297.51} = 9.52$$

The rejection region requires $\alpha = .01$ in the upper tail of the χ^2 distribution with df $= (r-1)(c-1) = (2-1)(2-1) = 1$. From Table VII, Appendix A, $\chi^2_{.01} = 6.63490$. The rejection region is $\chi^2 > 6.63491$.

Since the observed value of the test statistic falls in the rejection region ($\chi^2 = 9.52 > 6.63491$), H_0 is rejected. There is sufficient evidence to indicate that the genetic trait occurs at a higher rate in LDD patients than in the controls at $\alpha = .01$.

c. To construct a bar graph, we will first compute the proportion of LDD/Control patients that have the genetic trait. Of the LLD patients, $21 / 171 = .123$ have the trait. For the Controls, $15 / 321 = .043$. The bar graph is:

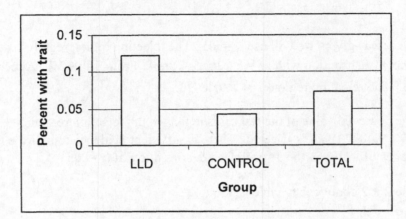

Because the bars are not close to being the same height, it indicates that the genetic trait appears at a higher rate among LLD patients than among the controls. This supports the conclusion of the test in part **b**.

13.49 a. No. In a multinomial experiment, each trial results in one of $k = 8$ possible outcomes. In this experiment, each trial cannot result in one of all of the eight possible outcomes. The people are assigned to a particular diet before the experiment. Once the people are assigned to a particular diet, there are only two possible outcomes, cancer tumors or no cancer tumors.

b. The expected cell counts are:

$$E_{11} = \frac{R_1 C_1}{n} = \frac{80(30)}{120} = 20 \quad E_{21} = \frac{R_2 C_1}{n} = \frac{40(30)}{120} = 10$$

$$E_{12} = \frac{R_1 C_2}{n} = \frac{80(30)}{120} = 20 \quad E_{22} = \frac{R_2 C_2}{n} = \frac{40(30)}{120} = 10$$

$$E_{13} = \frac{R_1 C_3}{n} = \frac{80(30)}{120} = 20 \quad E_{23} = \frac{R_2 C_3}{n} = \frac{40(30)}{120} = 10$$

$$E_{14} = \frac{R_1 C_4}{n} = \frac{80(30)}{120} = 20 \quad E_{24} = \frac{R_2 C_4}{n} = \frac{40(30)}{120} = 10$$

c. The test statistic is $\chi^2 = \sum \sum \frac{[n_{ij} - E_{ij}]^2}{E_{ij}}$

$$= \frac{(27-20)^2}{20} + \frac{(20-20)^2}{20} + \frac{(19-20)}{20} + \frac{(14-20)^2}{20} + \frac{(3-10)^2}{10} + \frac{(10-10)^2}{10}$$

$$+ \frac{(11-10)^2}{10} + \frac{(16-10)^2}{10} = 12.9$$

d. To determine if diet and presence/absence of cancer are independent, we test:

H_0: Diet and presence/absence of cancer are independent
H_a: Diet and presence/absence of cancer are dependent

The test statistic is $\chi^2 = 12.9$ (from part c).

The rejection region requires $\alpha = .05$ in the upper tail of the χ^2 distribution with df = $(r-1)(c-1) = (2-1)(4-1) = 3$. From Table VII, Appendix A, $\chi^2_{.05} = 7.81473$. The rejection region is $\chi^2 > 7.81473$.

Since the observed value of the test statistic falls in the rejection region ($\chi^2 = 12.9 > 7.81473$), H_0 is rejected. There is sufficient evidence to indicate that diet and presence/absence of cancer are dependent for $\alpha = .05$.

e. $\hat{p}_1 = 27/30 = .9$; $\hat{p}_2 = 20/30 = .667$

For confidence coefficient .95, $\alpha = .05$ and $\alpha/2 = .05/2 = .025$. From Table IV, Appendix A, $z_{.025} = 1.96$. The confidence interval is:

$$(\hat{p}_1 - \hat{p}_2) \pm z_{.025} \sqrt{\frac{\hat{p}_1 \hat{q}_1}{n_1} + \frac{\hat{p}_2 \hat{q}_2}{n_1}} \Rightarrow (.900 - .667) \pm 1.96 \sqrt{\frac{.900(.100)}{30} + \frac{.667(.333)}{30}}$$

$$\Rightarrow .233 \pm .200 \Rightarrow (.033, .433)$$

We are 95% confident that the difference in the percentage of rats on a high fat/no fiber diet with cancer and the percentage of rats on a high fat/fiber diet with cancer is between 3.3% and 43.3%.

13.51   a.   Some preliminary calculations:

$$E_1 = np_{1,0} = 28(.25) = 7$$
$$E_2 = np_{2,0} = 28(.25) = 7$$
$$E_3 = np_{3,0} = 28(.25) = 7$$
$$E_4 = np_{4,0} = 28(.25) = 7$$

To determine if the true proportions of intellectually disabled elderly patients in each of the hearing loss categories differ, we test:

H_0: $p_1 = p_2 = p_3 = p_4 = .25$
H_a: At least one p_i differs from .25, $i = 1, 2, 3, 4$

The test statistic is $\chi^2 = \sum \frac{[n_i - E_i]^2}{E_i} = \frac{(7-7)^2}{7} + \frac{(7-7)^2}{7} + \frac{(9-7)^2}{7} + \frac{(5-7)^2}{7}$

$$= 1.143$$

The rejection region requires $\alpha = .05$ in the upper tail of the χ^2 distribution with df = $k - 1 = 4 - 1 = 3$. From Table VII, Appendix A, $\chi^2_{.05} = 7.81473$. The rejection region is $\chi^2 > 7.81473$.

Since the observed value of the test statistic does not fall in the rejection region ($\chi^2 = 1.143 \not> 7.81473$), H_0 is not rejected. There is insufficient evidence to indicate that at least one of the true proportions of intellectually disabled elderly patients in each of the hearing loss categories differs at $\alpha = .05$.

b. $\hat{p} = \dfrac{x}{n} = \dfrac{5}{28} = .179$

For confidence coefficient .90; $\alpha = .10$ and $\alpha/2 = .10/2 = .05$. From Table IV, Appendix A, $z_{.05} = 1.645$. The confidence interval is

$$\hat{p} \pm z_{.05}\sqrt{\dfrac{\hat{p}\hat{q}}{n}}$$

$$\Rightarrow .179 \pm 1.645\sqrt{\dfrac{(.179)(1-.179)}{28}}$$

$$\Rightarrow .179 \pm .119$$

$$\Rightarrow (.060, .298)$$

We are 90% confident that the true proportion of disabled elderly patients with severe hearing loss is between .060 and .298.

13.53   First, the values of the variables will be defined.

1.   The variable WAR has values 1, 2, and 9. 1 = Support, 2 = Oppose, and 9 = Don't know or refused to answer.
2.   The variable INTERNET has values 1, 2, and 9. 1 = Yes, 2 = No, and 9 = Don't know or refused to answer.
3.   The variable PARTY has values 1, 2, 3, 4, 5, and 9. 1 = Republican, 2 = Democrat, 3 = Independent, 4 = No preference, 5 = Other, and 9 = Don't know or refused to answer.
4.   The variable VET has values 1, 2, 3, 4, and 9. 1 = Yes, I have, 2 = Yes, other household member, 3 = Yes, both, 4 = No, and 9 = Don't know or refused to answer.
5.   The variable IDEO has values 1, 2, 3, 4, 5, and 9. 1 = Very conservative, 2 = Conservative, 3 = Moderate, 4 = Liberal, 5 = Very liberal, and 9 = Don't know or refused to answer.
6.   The variable RACE has values 1, 2, 3, 4, 5, and 9. 1 = White, 2 = Black or African American, 3 = Asian or Pacific Islander, 4 = Mixed, 5 = Native American, 6 = Other and 9 = Don't know or refused to answer.
7.   The variable INCRANGE has values 1, 2, 3, 4, 5, 6, 7, 8, and 9. 1 = <$10,000, 2 = $10,000-under $20,000, 3 = $20,000-under $30,000, 4 = $30,000-under $40,000, 5 = $40,000-under $50,000, 6 = $50,000-under $75,000, 7 = $75,000 to under $100,000, 8 = Over $100,000, and 9 = Don't know or refused to answer.
8.   The variable CUMMUNIT has values 1, 2, 3, and 9. 1 = Urban, 2 = Suburban, 3 = Rural, and 9 = Don't know or refused to answer.

To do the analysis, all values of 9 were converted to missing values and are not included in the analysis. Alpha equal to .05 was used for all tests.

WAR versus INTERNET. Using MINITAB, the results of the analyses are:

Tabulated statistics: War, Internet

```
Rows: War    Columns: Internet

                 1         2   Missing      All

1              703       399         4     1102
             77.94     76.29         *    77.33

2              199       124         1      323
             22.06     23.71         *    22.67

Missing         27        38         0        *
                 *         *         *        *

All            902       523         *     1425
            100.00    100.00         *   100.00

Cell Contents:        Count
                      % of Column


Pearson Chi-Square = 0.512, DF = 1, P-Value = 0.474
Likelihood Ratio Chi-Square = 0.510, DF = 1, P-Value = 0.475
```

To determine if Support of Iraq War and access of internet are dependent, we test:

H_0: Support of the Iraq War and access of internet are independent
H_a: Support of the Iraq War and access of internet are dependent

The test statistic is $\chi^2 = .512$ and the p-value is $p = .474$. Since the p-value is not less than $\alpha = .05$, H_0 is not rejected. There is insufficient evidence to indicate that Support of Iraq war and access of internet are dependent at $\alpha = .05$.

WAR versus PARTY. Since only 3 respondents selected "Other" I combined the "Other" group with the "No preference" group" so that the expected counts in all cells are at least 5. Using MINITAB, the results of the analyses are:

Tabulated statistics: War, Party

```
Rows: War    Columns: Party

              1       2       3       4  Missing      All

1           451     262     315      53       25     1081
          95.55   60.23   73.77   80.30        *    77.21

2            21     173     112      13        5      319
           4.45   39.77   26.23   19.70        *    22.79

Missing       8      27      21       5        4        *
              *       *       *       *        *        *

All         472     435     427      66        *     1400
         100.00  100.00  100.00  100.00        *   100.00

Cell Contents:       Count
                     % of Column


Pearson Chi-Square = 164.761, DF = 3, P-Value = 0.000
Likelihood Ratio Chi-Square = 189.289, DF = 3, P-Value = 0.000
```

To determine if Support of Iraq War and Party are dependent, we test:

H_0: Support of the Iraq War and Party are independent
H_a: Support of the Iraq War and Party are dependent

The test statistic is $\chi^2 = 164.761$ and the p-value is $p = 0.000$. Since the p-value is less than $\alpha = .05$, H_0 is rejected. There is sufficient evidence to indicate that Support of Iraq war and Party are dependent at $\alpha = .05$.

If you look at the column percents in the above table, we see that 95.55% of the Republicans support the Iraq War, while only 60.23% of Democrats support the Iraq War. In addition, 73.33% of the Independents support the war and 80.3% of those with no preference or other support the war.

WAR versus VET. Using MINITAB, the results of the analyses are:

Tabulated statistics: War, vet

```
Rows: War    Columns: vet

                 1        2        3        4   Missing      All

1              209      237       27      631         2     1104
             84.27    76.45    77.14    75.66         *    77.37

2               39       73        8      203         1      323
             15.73    23.55    22.86    24.34         *    22.63

Missing          6       15        0       44         0        *
                 *        *        *        *         *        *

All            248      310       35      834         *     1427
            100.00   100.00   100.00   100.00         *   100.00

Cell Contents:         Count
                       % of Column


Pearson Chi-Square = 8.295, DF = 3, P-Value = 0.040
Likelihood Ratio Chi-Square = 8.858, DF = 3, P-Value = 0.031
```

To determine if Support of Iraq War and Veteran status are dependent, we test:

H_0: Support of the Iraq War and Veteran status are independent
H_a: Support of the Iraq War and Veteran status are dependent

The test statistic is $\chi^2 = 8.295$ and the p-value is $p = 0.040$. Since the p-value is less than $\alpha = .05$, H_0 is rejected. There is sufficient evidence to indicate that Support of Iraq War and Veteran status are dependent at $\alpha = .05$.

If you look at the column percents in the above table, we see that 84.27% of the veterans support the Iraq War, while 76.45% of those who have a family member who is a veteran support the Iraq War. In addition, 77.14% of those who are vets and have a family member who is a vet support the war and 75.66% of those who are not veterans support the war.

WAR versus IDEO. Using MINITAB, the results of the analyses are:

Tabulated statistics: War, ideo

```
Rows: War   Columns: ideo

               1       2       3       4       5  Missing      All

1             52     453     415     102      21       63     1043
           75.36   89.88   78.75   52.85   34.43        *    77.03

2             17      51     112      91      40       13      311
           24.64   10.12   21.25   47.15   65.57        *    22.97

Missing        2      17      19      11       2       14        *
               *       *       *       *       *        *        *

All           69     504     527     193      61        *     1354
          100.00  100.00  100.00  100.00  100.00        *   100.00

Cell Contents:       Count
                     % of Column


Pearson Chi-Square = 174.386, DF = 4, P-Value = 0.000
Likelihood Ratio Chi-Square = 161.296, DF = 4, P-Value = 0.000
```

To determine if Support of Iraq War and Political views are dependent, we test:

H_0: Support of the Iraq War and Political views are independent
H_a: Support of the Iraq War and Political views are dependent

The test statistic is $\chi^2 = 174.386$ and the p-value is $p = 0.000$. Since the p-value is less than $\alpha = .05$, H_0 is rejected. There is sufficient evidence to indicate that Support of Iraq War and Political views are dependent at $\alpha = .05$.

If you look at the column percents in the above table, we see that 75.36% of the Very Conservatives support the Iraq War, while 89.88% of the Conservatives support the Iraq War. In addition, 78.75% of Moderates support the war, 52.85% of Liberals support the war, and 34.43% of the Very Liberals support the war.

WAR versus RACE. Because there were less than 24 respondents for the race categories Asian/Pacific Islander, Mixed, Native American, and Other, these 4 categories were combined. Using MINITAB, the results of the analyses are:

Tabulated statistics: War, Race

```
Rows: War    Columns: Race

                  1          2          3    Missing        All

1               989         51         54         12       1094
              81.00      46.79      72.97          *      77.92

2               232         58         20         14        310
              19.00      53.21      27.03          *      22.08

Missing          50          9          4          2          *
                  *          *          *          *          *

All            1221        109         74          *       1404
             100.00     100.00     100.00          *     100.00

Cell Contents:        Count
                      % of Column


Pearson Chi-Square = 69.181, DF = 2, P-Value = 0.000
Likelihood Ratio Chi-Square = 57.984, DF = 2, P-Value = 0.000
```

To determine if Support of Iraq War and Race are dependent, we test:

H_0: Support of the Iraq War and Race are independent
H_a: Support of the Iraq War and Race are dependent

The test statistic is $\chi^2 = 69.181$ and the p-value is $p = 0.000$. Since the p-value is less than $\alpha = .05$, H_0 is rejected. There is sufficient evidence to indicate that Support of Iraq War and Race are dependent at $\alpha = .05$.

If you look at the column percents in the above table, we see that 81.00% of the Whites support the Iraq War, while 46.79% of the Blacks support the Iraq War. In addition, 72.96% of the Others support the war.

WAR versus INCRANGE Using MINITAB, the results of the analyses are:

Tabulated statistics: War, incrange

```
Rows: War    Columns: incrange

                 1        2        3        4        5        6        7        8

1               37       79      122      112      103      196      132      134
             61.67    73.15    77.71    77.78    83.06    82.35    77.19    74.03

2               23       29       35       32       21       42       39       47
             38.33    26.85    22.29    22.22    16.94    17.65    22.81    25.97

Missing         12       10        4        9        2        5        2        3
                 *        *        *        *        *        *        *        *

All             60      108      157      144      124      238      171      181
            100.00   100.00   100.00   100.00   100.00   100.00   100.00   100.00

          Missing      All

1             191      915
                *    77.35

2              56      268
                *    22.65

Missing        18        *
                *        *

All             *     1183
                *   100.00

Cell Contents:       Count
                     % of Column


Pearson Chi-Square = 16.387, DF = 7, P-Value = 0.022
Likelihood Ratio Chi-Square = 15.684, DF = 7, P-Value = 0.028
```

To determine if Support of Iraq War and Income range are dependent, we test:

H_0: Support of the Iraq War and Income range are independent
H_a: Support of the Iraq War and Income range are dependent

The test statistic is $\chi^2 = 16.387$ and the p-value is $p = 0.022$. Since the p-value is less than $\alpha = .05$, H_0 is rejected. There is sufficient evidence to indicate that Support of Iraq War and Income range are dependent at $\alpha = .05$.

If you look at the column percents in the above table, we see that those in the income ranges of $40,000 to under $50,000 and from $50,000 to under $75,000 support the Iraq War the most. Those in the lowest income range and the highest income range tend to not support the Iraq War as much as those in the middle income brackets.

WAR versus COMMUNIT Using MINITAB, the results of the analyses are:

Tabulated statistics: War, communit

```
Rows: War    Columns: communit

            1        2        3      All

1         235      605      266     1106
        69.73    78.27    83.13    77.34

2         102      168       54      324
        30.27    21.73    16.88    22.66

Missing    15       35       15        *
            *        *        *        *

All       337      773      320     1430
       100.00   100.00   100.00   100.00

Cell Contents:      Count
                    % of Column


Pearson Chi-Square = 17.618, DF = 2, P-Value = 0.000
Likelihood Ratio Chi-Square = 17.309, DF = 2, P-Value = 0.000
```

To determine if Support of Iraq War and Community are dependent, we test:

H_0: Support of the Iraq War and Community are independent
H_a: Support of the Iraq War and Community are dependent

The test statistic is $\chi^2 = 17.618$ and the p-value is $p = 0.000$. Since the p-value is less than $\alpha = .05$, H_0 is rejected. There is sufficient evidence to indicate that Support of Iraq War and Community are dependent at $\alpha = .05$.

If you look at the column percents in the above table, we see that 69.73% of those who live in the urban areas support the Iraq War, while 78.27% of those who live in suburban areas support the Iraq War. In addition, 83.13% of those who live in Rural areas support the war.

13.55 a. Some preliminary calculations are:

$E_1 = np_{1,0} = 85(.26) = 22.1$
$E_2 = np_{2,0} = 85(.30) = 25.5$
$E_3 = np_{3,0} = 85(.11) = 9.35$
$E_4 = np_{4,0} = 85(.14) = 11.9$
$E_5 = np_{5,0} = 85(.19) = 16.15$

To determine of probabilities differ from the hypothesized values, we test:

H_0: $p_1 = .26, p_2 = .30, p_3 = .11, p_4 = .14, p_5 = .19$
H_a: At least one of the probabilities differs from its hypothesized value.

The test statistic is $\chi^2 = \sum \dfrac{[n_i - E_i]^2}{E_i}$

$$= \frac{(32-22.1)^2}{22.1} + \frac{(26-25.5)^2}{25.5} + \frac{(15-9.35)^2}{9.35} + \frac{(6-11.9)^2}{11.9} + \frac{(6-16.15)^2}{16.15}$$

$$= 17.16$$

The rejection region requires $\alpha = .05$ in the upper tail of the χ^2 distribution with $df = k - 1 = 5 - 1 = 4$. From Table VII, Appendix A, $\chi^2_{.05} = 11.0705$. The rejection region is $\chi^2 > 11.0705$.

Since the observed value of the test statistic falls in the rejection region ($\chi^2 = 17.16 > 11.0705$), reject H_0. There is sufficient evidence to indicate the probabilities differ from their hypothesized values at $\alpha = .05$.

b. $\hat{p}_1 = \dfrac{32}{85} = .376$

For confidence coefficient .95, $\alpha = .05$ and $\alpha/2 = .05/2 = .025$. From Table IV, Appendix A, $z_{.025} = 1.96$. The 95% confidence interval is:

$$\hat{p}_i \pm z_{.025}\sqrt{\frac{\hat{p}_i \hat{q}_i}{n}} \Rightarrow .376 \pm 1.96\sqrt{\frac{.376(.624)}{85}} \Rightarrow .376 \pm .103 \Rightarrow (.273, \ .479)$$

We are 95% confident that the true proportion of Avonex MS patients who are exacerbation-free during a 2-year period is between .273 and .479.

c. From previous studies, it is known that 26% of the MS patients on a placebo experienced no exacerbations in a 2-year period. Since .26 does not fall in the 95% confidence interval, there is evidence that Avonex patients are more likely to have no exacerbations than the placebo patients at $\alpha = .05$.

13.57 Some preliminary calculations are:

$$E_{11} = \frac{R_1 C_1}{n} = \frac{12(10)}{24} = 5 \qquad\qquad E_{12} = \frac{R_1 C_2}{n} = \frac{12(14)}{24} = 7$$

$$E_{21} = \frac{R_2 C_1}{n} = \frac{12(10)}{24} = 5 \qquad\qquad E_{22} = \frac{R_2 C_2}{n} = \frac{12(14)}{24} = 7$$

To determine if a relationship exists between food choice and whether or not chickadees fed on gypsy moth eggs, we test:

H_0: Food choice and whether or not chickadees fed on gypsy moth eggs are independent
H_a: Food choice and whether or not chickadees fed on gypsy moth eggs are dependent

The test statistic is $\chi^2 = \sum\sum \dfrac{[n_{ij} - E_{ij}]^2}{E_{ij}}$

$$= \frac{(2-5)^2}{5} + \frac{(10-7)^2}{7} + \frac{(8-5)^2}{5} + \frac{(4-7)^2}{7} = 6.171$$

The rejection region requires $\alpha = .10$ in the upper tail of the χ^2 distribution with df $= (r-1)(c-1) = (2-1)(2-1) = 1$. From Table VII, Appendix A, $\chi^2_{.10} = 2.70554$. The rejection region is $\chi^2 > 2.70554$.

Since the observed value of the test statistic falls in the rejection region ($\chi^2 = 6.171 > 2.70554$), H_0 is rejected. There is sufficient evidence to indicate that a relationship exists between food choice and whether or not chickadees fed on gypsy moth eggs at $\alpha = .10$.

13.59   a.   Calculate $E_{ij} = \dfrac{R_i C_j}{n}$ for $i = 1, 2, 3, 4, 5,$ and $j = 1, 2, 3, 4, 5$.

The expected values for each cell in the table is summarized below.

Observed and Estimated Expected (in Parentheses) Counts

		Wife's Lifestyle					
		Pleasing	Outdoing	Avoiding	Control	Achieving	Totals
	Pleasing	9 (5.88)	6 (6.70)	6 (6.21)	3 (8.17)	9 (6.04)	33
	Outdoing	7 (8.20)	11 (9.34)	12 (8.65)	11 (11.39)	5 (8.43)	46
	Avoiding	8 (6.77)	8 (7.71)	6 (7.15)	11 (9.41)	5 (6.96)	38
Husband's Lifestyle	**Control**	8 (9.09)	10 (10.35)	7 (9.59)	15 (12.62)	11 (9.34)	51
	Achieving	4 (6.06)	6 (6.90)	7 (6.40)	10 (8.42)	7 (6.23)	34
	Totals	36	41	38	50	37	202

The test is: H_0: The wife's lifestyle types and the husband's lifestyle types are independent.
H_a: The wife's lifestyle types and the husband's lifestyle types are dependent.

The test statistic is $\chi^2 = \sum\sum \dfrac{[n_{ij} - E_{ij}]^2}{E_{ij}}$

$$= \frac{(9-5.88)^2}{5.88} + \frac{(6-6.70)^2}{6.70} + \cdots + \frac{(10-8.42)^2}{8.42} + \frac{(7-6.23)^2}{6.23}$$
$$= 13.715$$

Since no α is given we will use $\alpha = .05$. The rejection region requires $\alpha = .05$ in the upper tail of the χ^2 distribution with df $= (r-1)(c-1) = 4(4) = 16$. From Table VII, Appendix A, $\chi^2_{.05} = 26.2962$. The rejection region is $\chi^2 > 26.2962$.

Since the observed value of the test statistic does not fall in the rejection region ($\chi^2 = 13.715 \not> 26.2962$), H_0 is not rejected. There is insufficient evidence to indicate that the wife's lifestyle types and the husband's lifestyle types are dependent at $\alpha = .05$.

b.   There are two tests:

Wife

To determine if there are differences in the proportions of wives who fall into the five personality profiles, we test:

H_0: $p_1 = p_2 = p_3 = p_4 = p_5 = .20$
H_a: At least one of the proportions differs from .20.

where
p_1 = proportion of all wives that are pleasing
p_2 = proportion of all wives that are outstanding
p_3 = proportion of all wives that are avoiding
p_4 = proportion of all wives that are controlling
p_5 = proportion of all wives that are achieving

$E_1 = E_2 = E_3 = E_4 = E_5 = 202(.2) = 40.4$

The test statistic is $\chi^2 = \sum \dfrac{[n_i - E_i]^2}{E_i}$

$$= \frac{(36-40.4)^2}{40.4} + \frac{(41-40.4)^2}{40.4} + \frac{(38-40.4)^2}{40.4} + \frac{(50-40.4)^2}{40.4} + \frac{(37-40.4)^2}{40.4}$$
$$= 3.198$$

Since no α is given, we will use $\alpha = .05$. The rejection region is $\chi^2 > 9.48773$ for df = $k - 1 = 5 - 1 = 4$. Since the observed value of the test statistic does not fall in the rejection region ($\chi^2 = 3.198 \not> 9.48773$), do not reject H_0. There is insufficient evidence to indicate that a difference exists in the proportions of wives who fall in the five personality profiles at $\alpha = .05$.

Husband

To determine if there are differences in the proportions of wives who fall into the five personality profiles, we test:

H_0: $p_1 = p_2 = p_3 = p_4 = p_5 = .20$
H_a: At least one of the proportions differs from .20.

where
p_1 = proportion of all husbands that are pleasing
p_2 = proportion of all husbands that are outdoing
p_3 = proportion of all husbands that are avoiding
p_4 = proportion of all husbands that are controlling
p_5 = proportion of all husbands that are achieving

$E_1 = E_2 = E_3 = E_4 = E_5 = 202(.2) = 40.4$

The test statistic is $\chi^2 = \sum \dfrac{[n_i - E_i]^2}{E_i}$

$$= \frac{(33 - 40.4)^2}{40.4} + \frac{(46 - 40.4)^2}{40.4} + \frac{(38 - 40.4)^2}{40.4} + \frac{(51 - 40.4)^2}{40.4} + \frac{(34 - 40.4)^2}{40.4}$$
$$= 6.069$$

Using $\alpha = .05$, in the rejection region is $\chi^2 > 9.48773$ for df $= k - 1 = 5 - 1 = 4$. Since the observed value of the test statistic does not fall in the rejection region ($\chi^2 = 6.069 \not> 9.48773$), do not reject H_0. There is insufficient evidence to indicate that a difference exists in the proportions of husbands who fall in the five personality profiles at $\alpha = .05$.

13.61   a.   The observed frequencies for the 3 groups are:

Mail-only – 262
Internet-only – 43
Both – 135

Some preliminary calculations are:

$E_1 = E_2 = E_3 = np_{1,0} = 440(1/3) = 146.67$

To determine if the proportions of mail-only, internet-only and both users are different, we test:

H_0: $p_1 = p_2 = p_3 = 1/3$
H_a: At least one p_i differs from it hypothesized value

The test statistic is

$$\chi^2 = \sum \frac{[n_i - E_i]^2}{E_i} = \frac{[262 - 146.67]^2}{146.67} + \frac{[43 - 146.67]^2}{146.67} + \frac{[135 - 146.67]^2}{146.67} = 164.89$$

The rejection region requires $\alpha = .05$ in the upper tail of the χ^2 distribution with df $= k - 1 = 3 - 1 = 2$. From Table VII, Appendix A, $\chi^2_{.05} = 5.99147$. The rejection region is $\chi^2 > 5.99147$.

Since the observed value of the test statistic falls in the rejection region ($\chi^2 = 164.89 > 5.99147$), H_0 is rejected. There is sufficient evidence to indicate the proportions of mail-only, internet-only and both users are different at $\alpha = .05$.

Using MINITAB, a graph of the categories is:

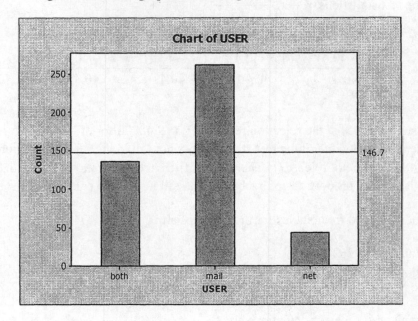

b. **Type of User vs Gender:**

Using MINITAB, the contingency table is:

Tabulated statistics: USER, GENDER

```
Rows: USER    Columns: GENDER

          Female    Male      All

both         104      31      135
           77.04   22.96   100.00
           32.70   25.41    30.68

mail         178      84      262
           67.94   32.06   100.00
           55.97   68.85    59.55

net           36       7       43
           83.72   16.28   100.00
           11.32    5.74     9.77

All          318     122      440
           72.27   27.73   100.00
          100.00  100.00   100.00

Cell Contents:        Count
                      % of Row
                      % of Column


Pearson Chi-Square = 6.797, DF = 2, P-Value = 0.033
Likelihood Ratio Chi-Square = 7.105, DF = 2, P-Value = 0.029
```

To determine if coupon user type is related to gender, we test:

H_0: Type of coupon user and gender are independent
H_0: Type of coupon user and gender are dependent

From the printout, the test statistic is $\chi^2 = 6.797$ and the p-value is p = .003.

Since the observed p-value is less than $\alpha = .05$ (p = .003 < .05), H_o is rejected. There is sufficient evidence to indicate the type of coupon user is related to gender at $\alpha = .05$.

Using MINITAB, a graph of the results is:

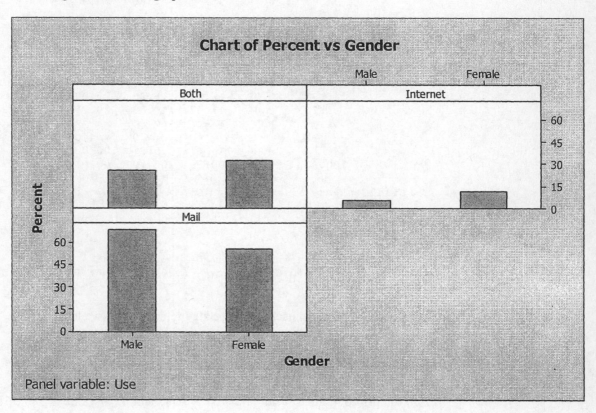

The majority of male and female users use coupons distributed through the mail. However, the proportion of male users who use coupons distributed through the mail is larger than the proportion of females. Very few users use coupons distributed via the Internet, but a higher proportion of females use coupons distributed via the Internet than males.

Type of User vs Education:

Using MINITAB, the contingency table is:

Tabulated statistics: USER, EDUC

```
Rows: USER   Columns: EDUC

          COLL     GRAD       HS     PROF      All

both        62       38       19       16      135
         45.93    28.15    14.07    11.85   100.00
         34.83    28.79    31.67    22.86    30.68

mail        96       85       34       47      262
         36.64    32.44    12.98    17.94   100.00
         53.93    64.39    56.67    67.14    59.55

net         20        9        7        7       43
         46.51    20.93    16.28    16.28   100.00
         11.24     6.82    11.67    10.00     9.77

All        178      132       60       70      440
         40.45    30.00    13.64    15.91   100.00
        100.00   100.00   100.00   100.00   100.00

Cell Contents:       Count
                     % of Row
                     % of Column


Pearson Chi-Square = 6.587, DF = 6, P-Value = 0.361
Likelihood Ratio Chi-Square = 6.786, DF = 6, P-Value = 0.341
```

To determine if coupon user type is related to education, we test:

H_0: Type of coupon user and education are independent
H_0: Type of coupon user and education are dependent

From the printout, the test statistic is $\chi^2 = 6.587$ and the p-value is $p = .361$.

Since the observed p-value is not less than $\alpha = .05$ ($p = .361 \not< .05$), H_0 is not rejected. There is insufficient evidence to indicate the type of coupon user is related to education at $\alpha = .05$.

Using MINITAB, a graph of the results is:

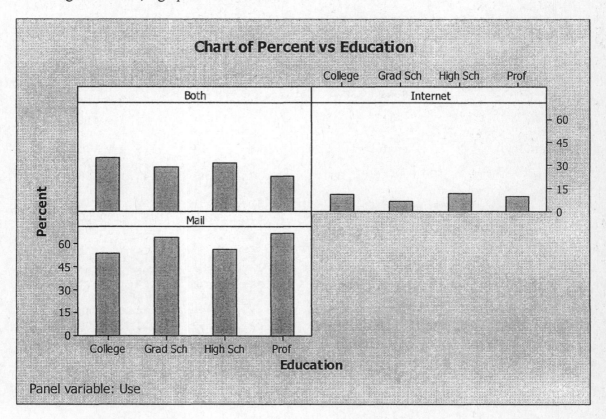

For each category of use, the proportions of each education level are about the same. This corresponds to the conclusion above that coupon use is not related to education level.

Type of User vs Work Status:

Using MINITAB, the contingency table is:

Tabulated statistics: USER, WORK

```
Rows: USER    Columns: WORK

              1         2         3         4       All

both         90        13        17        15       135
          66.67      9.63     12.59     11.11    100.00
          33.71     25.00     33.33     21.43     30.68

mail        148        31        31        52       262
          56.49     11.83     11.83     19.85    100.00
          55.43     59.62     60.78     74.29     59.55

net          29         8         3         3        43
          67.44     18.60      6.98      6.98    100.00
          10.86     15.38      5.88      4.29      9.77

All         267        52        51        70       440
          60.68     11.82     11.59     15.91    100.00
         100.00    100.00    100.00    100.00    100.00

Cell Contents:      Count
                    % of Row
                    % of Column


Pearson Chi-Square = 11.687, DF = 6, P-Value = 0.069
Likelihood Ratio Chi-Square = 12.208, DF = 6, P-Value = 0.057

* NOTE * 1 cells with expected counts less than 5
```

To determine if coupon user type is related to work status, we test:

H_0: Type of coupon user and work status are independent
H_0: Type of coupon user and work status are dependent

From the printout, the test statistic is $\chi^2 = 11.687$ and the p-value is p = .069.

Since the observed p-value is not less than $\alpha = .05$ ($p = .069 \nless .05$), H_0 is not rejected. There is insufficient evidence to indicate the type of coupon user is related to work status at $\alpha = .05$.

Using MINITAB, a graph of the results is:

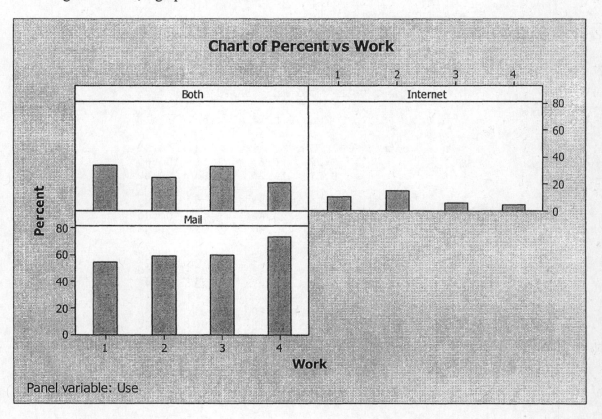

Again, for each category of use, the proportions of each work level are about the same. This corresponds to the conclusion above that coupon use is not related to work level.

Type of User vs Coupon satisfaction:

Using MINITAB, the contingency table is:

Tabulated statistics: USER, SATISF

```
Rows: USER    Columns: SATISF

            No      Some      Yes      All

both         3         9      123      135
          2.22      6.67    91.11   100.00
          8.57     11.25    37.85    30.68

mail        28        62      172      262
         10.69     23.66    65.65   100.00
         80.00     77.50    52.92    59.55

net          4         9       30       43
          9.30     20.93    69.77   100.00
         11.43     11.25     9.23     9.77

All         35        80      325      440
          7.95     18.18    73.86   100.00
        100.00    100.00   100.00   100.00

Cell Contents:      Count
                    % of Row
                    % of Column


Pearson Chi-Square = 30.418, DF = 4, P-Value = 0.000
Likelihood Ratio Chi-Square = 34.934, DF = 4, P-Value = 0.000

* NOTE * 1 cells with expected counts less than 5
```

To determine if coupon user type is related to coupon satisfaction, we test:

H_0: Type of coupon user and coupon satisfaction are independent
H_0: Type of coupon user and coupon satisfaction are dependent

From the printout, the test statistic is $\chi^2 = 30.418$ and the p-value is $p = .000$.

Since the observed p-value is less than $\alpha = .05$ ($p = .000 < .05$), H_0 is rejected. There is sufficient evidence to indicate the type of coupon user is related to coupon satisfaction at $\alpha = .05$.

Using MINITAB, a graph of the results is:

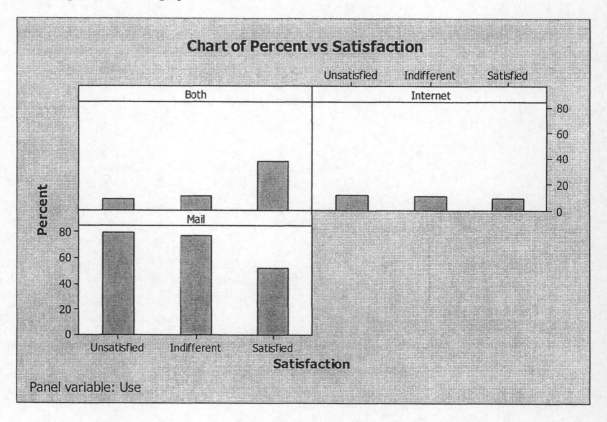

Again, very few users use coupons distributed via the Internet. The proportions of the different levels of satisfaction are about the same for this category of coupon use. For coupons distributed through the mail, the proportions of users who were unsatisfied or indifferent are larger than the proportion satisfied. However, for users of coupons distributed through both the mail and the Internet, the proportion of users who were satisfied is greater then the proportions of users unsatisfied or indifferent.

13.63   a. $\chi^2 = \sum \dfrac{[n_i - E_i]^2}{E_i}$

$= \dfrac{(26-23)^2}{23} + \dfrac{(146-136)^2}{136} + \dfrac{(361-341)^2}{341} + \dfrac{(143-136)^2}{136} + \dfrac{(13-23)^2}{23}$

$= 9.647$

b.   From Table VII, Appendix A, with df = 5, $\chi^2_{.05} = 11.0705$

c.   No. $\chi^2 = 9.647 \not> 11.0705$. Do not reject H_0. There is insufficient evidence to indicate the salary distribution is nonnormal for $\alpha = .05$.

d.   The p-value $= P(\chi^2 \geq 9.647)$.

Using Table VII, Appendix A, with df = 5,

$.05 < P(\chi^2 \geq 9.647) < .10$.

13.65　Using MINITAB, the results of the analyses are:

Tabulated statistics: Candidate, Time

```
Using frequencies in Fr


Rows: Candidate    Columns: Time

              1        2        3        4        5        6      All

Coppin       55       51      109       98       88      104      505
          13.89    13.56    13.37    13.98    14.08    14.04    13.82
           54.7     52.0    112.6     96.9     86.4    102.4    505.0


Montes      133      117      255      211      186      227     1129
          33.59    31.12    31.29    30.10    29.76    30.63    30.90
          122.4    116.2    251.8    216.6    193.1    229.0   1129.0


Smith       208      208      451      392      351      410     2020
          52.53    55.32    55.34    55.92    56.16    55.33    55.28
          218.9    207.9    450.5    387.5    345.5    409.6   2020.0


All         396      376      815      701      625      741     3654
         100.00   100.00   100.00   100.00   100.00   100.00   100.00
          396.0    376.0    815.0    701.0    625.0    741.0   3654.0

Cell Contents:        Count
                      % of Column
                      Expected count

Pearson Chi-Square = 2.284, DF = 10, P-Value = 0.994
Likelihood Ratio Chi-Square = 2.272, DF = 10, P-Value = 0.994
```

To determine if time period and votes received by the candidates are independent, we test:

H_0: Votes received by candidates and time period are independent
H_a: Votes received by candidates and time period are dependent

The test statistic is $\chi^2 = 2.284$.　(from printout)

The p-value is $p = .994$. Since the p-value is so large, H_0 would not be rejected for any reasonable value of α. There is insufficient evidence to indicate that the votes received by the candidates are dependent on the time period. In other words, there is no evidence that the percentages of votes received by the candidates change over time. If one looks at the column percentages (the second number in each cell above), the values are very similar for each time period.

A graph of the column percents is:

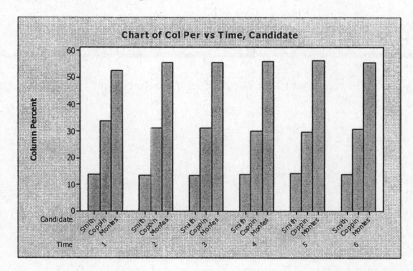

As one can see from the graph, the percent of votes received by the candidates are essentially the same for each time period.

All of the above indicates that the probability of a candidate receiving votes is independent of the time period. However, this does not necessarily imply a rigged election. If each time period is considered a random sample of voters, then the percentage of votes received by each candidate should be similar to the actual percentage of voters who favor each candidate.

Nonparametric Statistics

14.1   The sign test is preferred to the *t*-test when the population from which the sample is selected is not normal.

14.3   a. $P(x \geq 6) = 1 \leq P(x - 5) = 1 - .937 = .063$

  b. $P(x \geq 5) = 1 \leq P(x - 4) = 1 - .500 = .500$

  c. $P(x \geq 8) = 1 \leq P(x - 7) = 1 - .996 = .004$

  d. $P(x \geq 10) = 1 \leq P(x - 9) = 1 - .849 = .151$

    $\mu = np = 15(.5) = 7.5$ and $\sigma = \sqrt{npq} = \sqrt{15(.5)(.5)} = 1.9365$

    $P(x \geq 10) \approx P\left(z \geq \dfrac{(10 - .5) - 7.5}{1.9365} \right) = P(z \geq 1.03) = .5 - .3485 = .1515$

  e. $P(x \geq 15) = 1 \leq P(x \leq 14) = 1 - .788 = .212$

    $\mu = np = 25(.5) = 12.5$ and $\sigma = \sqrt{npq} = \sqrt{25(.5)(.5)} = 2.5$

    $P(x \geq 15) \approx P\left(z \geq \dfrac{(15 - .5) - 12.5}{2.5} \right) = P(z \geq .80) = .5 - .2881 = .2119$

14.5   To determine if the median is greater than 80, we test:

H_0: $\eta = 80$
H_a: $\eta > 80$

The test statistic is $S =$ number of measurements greater than $80 = 16$.

The *p*-value $= P(x \geq 16)$ where x is a binomial random variable with $n = 25$ and $p = .5$. From Table II,

    p-value $= P(x \geq 16) = 1 - P(x \leq 15) = 1 - .885 = .115$

Since the *p*-value $= .115 > \alpha = .10$, H_0 is not rejected. There is insufficient evidence to indicate the median is greater than 80 at $\alpha = .10$.

We must assume the sample was randomly selected from a continuous probability distribution.

Note: Since $n \geq 10$, we could use the large-sample approximation.

14.7   To determine if the median cesium amount of lichen differs from .003, we test:

$$H_0: \eta = .003$$
$$H_a: \eta \neq .003$$

The test statistic is $S = \{$number of observations greater than .003$\} = 8$ (from printout). The p-value is $p = .0391$. Since the p-value is less than $\alpha = .10$, H_0 is rejected. There is sufficient evidence to indicate the median cesium amount of lichen differs from .003 at $\alpha = .10$.

This result agrees with the t-test in Exercise 8.65.

14.9   a.   To determine if the median daily ammonia concentration for all afternoon drive-time days exceeds 1.5 ppm, we test:

$$H_0: \eta = 1.5$$
$$H_a: \eta > 1.5$$

b.   The test statistic is $S =$ number of measurements greater than $1.5 = 3$.

c.   The p-value $= P(x \geq 3)$ where x is a binomial random variable with $n = 8$ and $p = .5$. From Table II, Appendix A,

$$P(x \geq 3) = 1 - P(x \leq 2) = 1 - .145 = .855.$$

d.   Sine the p-value is greater than α ($p = .855 > .05$), H_0 is not rejected. There is insufficient evidence to indicate that the median daily ammonia concentration for all afternoon drive-time days exceeds 1.5 ppm at $\alpha = .05$.

14.11   a.   To determine whether the median biting rate is higher in bright, sunny weather, we test:

$$H_0: \eta = 5$$
$$H_a: \eta > 5$$

b.   The test statistic is $z = \dfrac{(S - .5) - .5n}{.5\sqrt{n}} = \dfrac{(95 - .5) - .5(122)}{.5\sqrt{122}} = 6.07$
(where $S =$ number of observations greater than 5)

The p-value is $p = P(z \geq 6.07)$. From Table IV, Appendix A, $p = P(z \geq 6.07) \approx 0.0000$.

c.   Since the observed p-value is less than α ($p = 0.0000 < .01$), H_0 is rejected. There is sufficient evidence to indicate that the median biting rate in bright, sunny weather is greater than 5 at $\alpha = .01$.

14.13   To determine if 50% of the ingots have a freckle index of 10 or higher, we test:

H_o: $\eta = 10$
H_a: $\eta > 10$

S = number of measurements greater than 10 = 10.

The test statistic is $z = \dfrac{(S - .5) - .5n}{.5\sqrt{n}} = \dfrac{(10 - .5) - .5(18)}{.5\sqrt{18}} = 0.24$

The p-value = $P(z \geq 0.24)$. From Table IV, Appendix A, $p = P(z \geq 0.24)$ = .
5 − .0948 = .4052.

Since the p-value is greater than $\alpha = .01$ ($p = .4052 > .01$), H_0 is not rejected. There is insufficient evidence to indicate that 50% of the ingots have a freckle index of 10 or higher at $\alpha = .01$.

14.15   To determine if the population median skidding distance is more than 400 meters, we test:

H_0: $\eta = 400$
H_a: $\eta > 400$

The test statistic is S = number of measurements greater than 400 = 8.

The p-value = $P(x \geq 8)$ where x is a binomial random variable with $n = 20$ and $p = .5$. From Table II, Appendix A,

$P(x \geq 8) = 1 - P(x \leq 7) = 1 - .132 = .868$

Since the p-value is greater than $\alpha = .10$ (p = .868 > .10), H_0 is not rejected. There is insufficient evidence to indicate the population median skidding distance is more than 400 meters at $\alpha = .10$.

14.17   The statement "If the rank sum for sample 1 is much larger than the rank sum for sample 2 when $n_1 = n_2$, then the distribution of population1 is likely to be shifted to the right of the distribution of population 2." is true.

14.19   a.   The test statistic is T_1, the rank sum of population A (because $n_1 < n_2$).

The rejection region is $T_1 \leq 41$ or $T_1 \geq 71$, from Table XII, Appendix A, with $n_1 = 7$, $n_2 = 8$, and $\alpha = .10$.

b.   The test statistic is T_1, the rank sum of population A (because $n_1 = n_2$).

The rejection region is $T_1 \geq 50$, from Table XII, Appendix A, with $n_1 = 6$, $n_2 = 6$, and $\alpha = .05$.

c. The test statistic is T_1, the rank sum of population A (because $n_1 < n_2$).

The rejection region is $T_1 \leq 43$, from Table XII, Appendix A, with $n_1 = 7$, $n_2 = 10$, and $\alpha = .025$.

d. Since $n_1 = n_2 = 20$, the test statistic is $z = \dfrac{T_1 - \dfrac{n_1(n_1 + n_2 + 1)}{2}}{\sqrt{\dfrac{n_1 n_2 (n_1 + n_2 + 1)}{12}}}$

The rejection region is $z < -z_{\alpha/2}$ or $z > z_{\alpha/2}$. For $\alpha = .05$ and $\alpha/2 = .05/2 = .025$, $z_{.025} = 1.96$ from Table IV, Appendix A. The rejection region is $z < -1.96$ or $z > 1.96$.

14.21 H_0: The probability distributions of treatments A and B are identical
H_a: The probability distribution of treatment A lies to the left of that for treatment B

The test statistic is

$$z = \frac{T_1 - \dfrac{n_2(n_1 + n_2 + 1)}{2}}{\sqrt{\dfrac{n_1 n_2 (n_1 + n_2 + 1)}{12}}} = \frac{173 - \dfrac{15(15 + 15 + 1)}{2}}{\sqrt{\dfrac{15(15)(15 + 15 + 1)}{12}}} = \frac{-59.5}{24.1091} = -2.47$$

The rejection region requires $\alpha = .05$ in the lower tail of the z distribution. From Table IV, Appendix A, $z_{.05} = 1.645$. The rejection region is $z < -1.645$.

Since the observed value of the test statistic falls in the rejection region ($-2.47 < -1.645$), H_0 is rejected. There is sufficient evidence to conclude that distribution A is shifted to the left of distribution B for $\alpha = .05$.

14.23

Sample from Population 1 (A)	Rank	Sample from Population 2 (B)	Rank
15	13	5	2.5
10	8.5	12	10.5
12	10.5	9	6.5
16	14	9	6.5
13	12	8	4.5
8	4.5	4	1
	$T_1 = 62.5$	5	2.5
		10	8.5
			$T_2 = 42.5$

a. H_0: The two sampled populations have identical probability distributions
   H_a: The probability distribution for population A is shifted to the left or to the right of that for B

   The test statistic is $T_1 = 62.5$ since sample A has the smallest number of measurements.

   The null hypothesis will be rejected if $T_1 \le T_L$ or $T_1 \ge T_U$ where T_L and T_U correspond to $\alpha = .05$ (two-tailed), $n_1 = 6$ and $n_2 = 8$. From Table XII, Appendix A, $T_L = 29$ and $T_U = 61$.

   Reject H_0 if $T_1 \le 29$ or $T_1 \ge 61$.

   Since $T_1 = 62.5 \ge 61$, we reject H_0 and conclude there is sufficient evidence to indicate population A is shifted to the left or right of population B at $\alpha = .05$.

b. H_0: The two sampled populations have identical probability distributions
   H_a: The probability distribution for population A is shifted to the right of population B

   The test statistic remains $T_1 = 62.5$.
   The null hypothesis will be rejected if $T_1 \ge T_U$ where T_U corresponds to $\alpha = .05$ (one-tailed), $n_1 = 6$ and $n_2 = 8$. From Table XII, Appendix A, $T_U = 58$.

   Reject H_0 if $T_1 \ge 58$.

   Since $T_1 = 62.5 \ge 58$, we reject H_0 and conclude there is sufficient evidence to indicate population A is shifted to the right of population B at $\alpha = .05$.

14.25   a. To determine if the distribution of FNE scores for bulimic students is shifted above the corresponding distribution for female students with normal eating habits, we test:

   H_0: The distribution of FNE scores for bulimic students is the same as the corresponding distribution for female students with normal eating habits

   H_a: The distribution of FNE scores for bulimic students is shifted above the corresponding distribution for female students with normal eating habits

b.   The data ranked are:

Bulimic Students	Rank	Normal Students	Rank
21	21.5	13	8.5
13	8.5	6	1
10	4.5	16	13.5
20	19.5	13	8.5
25	25	8	3
19	17	19	17
16	13.5	23	23
21	21.5	18	15
24	24	11	6
13	8.5	19	17
14	11	7	2
		10	4.5
	$T_1 = 174.5$	15	12
		20	19.5

$$T_2 = 150.5$$

c.   The sum of the ranks of the 11 FNE scores for bulimic students is $T_1 = 174.5$.

d.   The sum of the ranks of the 14 FNE scores for normal students is $T_2 = 150.5$.

e.   The test statistic is $z = \dfrac{T_1 - \dfrac{n_1(n_1 + n_2 + 1)}{2}}{\sqrt{\dfrac{n_1 n_2 (n_1 + n_2 + 1)}{12}}} = \dfrac{174.5 - \dfrac{11(11 + 14 + 1)}{2}}{\sqrt{\dfrac{11(14)(11 + 14 + 1)}{12}}} = 1.72$

The rejection region requires $\alpha = .10$ in the upper tail of the z distribution. From Table IV, Appendix A, $z_{.10} = 1.28$. The rejection region is $z > 1.28$.

f.   Since the observed value of the test statistic falls in the rejection region ($z = 1.72 > 1.28$), H_0 is rejected. There is sufficient evidence to indicate that the distribution of FNE scores for bulimic students is shifted above the corresponding distribution for female students with normal eating habits at $\alpha = .10$.

14.27   a.

Deaf Children	Rank	Hearing Children	Rank
2.75	18	1.15	1
3.14	19	1.65	6
3.23	20	1.43	4
2.30	15	1.83	8.5
2.64	17	1.75	7
1.95	10	1.23	2
2.17	13	2.03	12
2.45	16	1.64	5
1.83	8.5	1.96	11
2.23	14	1.37	3
	$T_1 = 150.5$		$T_2 = 59.5$

H_0: The probability distributions of eye movement rates are identical for deaf and hearing children

H_a: The probability distribution of eye movement rates for deaf children lies to the right of that for hearing children

The test statistic can be either T_1 or T_2 since the sample sizes are equal. Use $T_1 = 150.5$.

The null hypothesis will be rejected if $T_1 \geq T_U$ where $\alpha = .05$ (one-tailed), $n_1 = 10$ and $n_2 = 10$. From Table XII, Appendix A, $T_U = 127$.

Reject H_0 if $T_1 \geq 127$.

Since $T_1 = 150.5 \geq 127$, we reject H_0. There is sufficient evidence to indicate that deaf children have greater visual acuity than hearing children at $\alpha = .05$.

b. H_0: The probability distributions of eye movement rates are identical for deaf and hearing children

H_a: The probability distribution of eye movement rates for deaf children lies to the right of that for hearing children

The test statistic is $z = \dfrac{T_1 - \dfrac{n_1(n_1 + n_2 + 1)}{2}}{\sqrt{\dfrac{n_1 n_2 (n_1 + n_2 + 1)}{12}}} = \dfrac{150.5 - \dfrac{10(10 + 10 + 1)}{2}}{\sqrt{\dfrac{10(10)(10 + 10 + 1)}{12}}} = 3.44$

The rejection region requires $\alpha = .05$ in the upper tail of the z distribution. From Table IV, Appendix A, $z_{.05} = 1.645$. The rejection region is $z > 1.645$.

Since the observed value of the test statistic falls in the rejection region ($z = 3.44 > 1.645$), H_0 is rejected. There is sufficient evidence to support the psychologist's claim that deaf children have greater visual acuity than hearing children at $\alpha = .05$.

The results agree with those found in part a.

14.29   a.   Since the data are probably not from a normal distribution the Wilcoxon rank sum test should be used.

   b.   To determine if the relational intimacy scores for the participants in the CMC group are lower than the relational intimacy scores for the participants in the FTF group, we test:

   H_0: The probability distributions of those in the CMC group and those in the FTF group are identical
   H_a: The probability distribution of the CMC group is shifted to the left of that for the FTF group

   c.   The rejection region requires $\alpha = .10$ in the lower tail of the z distribution. From Table IV, Appendix A, $z_{.10} = 1.28$. The rejection region is $z < -1.28$.

   d.   First, we rank the data:

CMC Group	Rank	FTF Group	Rank
4	34.5	5	
3	13.5	4	34.5
3	13.5	4	34.5
4	34.5	4	34.5
3	13.5	3	13.5
3	13.5	3	13.5
3	13.5	3	13.5
3	13.5	4	34.5
4	34.5	3	13.5
4	34.5	3	13.5
3	13.5	3	13.5
4	34.5	3	13.5
3	13.5	4	34.5
3	13.5	4	34.5
2	2	4	34.5
4	34.5	4	34.5
2	2	4	34.5
4	34.5	3	13.5
5	47	3	13.5
4	34.5	3	13.5
4	34.5	4	34.5
4	34.5	4	34.5
5	47	2	2
3	13.5	4	34.5
	$T_1 = 578$		$T_2 = 598$

The test statistic is

$$z = \frac{T_1 - \frac{n_2(n_1 + n_2 + 1)}{2}}{\sqrt{\frac{n_1 n_2 (n_1 + n_2 + 1)}{12}}} = \frac{578 - \frac{24(24 + 24 + 1)}{2}}{\sqrt{\frac{24(24)(24 + 24 + 1)}{12}}} = \frac{-10}{48.4974} = -.21$$

Since the observed value of the test statistic does not fall in the rejection region ($z = -.21 \not< -1.28$), H_0 is not rejected. There is insufficient evidence to indicate the relational intimacy scores for the participants in the CMC group are lower than the relational intimacy scores for the participants in the FTF group at $\alpha = .10$.

14.31   a.   The Students' t procedure requires that the populations sampled from be normal. The distributions of the zinc measurements in the 3 locations may not be normal.

b.   The ranks of the observations and the sums for the two groups are:

Text-line	Rank	Intersection	Rank
.335	4	.393	7
.374	6	.353	5
.440	8	.285	1
		.295	2
		.319	3
	$T_1 = 18$		$T_2 = 18$

To determine if the distributions of zinc measurements for the text-line and the intersection have different centers of location, we test:

H_0: The distributions of zinc measurements for the text-line and the intersection are identical

H_a: The distribution of zinc measurements for the text-line is shifted to the right or left of that for the intersection

The test statistic is $T_1 = 18$.

The null hypothesis will be rejected if $T_1 \leq T_L$ or $T_1 \geq T_U$ where $\alpha = .05$ (two-tailed), $n_1 = 3$ and $n_2 = 5$. From Table XII, Appendix A, $T_L = 6$ and $T_U = 21$. The rejection region is $T_1 \leq 6$ or $T_1 \geq 21$.

Since the observed value of the test statistic does not fall in the rejection region ($T_1 \not\leq 6$ and $T_1 \not\geq 21$), H_0 is not rejected. There is insufficient evidence to indicate the distributions of zinc measurements for the text-line and the intersection have different centers of location at $\alpha = .05$.

c. The ranks of the observations and the sums for the two groups are:

Witness-line	Rank	Intersection	Rank
.210	2	.393	9
.262	3	.353	8
.188	1	.285	4
.329	7	.295	5
.439	11	.319	6
.397	10		
	$T_1 = 34$		$T_2 = 32$

To determine if the distributions of zinc measurements for the witness-line and the intersection have different centers of location, we test:

H_0: The distributions of zinc measurements for the witness-line and the intersection are identical

H_a: The distribution of zinc measurements for the witness-line is shifted to the right or left of that for the intersection

The test statistic is $T_2 = 32$.

The null hypothesis will be rejected if $T_2 \leq T_L$ or $T_2 \geq T_U$ where $\alpha = .05$ (two-tailed), $n_1 = 6$ and $n_2 = 5$. From Table XII, Appendix A, $T_L = 19$ and $T_U = 41$. The rejection region is $T_2 \leq 19$ or $T_2 \geq 41$.

Since the observed value of the test statistic does not fall in the rejection region ($T_2 \not\leq 19$ and $T_2 \not\geq 41$), H_0 is not rejected. There is insufficient evidence to indicate the distributions of zinc measurements for the witness-line and the intersection have different centers of location at $\alpha = .05$.

d. The median zinc score for the Text-line is .374. The median zinc score for the Witness-line is .296. The median zinc score for the Intersection is .319. We found no difference in the distributions of zinc scores between the Text-line and the Intersection. We also found no difference in the distributions of zinc scores between the Witness-line and the Intersection. However, we did not compare the Text-line and the Witness-line zinc scores. These two distributions appear to be the furthest apart.

14.33   a.   Since the data can take on a very limited number of values, the data are probably not from a normal distribution.

   b.   Using SAS, the output for the Wilcoxon test is:

```
                    The NPAR1WAY Procedure

         Wilcoxon Scores (Rank Sums) for Variable GSHWS
                   Classified by Variable COND

                    Sum of      Expected       Std Dev        Mean
   COND       N     Scores      Under H0       Under H0       Score
   ------------------------------------------------------------------------
   ATIPS     98     8039.0      11123.0        476.284142     82.030612
   TIPS     128    17612.0      14528.0        476.284142    137.593750

              Average scores were used for ties.


                    Wilcoxon Two-Sample Test

         Statistic                   8039.0000

         Normal Approximation
         Z                             -6.4741
         One-Sided Pr <  Z             <.0001
         Two-Sided Pr > |Z|            <.0001

         t Approximation
         One-Sided Pr <  Z             <.0001
         Two-Sided Pr > |Z|            <.0001

      Z includes a continuity correction of 0.5.


                    Kruskal-Wallis Test

         Chi-Square                  41.9273
         DF                                1
         Pr > Chi-Square              <.0001
```

To determine if there was a difference in the level of involvement in science homework assignments between TIPS and ATIPS students, we test:

H_0: The distributions of the involvement scores for science homework for the two groups have the same location

H_a: The distributions of the involvement scores for science homework for the TIPS students is shifted to the right or left of that for the ATIPS students

From the printout, the test statistic is $z = -6.4741$ and the p-value is $p < .0001$. Since the p-value is less than α ($p < .0001 < .05$), H_0 is rejected. There is sufficient evidence to indicate that there is a difference in the level of involvement in science homework assignments between TIPS and ATIPS students at $\alpha = .05$.

c. Using SAS, the output for the Wilcoxon test is:

```
                  The NPAR1WAY Procedure

       Wilcoxon Scores (Rank Sums) for Variable MTHHWS
                 Classified by Variable COND

                     Sum of     Expected      Std Dev        Mean
COND        N        Scores     Under H0      Under H0       Score
---------------------------------------------------------------------
ATIPS       98      10935.0     11123.0       472.485707    111.591837
TIPS       128      14715.0     14528.0       472.485707    114.960938

            Average scores were used for ties.


                    Wilcoxon Two-Sample Test

          Statistic                 10936.0000

          Normal Approximation
          Z                            -0.3947
          One-Sided Pr <  Z             0.3465
          Two-Sided Pr > |Z|            0.6930

          t Approximation
          One-Sided Pr <  Z             0.3467
          Two-Sided Pr > |Z|            0.6934

       Z includes a continuity correction of 0.5.


                    Kruskal-Wallis Test

          Chi-Square                    0.1566
          DF                                 1
          Pr > Chi-Square               0.6923
```

To determine if there was a difference in the level of involvement in mathematics homework assignments between TIPS and ATIPS students, we test:

H_0: The distributions of the involvement scores for mathematics homework for the two groups have the same location

H_a: The distributions of the involvement scores for mathematics homework for the TIPS students is shifted to the right or left of that for the ATIPS students

From the printout, the test statistic is $z = -.3947$ and the p-value is $p = .6930$. Since the p-value is not less than α ($p = .6930 > .05$), H_0 is not rejected. There is insufficient evidence to indicate that there is a difference in the level of involvement in mathematics homework assignments between TIPS and ATIPS students at $\alpha = .05$.

d. Using SAS, the output for the Wilcoxon test is:

```
               The NPAR1WAY Procedure

      Wilcoxon Scores (Rank Sums) for Variable LAHWS
               Classified by Variable COND

                    Sum of      Expected      Std Dev        Mean
  COND        N     Scores      Under H0      Under H0       Score
  -------------------------------------------------------------------
  ATIPS      98    10496.0      11123.0      464.844149    107.102041
  TIPS      128    15155.0      14528.0      464.844149    118.398438

            Average scores were used for ties.


                 Wilcoxon Two-Sample Test

          Statistic              10496.0000

          Normal Approximation
          Z                         -1.3478
          One-Sided Pr <  Z          0.0889
          Two-Sided Pr > |Z|         0.1777

          t Approximation
          One-Sided Pr <  Z          0.0895
          Two-Sided Pr > |Z|         0.1791

     Z includes a continuity correction of 0.5.


                  Kruskal-Wallis Test

          Chi-Square                 1.8194
          DF                              1
          Pr > Chi-Square            0.1774
```

To determine if there was a difference in the level of involvement in language arts homework assignments between TIPS and ATIPS students, we test:

H_0: The distributions of the involvement scores for language arts homework for the two groups have the same location

H_a: The distributions of the involvement scores for language arts homework for the TIPS students is shifted to the right or left of that for the ATIPS students

From the printout, the test statistic is $z = -1.3478$ and the p-value is $p = .1777$. Since the p-value is not less than α ($p = .1777 > .05$), H_0 is not rejected. There is insufficient evidence to indicate that there is a difference in the level of involvement in language arts homework assignments between TIPS and ATIPS students at $\alpha = .05$.

14.35 We assume that the probability distribution of differences is continuous so that the absolute differences will have unique ranks. Although tied (absolute) differences can be assigned average ranks, the number of ties should be small relative to the number of observations to assure validity.

14.37 a. The hypotheses are:

H_0: The two sampled populations have identical probability distributions

H_a: The probability distributions for population A is shifted to the right of that for population B

b. Some preliminary calculations are:

Treatment A	Treatment B	Difference A − B	Rank of Absolute Difference
54	45	9	5
60	45	15	10
98	87	11	7
43	31	12	9
82	71	11	7
77	75	2	2.5
74	63	11	7
29	30	−1	1
63	59	4	4
80	82	−2	2.5
			$T_- = 3.5$

The test statistic is $T_- = 3.5$

The rejection region is $T_- \leq 8$, from Table XIII, Appendix A, with $n = 10$ and $\alpha = .025$.

Since the observed value of the test statistic falls in the rejection region ($T_- = 3.5 \leq 8$), H_0 is rejected. There is sufficient evidence to indicate the responses for A tend to be larger than those for B at $\alpha = .025$.

14.39   a. H_0:   The two sampled populations have identical probability distributions
H_a: The probability distribution for population A is located to the right of that for population B

   b.   The test statistic is:

$$z = \frac{T_+ - \dfrac{n(n+1)}{4}}{\sqrt{\dfrac{n(n+1)(2n+1)}{24}}} = \frac{354 - \dfrac{30(30+1)}{4}}{\sqrt{\dfrac{30(30+1)(60+1)}{24}}} = \frac{121.5}{48.6184} = 2.499$$

The rejection region requires $\alpha = .05$ in the upper tail of the z distribution. From Table IV, Appendix A, $z = 1.645$. The rejection region is $z > 1.645$.

Since the observed value of the test statistic falls in the rejection region ($z = 2.499 > 1.645$), H_0 is rejected. There is sufficient evidence to indicate population A is located to the right of that for population B at $\alpha = .05$.

   c.   The p-value = $P(z \geq 2.499) = .5 - .4938 = .0062$ (using Table IV, Appendix A).

   d.   The necessary assumptions are:

   1.   The sample of differences is randomly selected from the population of differences.
   2.   The probability distribution from which the sample of paired differences is drawn is continuous.

Nonparametric Statistics   **499**

14.41 a. Since two measurements were obtained from each patients (before and after), the measurements are not independent of each other. Each "before" measurement is paired with an "after" measurement.

b. The distributions for both before and after treatment are skewed to the right and the spread of the distribution before treatment is much larger than the spread after treatment.

c. Since the p-value is so small ($p < .0001$), H_0 would be rejected. There is sufficient evidence to indicate the ichthyotherapy was effective in treating psoriasis for any reasonable value of α.

14.43 a. To determine if the distributions of the FSI scores for good and average readers differ in location, we test:

H_0: The distribution of the FSI scores of good readers is identical to the distribution of the FSI scores of average readers

H_a: The distribution of the of the FSI scores of good readers is shifted to the right or left of the distribution of the FSI scores of average readers

b, c.

| STRATEGY | FSI Scores | | Difference | Rank of Absolute Difference |
	Good Readers	Average Readers		
Word meaning	.38	.32	.06	2
Words in context	.29	.25	.04	1
Literal comprehension	.42	.25	.17	5
Draw inference from single string	.60	.26	.34	8
Draw inference from multiple string	.45	.31	.14	4
Interpretation of metaphor	.32	.14	.18	6.5
Find salient or main idea	.21	.03	.18	6.5
Form judgment	.73	.80	−.07	3

d. Positive rank sum $T_+ = 33$. Negative rank sum $T_- = 3$. The test statistic is the smaller of of T_+ and T_- and is $T_- = 3$.

e. The rejection region is $T_+ \leq 4$, from Table XIII, Appendix A, with $n = 8$ and $\alpha = .05$ (two-tails).

f. Since the observed value of the test statistic falls in the rejection region ($T_+ = 3 \leq 4$), H_0 is rejected. There is sufficient evidence to indicate the distributions of the FSI scores for good and average readers differ in location at $\alpha = .05$.

14.45   Some preliminary calculations are:

Day	Change in Temp.	Change in Transverse Strain Field Meas.	Change in Transverse Strain 3D Model	Diff.	Rank of Absolute Difference
Oct. 24	-6.3	-58	-52	-6	3
Dec. 3	13.2	69	59	10	5
Dec. 15	3.3	35	32	3	2
Feb. 2	-14.8	-32	-24	-8	4
Mar. 25	1.7	-40	-39	-1	1
May 24	-.2	-83	-71	-12	6
					$T_+ = 7$

To determine if there is a shift in the change in transverse strain distributions between field measurements and the 3D model, we test:

H_0: The change in transverse strain distribution for field measurements is identical to the change in transverse strain distribution for the 3D model

H_a: The change in transverse strain distribution for field measurements is shifted to the right or left of the change in transverse strain distribution for the 3D model

The test statistic is T = smaller of T_- and T_+ which is $T_+ = 7$.

The rejection region is $T_+ \leq 1$, from Table XIII, Appendix A, with n = 6 and $\alpha = .05$ (two-tails).

Since the observed value of the test statistic does not fall in the rejection region ($T_+ = 7 \not\leq 1$), H_0 is not rejected. There is insufficient evidence to indicate there is a shift in the change in transverse strain distributions between field measurements and the 3D model at $\alpha = .05$.

14.47   Some preliminary calculations are:

Patient	Before Thickness	After Thickness	Difference	Rank of Absolute difference
1	11.0	11.5	−0.5	6
2	4.0	6.4	−2.4	20
3	6.3	6.1	0.2	3
4	12.0	10.0	2.0	15.5
5	18.2	14.7	3.5	24
6	9.2	7.3	1.9	14
7	7.5	6.1	1.4	12
8	7.1	6.4	0.7	8.5
9	7.2	5.7	1.5	13
10	6.7	6.5	0.2	3
11	14.2	13.2	1.0	10
12	7.3	7.5	−0.2	3
13	9.7	7.4	2.3	18.5
14	9.5	7.2	2.3	18.5
15	5.6	6.3	−0.7	8.5
16	8.7	6.0	2.7	21
17	6.7	7.3	−0.6	7
18	10.2	7.0	3.2	23
19	6.6	5.3	1.3	11
20	11.2	9.0	2.2	17
21	8.6	6.6	2.0	15.5
22	6.1	6.3	−0.2	3
23	10.3	7.2	3.1	22
24	7.0	7.2	−0.2	3
25	12.0	8.0	4.0	25

Negative rank sum $T_- = 50.5$
Positive rank sum $T_+ = 274.5$

To determine if the treatment for tendonitis tends to reduce the thickness of tendons, we test:

H_0: The distribution of the thickness before treatment has the same location as the distribution of the thickness after the treatment

H_a: The distribution of the thickness before the treatment is shifted to the right of that for the thickness after the treatment

The test statistics is T = smaller of T_- and T_+ which is $T_- = 50.5$.

Reject H_0 if $T_- \le T_0$ where T_0 is based on $\alpha = .10$ and $n = 25$ (one-tailed).

Reject H_0 if $T_- \le 101$ (From Table XIII, Appendix A)

(Note: There is no value for $\alpha = .10$ for a one-tailed test. However, if we reject H_0 for $\alpha = .05$, we would also reject H_0 for $\alpha = .10$.)

Since the observed value of the test statistic falls in the rejection region $(T_- = 50.5 \le 101)$, H_0 is rejected. There is sufficient evidence to indicate the treatment for tendonitis tends to reduce the thickness of tendons at $\alpha = .10$.

14.49   Some preliminary calculations are:

Bowler	After 4 Strikes	After 4 Non-Strikes	Difference	Rank of Absolute difference
1	.683	.432	.251	3
2	.684	.400	.284	4
3	.632	.421	.211	2
4	.610	.529	.081	1

Negative rank sum $T_- = 0$
Positive rank sum $T_+ = 10$

a.   To determine if the data support the "hot hand" theory in bowling, we test:

H_0: The distribution of the proportion of strikes after rolling 4 strikes has the same location as the distribution of the proportion of strikes after rolling 4 non-strikes

H_a: The distribution of the proportion of strikes after rolling 4 strikes is shifted to the right of the distribution of the proportion of strikes after rolling 4 non-strikes

The test statistic is the smaller of T_- and T_+ which is $T_- = 0$.

There are no table values in Table XII when $n = 4$. Even if all the observations support rejecting H_0, the p-value is greater than .05. Thus, we are unable to reject H_0. There is insufficient evidence to support the "hot hand" theory at $\alpha = .05$.

b.   When all 43 bowlers are included, the p-value is $p = 0$. Since this p-value is so small, we would reject H_0 for any reasonable value of α. There is sufficient evidence to support the "hot hand" theory at any reasonable value of α.

14.51   The distribution of H is approximately a χ^2 if the null hypothesis is true and if the sample sizes, n_j, for each of the k distributions is more than 5.

14.53   a.   A completely randomized design was used.

b.   The hypotheses are:

H_0: The three probability distributions are identical
H_a: At least two of the three probability distributions differ in location

c.   The rejection region requires $\alpha = .01$ in the upper tail of the χ^2 distribution with df $= k - 1 = 3 - 1 = 2$. From Table VII, Appendix A, $\chi^2_{.01} = 9.21034$. The rejection region is $H > 9.21034$.

d.   Some preliminary calculations are:

I Observation	Rank	II Observation	Rank	III Observation	Rank
34	5.5	24	2	72	14
56	10	18	1	101	20
65	12	27	3	91	19
59	11	41	7	76	16
82	18	34	5.5	80	17
70	13	42	8	75	15
45	9	33	4		
	$R_A = 78.5$		$R_B = 30.5$		$R_C = 101$

The test statistic is $H = \dfrac{12}{n(n+1)} \sum \dfrac{R_j^2}{n_j} - 3(n+1)$

$$= \dfrac{12}{20(20+1)}\left[\dfrac{78.5^2}{7} + \dfrac{30.5^2}{7} + \dfrac{101^2}{6}\right] - 3(20+1) = 77.5252 - 63 = 14.5252$$

Since the observed value of the test statistic falls in the rejection region ($H = 14.5252 >$ 9.21034), H_0 is rejected. There is sufficient evidence to indicate at least two of the three probability distributions differ in location at $\alpha = .01$.

14.55   a – d. The ranks and sums of ranks appear in the table:

Scopolamine	Rank	Placebo	Rank	No Drug	Rank
5	2.5	8	13	8	13
8	13	10	20.5	9	17
8	13	12	26.5	11	23.5
6	6.5	10	20.5	12	26.5
6	6.5	9	17	11	23.5
6	6.5	7	10	10	20.5
6	6.5	9	17	12	26.5
8	13	10	20.5	12	26.5
6	6.5				
4	1				
5	2.5				
6	6.5				
	$R_1 = 84$		$R_2 = 145$		$R_3 = 177$

e. $H = \dfrac{12}{n(n+1)} \sum \dfrac{R_j^2}{n_j} - 3(n+1) = \dfrac{12}{28(28+1)}\left(\dfrac{84^2}{12} + \dfrac{145^2}{8} + \dfrac{177^2}{8}\right) - 3(28+1) = 18.40$

f.   To determine if the distributions of the number of word pairs recalled for the three groups have different locations, we test:

H_0: The three probability distributions are identical
H_a: At least two of the three probability distributions differ in location

The test statistic is $H = 18.40$.

The rejection region requires $\alpha = .05$ in the upper tail of the χ^2 distribution with df $=$ $k - 1 = 3 - 1 = 2$. From Table VII, Appendix A, $\chi^2_{.05} = 5.99147$. The rejection region is $\chi^2 > 5.99147$.

Since the observed value of the test statistic falls in the rejection region ($H = 18.40 > 5.99147$), H_0 is rejected. There is sufficient evidence to indicate the distributions of the number of word pairs recalled for the three groups have different locations at $\alpha = .05$.

g.   Re-ranking the data for only groups 1 and 2, we get:

Scopolamine	Rank	Placebo	Rank
5	2.5	8	12.5
8	12.5	10	18
8	12.5	12	20
6	6.5	10	18
6	6.5	9	15.5
6	6.5	7	10
6	6.5	9	15.5
8	12.5	10	18
6	6.5		
4	1		
5	2.5		
6	6.5		
	$T_1 = 82.5$		$T_2 = 127.5$

To determine if the distributions of the number of word pairs recalled for the Scopolamine group is shifted to the left of that for the Placebo group, we test:

H_0: The two probability distributions are identical
H_a: The probability distribution of the Scopolamine group is shifted to the left of that for the Placebo group

The test statistic is the T_2 because $n_2 < n_1$.

There is no table value for this problem since $n_1 = 12$ and the largest n in Table XII is 10. We will use the normal approximation even though both sample sizes are not greater than or equal to 10 ($n_2 = 8$).

The test statistic is $z = \dfrac{T_1 - \dfrac{n_1(n_1 + n_2 + 1)}{2}}{\sqrt{\dfrac{n_1 n_2 (n_1 + n_2 + 1)}{12}}} = \dfrac{82.5 - \dfrac{12(12 + 8 + 1)}{2}}{\sqrt{\dfrac{12(8)(12 + 8 + 1)}{12}}} = -3.36$

The rejection region requires $\alpha = .05$ in the lower tail of the z distribution. From Table IV, Appendix A, $z_{.05} = 1.645$. The rejection region is $z < -1.645$.

Since the observed value of the test statistic falls in the rejection region ($z = -3.36 < -1.645$), H_0 is rejected. There is sufficient evidence to indicate the distributions of the number of word pairs recalled for the Scopolamine group is shifted to the left of that for the Placebo group at $\alpha = .05$.

14.57   a.   To determine if the Group B subjects exercise for longer durations than the Group A subjects, we test:

H_0:   The two sampled populations have identical probability distributions
H_a: The group B probability distribution is shifted to the right of that of the probability distribution of Group A

b.   The test statistic is $H = 5.1429$.

The p-value is $p = .0233$. Since the p-value is so small, H_0 would be rejected for any $\alpha > .0233$. There is sufficient evidence to indicate that the Group B subjects exercise for longer durations than the Group A subjects for $\alpha > .0233$.

c.   For the Kruskal-Wallis H-test to be valid, the following assumptions must be met:

1.   The two samples are random and independent
2.   There are five or more measurements in each sample.
3.   The two probability distributions from which the samples are drawn are continuous.

For this problem, the second assumption is not met. One of the samples contained only four observations.

Since there are only two populations, the Wilcoxon Rank Sum test can be used. For this test, the only necessary assumptions are:

1.   The two samples are random and independent.
2.   The two probability distributions from which the samples are drawn are continuous.

14.59 Some preliminary calculations are:

AR	Rank	AC	Rank	A	Rank	P	Rank
.51	38	.50	36	.16	11.5	.58	41.5
.58	41.5	.30	21	.10	7.5	.12	9
.52	39	.47	33.5	.20	15	.62	44
.47	33.5	.36	25	.29	20	.43	29.5
.61	43	.39	26	-.14	3	.26	19
.00	4	.22	18	.18	13	.50	36
.32	23	.20	15	-.35	1	.44	31
.53	40	.21	17	.31	22	.20	15
.50	36	.15	10	.16	11.5	.42	28
.46	32	.10	7.5	.04	6	.43	29.5
.34	24	.02	5	-.25	2	.40	27
	$R_1 = 354$		$R_2 = 214$		$R_3 = 112.5$		$R_4 = 309.5$

To determine if the distributions of the task scores differ in location among the 4 groups, we test:

H_0: The distributions of the task scores for the 4 groups are identical
H_a: At least two of the four distributions differ in location

The test statistic is

$$H = \frac{12}{n(n+1)} \sum \frac{R_j^2}{n_j} - 3(n+1)$$

$$= \frac{12}{44(44+1)} \left[\frac{354^2}{11} + \frac{214^2}{11} + \frac{112.5^2}{11} + \frac{309.5^2}{11} \right] - 3(44+1)$$

$$= 154.027 - 135 = 19.027$$

The rejection region requires $\alpha = .05$ in the upper tail of the χ^2 distribution with df $= k - 1 = 4 - 1 = 3$. From Table VII, Appendix A, $\chi^2_{.05} = 7.81473$. The rejection region is $H > 7.81473$.

Since the observed value of the test statistic falls in the rejection region ($H = 19.027 > 7.81473$), H_0 is rejected. There is sufficient evidence to indicate the distributions of the task scores differ in location among the 4 groups at $\alpha = .05$. We can infer that the task scores are affected by the amount of alcohol and reward presented.

14.61   a.   Using MINITAB, the histograms of the three data sets are:

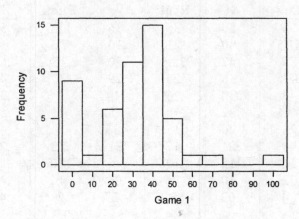

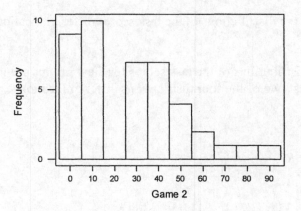

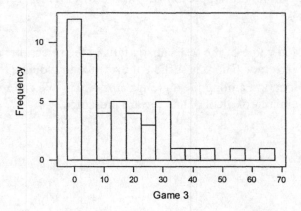

None of these three graphs look approximately normal. Thus, the assumption of normality necessary for ANOVA is likely violated.

b. Using MINITAB, the results of the Kruskal-Wallis analysis are:

```
Kruskal-Wallis Test on Percent

Game        N     Median    Ave Rank        Z
1           50    32.000    83.1          2.88
2           42    26.000    73.0          0.59
3           47     9.000    53.3         -3.49
Overall    139               70.0

H = 13.61  DF = 2  P = 0.001
H = 13.66  DF = 2  P = 0.001 (adjusted for ties)
```

To determine if the percentages of names recalled are different for the three retrieval methods, we test:

H_0: The distributions of the three retrieval methods are identical
H_a: At least two of the three retrieval methods have probability distributions that differ in location

From the printout, the test statistic is $H = 13.66$ and the p-value is $p = .001$. Since the p-value is less than α ($p = .001 < .05$), H_0 is rejected. There is sufficient evidence to indicate that the percentages of names recalled are different for the three retrieval methods at $\alpha = .05$.

14.63 To correctly rank the data in a randomized block design, we would use b - for each block, rank the data across the treatments from smallest to largest.

14.65 a. The number of blocks, b, is 6.

b. H_0: The probability distributions for the four treatments are identical
H_a: At least two of the probability distributions differ in location

c. The test statistic is $F_r = \dfrac{12}{bk(k+1)} \sum R_j^2 - 3b(k+1)$

$$= \dfrac{12}{6(4)(4+1)}[11^2 + 21^2 + 21^2 + 7^2] - 3(6)(4+1) = 105.2 - 90 = 15.2$$

The rejection region requires $\alpha = .10$ in the upper tail of the χ^2 distribution with df = $k - 1 = 4 - 1 = 3$. From Table VII, Appendix A, $\chi^2_{.10} = 6.25139$. The rejection region is $F_r > 6.25139$.

Since the observed value of the test statistic falls in the rejection region ($F_r = 15.2 > 6.25139$), reject H_0. There is sufficient evidence to indicate a difference in the location of at least two of the four treatments at $\alpha = .10$.

d. The p-value = $P(F_r \geq 15.2) = P(\chi^2 \geq 15.2)$. With 3 degrees of freedom, $F_r = 15.2$ falls above $\chi^2_{.005}$; therefore, p-value $< .005$.

e.   The means are:

$$\bar{R}_A = \frac{R_A}{b} = \frac{11}{6} = 1.833 \qquad\qquad \bar{R}_B = \frac{R_B}{b} = \frac{21}{6} = 3.5$$

$$\bar{R}_C = \frac{R_C}{b} = \frac{21}{6} = 3.5 \qquad\qquad \bar{R}_D = \frac{R_D}{b} = \frac{7}{6} = 1.167$$

$$\bar{R} = \frac{1}{2}(k+1) = \frac{1}{2}(4+1) = 2.5$$

$$F_r = \frac{12}{k(k+1)}\sum b(\bar{R}_j - \bar{R})^2$$

$$= \frac{12}{(4)(4+1)}[6(1.833-2.5)^2 + 6(3.5-2.5)^2 + 6(3.5-2.5)^2 + 6(1.167 - 2.5)^2]$$

$$= .6[25.3307] = 15.2$$

14.67 $R_1 = 16 \qquad R_2 = 7 \qquad R_3 = 23 \qquad R_4 = 14$

To determine if at least two of the treatment probability distributions differ in location, we test:

H_0: The probability distributions of the four treatments are identical
H_a: At least two of the probability distributions differ in location

The test statistic is $F_r = \dfrac{12}{bk(k+1)}\sum R_j^2 - 3b(k+1)$

$$= \frac{12}{6(4)(4+1)}[16^2 + 7^2 + 23^2 + 14^2] - 3(6)(4+1) = 103 - 90 = 13$$

The rejection region requires $\alpha = .05$ in the upper tail of the χ^2 distribution with df $= k - 1 = 4 - 1 = 3$. From Table VII, Appendix A, $\chi^2_{.05} = 7.84173$. The rejection region is $F_r > 7.81473$.

Since the observed value of the test statistic falls in the rejection region ($F_r = 13 > 7.81473$), reject H_0. There is sufficient evidence to indicate a difference in the location for at least two of the probability distributions at $\alpha = .05$.

14.69   a.   From the printout, the rank sums are: $R_1 = 27.0$, $R_2 = 32.5$, $R_3 = 29.0$, and $R_4 = 31.5$.

b. $F_r = \dfrac{12}{bk(k+1)}\sum R_j^2 - 3b(k+1) = \dfrac{12}{12(4)(4+1)}(27^2 + 32.5^2 + 29^2 + 31.5^2) - 3(12)(4+1) = .925$

c.   From the printout, $F_r = S = .93$ and the p-value is p $= .819$.

d.   To determine if the atlas theme ranking distributions of the four groups differ, we test:

H_0: The probability distributions of the atlas theme rankings are the same for the four groups
H_a: The probability distributions of the atlas theme rankings differ in location

The test statistic is $F_r = .925$ and the p-value is $p = .819$. Since the p-value is so large, there is no evidence to reject H_0 for any reasonable value of α. There is insufficient evidence to indicate that the atlas theme ranking distributions of the four groups differ.

14.71   a.   The rejection region requires $\alpha = .01$ in the upper tail of the χ^2 distribution with df = $k - 1 = 3 - 1 = 2$. From Table VII, Appendix A, $\chi^2_{.01} = 9.21034$. The rejection region is $\chi^2 > 9.21034$.

b.   Since the observed value of the test statistic falls in the rejection region ($\chi^2 = 19.16 > 9.21034$), H_0 is rejected. There is sufficient evidence to indicate the distributions of proportions of eye fixations on the interior mirror differ in location among the 3 treatments at $\alpha = .01$.

c.   Since the observed value of the test statistic does not fall in the rejection region ($\chi^2 = 7.80 \ngtr 9.21034$), H_0 is not rejected. There is insufficient evidence to indicate the distributions of proportions of eye fixations on the off-side mirror differ in location among the 3 treatments at $\alpha = .01$.

d.   Since the observed value of the test statistic falls in the rejection region ($\chi^2 = 20.67 > 9.21034$), H_0 is rejected. There is sufficient evidence to indicate the distributions of proportions of eye fixations on the speedometer differ in location among the 3 treatments at $\alpha = .01$.

14.73   Some preliminary calculations are:

Boxer	M1	Rank	R1	Rank	M5	Rank	R5	Rank
1	1243	1	1244	2	1291	4	1262	3
2	1147	2	1053	1	1169	3	1177	4
3	1247	1	1375	4	1309	2	1321	3
4	1274	2	1235	1	1290	4	1285	3
5	1177	2	1139	1	1233	3	1238	4
6	1336	2	1313	1	1366	4	1362	3
7	1238	1	1279	4	1275	3	1261	2
8	1261	2	1152	1	1289	4	1266	3
		$R_1 = 13$		$R_2 = 15$		$R_3 = 27$		$R_4 = 25$

To compare the punching power means of the four interventions, we test:

H_0: The four probability distributions of punching power are the same
H_a: At least two of the probability distributions of punching power differ in location

The test statistic is

$$F_r = \frac{12}{bk(k+1)} \sum R_j^2 - 3b(k+1) = \frac{12}{8(4)(4+1)}(13^2 + 15^2 + 27^2 + 25^2) - 3(8)(4+1) = 11.1$$

Since no α was given, we will use $\alpha = .05$. The rejection region requires $\alpha = .05$ in the upper tail of the χ^2 distribution with df = $k - 1 = 4 - 1 = 3$. From Table VII, Appendix A, $\chi^2_{.05} = 7.81473$. The rejection region is $F_r > 7.81473$.

Since the observed value of the test statistic falls in the rejection region ($F_r = 11.1 > 7.81473$), H_0 is rejected. There is sufficient evidence to indicate that the punching power means of the four interventions differ at $\alpha = .05$.

This is the same result that we obtained in Exercise 10.68.

14.75 The ranks and sums of ranks within each treatment are:

Week	Monday	Tuesday	Wednesday	Thursday	Friday
1	5	1	4	2	3
2	5	4	3	1	2
3	2.5	2.5	5	1	4
4	2	1	3.5	5	3.5
5	5	1	2	3	4
6	4	2	3	1	5
7	5	3.5	1.5	3.5	1.5
8	4	2	1	3	5
9	1	2	5	3	4
	$R_1 = 33.5$	$R_2 = 19$	$R_3 = 28$	$R_4 = 22.5$	$R_5 = 32$

To determine if the distributions of absentee rates differ in location fro the five days of the work week, we test:

H_0: The probabilities of the five days of the week are identical
H_a: The distributions of at least two days of the week differ in location

The test statistic is

$$F_r = \frac{12}{bk(k+1)} \sum R_j^2 - 3b(k+1)$$

$$= \frac{12}{9(5)(5+1)} \left(33.5^2 + 19^2 + 28^2 + 22.5^2 + 32^2\right) - 3(9)(5+1) = 6.778$$

Since no alpha level was given, we will use $\alpha = .05$. The rejection region requires $\alpha = .05$ in the upper tail of the χ^2 distribution with df $= k - 1 = 5 - 1 = 4$. From Table VII, Appendix A, $\chi^2_{.05} = 9.48773$. The rejection region is $\chi^2 > 9.48773$.

Since the observed value of the test statistic does not fall in the rejection region ($\chi^2 = 6.778 \not> 9.48773$), H_0 is not rejected. There is insufficient evidence to indicate the distributions of absentee rates differ in location fro the five days of the work week at $\alpha = .05$.

14.77 The conditions required for a valid Spearman's test are:

1. The sample of experimental units on which the two variables are measured is randomly selected.
2. The probability distributions of the two variables are continuous.

14.79 a. From Table XIV with $n = 10$, $r_{s,\alpha/2} = r_{s,.025} = .648$. The rejection region is $r_s > .648$ or $r_s < -.648$.

b. From Table XIV with $n = 20$, $r_{s,\alpha} = r_{s,.025} = .450$. The rejection region is $r_s > .450$.

c. From Table XIV with $n = 30$, $r_{s,\alpha} = r_{s,.01} = .432$. The rejection region is $r_s < -.432$.

14.81   a. H_0: $\rho_s = 0$
             H_a: $\rho_s \neq 0$

b.   The test statistic is $r_s = \dfrac{SS_{uv}}{\sqrt{SS_{uu}SS_{vv}}}$

x	Rank, u	y	Rank, v	u^2	v^2	uv
0	3	0	1.5	9	2.25	45
3	5.5	2	5	30.25	25	27.5
0	3	2	5	9	25	15
−4	1	0	1.5	1	2.25	1.5
3	5.5	3	7	30.25	49	38.5
0	3	1	3	9	9	9
4	7	2	5	49	25	35
	$\sum u = 28$		$\sum v = 28$	$\sum u^2 = 137.5$	$\sum v^2 = 137.5$	$\sum uv = 131$

$$SS_{uv} = \sum uv - \frac{\left(\sum u\right)\left(\sum v\right)}{n} = 131 - \frac{28(28)}{7} = 19$$

$$SS_{uu} = \sum u^2 - \frac{\left(\sum u\right)^2}{n} = 137.5 - \frac{(28)^2}{7} = 25.5$$

$$SS_{vv} = \sum v^2 - \frac{\left(\sum v\right)^2}{n} = 137.5 - \frac{(28)^2}{7} = 25.5$$

$$r_s = \frac{19}{\sqrt{25.5(25.5)}} = .745$$

Reject H_0 if $r_s < -r_{s,\alpha/2}$ or $r_s > r_{s,\alpha/2}$ where $\alpha/2 = .025$ and $n = 7$:

Reject H_0 if $r_s < -.786$ or $r_s > .786$ (from Table XIV, Appendix A).

Since the observed value of the test statistic does not fall in the rejection region, ($r_s = .745 \not> .786$), do not reject H_0. There is insufficient evidence to indicate x and y are correlated at $\alpha = .05$.

c.   The p-value is $P(r_s \geq .745) + P(r_s \leq -.745)$. For $n = 7$, $r_s = .745$ is above $r_{s,.025}$ where $\alpha/2 = .025$ and below $r_{s,.05}$ where $\alpha/2 = .05$. Therefore, $2(.025) = .05 < p\text{-value} < 2(.05) = .10$.

d.   The assumptions of the test are that the samples are randomly selected and the probability distributions of the two variables are continuous.

14.83 a, b. The ranks and Spearman correlation coefficient are:

Porosity	Rank, u	Diameter	Rank, v	u^2	v^2	uv
18.8	5	12.0	5	25	25	25
18.3	4	9.7	3	16	9	12
16.3	2	7.3	2	4	4	4
6.9	1	5.3	1	1	1	1
17.1	3	10.9	4	9	16	12
20.4	6	16.8	6	36	36	36
	$\sum u = 21$		$\sum v = 21$	$\sum u^2 = 91$	$\sum v^2 = 91$	$\sum uv = 90$

$$SS_{uv} = \sum uv - \frac{\left(\sum u\right)\left(\sum v\right)}{n} = 90 - \frac{21(21)}{6} = 16.5$$

$$SS_{uu} = \sum u^2 - \frac{\left(\sum u\right)^2}{n} = 91 - \frac{21^2}{6} = 17.5$$

$$SS_{vv} = \sum v^2 - \frac{\left(\sum v\right)^2}{n} = 91 - \frac{21^2}{6} = 17.5$$

$$r_s = \frac{SS_{uv}}{\sqrt{SS_{uu} SS_{vv}}} = \frac{16.5}{\sqrt{17.5(17.5)}} = .9429$$

Since r_s is very close to 1, apparent porosity and mean pore diameter are strongly, positively related.

c. To determine if apparent porosity and mean pore diameter have a positive rank correlation, we test:

H_o: $\rho_s = 0$
H_a: $\rho_s > 0$

The test statistic is $r_s = .943$.

Reject H_0 if $r_s > r_{s, \alpha}$ where $\alpha = .01$ and n = 6.

Reject H_0 if $r_s > .943$ (From Table XIV, Appendix A)

Since the observed value of r_s does not fall in the rejection region ($r_s = .9429 \not> .943$), H_0 is not rejected. There is insufficient evidence of a positive rank correlation between apparent porosity and mean pore diameter at $\alpha = .01$.

14.85   a.   Some preliminary calculations are:

Blood Lactate Level	Rank u	Perceived Recovery	Rank v	u^2	v^2	uv
3.8	3	7	1.5	9	2.25	4.5
4.2	5.5	7	1.5	30.25	2.25	8.25
4.8	7	11	3	49	9	21
4.1	4	12	5	16	25	20
5.0	8	12	5	64	25	40
5.3	9.5	12	5	90.25	25	47.5
4.2	5.5	13	7	30.25	49	38.5
2.4	1	17	9	1	81	9
3.7	2	17	9	4	81	18
5.3	9.5	17	9	90.25	81	85.5
5.8	12	18	11.5	144	132.25	138
6.0	14	18	11.5	196	132.25	161
5.9	13	21	14.5	169	210.25	188.5
6.3	15	21	14.5	225	210.25	217.5
5.5	12	20	13	121	169	143
6.5	16	24	16	256	256	256

$$\sum u = 136 \qquad \sum v = 136 \quad \sum u^2 = 1495 \quad \sum v^2 = 1490.5 \quad \sum uv = 1396.25$$

$$\text{SS}_{uv} = \sum uv - \frac{\left(\sum u\right)\left(\sum v\right)}{n} = 1396.25 - \frac{136(136)}{16} = 240.25$$

$$\text{SS}_{uu} = \sum u^2 - \frac{\left(\sum u\right)^2}{n} = 1495 - \frac{136^2}{16} = 339$$

$$\text{SS}_{vv} = \sum v^2 - \frac{\left(\sum v\right)^2}{n} = 1490.5 - \frac{136^2}{16} = 334.5$$

a.   The ranks of the blood lactate levels are stored in the "u" column above.

b.   The ranks of the perceived recovery values are stored in the "v" column above.

c. $r_s = \dfrac{\text{SS}_{uv}}{\sqrt{\text{SS}_{uu}\text{SS}_{vv}}} = \dfrac{240.25}{\sqrt{339(334.5)}} = .713$

Since r_s is relatively close to 1, blood lactate level and perceived recovery have a fairly strong, positive relationship.

d.   Reject H_0 if $r_s < -r_{s,\alpha/2}$ or if $r_s > r_{s,\alpha/2}$ where $\alpha/2 = .05$ and $n = 16$;
Reject H_0 if $r_s < -.425$ or $r_s > .425$ (from Table XIV, Appendix A.)

e. To determine if blood lactate level and perceived recovery are rank correlated, we test:

H_0: Blood lactate level and Perceived recovery are not rank correlated
H_a: Blood lactate level and Perceived recovery are rank correlated
The test statistic is $r_s = .713$.

Since the observed value of the test statistic falls in the rejection region, ($r_s = .713 >$ .425), H_0 is rejected. There is sufficient evidence to indicate that blood lactate level and perceived recovery are rank correlated at $\alpha = .10$.

14.87 a. Since $r_s = .643$ is not real close to 1, there is a moderate positive relationship between the public agenda score and the media agenda score.

b. Since $r_s = .714$ is somewhat close to 1, there is a positive relationship between the public agenda score and the media agenda score, once 'length of war' is removed.

c. To determine if there is a positive rank correlation between the public agenda score and the media agenda score, we test:

H_0: $\rho_s = 0$
H_a: $\rho_s > 0$

The test statistic is $r_s = .714$.

Reject H_0 if $r_s > r_{s, \alpha}$ where $\alpha = .01$ and $n = 235$. The largest n in Table XIV is $n = 30$. The critical value for $n = 30$ is .432. As n increases, the critical value decreases. Thus, if we reject H_0 with $n = 30$, we would also reject H_0 with $n = 235$.

Reject H_0 if $r_s > .432$ (From Table XIV, Appendix A, $n = 30$)

Since the observed value of r_s falls in the rejection region ($r_s = .714 > .432$), H_0 is rejected. There is sufficient evidence of a positive rank correlation between the public agenda score and the media agenda score at $\alpha = .01$.

14.89   a.   Some preliminary calculations are:

FCAT-Math	Rank, u	% Below Poverty	Rank, v	u^2	v^2	uv
166.4	7	91.7	22	49	484	154
159.6	4	90.2	21	16	441	84
159.1	3	86.0	20	9	400	60
155.5	1	83.9	19	1	361	19
164.3	5	80.4	18	25	324	90
169.8	9	76.5	17	81	289	153
155.7	2	76.0	16	4	256	32
165.2	6	75.8	15	36	225	90
175.4	13	75.6	14	169	196	182
178.1	16	75.0	13	256	169	208
167.1	8	74.7	12	64	144	96
177.0	15	63.2	11	225	121	165
174.2	11	52.9	10	121	100	110
175.6	14	48.5	9	196	81	126
170.8	10	39.1	8	100	64	80
175.1	12	38.4	7	144	49	84
182.8	20.5	34.3	6	420.25	36	123
180.3	18	30.3	4.5	324	20.25	81
178.8	17	30.3	4.5	289	20.25	76.5
181.4	19	29.6	3	361	9	57
182.8	20.5	26.5	2	420.25	4	41
186.1	22	13.8	1	484	1	22
	$\sum u = 253$		$\sum v = 253$	$\sum u^2 = 3794.5$	$\sum v^2 = 3794.5$	$\sum uv = 2133.5$

$$SS_{uv} = \sum uv - \frac{\sum u \sum v}{n} = 2133.5 - \frac{253(253)}{22} = -776$$

$$SS_{uu} = \sum u^2 - \frac{\left(\sum u\right)^2}{n} = 3794.5 - \frac{253^2}{22} = 885$$

$$SS_{vv} = \sum v^2 - \frac{\left(\sum v\right)^2}{n} = 3794.5 - \frac{253^2}{22} = 885$$

$$r_s = \frac{SS_{uv}}{\sqrt{SS_{uu}SS_{vv}}} = \frac{-776}{\sqrt{885(885)}} = -.877$$

Since r_s is very close to -1, there is a fairly strong, negative relationship between FCAT math scores and the percentage of students below the poverty level.

b.   Some preliminary calculations are:

FCAT-Read	Rank, u	% Below Poverty	Rank, v	u^2	v^2	uv
165.0	7.5	91.7	22	56.25	484	165
157.2	1	90.2	21	1	441	21
164.4	5	86.0	20	25	400	100
162.4	3	83.9	19	9	361	57
162.5	4	80.4	18	16	324	72
164.9	6	76.5	17	36	289	102
162.0	2	76.0	16	4	256	32
165.0	7.5	75.8	15	56.25	225	112.5
173.7	14	75.6	14	196	196	196
171.0	11	75.0	13	121	169	143
169.4	9	74.7	12	81	144	108
172.9	13	63.2	11	169	121	143
172.7	12	52.9	10	144	100	120
174.9	16	48.5	9	256	81	144
174.8	15	39.1	8	225	64	120
170.1	10	38.4	7	100	49	70
181.4	20	34.3	6	400	36	120
180.6	19	30.3	4.5	361	20.25	85.5
178.0	18	30.3	4.5	324	20.25	81
175.9	17	29.6	3	289	9	51
181.6	21	26.5	2	441	4	42
183.8	22	13.8	1	484	1	22
	$\sum u = 253$		$\sum v = 253$	$\sum u^2 = 3794.5$	$\sum v^2 = 3794.5$	$\sum uv = 2107$

$$SS_{uv} = \sum uv - \frac{\sum u \sum v}{n} = 2107 - \frac{253(253)}{22} = -802.5$$

$$SS_{uu} = \sum u^2 - \frac{\left(\sum u\right)^2}{n} = 3794.5 - \frac{253^2}{22} = 885$$

$$SS_{vv} = \sum v^2 - \frac{\left(\sum v\right)^2}{n} = 3794.5 - \frac{253^2}{22} = 885$$

$$r_s = \frac{SS_{uv}}{\sqrt{SS_{uu}SS_{vv}}} = \frac{-802.5}{\sqrt{885(885)}} = -.907$$

Since r_s is very close to -1, there is a fairly strong, negative relationship between FCAT reading scores and the percentage of students below the poverty level.

c. To determine whether the FCAT math scores and the percentage of students below the poverty level are negatively rank correlated, we test:

H_0: $\rho_s = 0$
H_a: $\rho_s < 0$

The test statistic is $r_s = -.877$.

Reject H_0 if $r_s < -r_{s,.01}$ with $n = 22$

Reject H_0 if $r_s < -.508$ (From Table XIV, Appendix A)

Since the observed value of the test statistic falls in the rejection region ($r_s = -.877 < -.508$), H_0 is rejected. There is sufficient evidence to indicate FCAT math scores and the percentage of students below the poverty level are negatively rank correlated at $\alpha = .01$.

d. To determine whether the FCAT reading scores and the percentage of students below the poverty level are negatively rank correlated, we test:

H_0: $\rho_s = 0$
H_a: $\rho_s < 0$

The test statistic is $r_s = -.907$.

Reject H_0 if $r_s < -r_{s,.01}$ with $n = 22$

Reject H_0 if $r_s < -.508$ (From Table XIV, Appendix A)

Since the observed value of the test statistic falls in the rejection region ($r_s = -.907 < -.508$), H_0 is rejected. There is sufficient evidence to indicate FCAT reading scores and the percentage of students below the poverty level are negatively rank correlated at $\alpha = .01$.

14.91 a. Some preliminary calculations are:

Change KSADS-MRS	Rank, u	Improve CGI-BP	Rank, v	u^2	v^2	uv
80	18	6	18	324	324	324
65	17	5	17	289	289	289
20	15.5	4	15	240.25	225	232.5
−15	13	4	15	169	225	195
−50	9	4	15	81	225	135
20	15.5	3	12	240.25	144	186
−30	11	3	12	121	144	132
−70	5.5	3	12	30.25	144	66
−10	14	2	6	196	36	84
−25	12	2	6	144	36	72
−35	10	2	6	100	36	60
−65	7.5	2	6	56.25	36	45
−65	7.5	2	6	26.25	36	45
−70	5.5	2	6	30.25	36	33
−80	4	2	6	16	36	24
−90	2.5	2	6	6.25	36	15
−95	1	2	6	1	36	6
−90	2.5	1	1	6.25	1	2.5
	$\sum u = 171$		$\sum v = 171$	$\sum u^2 = 2107$	$\sum v^2 = 2045$	$\sum uv = 1946$

$$SS_{uv} = \sum uv - \frac{\sum u \sum v}{n} = 1946 - \frac{171(171)}{18} = 321.5$$

$$SS_{uu} = \sum u^2 - \frac{\left(\sum u\right)^2}{n} = 2107 - \frac{171^2}{18} = 482.5$$

$$SS_{vv} = \sum v^2 - \frac{\left(\sum v\right)^2}{n} = 2045 - \frac{171^2}{18} = 420.5$$

$$r_s = \frac{SS_{uv}}{\sqrt{SS_{uu}SS_{vv}}} = \frac{321.5}{\sqrt{482.5(420.5)}} = .714$$

b.   To determine if there is a positive rank correlation between the two test score changes in the population of all pediatric patients with manic symptoms, we test:

H_0: $\rho_s = 0$
H_a: $\rho_s > 0$

The test statistic is $r_s = .714$.

Reject H_0 if $r_s > r_{s, .05}$ with $n = 18$

Reject H_0 if $r_s > .399$ (From Table XIV, Appendix A)

Since the observed value of the test statistic falls in the rejection region ($r_s = .714 > .399$), H_0 is rejected. There is sufficient evidence to indicate there is a positive rank correlation between the two test score changes in the population of all pediatric patients with manic symptoms at $\alpha = .05$.

14.93 It is appropriate to use the t and F tests for comparing two or more population means when the populations sampled are normal and the population variances are equal.

14.95 a.   To compare two populations with independent samples, one would use the Wilcoxon Rank Sum test.

b.   To make an inference about a population median, one would use the sign test.

c.   To compare three or more populations with independent samples, one would use the Kruskal-Wallis H-test.

d.   To make inferences about a rank correlation, one would use the Spearman's rank correlation coefficient test.

e.   To compare two populations with matched pairs, one would use the Wilcoxon Signed Rank Test.

f.   To compare three or more populations with a block design, one would use the Friedman's F_r-test.

14.97   a.   Some preliminary calculations are:

Pair	X	Rank u_i	Y	Rank v_i	u_i^2	v_i^2	$u_i v_i$
1	19	5	12	5	25	25	25
2	27	7	19	8	49	64	56
3	15	2	7	1	4	1	2
4	35	9	25	9	81	81	81
5	13	1	11	4	1	16	4
6	29	8	10	2.5	64	6.25	20
7	16	3.5	16	6	12.25	36	21
8	22	6	10	2.5	36	6.25	15
9	16	3.5	18	7	12.25	49	24.5
		$\sum u_i = 45$		$\sum v_i = 45$	$\sum u_i^2 = 284.5$	$\sum v_i^2 = 284.5$	$\sum u_i v_i = 248.5$

$$SS_{uv} = \sum u_i v_i - \frac{\sum u_i v_i}{n} = 248.5 - \frac{45(45)}{9} = 23.5$$

$$SS_{uu} = \sum u_i^2 - \frac{\left(\sum u_i\right)^2}{n} = 284.5 - \frac{45^2}{9} = 59.5$$

$$SS_{vv} = \sum v_i^2 - \frac{\left(\sum v_i\right)^2}{n} = 284.5 - \frac{45^2}{9} = 59.5$$

To determine if the Spearman rank correlation differs from 0, we test:

$H_0: \rho_s = 0$
$H_a: \rho_s \neq 0$

The test statistic is $r_s = \dfrac{SS_{uv}}{\sqrt{SS_{uu}SS_{vv}}} = \dfrac{23.5}{\sqrt{59.5(59.5)}} = .40$

Reject H_0 if $r_s < -r_{s,\alpha/2}$ or if $r_s > r_{s,\alpha/2}$ where $\alpha/2 = .025$ and $n = 9$:

Reject H_0 if $r_s < -.683$ or if $r_s > .683$ (from Table XIV, Appendix A)

Since the observed value of the test statistic does not fall in the rejection region ($r_s = .40$ $\not>$ .683), H_0 is not rejected. There is insufficient evidence to indicate that Spearman's rank correlation between x and y is significantly different from 0 at $\alpha = .05$.

b. Use the Wilcoxon signed rank test. Some preliminary calculations are:

Pair	X	Y	Difference	Rank of Absolute Difference
1	19	12	7	3
2	27	19	8	4.5
3	15	7	8	4.5
4	35	25	10	6
5	13	11	2	1.5
6	29	10	19	8
7	16	16	0	(eliminated)
8	22	10	12	7
9	16	18	−2	1.5
				$T_- = 1.5$

To determine if the probability distribution of x is shifted to the right of that for y, we test:

H_0: The probability distributions are identical for the two variables

H_a: The probability distribution of x is shifted to the right of the probability distribution of y

The test statistic is $T = T_- = 1.5$

Reject H_0 if $T \leq T_0$ where T_0 is based on $\alpha = .05$ and $n = 8$ (one-tailed):

Reject H_0 if $T \leq 6$ (from Table XIII, Appendix A).

Since the observed value of the test statistic falls in the rejection region ($T = 1.5 \leq 6$), reject H_0 at $\alpha = .05$. There is sufficient evidence to conclude that the probability distribution of x is shifted to the right of that for y.

14.99 Some preliminary calculations are:

Block	1	Rank	2	Rank	3	Rank	4	Rank	5	Rank
1	75	4	65	1	74	3	80	5	69	2
2	77	3	69	1	78	4	80	5	72	2
3	70	4	63	1.5	69	3	75	5	63	1.5
4	80	3.5	69	1	80	3.5	86	5	77	2
		$R_1 = 14.5$		$R_2 = 4.5$		$R_3 = 13.5$		$R_4 = 20$		$R_5 = 7.5$

To determine whether at least two of the treatment probability distributions differ in location, use Friedman F_r test.

H_0: The five treatments have identical probability distributions

H_a: At least two of the populations have probability distributions that differ in location

The test statistic is $F_r = \dfrac{12}{bk(k+1)} \sum R_j^2 - 3b(k+1)$

$$= \frac{12}{4(5)(6)} [(14.5)^2 + (4.5)^2 + (13.5)^2 + (20)^2 + (7.5)^2] - 3(4)(6) = 14.9$$

The rejection region requires $\alpha = .05$ in the upper tail of the χ^2 distribution with df $= k - 1$ $= 5 - 1 = 4$. From Table VII, Appendix A, $= 9.48773$. The rejection region is $F_r > 9.48773$.

Since the observed value of the test statistic falls in the rejection region ($F_r = 14.9 >$ 9.48773), H_0 is rejected. There is sufficient evidence to indicate that at least two of the treatment means differ in location at $\alpha = .05$.

14.101

Test A	Rank	Test B	Rank
90	12	66	4
71	6	78	8
83	11	50	1
82	10	68	5
75	7	80	9
91	13	60	2
65	3		$T_2 = 29$
	$T_1 = 62$		

H_0: The probability distributions of scores for tests A and B are identical
H_a: There is a shift in the locations of the probability distributions of scores for tests A and B

The test statistic is $T_2 = 29$.

The null hypothesis will be rejected if $T_2 \le T_L$ or $T_2 \ge T_U$ where $\alpha = .05$ (two-tailed), $n_1 = 6$ and $n_2 = 7$. From Table XII, Appendix A, $T_L = 28$ and $T_U = 56$.

Reject H_0 if $T_2 \le 28$ or $T_2 \ge 56$.

Since $T_2 = 29 \not\le 28$ and $T_2 = 29 \not\ge 56$, do not reject H_0. There is insufficient evidence to indicate a shift in location for tests A and B at $\alpha = .05$.

14.103 Some preliminary calculations are:

Location	Temephos	Rank	Malsathion	Rank	Fenitrothion	Rank	Fenthion	Rank	Chlorpyrifos	Rank
Anguilla	4.6	5	1.2	1	1.5	2.5	1.8	4	1.5	2.5
Antigua	9.2	5	2.9	3	2.0	1.5	7.0	4	2.0	1.5
Dominica	7.8	5	1.4	1	2.4	2	4.2	4	4.1	3
Guyana	1.7	2	1.9	4	2.2	5	1.5	1	1.8	3
Jamaica	3.4	3	3.7	4	2.0	2	1.5	1	7.1	5
St. Lucia	6.7	4	2.7	1.5	2.7	1.5	4.8	3	8.7	5
Suriname	1.4	1	1.9	3	2.0	4	2.1	5	1.7	2
		$R_1 = 13$		$R_2 = 15$		$R_3 = 18.5$		$R_4 = 22$		$R_5 = 22$

To determine if the resistance ratio distributions of the 5 insecticides differ, we test:

H_0: The distributions of the 5 insecticide ratios are the same
H_a: At least two of the distributions of insecticide ratios differ

The test statistic is $F_r = \dfrac{12}{bk(k+1)}\sum R_j^2 - 3b(k+1)$

$$= \dfrac{12}{7(5)(5+1)}(25^2 + 17.5^2 + 18.5^2 + 22^2 + 22^2) - 3(7)(5+1) = 2.086$$

Since no α was given, we will use $\alpha = .05$. The rejection region requires $\alpha = .05$ in the upper tail of the χ^2 distribution with df $= k - 1 = 5 - 1 = 4$. From Table VII, Appendix A, $\chi^2_{.05} = 9.48773$. The rejection region is $F_r > 9.48773$.

Since the observed value of the test statistic does not fall in the rejection region ($F_r = 2.086 \not> 9.48773$), H_0 is not rejected. There is insufficient evidence to indicate that the resistance ratio distributions of the 5 insecticides differ at $\alpha = .05$.

14.105 a.   Some preliminary calculations are:

Week	Number of Strikes	Rank u_i	Age of fish	Rank v_i	$d_i=u_i-v_i$	d_i^2
1	85	9	120	1	−8	64
2	63	8	136	2	−6	36
3	34	3	150	3	0	0
4	39	5	155	4	1	1
5	58	7	162	5	2	4
6	35	4	169	6	−2	4
7	57	6	178	7	−1	1
8	12	1	184	8	−7	49
9	15	2	190	9	−7	49
					$\sum d_i^2 =$	208

$$r_s = 1 - \dfrac{6\sum d_i^2}{n(n^2-1)} = 1 - \dfrac{6(208)}{9(9^2-1)} = 1 - 1.733 = -.733$$

b.   To determine whether the number of strikes and age are negatively correlated, we test:

H_0: $\rho_s = 0$
H_a: $\rho_s < 0$

The test statistic is $r_s = -.733$.

Reject H_0 if $r_s < -r_{s,\alpha}$ where $\alpha = .01$ and $n = 9$.

Reject H_0 if $r_s < -.783$ (From Table XIV, Appendix A)

Since the observed value of the test statistic does not fall in the rejection region ($r_s = -.733 \not< -.783$), H_0 is not rejected. There is insufficient evidence to indicate that the number of strikes and age are negatively correlated at $\alpha = .01$.

14.107 a.   To determine if the median level of TCDD in the fat tissue of Vietnam vets exceeds 3 ppt, we test:

H_0: $\eta = 3$
H_a: $\eta > 3$

The test statistic is S = number of measurements greater than 3 = 14.

The p-value = $P(x \geq 14)$ where x is a binomial random variable with n = 20 and p = .5. From Table II, Appendix A,

$P(x \geq 14) = 1 - P(x \leq 13) = 1 - .942 = .058$

Since the p-value is greater than $\alpha = .05$ ($p = .058 > .05$), H_0 is not rejected. There is insufficient evidence to indicate the median level of TCDD in the fat tissue of Vietnam vets exceeds 3 ppt at $\alpha = .05$.

b.   To determine if the median level of TCDD in the plasma of Vietnam vets exceeds 3 ppt, we test:

H_0: $\eta = 3$
H_a: $\eta > 3$

The test statistic is S = number of measurements greater than 3 = 12.

The p-value = $P(x \geq 12)$ where x is a binomial random variable with n = 20 and p = .5. From Table II, Appendix A,

$P(x \geq 12) = 1 - P(x \leq 11) = 1 - .748 = .252$

Since the p-value is greater than $\alpha = .05$ (p = .252 > .05), H_0 is not rejected. There is insufficient evidence to indicate the median level of TCDD in the plasma of Vietnam vets exceeds 3 ppt at $\alpha = .05$

b.   Some preliminary calculations are:

Fat	Plasma	Difference	Rank of Absolute Difference
4.9	2.5	2.4	12.5
6.9	3.5	3.4	17
10.0	6.8	3.2	16
4.4	4.7	-0.3	5
4.6	4.6	0	1
1.1	1.8	-0.7	9
2.3	2.5	-0.2	3.5
5.9	3.1	2.8	15
7.0	3.1	3.9	18
5.5	3.0	2.5	14
7.0	6.9	0.1	2
1.4	1.6	-0.2	3.5
11.0	20.0	-9.0	20
2.5	4.1	-1.6	10
4.4	2.1	2.3	11
4.2	1.8	2.4	12.5
41.0	36.0	5.0	19
2.9	3.3	-0.4	6
7.7	7.2	0.5	7.5
2.5	2.0	0.5	7.5

$$T_- = 57$$
$$T_+ = 153$$

To determine if the distribution of TCDD levels in fat is shifted above or below the distribution of TCDD levels in plasma, we test:

H_0: The distribution of TCDD levels in fat is identical the distribution of TCDD levels in plasma

H_a: The distribution of TCDD levels in fat is shifted above or below the distribution of TCDD levels in plasma

The test statistic is T = smaller of T_- and T_+ which is $T_- = 57$.

The rejection region is $T_- \leq 81$, from Table XIII, Appendix A, with $n = 24$ and $\alpha = .05$ (two-tails).

Since the observed value of the test statistic falls in the rejection region ($T_- = 57 < 81$), H_0 is rejected. There is sufficient evidence to indicate the distribution of TCDD levels in fat is shifted above or below the distribution of TCDD levels in plasma at $\alpha = .05$.

d. Some preliminary calculations are:

Fat	Rank, u	Plasma	Rank, v	u^2	v^2	uv
4.9	11	2.5	6.5	121	42.25	71.5
6.9	14	3.5	12	196	144	168
10.0	18	6.8	16	324	256	288
4.4	8.5	4.7	15	72.25	225	127.5
4.6	10	4.6	14	100	196	140
1.1	1	1.8	2.5	1	6.25	2.5
2.3	3	2.5	6.5	9	42.25	19.5
5.9	13	3.1	9.5	169	90.25	123.5
7.0	15.5	3.1	9.5	240.25	90.25	147.25
5.5	12	3.0	8	144	64	96
7.0	15.5	6.9	17	240.25	289	263.5
1.4	2	1.6	1	4	1	2
11.0	19	20.0	19	361	361	361
2.5	4.5	4.1	13	20.25	169	58.5
4.4	8.5	2.1	5	72.25	25	42.5
4.2	7	1.8	2.5	49	6.25	17.5
41.0	20	36.0	20	400	400	400
2.9	6	3.3	11	36	121	66
7.7	17	7.2	18	289	324	306
2.5	4.5	2.0	4	20.25	16	18
	$\sum u = 210$		$\sum v = 210$	$\sum u^2 = 2868.5$	$\sum v^2 = 2868.5$	$\sum uv = 2718.75$

$$SS_{uv} = \sum uv - \frac{\left(\sum u\right)\left(\sum v\right)}{n} = 2718.75 - \frac{210(210)}{20} = 513.75$$

$$SS_{uu} = \sum u^2 - \frac{\left(\sum u\right)^2}{n} = 2868.5 - \frac{210^2}{20} = 663.5$$

$$SS_{vv} = \sum v^2 - \frac{\left(\sum v\right)^2}{n} = 2868.5 - \frac{210^2}{20} = 663.5$$

$$r_s = \frac{SS_{uv}}{\sqrt{SS_{uu}SS_{vv}}} = \frac{513.75}{\sqrt{663.5(663.5)}} = .774$$

To determine if there is a positive association between the two TCDD measures, we test:

H_0: $\rho_s = 0$
H_a: $\rho_s > 0$

The test statistic is $r_s = .774$

From Table XIV, Appendix A, with $n = 20$ and $\alpha = .05$, the rejection region is $r_s > .377$.

Since the observed value of the test statistic falls in the rejection region ($r_s = .774 > .377$), H_0 is rejected. There is sufficient evidence to indicate that there is a positive association between the two TCDD measures at $\alpha = .05$.

14.109 a.

Private Sector	Rank	Private Sector	Rank
2.58%	10	5.40%	15
5.05	13	2.55	9
.05	1	9.00	16
2.10	5	10.55	17
4.30	12	1.02	2
2.25	6	5.11	14
2.50	8	12.42	18
1.94	4	1.67	3
2.33	7	3.33	11
	$T_1 = 66$		$T_2 = 105$

To determine if the two populations have identical distributions, we test:

H_0: The two sampled sectors have identical probability distributions
H_a: The probability distribution for the public sector is shifted to the right of the private sector.

The test statistic is $T_2 = 105$.

The rejection region is $T_2 \geq 105$ from Table XII, Appendix A, with $n_1 = 9$, $n_2 = 9$ and $\alpha = .05$. Since the observed value of the test statistic falls in the rejection region ($T_2 = 105 \geq 105$), H_0 is rejected. We can conclude that the probability distribution for the public sector is shifted to the right of the private sector at $\alpha = .05$.

b. The p-value is approximately equal to .05 because the test statistic is equal to the critical value T_U.

c. 1. The two samples are random and independent.
   2. The two probability distributions from which the samples are drawn are continuous.

14.111 Some preliminary calculations are:

Metal	I	Rank	II	Rank	III	Rank
1	4.6	2	4.2	1	4.9	3
2	7.2	3	6.4	1	7.0	2
3	3.4	1.5	3.5	3	3.4	1.5
4	6.2	3	5.3	1	5.9	2
5	8.4	3	6.8	1	7.8	2
6	5.6	2	4.8	1	5.7	3
7	3.7	1.5	3.7	1.5	4.1	3
8	6.1	1	6.2	2	6.4	3
9	4.9	3	4.1	1	4.2	2
10	5.2	3	5.0	1	5.1	2
		$R_1 = 23$		$R_2 = 13.5$		$R_3 = 23.5$

To determine if there is a difference in the probability distributions of the amounts of corrosion among the three types of sealers, we test:

H_0: The probability distributions of corrosion amounts are identical for the three types of sealers

H_a: At least two of the probability distributions differ in location

The test statistic is $F_r = \dfrac{12}{bk(k+1)}\sum R_j^2 - 3b(k+1)$

$$= \frac{12}{10(3)(3+1)}[23^2 + 13.5^2 + 23.5^2] - 3(10)(3+1) = 126.35 - 120 = 6.35$$

The rejection region requires $\alpha = .05$ in the upper tail of the χ^2 distribution with df $= k - 1$ $= 3 - 1 = 2$. From Table VII, Appendix A, $\chi^2_{.05} = 5.99147$. The rejection region is $F_r > 5.99147$.

Since the observed value of the test statistic falls in the rejection region ($F_r = 6.35 > 5.99147$), reject H_0. There is sufficient evidence to indicate a difference in the probability distributions among the three types of sealers at $\alpha = .05$.

14.113 Some preliminary calculations are:

List 1	Rank	List 2	Rank	List 3	Rank
48	19	41	14	18	2
43	16	36	10	42	15
39	12	29	5	28	4
57	20	40	13	38	11
21	3	35	9	15	1
47	18	45	17	33	8
58	21	32	7	31	6
	$R_1 = 109$		$R_2 = 75$		$R_3 = 47$

To determine if there is a difference between at least two of the probability distributions of the numbers of word associates that subjects can name for the three lists, we test:

H_0: The probability distributions of the numbers of word associates named are the same for the three lists

H_a: At least two of the three probability distributions differ in location

The test statistic is $H = \dfrac{12}{n(n+1)}\sum \dfrac{R_j^2}{n_j} - 3(n+1)$

$$= \frac{12}{21(21+1)}\left[\frac{(109)^2}{7} + \frac{(75)^2}{7} + \frac{(47)^2}{7}\right] - 3(21+1)$$

$$= 73.154 - 66 = 7.154$$

The rejection region requires $\alpha = .05$ in the upper tail of the χ^2 distribution with df $= k - 1 = 3$ $- 1 = 2$. From Table VII, Appendix A, $\chi^2_{.05} = 5.99147$. The rejection region is $H > 5.99147$.

Since the observed value of the test statistic falls in the rejection region ($H = 7.154 >$ 5.99147), reject H_0. There is sufficient evidence to indicate a difference in location for at least two of the probability distributions of the numbers of word associates at $\alpha = .05$.

14.115 a. The correlation between a variable and itself will always be 1.00. Therefore, the Spearman's rank correlations on the diagonals are always 1.00.

      b. The Spearman's rank correlation between EDSS and ARM is $-.439$. There is a moderately weak, negative relationship between the Expanded Disability Status Scale and the peak oxygen uptake during arm cranking exercise.

The Spearman's rank correlation between EDSS and LEG/ARM is $-.655$. There is a moderately strong, negative relationship between the Expanded Disability Status Scale and the peak oxygen uptake during leg cycling and arm cranking exercise.

The Spearman's rank correlation between EDSS and LEG is $-.512$. There is a moderate, negative relationship between the Expanded Disability Status Scale and the peak oxygen uptake during leg cycling exercise.

The Spearman's rank correlation between ARM and LEG/ARM is $.634$. There is a moderately strong, positive relationship between the peak oxygen uptake during arm cranking exercise and the peak oxygen uptake during combined leg cycling and arm cranking exercise.

The Spearman's rank correlation between ARM and LEG is $.722$. There is a moderately strong, positive relationship between the peak oxygen uptake during arm cranking exercise and the peak oxygen uptake during leg cycling exercise.

The Spearman's rank correlation between LEG/ARM and LEG is $.890$. There is a strong, positive relationship between the peak oxygen uptake during combined leg cycling and arm cranking exercise and peak oxygen uptake during leg cycling exercise.

      c. To determine if EDSS and peak oxygen uptake for arm cranking exercise are negatively correlated, we test:

H_0: $\rho_s = 0$
H_a: $\rho_s < 0$

The test statistic is $r_s = -.439$.

Reject H_0 if $r_s < -r_{s,\alpha}$ where $\alpha = .01$ and $n = 10$:

Reject H_0 if $r_s < -.745$ (using Table XIV, Appendix A)

Since the observed value of the test statistic does not fall in the rejection region ($r_s = -.439 \not< -.745$), H_0 is not rejected. There is insufficient evidence to indicate that there is a negative correlation between EDSS and peak oxygen uptake for arm cranking exercise at $\alpha = .01$.

To determine if EDSS and peak oxygen uptake for combined leg cycling and arm cranking exercise are negatively correlated, we test:

H_0: $\rho_s = 0$
H_a: $\rho_s < 0$

The test statistic is $r_s = -.655$.

Reject H_0 is $r_s < -r_{s,\alpha}$ where $\alpha = .01$ and $n = 10$:

Reject H_0 if $r_s < -.745$ (using Table XIV, Appendix A)

Since the observed value of the test statistic does not fall in the rejection region ($r_s = -.655 \not< -.745$), H_0 is not rejected. There is insufficient evidence to indicate that there is a negative correlation between EDSS and peak oxygen uptake for combined leg cycling and arm cranking exercise at $\alpha = .01$.

To determine if EDSS and peak oxygen uptake for leg cycling exercise are negatively correlated, we test:

H_0: $\rho_s = 0$
H_a: $\rho_s < 0$

The test statistic is $r_s = -.512$.

Reject H_0 is $r_s < -r_{s,\alpha}$ where $\alpha = .01$ and $n = 10$:

Reject H_0 if $r_s < -.745$ (using Table XIV, Appendix A)

Since the observed value of the test statistic does not fall in the rejection region ($r_s = -.512 \not< -.745$), H_0 is not rejected. There is insufficient evidence to indicate that there is a negative correlation between EDSS and peak oxygen uptake for leg cycling exercise at $\alpha = .01$.

14.117 Some preliminary calculations are:

Staff		Trainees		Undergraduates	
Observation	Rank	Observation	Rank	Observation	Rank
78	18.5	80	21.5	65	1
79	20	75	13	70	5
85	26	72	6.5	74	10.5
93	30	68	2.5	78	18.5
90	29	75	13	68	2.5
76	16	69	4	74	10.5
86	27	81	23.5	80	21.5
88	28	76	16	73	8.5
84	25	72	6.5	75	13
81	23.5	76	16	73	8.5
	$R_1 = 243$		$R_2 = 122.5$		$R_3 = 99.5$

To determine if there is a difference in the probability distributions of the number of correct identifications among the three types of judges, we test:

H_0: The probability distributions of the number of correct identifications are the same for the three types of judges

H_a: The probability distributions of the number of correct identifications differ in location for at least two of the types of judges

The test statistic is $H = \dfrac{12}{n(n+1)} \sum \dfrac{R_j^2}{n_j} - 3(n+1)$

$$= \frac{12}{30(30+1)} \left(\frac{243^2}{10} + \frac{122.5^2}{10} + \frac{99.5^2}{10} \right) - 3(30+1) = 15.330$$

The rejection region requires $\alpha = .05$ in the upper tail of the χ^2 distribution with df $= k - 1 = 3 - 1 = 2$. From Table VII, Appendix A, $\chi_{.05}^2 = 5.99147$. The rejection region is $H > 5.99147$.

Since the observed value of the test statistic falls in the rejection region ($H = 15.330 > 5.99147$), H_0 is rejected.

There is sufficient evidence to indicate a difference in the probability distributions of the number of correct identifications for at least two of the types of judges at $\alpha = .05$.

14.119 a.   Some preliminary calculations are:

Ear	Rank Spray A	Rank Spray B	Rank Spray C
1	2	3	1
2	2	3	1
3	1	3	2
4	3	2	1
5	2	1	3
6	1	3	2
7	2.5	2.5	1
8	2	3	1
9	2	3	1
10	2	3	1
	$R_1 = 19.5$	$R_2 = 26.5$	$R_3 = 14$

To determine whether the distributions of the levels of aflatoxins in corn differ for at least two of the three sprays, we test:

H_0: The three populations have probability distributions that are identical
H_a: At least two of the populations have probability distributions that differ in location

The test statistic is $F_r = \dfrac{12}{bk(k+1)} \sum R_j^2 - 3b(k+1)$

$$= \frac{12}{10(3)(4)} \left[(19.5)^2 + (26.5)^2 + (14)^2 \right] - 3(10)(4) = 127.85 - 120 = 7.85$$

The rejection region requires $\alpha = .05$ in the upper tail of the χ^2 distribution with df $= k - 1 = 3 - 1 = 2$. From Table VII, Appendix A, $\chi^2_{.05} = 5.99147$. The rejection region is $F_r > 5.99147$.

Since the observed value of the test statistic falls in the rejection region ($F_r = 7.85 > 5.99147$), H_0 is rejected. There is sufficient evidence to indicate that the distributions of the levels of aflatoxin in corn differ for at least 2 of the 3 sprays at $\alpha = .05$.